STAINING PROCEDURES

FOURTH EDITION

STAINING PROCEDURES

FOURTH EDITION

Edited by George Clark

GENERAL METHODS
George Clark
William J. Dougherty
Frederick H. Kasten
ANIMAL HISTOTECHNIQUE
George E. Cantwell
George Clark
Robert E. Coalson
John L. Mohr
S. S. Spicer
BOTANICAL SCIENCES
Ronald L. Phillips
Henry Schneider
MICROBIOLOGY
James W. Bartholomew

Published for the
BIOLOGICAL STAIN COMMISSION

by

WILLIAMS & WILKINS
Baltimore • London • Los Angeles • Sydney

First edition published by the BIOLOGICAL STAIN COMMISSION
Editions published by Williams & Wilkins:
    Second edition, 1960
    Third edition, 1973

Made in the United States of America

Library of Congress Cataloging in Publication Data

Main entry under title:

Staining procedures used by the Biological Stain Commission.

    Bibliography: p.
    Includes index.
    1. Stains and staining (Microscopy) I. Clark, George, 1905– II. Biological Stain Commission.
QH237.C66 1980   578'.64   79-29658
ISBN 0-683-01707-1

Composed and printed at the
Waverly Press, Inc.   86 87 88 89   10 9 8 7 6 5

PREFACE TO FOURTH EDITION

While some new technics have been added, some older procedures deleted, and some rearrangements made, in many respects the fourth edition is a new book. In view of the interest in and increased use of plastic embedded tissue for light microscopy, a new chapter has been added. This should be very useful to both the neophyte and to those already using this procedure. Fluorescence microscopy is also a rapidly growing field and has been greatly expanded. Although many of the procedures used for vertebrate histotechnic may be used on invertebrate material, there are many procedures that were devised primarily for invertebrate tissue. This is covered in a new section. The staining of endocrine glands also requires many special procedures. This section, too, has been completely revised. In the botany section, the chapters on plant cytogenetics and on pollen and pollen tubes are new.

PREFACE TO SECOND EDITION

This publication is sponsored by the Biological Stain Commission and is intended to represent staining methods used in microtechnic in the laboratories of Commission members. It is not intended to be a complete treatise in regard to staining, nor even to include the only way of getting good results with the various methods here included. Nor are the procedures given intended to be considered as *standard* or *official*. They are all put in a standardized *form*, but only for the convenience of users.

The first edition of this book was published in a loose-leaf form, the first leaflets thereof appearing in 1943. Eventually it consisted of nine leaflets. These leaflets were periodically revised during the period from 1943 to 1955 some of them going through 4 editions. The present publication, however, is the first in which all the material has been gathered together as a "hardcover" book, and is regarded as only the 2nd edition of the entire contents.

Members of the Stain Commission contributing to the first (loose-leaf) edition are listed in its introductory leaflet. When the senior author retired several years ago as an officer of the Commission, it was anticipated that this publication would be allowed to go out of print and that its revision would not be undertaken. However, a continuing steady demand for the material in it has resulted in a decision by the Commission trusted to sponsor a new and revised edition.

In the first edition the names of 28 collaborators are given on the page of Acknowledgements. On the title page of the current edition only those specially assisting in the revision are named; but acknowledgement is hereby given to the following whose assistance made the first edition possible (but who are not named on the present title page): C. E. Allen, Wanda Brentzel, Clyde Chandler, G. H. Chapman, E. V. Cowdry, A. B. Dawson, John Einset, J. O. Foley, M. F. Guyer, P. H. Hartz, Raphael Isaacs, M. W. Jennison, H. E. Jordan, R. R. Kudo, B. R. Nebel, L. F. Randolph, Ruth Rhines, L. W. Sharp, K. A. Stiles, W. D. Stovall, W. R. Taylor, J. M. Thuringer, W. F. Windle, Isidore Wodinsky, Conway Zirkle. Special mention should also be made of the fact that Chapter 4 (Neurological Methods) has been completely revised for the present edition by H. A. Davenport; while Chapters 6–8 (Plant Microtechnic) owe their new form largely to the valuable suggestions made by Frank H. Smith and Charlotte Pratt.

H. J. Conn
Geneva, N.Y.
Mary A. Darrow
Victor M. Emmel
Rochester, N.Y.

CONTRIBUTORS

James W. Bartholomew, Professor of Microbiology,
University of Southern California

George E. Cantwell, U. S. Department of Agriculture,
Environmental Bee Laboratory,
Beltsville, Maryland

George Clark, Consultant in Anatomy,
Medical University of South Carolina
and Research Associate,
Veterans Administration Medical Center,
Charleston, South Carolina

Robert E. Coalson, Professor of Anatomical Sciences and Assistant
Professor of Pathology,
University of Oklahoma

William J. Dougherty, Professor of Anatomy,
Medical University of South Carolina

Frederick H. Kasten, Professor of Anatomy,
Louisiana State University
School of Medicine and Visiting Professor of
Anatomy,
East Tennessee State University,
College of Medicine

John L. Mohr, Professor of Biology,
University of Southern California

Ronald L. Phillips, Professor of Genetics and Plant Breeding,
University of Minnesota

Henry Schneider, Plant Pathologist,
University of California Riverside

S. S. Spicer, Professor of Pathology,
Medical University of South Carolina

CONTENTS

GENERAL METHODS

CHAPTER 1

INTRODUCTION

George Clark

STATEMENT OF PURPOSE

Ever since the Biological Stain Commission was organized, its primary objective has been to secure uniformity and high quality in the dyes used in biological work. To a large measure this has been accomplished for many dyes by the certification procedures of the Biological Stain Commission and by *Biological Stains*, which now is in its ninth edition. A secondary objective has been improvement in staining technic. *Stain Technology*, now in its 54th volume, has partially accomplished this objective. The present book is a further attempt to satisfy this second objective. It is felt that while rigid standardization would be most undersirable, every effort should be made to make easier the utilization of various technics.

Many staining formulas are inaccurate, indefinite and omit crucial steps. These errors can take many forms: the dye or dyes may not be clearly defined, a reagent may be listed without statement as to whether or not the anhydrous or crystalline form is desired, the quantities listed may imply a much greater degree of exactness in measuring than is required (i.e., 3.1 N or 29%, etc.). Then, also, a variant of a technic may be published with no reference to the original. Checking inaccuracies of this sort has proved difficult. The original references are sometimes hard to get; and, if it proves, when they are located, that the original method differs from the one commonly used today under the same name, a further search through the literature is necessary to locate the source of the modification.

Wherever possible each procedure includes one or more references to show the source from which it is taken. Where two or more references are given, the oldest identifies first publication, usually with the authorship of the individual after whom it has been named. The others indicate the sources from which the procedure was taken (either without change or with slight emendations). Where possible the procedures were taken directly from the original. In some instances the method either has not been published or has not been published in the form given, or has been contributed either orally or by correspondence from some collaborator; in these cases a statement to this effect is included or is indicated in the bibliography.

PREPARATION OF SOLUTIONS

Throughout the literature relating to microtechnic, concentrations of solutions are expressed in more or less abbreviated form, as for example "2% NaCl" or "1:10 HCl." Sometimes the meanings of such expressions are obvious; sometimes they can be readily guessed by anyone familiar with the methods; but in many cases they are open to more than one possible interpretation. This can result in confusion. Nevertheless the use of such expressions as these is convenient and in a publication like this results in saving much paper and ink. So it has been decided to adopt some of these abbreviated forms throughout, but to define them so that there should be no misunderstanding as to what is intended. The following definitions are, therefore, formulated:

The expression "$n\%$ aqueous...," in the case of a solid or gaseous solute, indicates grams brought up to 100 ml with distilled water; in the case of a liquid, indicates milliliters brought up to 100 ml with distilled water.
The expression "$n\%$ alcoholic..." is used similarly, the solvent being 95% ethyl alcohol.

> NOTE: Whenever the term "alcohol" alone is used in this or any other context, *ethyl alcohol* is intended; if any other alcohol is intended, it is definitely specified; 95% alcohol is always meant unless some other strength (e.g. 100% or 80%) is specified.

Similarily the term water denotes deionized or distilled water. When tap water is desired, it is so mentioned.
The expressions "1:10 formalin," "1:10 HCl," etc., indicate 1 volume of the concentrated reagent to 9 volumes of distilled water; and "1:100" similarly means 1 volume to 99 volumes of distilled water, etc. In preparing these dilutions, the concentrated reagents are assumed to be the following:
Acid, acetic, glacial, $CH_3 \cdot COOH$, 99%.
Acid, hydrochloric, HCl, 35–37%, sp gr 1.18–1.19.
Acid, nitric, HNO_3, 68–70%, sp gr 1.40–1.42.
Acid, sulfuric, H_2SO_4, 96–98%, sp gr 1.84.
Ammonium hydroxide, NH_4OH, 28% NH_3, sp gr 0.90.
Formalin, 37–40% solution of formaldehyde, HCHO.

BUFFER TABLES

From Lillie (1965), pp. 655-674 (abridged)
The salts and acids used in preparing buffers should be of reagent grade, or those especially designated for buffer use. Since some salts are available in a variety of states of hydration, the molecular weight given on the label

should be used in preparing solutions. Sodium and potassium salts can be interchanged, providing that due account is taken of the difference in molecular weight.

Buffers containing sodium citrate and acetate are particularly liable to mold growth. This can be inhibited if stock buffers (M/10) are made up in 25% methanol. When diluted 1:20 or 1:25 in use the alcohol concentration becomes negligible.

Formulas and Molecular Weights of Buffer Ingredients

Citric acid, anhydrous	$C_3H_4(OH)(COOH)_3$	192.12
Citric acid, crystals	$C_3H_4(OH)(COOH)_3 \cdot H_2O$	210.14
Potassium acid phosphate	KH_2PO_4	136.09
Sodium acetate	CH_3COONa	82.04
Sodium acetate, crystals	$CH_3COONa \cdot 3H_2O$	136.09
Sodium acid phosphate	NaH_2PO_4	138.01
Sodium citrate, crystals	$C_3H_4(OH)(COONa)_3 \cdot 5\frac{1}{2}H_2O$	357.18
Sodium citrate, granular	$C_3H_4(OH)(COONa)_3 \cdot 2H_2O$	294.12
Sodium phosphate, dibasic	Na_2HPO_4	141.98

Acetate Buffer (Walpole)

pH	M/5 Acetic Acid	M/5 Sodium Acetate
2.80	19.9	0.1
2.91	19.8	0.2
2.99	19.7	0.3
3.08	19.6	0.4
3.15	19.5	0.5
3.20	19.4	0.6
3.32	19.2	0.8
3.42	19.0	1.0
3.59	18.5	1.5
3.72	18.0	2.0
3.90	17.0	3.0
4.05	16.0	4.0
4.16	15.0	5.0
4.27	14.0	6.0
4.36	13.0	7.0
4.45	12.0	8.0
4.53	11.0	9.0
4.63	10.0	10.0
4.71	9.0	11.0
4.80	8.0	12.0

pH	M/5 Acetic Acid	M/5 Sodium Acetate
4.90	7.0	13.0
4.99	6.0	14.0
5.11	5.0	15.0
5.23	4.0	16.0
5.38	3.0	17.0
5.57	2.0	18.0
5.89	1.0	19.0
6.21	0.5	19.5

Walpole HCl Buffer

1 N HCl (ml)	1 M Sodium Acetate (ml)	Distilled H$_2$O (ml)	pH
20	10	20	0.65
18	10	22	0.75
16	10	24	0.91
14	10	26	1.07
13	10	27	1.24
12	10	28	1.42
11	10	29	1.71
10.7	10	29.3	1.85
10.5	10	29.5	1.99
10.2	10	29.8	2.32
10	10	30	2.64
9.7	10	30.3	3.09
9.5	10	30.5	3.29
9.0	10	31	3.61
8.5	10	31.5	3.79
8	10	32	3.95
7	10	33	4.19
6	10	34	4.39
5	10	35	4.58
4	10	36	4.76
3	10	37	4.92
2	10	38	5.20

NOTE: In the use of this buffer, other solutions added must replace part of the water, so that the 50-ml volume is not increased.

Citrate Buffer (Horecker-Lillie)

Stock M/10 solutions made in 25% methanol. pH values are for final aqueous dilution of 1:25 (M/250).

pH	M/10 Citric Acid	M/10 Sodium Citrate
2.95	19	1
3.08	18	2
3.23	17	3
3.40	16	4
3.58	15	5
3.80	14	6
4.01	13	7
4.27	12	8
4.50	11	9
4.65	10	10
4.84	9	11
5.04	8	12
5.26	7	13
5.50	6	14
5.74	5	15
5.97	4	16
6.17	3	17
6.42	2	18
6.77	1	19

McIlvaine-Lillie

Stock Solutions (M/10, M/5) made in 25% methanol. pH values are for final aqueous dilution of 1:25.

pH	M/10 Citric Acid	M/5 Disodium Phosphate
2.6	19.5	0.5
2.65	19.0	1.0
2.7	18.5	1.5
2.75	18.0	2.0
2.8	17.5	2.5
2.9	17.0	3.0
3.0	16.5	3.5
3.05	16.0	4.0
3.1	15.5	4.5
3.2	15.0	5.0
3.3	14.5	5.5
3.45	14.0	6.0
3.6	13.5	6.5

pH	M/10 Citric Acid	M/5 Disodium Phosphate
3.75	13.0	7.0
3.95	12.5	7.5
4.1	12.0	8.0
4.3	11.5	8.5
4.5	11.0	9.0
4.75	10.5	9.5
4.95	10.0	10.0
5.3	9.5	10.5
5.5	9.0	11.0
5.7	8.5	11.5
6.0	8.0	12.0
6.1	7.5	12.5
6.3	7.0	13.0
6.4	6.5	13.5
6.5	6.0	14.0
6.6	5.5	14.5
6.8	5.0	15.0
6.9	4.5	15.5
7.0	4.0	16.0
7.1	3.5	16.5
7.2	3.0	17.0
7.3	2.5	17.5
7.4	2.0	18.0
7.5	1.5	18.5
7.7	1.0	19.0
8.0	0.5	19.5

Sörensen's Phosphate Buffers

KH_2PO_4 or $NaH_2PO_4 \cdot H_2O$	Na_2HPO_4	pH at Specified Dilutions		
		M/10	M/15	M/200
48	2	5.31	5.42	5.63
47	3	5.53	5.60	5.81
46	4	5.67	5.74	5.95
45	5	5.78	5.83	6.06
44	6	5.86	5.91	6.14
43	7	5.94	5.99	6.22
42	8	6.02	6.07	6.30
41	9	6.08	6.14	6.36

KH$_2$PO$_4$ or NaH$_2$PO$_4$ · H$_2$O	Na$_2$HPO$_4$	pH at Specified Dilutions		
		M/10	M/15	M/200
40	10	6.12	6.19	6.42
39	11·	6.17	6.24	6.47
38	12	6.23	6.28	6.51
37	13	6.28	6.32	6.56
36	14	6.33	6.37	6.61
35	15	6.37	6.41	6.65
34	16	6.41	6.45	6.68
33	17	6.45	6.49	6.72
32	18	6.49	6.53	6.75
31	19	6.53	6.56	6.79
30	20	6.55	6.59	6.82
29	21	6.58	6.63	6.86
28	22	6.61	6.68	6.91
27	23	6.65	6.72	6.95
26	24	6.70	6.76	7.00
25	25	6.76	6.81	7.05
24	26	6.81	6.86	7.09
23	27	6.84	6.91	7.13
22	28	6.87	6.94	7.16
21	29	6.89	6.96	7.18
20	30	6.91	6.98	7.20
19	31	6.94	7.01	7.22
18	32	6.97	7.03	7.25
17	33	7.00	7.05	7.28
16	34	7.02	7.07	7.31
15	35	7.06	7.11	7.34
14	36	7.10	7.15	7.38
13	37	7.14	7.20	7.41
12	38	7.19	7.24	7.45
11	39	7.24	7.28	7.51
10	40	7.30	7.33	7.59
9	41	7.36	7.40	7.65
8	42	7.42	7.47	7.70
7	43	7.49	7.54	7.75
6	44	7.57	7.61	7.80
5	45	7.65	7.69	7.87
4	46	7.73	7.77	7.96
3	47	7.81	7.85	8.05
2	48	7.92	7.97	8.15

Boric Acid-Borate Buffer (Holmes)

M/5 boric acid; H_3BO_3. Dissolve 12.4 gm in distilled water and make up to 1 liter.

M/20 sodium tetraborate (borax); $Na_2B_4O_7 \cdot 10\ H_2O$ Dissolve 19.1 gm in distilled water and make up to 1 liter.

Mix the following quantities of the two solutions as shown below:

pH	M/5 Boric Acid (ml)	M/20 Borate (ml)
7.4	180	20
7.6	170	30
7.8	160	40
8.0	140	60
8.2	130	70
8.4	110	90
8.7	80	120
9.0	40	160

DYES FOR STAINING SOLUTIONS

The staining solutions given in the great majority of procedures included here call for dyes that have been put on the certification basis by the Biological Stain Commission, and a considerable number of them are stains for which specifications have been drawn up by the American Pharmaceutical Association in cooperation with the Stain Commission. Those who consult this publication for staining formulas should understand that the stains which have been put on the certification basis are approved by the Biological Stain Commission on a batch basis; no batch is approved unless it only meets the chemical and optical tests imposed by the Commission, but also gives satisfactory results in the procedures for which it is ordinarily employed. This method of approval has been adopted because no purely chemical or physical tests have yet been devised that can, in every instance, distinguish between a satisfactory and an unsatisfactory dye.

In the procedures the majority of the dyes are already on the certification list. Dyes certified by the Biological Stain Commission, or dyes of equal purity, should be used. The list below shows the dyes which are currently certified. The list also gives the minimum dye content for samples admitted for certification and the Color Index Number. This number identifies the dye while the name may vary with different suppliers. As a further aid in identification, page references to Lillie (H. J. Conn's *Biological Stains*, 1977) are also listed.

Dyes Certified by Biological Stain Commission

	Minimum Dye Content (%)	2nd Ed. (1956) Color Index No.	9th Ed. (1977) Biological Stains Page Nos.
Alizarin red S	*	58005	432
Aniline blue, water soluble	*	42755	289
Auramine O	80	41000	238
Azocarmine G	†	50085	381
Azure A	55	52005	420
Azure B	†	52010	421
Azure C	†	52002	419
Bismarck brown Y	45	21000	145
Brilliant cresyl blue	50	51010	393
Brilliant green	85	42040	251
Carmine	*	75470	476
Chlorazol black E	*	30235	193
Congo red	75	22120	147
Cresyl violet acetate	†	—	412
Crystal violet	88	42555	274
Darrow red	*	—	414
Eosin B	85	45400	345
Eosin Y	80	45380	342
Erythrosin B	80	45430	349
Ethyl eosin	78	45386	344
Fast green FCF	85	42053	253
Fluorescin isothiocyanate	70	—	338
Fuchsin, acid	55	42685	285
Fuchsin, basic	88	42500, 42510	260
Fuchsin, basic (special for flagella)		42500	260
Giemsa stain	*	—	496
Hematoxylin	*	75290	468
Indigocarmine	80	73015	451
Janus green B	50	11050	86
Jenner's stain	*	—	606
Light green SF yellowish	65	42095	257
Malachite green	90	42000	248
Martius yellow	†	10315	75
Methyl green	65	42585	280
Methyl orange	85	13025	98
Methyl violet 2B	75	42535	271

	Minimum Dye Content (%)	2nd Ed. (1956) Color Index No.	9th Ed. (1977) *Biological Stains* Page Nos.
Methylene blue	82	52015	423
Methylene blue thiocyanate tablets ca. 9 mg		52015	
Methylene violet	†	52041	429
Neutral red	50	50040	377
Nigrosin	*	50420	390
Nile blue A	70	51180	409
Orange G	80	16230	121
Orange II	85	15510	112
Orcein, synthetic	*	—	400
Phloxine B	80	45410	347
Protargol-S	*	—	484
Pyronin B	30	45010	328
Pyronin Y	45	45005	328
Resazurin	†	—	396
Resazurin tablets ca. 11 mg		—	
Rose Bengal	80	45440	350
Safranin O	80	50240	385
Sudan III	75	26100	168
Sudan IV	80	26105	169
Sudan black B	*	26150	173
Tetrachrome stain (MacNeal)	*	—	606
Thionin	85	52000	417
Toluidine blue O	50	52040	428
Wright's stain	*	—	605

* *Dye content not determined.*
† *No minimum dye content established.*

GENERAL METHODS

This book is not intended as a textbook on microtechnic and anyone unfamiliar with the general methods in this field should consult one of the excellent textbooks on this subject. In a very few instances specimens are examined without fixation or any form of hardening. Especially for enzyme and ion localization tissue can be frozen, usually at a very low temperature, and sectioned. Far more often they are fixed and hardened, embedded in some material and sectioned. The hardening also kills the cells. Since specimens contain autolytic enzymes, material soluble in water, are often friable and subject to bacterial and fungal infection, fixation must be as rapid as possible. In some instances perfusion with the fixing fluid (see p. 133) is necessary but in material only 4–6 mm in thickness fixation by immersion is adequate. The fixed material is sometimes frozen and sectioned but is usually embedded in paraffin, celloidin or plastic. For practical purposes one may regard the paraffin method, as applied to fixed tissue, to be the general procedure called for in this book. When any of the other procedures are called for or special fixatives are necessary, reference to this effect is made under the technic in question.

Fixation is particularly important because the procedure employed must be adapted to the tissue to be examined and must be specially designed to bring out and preserve the structures which it is desired to demonstrate. Because of the prime importance of fixation, the staining procedure is often determined by the method of fixation or vice versa; hence each staining method given in the later chapters specifies the type of fixation necessary or at least recommended. An almost innumerable variety of fixatives have been proposed; but only a comparatively small number of these have come into routine use. Those listed here are the most commonly employed fixatives in laboratories of microtechnic. Each can be regarded as having its own advantages, and in any staining procedure the specific recommendations in regard to fixation should be followed, unless there are definite reasons for some different procedure. Some staining procedures require special fixation—these are given under the individual staining method.

Preparation of tissue for plastic embedding. This requires special procedures and solutions and is covered in Chapter 2.

Occasionally stains are applied to the fixed tissue in bulk, but by far the majority of cases call for sections. With such material, as is well known, there are five essential steps in handling the tissue before staining has begun, namely:

1. Fixation, followed by washing
2. Dehydration
3. Clearing
4. Embedding
5. Sectioning and attaching sections to slides.

The first two of these steps are essentially the same, whether embedding is in paraffin or celloidin, but the remaining steps differ.

FIXING FLUIDS

Alcohol

Recommended for: Glycogen in animal tissue; other fluids are preferable, however.

Formula: 95% or preferably absolute ethyl alcohol. Absolute alcohol (9 vol) with formalin (1 vol) is sometimes employed; or 80% at temperatures 0°C or lower.

Procedure: Lower concentrations of alcohol, at room temperature, are likely to dissolve glycogen. For enzymes in tissues, (*e.g.*, phosphatases) 80% alcohol may be used for 18–24 hr at 0°C or lower. This should be followed by two 4–8 hr changes of 95% and of 100% alcohol, still in the cold, then petroleum ether and paraffin, not over 10 min each *in vacuo*. To prevent migration (and loss) of glycogen, the alcohol is preferably applied at −75°C, to freeze the tissue immediately. Mix solid carbon dioxide (Dry Ice) and alcohol. Immerse thin slices of tissue. When frozen hard (1–2 min) trim to desired size and shape and return to the carbon dioxide alcohol and place in −20°C "Deep Freeze" to fix. Change (prechilled) alcohol daily for a week. Allow to warm slowly to room temperature. Embed in paraffin through alcohol chloroform sequence.

Although some glycogens are well conserved with aqueous formalin fixations, others, probably lower polymers which are more soluble in water, require anhydrous fixation. The same procedure is applicable for dextrans, but there alcoholic periodic acid should be used in the demonstration procedure.

Alcohol and Acetic Acid

Carnoy (1886)

Recommended for: Plant histology and cytology.

Formula: Acetic acid, glacial   1 vol
95% alcohol   1–3 vol

Procedure: 12–24 hr fixation is generally preferred, although much shorter periods can sometimes be used when speed is essential. Fixation is followed by washing in 50–70% alcohol.

Carnoy's Fluid

Carnoy (1887)

Recommended for: Animal tissue, in general; glycogen in animal tissue, especially if used at 3–5°C; plant tissue, cytological. Unusually penetrating and quick acting.

		Plain	With chloroform
Formula:	Alcohol, absolute	60 ml	60 ml
	Chloroform		30 ml
	Acetic acid, glacial	10 ml	10 ml

Procedure: Fix small pieces of animal tissue 1½-3 hr at room temperature or 18 hr at 0°C; wash in absolute alcohol over $CaCO_3$; for glycogen, fix for 24 hr at 3–5°C. With insect ova, wash in 2 changes of 95% alcohol and proceed as quickly as possible through 100% alcohol and chloroform to paraffin. Fix root tips, 15 min or anthers 1 hr.

Methacarn

Puchtler et al. (1970)

This is similar to Carnoy's fluid except that absolute methyl alcohol is substituted for ethyl alcohol. Shrinkage, especially of collagen, is less and myofibrils are prominent. Use similarly to Carnoy's.

Formalin Mixtures

Blum (1893)

Formalin (1:10)

Recommended for: Animal tissue, especially nervous tissue and material for frozen sectioning and for celloidin embedding.

Formula:

Formalin	10 ml
Distilled water	90 ml

To maintain a neutral reaction, $CaCO_3$; or a lead oxide or carbonate should be added in excess.

Procedure: At least 24 hr at room temperatures; or about 24 hr at 35°C, 12 hr at 45°C, or 6 hr at 55°C. If shorter fixation times are important, stronger formalin solutions may be used; *e.g.* 1 hr in 1:5 formalin at 35–60°C. Tissue may be kept in formalin for considerable period before sectioning.

Buffered Neutral Formalin

Lillie (1948)

Recommended for: General fixative for all tissues. Especially good for hemoglobin and homesiderin. Does not produce formalin precipitates.

Formula:

Formalin	100 ml
Na_2HPO_4 (anhyd.)	6.5 gm
$NaH_2PO_4 \cdot H_2O$	4.0 gm
Distilled water	900 ml

Procedure: Fix tissues in the above for 1–2 days or longer.

Alcoholic Formalin

Recommended for: Plant microtechnic, especially for pollen tubes in styles. Glycogen in animal tissue.

Formula:

Formalin	6–10 ml
70% alcohol	100 ml

NOTE: The 10% formula is specially recommended for glycogen.

Procedure: Material may be examined promptly or stored in the fixative for a considerable period.

Calcium Acetate Formalin
Lillie and Fullmer (1976)

Recommended for: Animal tissue, in general. Lipids are better preserved than with many other fixatives.

Formula:

Formalin	10 ml
Calcium acetate (monohydrate)	2 gm
H_2O q.s.	100 ml

Procedure: Fix tissues 1–2 days or longer.

Cajal's Formalin-Ammonium Bromide (FAB)

Recommended for: Animal tissue, central nervous system.

Formula:

Formalin	15 ml
Distilled water	85 ml
NH_4Br	2 gm

Procedure: Fix material designed for Cajal's gold-sublimate procedure in this fluid for 2–25 days—5 days about optimum.

NOTE: This is a very acid fixative, since formaldehyde reacts with the ammonia to form hexamethylene tetramine and liberate hydrobromic acid. pH levels around 1.5 are observed. Mammalian erythrocytes are destroyed.

Formalin-Aceto-Alcohol (FAA)

Recommended for: Plant tissue, in general. Not suitable for plant cytology. Glycogen in animal tissue; central nervous system.

Formula:

50 or 70% ethyl alcohol	90 ml or less
Acetic acid, glacial	5 ml or less
Formalin	5 ml or more

Procedure: For delicate materials use the weaker (50%) alochol. For hard, woody materials decrease the quantity of acetic acid and increase the amount of formalin. Fix tissue for at least 18 hr. If dehydrating in tertiary butyl alcohol, do not wash, but go direct to the 50% stage of that method. With other methods of dehydration, wash ordinary plant material in 2 changes of 50% ethyl alcohol. Woody materials may be softened by washing 2 days in running water and 6 weeks in 50% aqueous HF.

Gum acadia formalin of Koenig et al., see p. 134.
Sugar formalin, see p. 134.

Gendre's Fluid

Gendre (1937)

Recommended for: Fixation of glycogen in liver.

Formula:

Saturated alcoholic picric acid	8 vol
Formalin	1.5 vol
Acetic acid, glacial	0.5 vol

Procedure: Fix tissue 1–4 hr.

Bouin's Fluid

Bouin (1897)

Recommended for: Animal and plant tissue in general.

Formula:

Saturated aqueous picric acid (about 1.22%)	75 ml
Formalin	25 ml
Acetic acid, glacial	5 ml

NOTE: The formalin, if desired, can be reduced to 20 ml.

Procedure: According to Lee (1937)—Fix up to 18 hr, then wash in 50% and then in 70% alcohol until picric acid is almost removed. According to Masson (1929)—Fix up to 3 days, then pour off the solution and cover the tissue with water.

Duboscq-Brazil Fluid

Gatenby and Cowdry (1928)

Recommended for: This is said to penetrate better than Bouin's and is especially recommended for arthropods containing parasites.

Formula:

Ethanol (80%)	160 ml
Formalin (40%)	60 ml
Glacial acetic acid	15 ml
Picric acid	1 gm

Procedure: Overnight fixation is adequate.

Randolph's CRAF

Randolph (1935)

Recommended for: Plant tissue, cytological.

Formula:

A. CrO_3	1.0 gm
Acetic acid, glacial	7 ml
Distilled water	92 ml
B. Formalin, neutral	30 ml
Distilled water	70 ml

Just before using, mix A and B.

Procedure: Fix cytological material in this fluid for 12–24 hr. Without washing transfer directly to 70 or 75% alcohol, changing several times at half-hour intervals. After thus removing most of the fixing fluid, transfer to 85% alcohol.

Chrome-Acetic Fluids

Chrome Acetic Fluid, Weak

Recommended for: Easily penetrated plant tissue, e.g., algae, fungi, Bryophyta, prothallia of the Pteridophyta, moss capsules, etc.

Formula:

10% aqueous CrO_3	2.5 ml
10% aqueous acetic acid	5.0 ml
Distilled water q.s.	100 ml

Procedure: Fix tissue 24 hr or more.

Chrome-Acetic Fluid, Medium

Recommended for: Plant tissue, such as root tips, small ovaries, etc.

Formula:

10% aqueous CrO_3	7 ml
10% aqueous acetic acid	10 ml
Distilled water q.s.	100 ml

Procedure: To facilitate penetration, 2% maltose, 2% urea, or 0.3–0.5% saponin may be added. Fix tissue for 24 hr or more.

Chrome-Acetic Fluid, Strong

Recommended for: Plant tissue, such as wood, tough leaves, etc.

Formula:

10% aqueous CrO_3	10 ml
10% aqueous acetic acid	30 ml
Distilled water q.s.	100 ml

Procedure: Fix tissue 24 hr or more, adding maltose, urea, or saponin to the fluid if desired.

Chrome-Osmo-Acetic Fluids

Various chrome-osmo-acetic mixtures have been proposed; and two methods of preparing them are described in the literature. The original Flemming fluid called for aqueous OsO_4 mixed, just before using, with CrO_3 dissolved in acetic acid. Because of the instability of OsO_4, either dry or in aqueous solution, this method of preparing the fluid is wasteful. It has long been known, however, that the mixed Flemming fluid does not change rapidly, and can be kept for some time in a black bottle. In certain types of work, fluids thus preserved are all right; but it is ordinarily preferable to keep the OsO_4 in 2% CrO_3, in which it is fairly stable, and to make up the fixing fluid just before using. Taylor (in McClung's *Microscopical Technique*, 2nd Ed., p. 218) gives various formulas recalculated on this basis; some of them are given below, following the original Flemming formula.

Flemming Fluid, Strong

Recommended for: Plant tissue.

Formula:

A. 1% aqueous CrO_3	45 ml
Acetic acid, glacial	3 ml
B. 2% aqueous OsO_4	12 ml

Preparation: Preferably mix A and B just before using, but the mixture may be kept some time in a black bottle.

Procedure: Fix tissue for 12–24 hr.

Flemming's Fluids, Prepared According to Taylor

Recommended for: Plant tissue.

Formula:

	Medium (ml)		
	Strong	"Bonn Formula"	Weak
10% aqueous CrO_3	3.1	0.33	1.5
2% OsO_4 in 2% aqueous CrO_3	12.0	0.62	5.0
10% aqueous acetic acid	30.0	3.0	1.0
Distilled water	11.9	6.27	96.5

Preparation: Prepare fresh just before using, as the acetic acid spoils the keeping qualities of the chrome-osmic solution.

Procedure: Fix tissue 24 hr or more.

Taylor's Modification for Plant Cytology

Recommended for: Root-tips and smears.

Formula:	10% acqueous CrO_3	0.2 ml
	2% OsO_4 in 2% aqueous CrO_3	1.5 ml
	10% aqueous acetic acid	2.0 ml
	Maltose	0.15 gm
	Distilled water	8.3 ml

Preparation: Make up in small quantities as desired, with the use of a pipette graduated in 0.01 ml. The amount of maltose may be varied.

Mercuric Chloride Mixtures

Schaudinn's Fluid

Schaudinn (1902)

Recommended for: Animal tissue in general

| **Formula:** | A. Saturated aqueous $HgCl_2$ | 2 parts |
| | B. Absolute Alcohol | 1 part |

Mix just before use.

Procedure: Fix tissue 6–16 hours. Wash in several changes of 50–70% alcohol, then preserve in either 70 or 80% alcohol. This procedure is repeated from the previous edition. Schaudinn was not specific. There have been many variants of this technic. (see pp. 000–000).

Heidenhain's SUSA Fluid

Heidenhain (1916)

Recommended for: Animal tissue

Formula:	$HgCl_2$	4.5 gm
	NaCl	0.5 gm
	Formalin	20 ml
	Trichloracetic acid	4 ml
	Distilled water	80 ml

Procedure: Fix about 12 hr; wash in 95% alcohol.

Helly's Fluid

Helly (1903)

Recommended for: Animal tissue, in general.

Formula:	$K_2Cr_2O_7$	2.5 gm
	$HgCl_2$, to saturation	5–8 gm
	Distilled water	100 ml

Just before using, add 5% of neutralized formalin.

Maximow's fluid is the same but calls for 10% formalin. This is, however, properly called Spuler's Fluid, publication of which antedated Helly's publication.

Procedure: Fix tissues 2–4 hr. Wash 12–24 hr in several changes of 70% alcohol, then transfer to 80% alcohol.

Zenker's Fluid

Zenker (1894)

Recommended for: Animal tissue, in general; specially valuable for sharp histological detail.

Formula:		
$K_2Cr_2O_7$	2.5 gm
$HgCl_2$	5 gm
Distilled water q.s.	100 ml
Acetic acid, glacial (added at time of use)	5% by volume

Dissolve the salts in the water with the aid of heat. Add the acetic acid only to that portion of this stock solution which is required on any given occasion.

Procedure: Fix tissue in this fluid 2–4 hr. Wash in several changes of 70% alcohol; then preserve in 80% alcohol.

Lillie's Buffered Sublimate

Lillie and Fullmer (1976)

Recommended for: Zymogen granules and other acid soluble structures, and where the chemical action of formalin on proteins is to be avoided.

Formula:		
$HgCl_2$	6 gm
$NaCH_3CO_2 \cdot 3H_2O$	2 gm
H_2O q.s.	100 ml

Procedure: The mixture is relatively stable. Fix tissue for 18–24 hr, preferably at 4°C.

Lillie's Acidified Sublimate

Lillie and Fullmer (1976)

Recommended for: An excellent general fixative especially for pancreatic islands.

Formula:		
$HgCl_2$	6 gm
H_2O	85 ml

Just before use, add 10 ml formalin and 5 ml glacial acetic acid.

Procedure: Fix for 12–24 hr perferably at 4°C.

Müller's Fluid

Müller (1859)

(An old formula in little use today)

Recommended for: Perfusion of animals in certain neurological work.

Formula:

$K_2Cr_2O_7$	2.0–2.5 gm
$Na_2SO_4 \cdot 10H_2O$	1.0 gm
Distilled water	100 ml

Procedure: Fix tissues in this fluid for 6–8 weeks, changing fluid daily during the first week, once a week thereafter. Ordinarily wash overnight in running water before placing in alcohol.

Marchi's Fluid

Marchi (1886)

Recommended for: Animal and plant tissue in general.

Formula: Mix 2 volumes of Müller's fluid with 1 vol of 1% aqueous OsO_4 just before use.

Procedure: Fix thin blocks of tissue not over 2 mm in thickness for 5–8 days. Wash in running water overnight and place in 70% alcohol.

Zirkle-Erliki Fluid

Recommended for: Plant tissue, for demonstration of mitochondria (all chromatin is dissolved).

Formula:

$K_2Cr_2O_7$	1.25 gm
$(NH_4)_2Cr_2^{2}O_7$	1.25 gm
$CuSO_4 \cdot 5H_2O$	1.0 gm
Distilled water	200 ml

Procedure: Fix tissue 48 hr; wash with running water.

Washing out: With many fixatives, complete removal of the ingredients of the fixing fluid must be accomplished before proceeding further with the technic. This is ordinarily done with either water or alcohol, depending upon the procedure in question. When alcohol or formaldehyde are the fixing agents, washing out is unnecessary; and it can sometimes be dispensed with in the case of acetic acid, picric acid, or even $HgCl_2$.

Baker's Formol Calcium

Recommended for: Lipids.

Formula:	10% CaCl$_2$	10 ml
	Conc. formalin	10 ml
	Distilled water	90 ml

Procedure: Tissues may remain in fixative indefinitely. Sometimes, CdCl$_2$ is added at the same concentration as the CaCl$_2$.

Newcomer's Fixative

Recommended for: Mucopolysaccharides and nucleoproteins.

Formula:	Isopropyl alcohol	60 ml
	Propionic acid	30 ml
	Petroleum ether	10 ml
	Acetone	10 ml
	Dioxane	10 ml

Procedure: For tissues 3 cm × 1 × 1, fix 12 to 18 hr at 4°C. For tissues 3 mm or less, fix 2 to 3 hr. This is a rapid fixative and is excellent for chromatin. It does not require more than 1 change of alcohol before clearing. However, there is excessive shrinkage and only small pieces are recommended.

Formalin-Ethanol-Propionic Acid

Recommended for: Plant tissues.

Formula:	Formalin conc.	10 ml
	95% ethanol	50 ml
	Propionic acid	5 ml
	Distilled water	35 ml

Procedure: For pollen tubes, fix flowers at least 24 hr. Used in the staining of callose.

Salt Solutions

Krebs-Ringer-Phosphate Solution

1. 0.9% NaCl (0.154 M)
2. 1.15% KCl (0.154 M)
3. 1.22% CaCl$_2$ (0.11 M)
4. 3.82% MgSO$_4$·7H$_2$O (0.154 M)
5. 0.1 M phosphate buffer, pH 7.4 (17.8 gm Na$_2$HPO$_4$·2H$_2$O plus 20 ml 1 N HCl; dilute to 1 liter).

To prepare the Krebs-Ringer-phosphate solution, the following amounts of the above are mixed:

100 parts of solution 1
4 parts of solution 2
3 parts of solution 3
1 part of solution 4
20 parts of solution 7

The solution is gassed with either air or oxygen as desired. A precipitate of calcium phosphate forms; this is suspended by shaking before using.

Phosphate Buffered Saline (PBS)-Dulbecco's Modified

1. $CaCl_2$ 100 mg/liter
2. KCl   200 mg/liter
3. KH_2PO_4 200 mg/liter
4. $MgSO_4$ 59 mg/liter
5. NaCl   8000 mg/liter
6. Na_2HPO_4 1150 mg/liter

The above formulation is referred to as PBS complete. When used during tissue dissociation to single cells, omit the calcium and magnesium salts; it is then referred to as PBS Incomplete.

Hanks' Balanced Salt Solution (Hanks' BSS)

1. $CaCl_2 \cdot 2H_2O$ 185.5 mg/liter
2. KCl   400.0 mg/liter
3. KH_2PO_4 60.0 mg/liter
4. $MgSO_4 \cdot 7H_2O$ 200.0 mg/liter
5. NaCl   8000.0 mg/liter
6. $NaHCO_3$ 350.0 mg/liter
7. Na_2HPO_4 47.5 mg/liter
8. Dextrose   1000.0 mg/liter
9. Phenol red, Na   17.0 mg/liter

Dryl's Salt Solution

1. Sodium citrate 0.1 M   20 ml
2. Sodium phosphate monobasic 0.1 M   10 ml
3. Sodium phosphate dibasic 0.1 M   10 ml
4. Redistilled water   945 ml
5. Calcium chloride   15 ml

In order to avoid precipitation of calcium salts, the $CaCl_2$ must be added last in preparing the salt mixture. This solution is recommended by Dryl (1959) in working with Paramecium. It is said to avoid the erratic toxicity of

Ringer's solution. Paramecia and a number of other protozoa live in this solution for many days without visible signs of damage.

WASHING

After fixation is complete it is necessary to wash out the excess fixative before proceeding with dehydration. The following rules are given for convenience.

1. All aqueous fluids, particularly those containing chromic acid or chromates, are washed out with water.
2. Alcoholic solutions should be washed out with plain alcohol of approximately the same strength as used in the fixative.
3. Fixatives containing picric acid, whether in aqueous or alcoholic solution, should always be washed out with alcohol, never with water unless the fixative contains some substance which fixes chromatin indissolubly.
4. If mercuric chloride is used in aqueous solution, it should be washed out with water, but if employed in an alcoholic solution, it should be washed out with 70% alcohol.

PARAFFIN METHOD

Dehydration

Little needs to be said about dehydration as the alcohol method is most common and is quite generally understood. Except for very delicate and friable materials steps of 70%, 95%, and either absolute or n-butyl alcohol are ample. When n-butyl alcohol is used as the final step in the alcohol series the timing is not critical and tissue can be stored in the n-butyl alcohol for a few days. The time is largely a function of block size and density of the tissue, but 2–6 hr in each is about average. Acetone is also a safe dehydrating agent and can replace alcohol in this procedure. It is the most rapid of the dehydrating agents in common use, but 3–4 changes must be used to ensure adequate dehydration. Dioxane, while expensive, is an excellent dehydrating agent. It is miscible with water, alcohol and most oils. It dehydrates and clears at the same time. However, it is very volatile and the fumes are toxic so a hood should be used.

Clearing

After alcoholic dehydration it is necessary to replace the alcohol with some agent miscible with paraffin. Since the ones commonly used have a high refractive index and render the tissue more or less transparent these substances are called clearing agents. Usually xylene, toluene or benzene are used but some prefer cedar wood oil, clove oil, carbon tetrachloride or methyl benzoate. Many make the change from alcohol to clearing agent gradual by interposing an intermediate bath consisting of equal parts of alcohol and the clearing agent. With dioxane and *n*-butyl alcohol no clearing agent is needed except that with *n*-butyl alcohol clearing of nervous tissue is essential to prevent loss of sections in staining.

NOTE: For a discussion of many additional methods for dehydration, clearing and embedding, see Lillie and Fullmer (1976).

The material is now ready for infiltration. It is well to have three dishes of paraffin and to shift the object from one to the other at equal intervals. The duration of the paraffin bath varies according to the size and density of the object. Many objects of from 3 to 5 mm in thickness are thoroughly saturated in an hour or less; others which are more impervious or which have impenetrable coverings may require several hours or even days. Some objects difficult to infiltrate are embedded most quickly and satisfactorily if the infiltration is allowed to take place in partial vacuum. This is most simply accomplished by use of a vacuum oven or by putting the paraffin and tissue in a test tube which is kept in a heated water bath, with a connection between the test tube and a faucet suction pump.

Paper boxes are prepared and, just as the material is ready, sufficient melted paraffin is transferred to the box, and the tissue placed in it. The box is filled with melted paraffin. This must be added before the surface of the underlying paraffin has congealed. The object is oriented with heated needles if necessary. As soon as the paraffin has congealed sufficiently to form a thin surface film, it is cooled rapidly by plunging it into cold, preferably running water; otherwise the paraffin will crystallize and become unsuited for sectioning.

Sectioning and Attaching Sections to Slides

Today, sectioning of paraffin embedded material is almost always done with a rotary microtome. Various types of microtomes are on the market and little need be said here in regard to their use. Such points as trimming the paraffin block so as to give continuous, straight ribbons of serial sections, maintaining the paraffin at a suitable temperature for sectioning, and setting the microtome to give proper thickness of section for the material in question, are all important; but this is hardly the place for discussing them in detail.

There are various methods for attaching sections to slides. Ordinarily some form of adhesive is applied to the slide before mounting the sections. The most widely used are Mayer's albumen and gelatin. The former may be purchased from most supply houses but is easily prepared. To 50 ml fresh egg albumen (white of egg) add 50 ml glycerin and 1 gm sodium salicylate. Place in a tall cylinder and shake until a large number of air bubbles are in suspension. Allow the bubbles to rise to the surface carrying with them the larger fragments of the white of the egg. Allow to stand and remove upper layer. The remaining clear solution should be filtered. Place a tiny drop of this adhesive on a slide and with the tip of the finger spread over the surface of the slide leaving a very thin film. Put slide on warming plate and place a drop or two of water on the slide and spread the section on this drop. When the section is flattened, position it with a fine brush and drain the water from the slide. Then place the slide in a rack and leave overnight on the warming plate. A simpler method is to use a solution of gelatin in a flotation bath (a thermostatically controlled water bath holding about 1000 ml). Gelatin (0.8–1.0 gm) is dusted on the surface of the water and, after the gelatin goes into solution, add 3–5 ml of "Photo-Flo" (from the photographer). The sections are floated on the warm water (about 40–45°C). Then a slide can be dipped under the section and on lifting the section will adhere to the slide. Place the slide in a rack and leave overnight on the warming plate.

Even though "precleaned" slides are available when the procedure to be used causes sections to fall off the slide it is best to use "subbed" slides. Place slides in a glass staining rack and immerse in sulfuric acid dichromate cleaning solution for a few minutes. Wash thoroughly in running water. Immerse slides in the "subbing" solution—3 gm gelatin, 0.3 gm $Cr_2(SO_4)_3 \cdot K_2SO_4 \cdot 24\ H_2O$ in 300 ml H_2O—for a few minutes. Then place rack in staining dish, loosely covered, in paraffin oven overnight. Sections can be mounted on these slides using the usual flotation bath as given above.

CELLOIDIN METHOD

When thick sections or when maintenance of exact relations are desired, the celloidin method, although time consuming, is the method of choice. After preliminary treatment similar to that for paraffin embedding, the tissue is brought through the lower grades of alcohol to absolute alcohol and then to ether-alcohol (absolute alcohol: one part; diethyl ether: one part). A block the size of a cat brain should remain in absolute alcohol (one change) for at least two weeks and in ether-alcohol (one change) for not less than two weeks; then into 5% celloidin (Parloidin, purified pyroxylin) for at least another week followed by 10% celloidin for another week. In the preparation of the celloidin solutions, the thin strips of celloidin should be briefly washed in absolute alcohol and then dissolved in ether-alcohol. To promote solution of the celloidin, the bottle should be upended daily.

Embed in appropriately sized paper boats. The paper must be sufficiently porous to allow slow passage of the ether-alcohol and sufficiently dense to retain the celloidin. The boats and a vial half filled with concentrated H_2SO_4 are placed in a small container that can be securely capped. The acid takes up the vapors of alcohol and ether that evaporate from the celloidin solution. The volume of acid increases and it becomes brownish in color. If a film forms on the celloidin, the container is either too large or the cap is not secure. Add additional 10% celloidin to maintain volume.

When the celloidin is firm enough to indent with finger and not stick to the finger, the block can be hardened by removal of the vial containing H_2SO_4 and pouring a small amount of chloroform into the container with the block. After a day or so the paper boat can be peeled off and the block cut to size. The block should be firm; if not, allow it to remain in the original container with some added chloroform. The block can be mounted immediately or stored in 70% alcohol. If stored, place the block in n-butyl alcohol overnight before attempting to mount. Fiber blocks are available in various sizes and are convenient for mounting of the tissue blocks. Place the surface of the fiber block into ether-alcohol (a Petri dish is convenient) and then into 10% celloidin. The same is done with the tissue block. The two are pressed together and additional celloidin applied to the basal sides of the tissue block. Continue pressing the two together and allow to air dry. When a film forms on the applied celloidin, the mounted block can be placed in a container with a small amount of chloroform.

When hard, the block can be mounted on a sliding or sledge microtome and sectioned. The block and knife must be kept damp with 70% alcohol throughout cutting. Sections can be cut from less than 10 μm to over 100 μm in thickness. The cut sections can be stained immediately or stored in 70% alcohol. After the sections are stained, they can be partially dehydrated in graded alcohols. Absolute alcohol must be avoided, for even in a short time the celloidin will soften and become sticky. However, since celloidin is not soluble in n-butyl alcohol, this may be used. For complete dehydration, sections should be left in terpineol-xylene (terpineol: 1 part; xylene: 3–4 parts) overnight and then cleared in xylene and mounted.

OTHER EMBEDDING METHODS

Tissues can be embedded in a variety of substances other than celloidin and paraffin. For example, the original Nissl method used tissue embedded in soap. Where preservation of all lipids is required the various water-soluble waxes are feasible embedding media. Also agar and gelatin have been used to hold friable tissue in place for sectioning on the freezing microtome. These methods are described at length by Steedman (1960) and to a much lesser extent by Lillie and Fullmer (1976). These methods are so rarely used that no discussion will be given here.

CHAPTER 2

PREPARATION OF SEMI-THIN SECTIONS OF TISSUES EMBEDDED IN WATER-SOLUBLE METHACRYLATE FOR LIGHT MICROSCOPY

William J. Dougherty

The advent of electron miscoscopy prompted a continuing search for embedding media that on the one hand could withstand electron bombardment and high vacuum conditions and on the other would not greatly distort molecular, cellular and tissue architecture. As a result, a variety of polyester and epoxy resins have found routine use in ultrastructural studies of prokaryotic and eukaryotic cells. These plastics have also proved to be useful for light microscopic observations and orientation of cells and tissues that are to be studied subsequently in the electron microscope. The number and variety of stains that can be applied to sections of epoxy- and polyester-embedded tissues, however, is strictly limited and this thwarts efforts to make routine the preparation of plastic-embedded tissue sections for diagnostic or teaching purposes.

In recent years, however, it has been demonstrated that sections of tissues embedded in water-soluble methacrylates, e.g., glycol methacrylate, can be stained very effectively with dyes and staining procedures that have been used commonly with paraffin-embedded tissues. One biological supply company has been offering for sale for several years hemotoxylin and eosin stained sections (1.5 μm thick) of glycol methacrylate-embedded tissues. In addition, a few newer stain combinations have been developed to reveal cytologic and histologic details that are better preserved in tissues embedded in water soluble methacrylates. It is the purpose of this chapter to present the procedures for embedding tissues in water-soluble methacrylates and for sectioning and staining such tissues.

Embedding tissues in glycol methacrylate offers several advantages. Either commonly used or special fixative mixtures can be employed prior to glycol methacrylate embedding and tissues may be dehydrated either in organic solvents such as alcohol or acetone or directly in aqueous dilutions of glycol methacrylate. Intermediate organic solvents such as xylol, toluene, benzene, etc., need not be employed. The plastic monomer is liquid at room temperature and infiltration of tissues with monomer can be carried out at room or lower temperatures. Polymerization of the plastic is routinely accomplished at room temperature within a period of ½ to 3 hr, but can be performed also at temperatures as low as $-20°C$ under ultraviolet light or as high as $60°C$ in an oven. Sectioning of tissues embedded in glycol methacrylate is done at room temperature and sections as thin as 0.5 μm or as thick as 10 μm can be cut with glass knives. The glass knives can vary in width from 6.5 to 38 mm making it possible to section either very small or fairly large specimens or samples of tissue. Steel knives can be used for sectioning but are prone to being nicked and causing sectioning artifacts and, in addition to being initially expensive, tend to have a high down time. Plastic sections are mounted directly on glass slides without albumen fixative, and do not require removal of the plastic prior to staining. Staining is accomplished by dipping dried sections directly into the staining solution without the prior need for running sections up to water or dehydrating sections through alcohols after staining. In general, the major practical advantages offered by embedding tissues in glycol methacrylate are (a) a shortened period between sampling of tissue and the final, stained section, and (b) the need for less elaborate and less expensive infiltrating and embedding equipment. Microtomes currently in use in laboratories cutting paraffin sections can be used for sectioning glycol methacrylate embedded tissues using a simple adapter for mounting glass knives.

Other advantages accrue from embedding tissues in glycol methacrylate. The thinner sections which can be cut provide better optical resolution and clarity. Optical effects due to superimposition of the image are lessened in 0.5–1.5-μm thick sections. Improved radioautographic and cytochemical resolution result also. Because the plastic need not be removed from the sections, the sections are stronger and the dimensions of cellular and tissue components are less affected by physical forces such as surface tension during drying of the sections.

INFILTRATION AND EMBEDDING MIXTURES AND PROCEDURES

The infiltration and embedding mixtures containing glycol methacrylate are liquid at room temperature and usually consist of three components, viz., glycol methacrylate monomer, a polymerization initiator or catalyst, and a plasticizer or filler. The glycol methacrylate monomer is supplied with hydroquinone or similar substance which serves to inhibit spontaneous polymerization during synthesis or storage. These inhibitors should be removed with activated charcoal prior to preparation of the infiltration and embedding mixtures. This can be done by adding 4 gm of activated charcoal to each 100 ml of glycol methacrylate monomer and shaking or stirring for about 1 hr prior to filtering through filter paper. This procedure should be repeated and the clear filtrate should be stored in a refrigerator or freezer prior to use.

A list of infiltration and embedding mixtures based on glycol methacrylate includes the following.

1. Feder and O'Brien (1968)

Glycol methacrylate*	94.5 ml
2,2'-Azobis(2-methylproprionitrile)†	0.5 gm
Polyethylene glycol 400†	5.0 ml

Dissolve the azobis and polyethylene glycol 400 in glycol methacrylate monomer. The mixture is filtered if it is not clear and can be stored for several months in the refrigerator.

The composition of the embedding mixture may have to be modified if it produces plastic blocks that are too hard or soft. If the blocks are too hard and brittle, more plasticizer, polyethylene glycol 400, should be added. A concentration of up to 10% may be used.

The differences between Mixtures 1 and 5 below are minor. One is measured volumetrically, while the other is prepared gravimetrically. Gravimetric preparation is extremely convenient and less hazardous to one's health if the laboratory has a top-loading balance.

2. Ruddell (1971)

Glycol methacrylate*	50 ml
2-Butoxyethanol	12 ml
(ethylene glycol monobutyl ether)†	
Benzoyl peroxide†	0.25 gm
Pyridine†	0.75 ml

Dissolve the benzoyl peroxide in the glycol methacrylate butoxyethanol mixture. The benzoyl peroxide will dissolve within 1–2 hr at room temperature. When solution is complete, add the pyridine to the mixture and chill to inhibit polymerization at room temperature. Infiltration of tissue with this embedding mixture should be done at 4–8°C.

3. Cole and Sykes (1974)

Glycol methacrylate*	100.0 ml
(prepolymerized)	
Benzoyl peroxide†	0.15 gm
Polyethylene glycol 400†	5.0 ml

---

* *Hartung Associates, 803 North 33rd St. & Cleveland Ave., Camden, NJ 08105.*
† *Fisher Scientific Co., 2775 Pacific Dr., Norcross, GA 30091.*

Dissolve the benzoyl peroxide in 100 ml of glycol methacrylate. Prepare "prepolymerized" glycol methacrylate by heating small portions of this mixture carefully with constant stirring. A hot plate-stirrer is convenient for this purpose, and a thermometer should be kept at the ready. As the temperature of the mixture approaches approximately 100°C, the mixture starts to become syrupy in consistency and an exothermic polymerization reaction occurs. At this point, the reaction vessel should be plunged quickly into an ice-slush or a dry ice-acetone slurry. When completely cooled, the viscosity of the partially polymerized mixture can be tested. If the proper viscosity is not reached after the initial heating and cooling sequence, this may be repeated as many times as is necessary to obtain proper viscosity. Once proper viscosity is achieved, 5.0 ml of polyethylene glycol 400 is stirred into each 100 ml of prepolymerized solution and the final mixture is stored at −20°C until needed.

4. Sims (1974)

A. Infiltrating solution:	
Glycol methacrylate*	80 ml
2-Butoxyethanol†	8 ml
Benzoyl peroxide†	0.5 gm
B. Promoter solution:	
Polyethylene glycol 400†	8 ml
N,N-dimethylaniline†	1 ml
C. Embedding mixture:	
Infiltrating solution	42 vol
Promoter solution	1 vol

Three separate solutions are used in this procedure. The infiltrating and promoter solutions can be prepared beforehand and stored at 0–4°C for several months. Once the embedding mixture is prepared, it is best to make use of it shortly, although polymerization can be inhibited by keeping it in the refrigerator or several degrees below room temperature.

5. Bennett, Wyrick, Lee and McNeil (1976)

Glycol methacrylate*	95 gm
2,2′-Azobis(2-methylproprionitrile)†	0.4 gm
Polyethylene glycol 400†	5 gm

Similar to directions for Mixture 1 above, except that components are measured by weight.

Infiltration and embedding procedures with any of the above listed plastic mixtures are simple and similar. Transfer fixed and dehydrated specimens from absolute alcohol to a 1:1 mixture of absolute alcohol and monomer mixture for ½ to 24 hr, depending upon the size of the tissue sample. Tissues are then transferred through at least two changes of the monomer mixture, allowing ½ to 24 hr for each change, depending upon the size of the tissue, before placing and orienting the tissues in suitable molds filled with fresh monomer mixture for final polymerization. Embedding molds may be gelatin capsules,‡ BEEM capsules,§ plastic block molding cups,‖ or other suitable and commonly available molds. Paper molds or wax-coated cardboard molds should be avoided, however.

Polymerization is usually accomplished at room temperature with Mixtures 2 and 4. With Mixtures 1, 3 and 5, polymerization is facilitated by heating in an oven at 37–40°C. Air should be restricted from the open end of the embedding mold during polymerization, because of the inhibition of polymerization by oxygen. The caps of gelatin or BEEM capsules are convenient for excluding oxygen after the capsules have been filled to the brim with monomer. Aluminum stubs‖ to which the polymerized plastic will adhere and which fit into the microtome feed assembly can be used to exclude oxygen when plastic block molding cups are used.

TYPICAL FIXATION, DEHYDRATION AND EMBEDDING SCHEDULE FOR GLYCOL METHACRYLATE EMBEDDING

Fixation: Fix tissues to be studied by light microscopy by immersion in or by vascular perfusion of an appropriate fixative mixture. Fixation time may vary from 10 min to overnight depending upon the specimen and its size. Fixation temperature may be at room temperature prior to most histological staining procedures or at 0–4°C prior to histochemical procedures designed to reveal enzymatic activity. Electron microscopic studies indicate that buffered fixatives using well purified fixative reagents give the best preservation of tissue, cellular and subcelluar architecture. Purified aldehydes such as glutaraldehyde and paraformaldehyde and buffer salts are commercially available.

Dehydration, infiltration and embedding: Usual dehydration practices can be followed prior to glycol methacrylate embedding. If the fixative mixture was aqueous, specimens or tissue samples are washed well in the buffer that was used to prepare the fixative mixture or in tap water prior to immersion in solutions of alcohol or acetone of increasing concentrations. If nonaqueous fixative mixtures are used, specimens or tissue samples are transferred to absolute alcohol or acetone. Samples are then immersed in a mixture of equal volumes of alcohol (or acetone) and glycol methacrylate embedding mixture prior to immersion in the glycol methacrylate embedding mixture alone.

---

‡ Eli Lilly and Company, Indianapolis, IN 46206.
§ Electron Microscopy Sciences, P.O. Box 251, Fort Washington, PA 19034.
‖ E. I. duPont deNemours and Co., Inc., Instruments Products, Biomedical Division, Newtown, CT 06470.

A typical schedule after aqueous fixatives is as follows:

Rinse	5 min to 1 hr
50% alcohol	5 min to 1 hr
70% alcohol	5 min to 1 hr
80% alcohol	5 min to 1 hr
90% alcohol	5 min to 1 hr
95% alcohol	5 min to 1 hr
100% alcohol	5 min to 1 hr
1:1 100% alcohol/ glycol methacrylate (GMA) mixture	½ hr to overnight
GMA mixture	½ hr to overnight, 2×
GMA mixture	Embed and polymerize

Alternatively, *small* specimens or samples of tissue may be dehydrated in aqueous dilutions of glycol methacrylate. Adequate preservation of histological architecture is not maintained if *large* tissue samples are dehydrated with glycol methacrylate. After rinsing in buffer or tap water, specimens or tissue samples are transferred directly to 85% glycol methacrylate in water. A typical schedule for glycol methacrylate dehydration is as follows:

rinse	5 min to 1 hr
85% GMA	2 changes, 1 hr each
95% GMA	2 changes, 1 hr each
GMA embedding mixture	1 hr
GMA embedding mixture	Polymerize

PREPARATION OF GLASS KNIVES AND SECTIONING

Sectioning tissues embedded in glycol methacrylate is done most satisfactorily with glass knives. Steel knives may be used for sectioning plastic-embedded tissues, but the knife must be sharpened to a keen and near-perfect edge. Both steel and glass knives are dulled by cutting plastic embedded tissues, but glass knives with near-perfect edges are relatively simple and quick to prepare, are inexpensive and easily discarded and replaced.

Glass knives may be prepared from strips of glass that are 6.5 mm thick and either 25 or 38 mm in width.¶ These glass strips can be broken into squares either manually with the aid of pliers or mechanically with the aid of commercially available glass knife makers.‖,¶ These squares are then

---

¶ *LKB Instruments, Inc., 12221 Parklawn Dr., Rockville, MD 20852.*

broken into triangles to provide Latta-Hartmann glass knives with a cutting edge that is at most 6 mm long. Squares can be prepared from plate glass that is as thick as 10 mm and from these Latta-Hartmann knives with 8–9-mm long cutting edges can be prepared. These glass knives are mounted in commercially available glass knife holders that are supplied with or are adaptable to most microtomes.

For specimens or tissues with dimensions that are greater than 8–9 mm, Ralph-type glass knives can be prepared from strips of glass that are of the same dimensions used for preparing Latta-Hartmann glass knives, i.e., 6.5 mm thick by 25 or 38 mm wide. These strips are broken into 25 or 38 mm squares either manually as described in detail by Bennett et al. (1976), or mechanically using commercially available Ralph-type glass knife makers.¶ Ralph knives provide near-perfect cutting edges that are as long as 30 mm, thus making it possible to section specimens that are approximately this dimension. Ralph knives are attached to 10–15-mm long strips of metal with either molten wax, double-sided Scotch tape, or other types of cement and are mounted in standard steel knife holders of most microtomes.

Trimming the specimen is performed much like that for paraffin blocks. Although trimmed blocks can be clamped directly in the microtome chuck, sectioning is generally more satisfactory when blocks are attached to a metal or wood carrier. If a metal stub to which the plastic will adhere is not used to exclude oxygen during polymerization, tissue blocks can be attached to a carrier with a rapid setting epoxy glue.

After installing a specimen block in the advance mechanism and a glass knife in the knife holder of the microtome, sectioning is begun. Since the sections do not form ribbons, their handling is like that used for nitrocellulose sections. Sections 0.5–1.5 μm in thickness are generally most useful and are made with a slow cutting speed. As the knife cuts into the block, the newly formed edge of the section is lifted from the dry knife edge with a pair of very fine watchmaker's forceps and light tension is maintained to reduce folding of the section. Sections are lifted very gently one at a time from the knife edge and transferred quickly either to a drop of water in the center of a clean glass slide or to a finger bowl filled with clean water where sections can be accumulated conveniently prior to being maneuvered onto glass slides. Sections usually stretch out when contact with water is made. Folding of sections, especially of the thinner and larger ones, is sometimes encountered. Folds can often be pulled flat with brushes or forceps while the section is on the water surface. Other measures may faciliate the stretching of sections and the removal of folds, including the use of warm water or water to which a drop or two of concentrated ammonium hydroxide has been added.

Sections are mounted on clean glass slides without the use of Mayer's albumen or other adhesive. To pick up sections which are accumulated on the surface of water in a finger bowl, a glass slide is partially inserted at a slant into the water. The slide is then brought up under the section or sections to be mounted, the orientation and position of each section is adjusted with a brush, and the slide is removed quickly from the water. Slides are placed on a hot plate at 30–40°C and allowed to dry. Thorough drying may take 10–15 min. Thereafter, sections are ready for staining.

Staining is accomplished without prior removal of the polymerized plastic. Dried sections are simply covered with or immersed in the dye solution of choice, rinsed briefly and blown dry with a quick blast of clean dry gas. The section is then ready for mounting under a coverslip in a suitable mounting medium.

STAINING METHODS FOR TISSUES EMBEDDED IN GLYCOL METHACRYLATE

Five procedures which we have found most useful are Bennett's Methylene Blue/Basic Fuchsin, Sidman's Acid Fuchsin-Toluidine Blue, Lockard's Luxol Fast Blue-Neutral Red, Hematoxylin and Eosin, and Van Gieson. The methods in use are given in this section. Actually most procedures given elsewhere in this volume can be used, usually with no or only slight modifications. Among those which we have used are: Buffered eosin-azure (p. 209), Feulgen (p. 201), Periodic acid Schiff (p. 200), Methyl green-pyronin (p. 199), Alcian blue (p. 201) either with or without PAS, Giemsa (p. 173), Masson (p. 118), and Foot's modification of Bielshowsky's method for collagen and reticular fibers (p. 126).

Methylene Blue/Basic Fuchsin
From Bennet et al. (1976)

Stock solutions:

Methylene blue, C.I. 52015	0.13 gm
Distilled water *q.s.*	100 ml
Basic fuchsin, C.I. 42500	0.13 gm
Distilled water *q.s.*	100 ml
0.2 M phosphate buffer, pH 7.2–8.0	

Staining mixture:

Methylene blue solution	12 ml
Basic fuchsin solution	12 ml
0.2 M phosphate buffer	21 ml
Ethanol (95% or absolute)	15 ml
Filter; may be used for 4–5 days	

Staining procedure:
1. Immerse sections in stain mixture for 10–20 sec.
2. Remove slides, dip briefly (10–30 sec) in distilled water and blow dry with a blast of clean, inert gas.

3. Mount dried sections under a cover slip using Permount.

Results: Nuclei stain blue, cytoplasm magenta with patches of blue ergastoplasm.

Acid Fuchsin-Toluidine Blue

From Sidman et al. (1961), Ashley and Feder (1966), Feder and O'Brien (1968)

Staining solutions:
A. Toluidine blue, C.I. 52040   0.05 gm
   Acetate buffer (pH 4.4) q.s.   100 ml
B. Acid fuchsin, C.I. 42685   1.0 g
   Distilled water q.s.   100 ml

Staining procedures:
1. Immerse sections in 1% acid fuchsin solution for 15 sec to 5 min.
2. Rinse briefly in distilled water.
3. Immerse sections in toluidine blue solution for 30 sec to 2 min.
4. Rinse briefly in distilled water and blow dry with a blast of clean, inert gas.
5. Sections may be restained with either or both stain solutions, as appropriate or necessary.

Results: Nuclei blue, cytoplasm red, red blood cells bright red.

Luxol Fast Blue-Neutral Red
From Lockard and Reers (1962)

Stock solutions:
A. Luxol fast blue in acidified alcohol
   10% acetic acid   20 ml
   Luxol fast blue, C.I. 00000   0.25 gm
   95% ethanol   380 ml
B. Lithium carbonate-lithium hydroxide solution
   $LiCO_3$ 5.0 gm
   $LiOH \cdot H_2O$ 0.01 gm
   Distilled water   1000 ml
C. Neutral red solution
   Neutral red, C.I. 00000   0.5 gm
   Distilled water   1000 ml
D. 0.2% $NaHSO_3$
   $NaHSO_3$ 2.0 gm
   Distilled water   1000 ml
E. 0.1 M acetate buffer, pH 5.6
   0.1 M Na acetate   9 vol
   0.1 M acetic acid   1 vol

F. Copper sulfate-chrome alum solution
$CuSO_4 \cdot 5H_2O$ 0.5 gm
$CuK(SO_4)_2 \cdot 12H_2O$ 0.5 gm
10% acetic acid   3.0 ml
Distilled water   250 ml

Staining solutions:
1. Luxol fast blue stain
   Luxol fast blue stock solution   1 vol
   95% ethanol   3 vol
2. Differentiating solution
   Stock lithium solution   1 vol
   Distilled water   9 vol
3. Neutral red stain
   Neutral red stock solution   3 vol
   0.1 M acetate buffer, pH 5.6   2 vol

Staining procedure:
1. Immerse sections in Luxol fast blue staining solution for 2 hr at 57°C, then 30 min at room temperature; rinse in 95% ethanol.
2. Differentiate in lithium carbonate differentiating solution at 0–4°C for as long as necessary; rinse in 70% ethanol for 1 min or less and then in distilled water for 1 min.
3. Treat with 0.2% $NaHSO_3$ for 1 min.
4. Immerse sections in 0.1 M acetate buffer, pH 5.6, for 1 min.
5. Stain in Neutral red staining solution for 15 min at room temperature; rinse in distilled water.
6. Immerse in copper sulfate-chrome alum solution for 1 sec; rinse in distilled water.
7. Blow dry with blast of clean, inert gas.

Results: Myelin sheaths are blue. Extracellular connective tissue elements are blue to light purple. Nuclei and cytoplasmic basophilic structures are red.

Hematoxylin and Eosin

Stock solutions:

A. Gill's triple hematoxylin
    Hematoxylin, C.I. 75290   6 gm
    Sodium iodate   0.6 gm
    Aluminum sulfate   52.8 gm
    Distilled water   690 ml
    Ethylene glycol   250 ml
    Acetic acid, glacial   60 ml

B. 1% Eosin

Eosin Y, C.I. 45380	1 g
Distilled water	20 ml
95% ethanol	80 ml

Staining solutions:
A. Gill's triple hematoxylin, full strength
B. Eosin:

Eosin stock solution	1 part
80% ethanol	3 parts
Acetic acid, glacial	0.5 ml*

* *Added to each 100 ml of staining solution just prior to use.*
C. Acid/alcohol:

HCl, conc.	20 ml
70% ethanol	1000 ml

Staining procedure:
1. Dip slides in distilled water, 10 sec.
2. Stain in Gill's triple hematoxylin, 1–2 hr.
3. Rinse in tap water, briefly.
4. Differentiate in acid/alcohol, 5–10 sec.
5. Rinse in tap water, 15 min.
6. Blow dry with blast of clean, inert gas.
7. Stain in eosin, 5–10 min.
8. Rinse in 100% ethanol, fast dip.
9. Rinse in acetone × 2, 1 min.
10. Blow dry with blast of clean, inert gas.
11. Mount dried sections under coverslip.

Results: Nuclei stain blue, cytoplasm pink.

Van Gieson Stain

Stock solutions:
A. Celestin blue B, C.I. 51050

Ferric ammonium sulfate	0.5 gm
Distilled water	5.0 gm
	100 ml

B. Weigert's hematoxylin

(1) Hematoxylin, C.I. 75290	10 gm
95% ethanol	100 ml
(2) Ferric chloride	11.6 gm
Distilled water	980 ml
HCl, conc.	10 ml

C. Van Gieson solution
    (1) Acid fuchsin, C.I. 42685   1 gm
        Distilled water   100 ml
    (2) Picric acid   3 ml
        Distilled water   100 ml
D. Acid alcohol solution
    HCl (7%)   20 ml
    Ethanol   1000 ml

Staining solutions:
A. Stock celestin blue B solution
B. Weigert's hematoxylin
    Solution A   10 ml
    Solution B   50 ml
    Absolute ethanol   40 ml
C. Van Gieson solution
    1% acid fuchsin   20 ml
    Picric acid   25 ml

Staining procedure:
1. Stain sections in stock Celestin blue B solution, 10 min.
2. Stain in Weigert's hematoxylin, 30 min.
3. Rinse in tap water.
4. Differentiate in acid/alcohol and rinse in water, quick dip.
5. Dilute ammonium hydroxide, dip.
6. Stain in Van Gieson solution, 7 min.
7. Rinse in water and dry sections with blast of clean, inert gas.
8. Mount sections under coverslip.

Results: Nuclei stain black, muscle yellow and collagen red.

CHAPTER 3

METHODS FOR FLUORESCENCE MICROSCOPY

Frederick H. Kasten

SECTION A: INTRODUCTION

This chapter has been greatly expanded because of the large number of fluorescence staining methods in current use and their growing importance. In the previous edition of this volume, less than a dozen fluorescent methods were described which were scattered throughout the book. The list is now enlarged to include more than 50 methods, which are largely placed together in this chapter for convenience. In addition to picking up some of the older technics which are used frequently in pathology, histology, and microbiology laboratories, there are many new ones included which are employed for chromosome identification (Q-banding methods) and molecular analysis in cytofluorometry (cf. Kasten's fluorescent Schiff reagents for DNA, cell surface labeling with fluorescamine). The lack of space precludes the inclusion of certain nucleic acid fluorophores and protein-binding agents which are important research tools in molecular cytochemistry, especially for cytofluorometry and cell sorting applications (cf. ethidium bromide, propidium bromide, actinomycin D, bromodeoxyuridine substitution, Fluorescein isothiocyanate, Sulfaflavine, and Sulpho-rhodamine 101). Some fluorochromes are ideal for supravital staining of living cells because innocuous dye concentrations are sufficient to impart visible or photodetectable fluorescence, as for example Acridine Orange, SITS, Fluorescein diacetate (FDA), and DAPI. Xylenol orange, a substitute for tetracyclines, is injected intravitally to localize newly deposited calcium salts in bone. As a rapid screening technique for acid-fast bacilli and fungi, a new Auramine O-Acridine Orange method exploits the brilliance of fluorescence to demonstrate these organisms. Excellent detail is revealed in semi-thick sections of epoxy-embedded tissues when the sections are stained with fluorochromes specific for carbohydrates in the fluorescent-PAS Method. Mycoplasmas and DNA viral infections of cell cultures are now detectable with a simple and sensitive fluorochroming technique using DAPI or Hoechst 33258. Certain naturally occurring substances in tissues are detectable by their autofluorescing properties, such as thiamine or vitamin B_1 and vitamin A. However, these and other methods for natural fluorescing compounds require considerable care in interpretation. The availability of high-class fluorescence microscopic equipment and light sources has helped make it possible to advance the study of the nervous system. Living nerve cells can now be studied using fluorescent Procion dyes. A standard procedure has been selected for this volume using Procion Yellow M4RS. Similar advances are occurring in the histochemical detection of neurotransmitters in the nervous system, starting with the Falck-Hillarp method. It has not been possible to include certain cytofluorometric applications, despite their great interest, such as the resolving and separating of X and Y spermatids (Meistrich et al., 1978) and the sorting of human chromosomes (Carrano et al., 1979).

The methods included are a representative cross section of those in use by biologists and cytochemists. To include all methods would require a separate book devoted entirely to fluorescence microscopy. Hopefully, the methods selected on the basis of their general applicability and present-day or potential importance will be adequate for the majority of readers.

SECTION B: THEORY AND PRACTICAL ASPECTS OF FLUORESCENCE MICROSCOPY

Fluorescence is an optical phenomenon in which UV or short wavelength visible light energy (violet, blue) is absorbed by a molecule (fluorophore) and rapidly reemitted as light of a longer wavelength. For example, blue light is used to irradiate a tissue section previously stained with Acridine Orange. Some of the stained structures will emit green light and others fluoresce red. The reason for fluorescence at two different wavelengths in this particular example (fluorochromasia) has to do with dye-dye interaction in the presence of different cell components (DNA, RNA).

Because the emitted light is of a longer wavelength than that used for excitation, the conventional transmission microscope simply requires two additional filters and a different light source in order to be converted into a fluorescence microscope. A special primary or excitation filter is inserted between the light source and the specimen. This filter can be violet or blue, according to whether UV (below 400 nm) or blue light (400–500 nm) is to isolated. Commonly, a train of primary filters is placed directly in front of the light source to permit various filter combinations to be selected. The other essential component of the microscope is the secondary or barrier filter, which is inserted between the specimen and the eye. Its purpose is to block the excitation wavelength from reaching the eyes and causing UV radiation damage. Glass lenses cannot be counted on to block potentially damaging UV light since glass cuts off light rays only below 300 nm. The barrier filter is frequently pale yellow although other colored filters may be used, according to the color of fluorescent light. In practice, the barrier filter is placed in the eyepiece if no provision is made in the microscope tube for permanently placed barrier filters. It should be emphasized that a special light source is required with high light intensity in the long UV-visible spectra. High pressure mercury vapor lamps are ideal because of the variety of strong emission lines which can be isolated. Recently, tungsten-halogen lamps have become available which are suitable for most purposes. For routine illumination of the specimen, a substage bright-field condenser is used (dia-illumination), which also permits observation by absorption and phase-contrast if desired. It is the simplest to use if a special fluorescence microscope with epi-illumination is not available. The results are relatively good for routine applications. However, dia-illumination cannot be used on thick or opaque objects, is difficult to use for fluorescence quantitation, and provides relatively weak emission at high magnifications. Epi-illumination is a system in which the objective is also the condenser. In addition, a dichroic mirror (preferred) or beamsplitter is employed to reflect the exciting light to the specimen and simultaneously transmit the fluorescent light to the eyepiece. All of the disadvantages mentioned above for dia-illumination are overcome in the epi-illumination system. The only disadvantage with the latter is that the emission intensity is relatively weak with very low power objectives. If a person has available a fluorescence microscope with both dia- and epi-illumination capabilities, this is an ideal setup. A third system of illumination known as dark-ground is not worth discussing. It is beyond the scope of this discussion to consider a specific choice of filters. If one has available a variety of primary and secondary filters, then he will be in a position to determine the correct combination for a particular fluorophore.

SECTION C: EMPLOYMENT OF FLUORESCENCE METHODS

Some tissue components are naturally fluorescent (autofluorescence), others can be altered to fluorophores by chemical treatment (induced fluorescence), and many tissue components can be bound to fluorescent dyes (fluorochromy). All unstained tissues retain some degree of autofluorescence, especially due to proteins. In animal tissues, extracellular connective tissue fibers (collagen, elastin), ceroid, lipofuscin, vitamins A and B_2, and porphyrins exhibit pronounced autofluorescence. Certain mitochondrial enzymes (NADH, flavoproteins) are believed to be responsible for much of the intracellular self-fluorescence. In plant tissues, natural fluorescence is exhibited by carotenes, chlorophyll, and a number of alkaloids. Generally, autofluorescence is undesirable since it may mask or cause confusion in interpreting a specific fluorescence resulting from dye binding. Also, natural fluorescence reduces the inherent sensitivity of the staining technic. By judicious selection of primary and secondary filters, undesired fluorescence can frequently be minimized. Fixatives containing heavy metals, such as mercury and iron, quench natural fluorescence as well as secondary fluorescence with added dyes. Aldehyde fixation is frequently avoided, as condensation reactions occur with tissue proteins which causes ring formation and increased autofluorescence. Formalin-fixed paraffin sections are said to be suitable for the fluorescence detection of amyloid, mucins, and acid-fast bacilli. However, formaldehyde-induced fluorescence is an extremely important technic in the histochemical detection of biogenic amines such as catecholamines and 5-hydroxytryptamine. Special nonfluorescent mounting media are utilized such as Fluormount. Others are listed on page 100. Some fluorescent trypanocidal agents (Acriflavine) and carcinogenic hydrocarbons (benzpyrene) have been detected by fluorescence microscopy after their introduction into tissues. Fluorochromy, which utilizes fluorescent dyes as stains, is the most widely used application of fluorescence microscopy. Many yellow, orange, and red dyes are fluorescent and are derived from various chemical groups of dyes, including acridine (Quinacrine), arylmethane (Auramine O, Pararosaniline), azo (Congo red), fluorone (Eosin, Fluorescein), quinone-imine (Neutral red), quinoline (Pseudoisocyanin), rosamine (Rhodamine 3G), thiazole (Thioflavine T), and xanthene (Pyronin Y). A list of fluorochromes and their characteristics is given in Section Q. The advantages of using fluorochromes in cellular binding are that the sensitivity of the staining reaction is increased (fluorescent structures are viewed against a dark background), small, nontoxic concentrations of fluorescent material yield visible fluorescence, cytologic detail is often improved (even in the same stained structure viewed in transmitted light), and fluorescent dyes may be used which show weak colors by routine microscopy (Thioflavine T). An additional advantage for quantitative purposes is the absence of distributional error, so that scanning or two-wavelength technics are unnecessary. On the other hand, some fluorochromed preparations are sensitive to small changes in pH, as in Acridine Orange fluorochroming for nucleic acids, and must be viewed in buffer solution to achieve proper color discrimination; these preparations are impermanent.

As is true for many other dyes, commercially available fluorescent dyes are frequently contaminated with other dyes (see review of this problem dealing with Basic fuchsin, Acridine Orange, and Pyronin B and Y by Kasten (1967a)). Fortunately, in many cases of dye contamination, qualitative staining results are not appreciably affected. This good fortune comes about in part because dyes which are certified by the Biological Stain Commission must always pass use tests. The contaminating components are closely related chemically to the dye in question or the degree of contamination is too low to do more than alter slightly the final color. Hopefully, purer dyes will become more available commercially for application to cytofluorometry where dye standardization between laboratories is important. Preparations which have been fluorochromed are subject to fading, especially on exposure to UV light. Fading is minimized in the absence of oxidizing agents, by exciting at long wavelengths, reducing unneccessary exposure to excitation, careful selection of mounting medium, and addition of reducing agents such as dithionite to the mounting medium (Gill, 1979). Fluorochromed sections which have faded can often be rejuvenated by storage in the dark at a low temperature.

It is hoped that this introduction to fluorescence methods will encourage readers to go to the original sources and to more intensive reviews for further information. To aid here, background references are listed and divided according to topic in Section D. Many of the older references are included for the sake of historical importance and completeness. For the methods described on subsequent pages, specific citations are given there.

SECTION D: GENERAL REFERENCES

Fluorescence Theory, Technical Aspects of Fluorescence Microscopy, and Applications

Haitinger (1934), Hamperl (1934), Barnard and Welch (1936), Haitinger (1938a), Ellinger (1940), Pringsheim (1948), Eppinger (1949), Bräutigam and Grabner (1949), Popper and Szanto (1950), Gurr (1951), Reichert (1952), Krieg (1953), Gottschewski (1954), Manigault (1955), Hamperl (1955), Richards (1955), Price and Schwartz (1956), Perner (1957), deLerma (1958), Gottschewski (1958), Haitinger (1959), Richards (1960), Romeis (1968), Pearse (1972), and Lillie and Fullmer (1976).

Fluorochromes

Harms (1965), Kasten (1967a), Matzke and Thiessen (1976), and Lillie (1977).

Intravital and Supravital Fluorochroming

Ellinger and Hirt (1932), Strugger (1940), Strugger (1947), Schümmelfeder (1950a), Zeiger and Harders (1951), Kosenow (1952), Stockinger (1964), West (1969), and Allison and Young (1969).

Fluorescence Microscopy of Plants

Frey-Wyssling (1947), Drawert (1951), Goodwin (1953), Stadelmann and Kinzel (1972), and Kapil and Tiwari (1978).

Fluorescence Microscopy of the Nervous System

Schümmelfeder (1950b), Scharf (1956a), Scharf (1956b), Kater and Nicholson (1973), Moore and Loy (1978), and Tweedle (1978).

Fluorescence Microscopy of Insects

Metcalf and Patton (1944).

Fluorescence Microscopy in Microbiology

Hagemann (1937), Haitinger (1938b), Strugger (1949), and Meisel et al. (1961).

Fluorochromes in Membrane Physiology

Waggoner (1979).

Fluorescence Spectroscopy

Zanker (1952), Morthland et al. (1954), DeBruyn and Smith (1959), Bradley and Wolf (1959), Udenfriend (1962), Ruch (1964), Rigler (1965), Caspersson et al. (1966), Rigler (1966), and Rigler (1969).

Fluorescence Probes of Chromosome Structure

Lubs et al. (1973), Uchida and Lin (1974), Jalal et al. (1974), Schweiger (1976), Latt (1976), Latt (1977), and Comings (1978).

Fluorescence Histochemistry

Ornstein et al. (1957) Kasten (1959), Kasten (1960a), Kasten (1964), Sandritter and Kasten (1964), Ruch (1966), Kasten (1967b), Zelenin (1967), Prenna (1968), Ruch (1970), Ganter and Jollès (1970), Pearse (1972), Chayen et al. (1973), Lillie and Fullmer (1976), Geyer and Luppa (1977), and Bancroft and Stevens (1977).

Cytofluorometry

Böhm and Sprenger (1968), Thaer and Sernetz (1973), Crissman et al. (1975), Thiessen and Thiessen (1977), Bloch et al. (1978), Arndt-Jovin and Jovin (1978), Mayall and Gledhill (1979), and Krug (1979).

SECTION E: METHODS FOR PLANT TISSUES

Combined Fluorescence-Light Microscopic Detection of Callose in Plant Organs by Aniline Blue-Lacmoid Staining

From Cheadle et al. (1953); Currier and Strugger (1956) and modified by Ramming et al. (1973); Peterson and Fletcher (1973)

Application: Plant material containing callose.

Designed to show: Sieve tubes and callose plugs in pollen tubes.

Fixation and tissue preparation: For pollen tubes, fix flowers in formalin ethanol-propionic acid (see p. 21) for at least 24 hr. Individual flowers are dehydrated through a tertiary butanol series and embedded in paraffin. Paraffin blocks are softened in Gifford's solution or in 4% phenol a few hours or in Mollifix* to facilitate sectioning. For sieve tubes (and pollen tubes) young plant parts are fixed in Carnoy's acetic-alcohol (1:3) for 24 hr. Segments are rinsed in water, cleared in 85% lactic acid in a boiling water bath for up to an hour, washed in water and stained directly. To prevent staining of pit callose formed in parenchyma cells during fixation, Peterson and Fletcher (1973) recommend quick killing the tissues by plunging them into boiling 95% ethanol for 10 min prior to acetic-alcohol fixation. Another quick-killing method is to fix in acetic-alcohol precooled to −74°C (Eschrich and Currier, 1964).

Preparation of staining solutions:

A. Aniline Blue (C.I. 42755) solution: Solution is prepared at a concentration of 0.005% in 0.15 M K_2HPO_4 at pH 8.6.

B. Sodium bicarbonate Solution 1: 1% $NaHCO_3$ in 25% ethanol.

C. Lacmoid (C.I. 51400, also known as Resorcinol Blue) solution: 0.25% in 30% ethanol containing a few milliliters of 1% $NaHCO_3$.

D. Sodium bicarbonate Solution 2: 1% $NaHCO_3$ in 50% ethanol.

Staining schedule:

1. Sections or squashes are brought to water and stained in Aniline Blue for 10 min.
2. Temporary mounts are made and observed directly with a fluorescence microscope.
3. Permanent preparations are made of tissue sections by first rinsing in sodium bicarbonate Solution 1 for 30 min.
4. Stain in lacmoid for 12–24 hr.
5. Rinse 1 min in sodium bicarbonate Solution 2.
6. Bring to absolute ethanol rapidly to avoid excessive loss of lacmoid.
7. Clear in xylene and mount.

Results: Callose plugs in pollen tubes fluoresce bright yellow-green, pollen tube wall is light yellow-green, and the cell walls and nuclei are brown. In other plant parts there is strong fluorescence of sieve tubes with some autofluorescence from other components (cuticle, plant hairs, lignified sites around xylem and phloem, some parenchyma cells of the stigma and style, and some root epidermal cells). The permanent preparations show the callose to be stained blue (bright-field) while other structures are unstained or stained (if a counterstain like safranin is used).

NOTE: Callose occurs in many different plant organs. It occurs within pollen tubes in a form known as callose plugs. It is also associated with sieve tubes in many plant regions. Increased callose formation occurs in response to injury or infection. Some workers believe that callose is a by-product of disturbed metabolism in the cell wall-cytoplasm interface and others believe that callose functions to plug tissues and constrict plasmo-

---

* *British Drug House, Chadwell Heath, Essex, England.*

desmata in pits and sieve areas. Fluorescence microscopy using Aniline Blue as a fluorochrome was introduced by Arens (1949) quoted by Ulrychova et al. (1976)), and popularized by Currier and Strugger (1956), and Martin (1959). Aniline blue fluorochroming can be used for fresh or fixed tissues but does not ordinarily produce permanent slides. Lacmoid is a selective blue nonfluorescent stain for callose (Cheadle et al., 1953); although not as sensitive as the fluorescence method, lacmoid is usable for permanent preparations. The technic described here is a combination of both technics. According to Ramanna (1973), aniline blue-induced fluorescence can be preserved by mounting the fluorochromed preparation in Euparal. A number of other dyes (in addition to lacmoid) have been proposed in staining callose for permanent preparations (Sprau, 1955; Ulrychova et al., 1976). These include coralline, benzoazurine (C.I. 24140), azurine, and Ponceau S (C.I. 27195).

An Optical Brightener for Studies of Plant Structure by Fluorescence Microscopy

From Hughes and McCully (1975)

Application: General study of plant cytology.

Designed to show: Cellulose, carboxylated polysaccharides, and β-1,3-glucans in plant walls, pectin, and certain other polysaccharides in extracellular mucilages of root caps.

Tissue preparation: Sectioned tissues are examined in two ways: a) Hand sections are cut with a razor blade and washed at least 2 min in tap water before staining. These preparations can be kept for several days by sealing coverslips with nail polish; b) 3% glutaraldehyde-fixed tissues are embedded in glycol methacrylate and sectioned at 1–2 μm. Material embedded in epoxy resins (Epon, Araldite) should have the plastic removed from sections before staining using the sodium methoxide technic (24 hr in methoxide-saturated methanol). Postfixation with osmium tetroxide reduces the staining but masks aldehyde-induced autofluorescence. For in vivo vital staining of intact plants, roots are maintained in darkness for several days in a nutrient solution containing optical brightner. Roots are fixed in 3% glutaraldehyde, dehydrated in a Methyl Cellosolve, ethanol, propanol, and butanol series, embedded in glycol methacrylate, sectioned at 1–2 μm, and mounted in media of low fluorescence.

Preparation of staining solutions: Calcofluor White M2R* optical brightener (disodium salt of 4,4-bis-(4-anilino-bis-diethylamino-S-triazin-2-ylaminol)-2,2-stilbene-disulfonic acid): 0.01% solution in distilled water.

---

* American Cyanamid Co., Montreal.

Staining Schedule A (hand sections):
1. Stain in optical brightener for 20–60 sec.
2. Wash in water 1 min.
3. Mount in tap water.

Staining Schedule B (embedded material):
1. Sections are stained in optical brightener for 1 min (or longer to obtain more fluorescence).
2. Wash in water 1 min.
3. Let sections air-dry.
4. Mount in immersion oil or Permount.

Staining Schedule C (in vivo staining of intact plants): Slides are observed without further staining.

Results: Fluorescence is green-yellow and difficult to distinguish from aldehyde-induced fluorescence of proteins and autofluorescence of lignins and phenolic compounds unless excitation is in the UV (365 nm peak) and fluorescence transmission is restricted to longer visible wavelengths. The original paper should be consulted for the detailed results, which are briefly summarized here. The brightener fluorochromes almost all types of cell walls but usually not cell contents. Fine microfibrillar detail of the walls is seen at high resolution and fluorescence dichroism occurs (2 distinct colors are seen, according to wall orientation with respect to the polarizer). Fluorescence is produced in a variety of extracellular components on various cells, including cellulose, callose, carboxylated polysaccharides, β-linked glucans, and pectin. There may be some binding to lignin. Anomalous fluorescence is seen in some regions of starch grains. A bright fluorescence is generated in nucleoli and cytoplasm when stained sections are exposed to light. Results are the same for hand sections of fixed or unfixed material, and for embedded sections. At low concentration of brightener, there is no effect on root growth. Growth inhibition occurs at high concentration, which suggests that cellular uptake can occur. Embedded section of vitally stained roots exhibit strong fluorescence in root cap cell walls, root cap mucilage, walls of epidermal and subepidermal cells, and mucilage produced by epidermal cells.

NOTE: Brightener-induced fluorescence fades with prolonged viewing, especially in thin sections, but the fluorescence can be restored in intensity by restaining. There is no loss of fluorescence when mounted sections are stored in the dark for 6 months. Optical brighteners are colorless dyes widely used as whitening agents and added to commercial detergents. This commercial use depends upon dye binding to plant fibers in fabrics, which is in agreement with morphologic studies on plant tissues. Brighteners generally exhibit little or no toxicity to bacteria, fungi, algae, higher plants, and humans although there is some controversy as to whether or not they produce mutations in microorganisms (yeast). Optical or fluorescence brighteners have been used to stain bacteria (see Section F, under "Fluorescence Brightener Method for Detecting Bacteria in Plant Tissues") fungal cell walls (Tsao, 1970), and pollen tubes (Jeffries and Belcher, 1974). A fluorescent brightener has been employed to demonstrate cellulose in different stages of the life cycle of cellular slime molds (Harrington and Raper, 1968).

Acridine Orange Method for Virus-infected Plant Cells

Modified from Hooker and Summanwar (1964)

Application: Virus-infected plant cells.

Designed to show: RNA inclusions in virus infected tissue. Potato virus X, tobacco etch virus, potato virus Y, tobacco necrosis virus.

Fixation: 50% ethyl alcohol.

No Embedding: Fresh cross sections of leaves or paradermal sections of the leaf epidermis.

Preparation of staining solutions:
A. Acridine orange (AO, C.I. 46005) stock solution: 0.1 gm in 100 ml distilled water.
B. M/15 Sörensen's phosphate buffer, pH 6.3 (see p. 6).
C. Working AO solution: Just before use, take 1 part of stock AO Solution A and dilute with 9 parts of buffer Solution B.
D. 0.1 M calcium chloride.

Staining schedule:
1. Place freehand sections of turgid fresh tissue from lower surface of virus-infected leaves into 50% ethanol for 20 min.
2. Rinse in 2 changes of distilled water (10 min each).
3. Stain in Solution C for 5 min.
4. Two distilled water rinses of 3-4 min each.
5. Place in Solution D for 5 min.
6. Two distilled water rinses of 2-3 min each.
7. One change of phosphate buffer (Solution B). Mount in buffer.
8. Examine immediately with fluorescence microscope as stain intensity fades rapidly.

Results: x bodies and striate bodies—red, nucleus—greenish white, mucoproteins—red.

NOTE: The AO method was applied originally to animal cell cultures infected with viruses by Armstrong (1956) using a method similar to that described in this section for AO fluorochroming of nucleic acids. The method has proved valuable in discriminating the various nucleic acid alterations undergone by cells during viral infection and the development of inclusion bodies. Such studies were reported, for example, of cultures infected with adenovirus (Armstrong and Hopper, 1959), psittacosis virus (Starr et al., 1960), poliovirus (Mayor, 1961a), rabies virus (Love et al., 1964), and bovine parainfluenza 3 virus (Kasten and Churchill, 1966). Through the elegant work of Mayor, it was demonstrated that by combining the fluorochromatic properties of AO with the use of proteases and nucleases on purified viruses, one can discriminate between single-stranded and double-stranded viruses containing either nucleic acid in their cores. Using these techniques, she and Hill showed that the DNA of bacteriophage ϕX174 is in the single-stranded form (Mayor and Hill, 1961). Similar approaches were used to analyze polyoma virus (Mayor, 1961b), poliovirus, and tobacco mosaic virus (Mayor and Diwan, 1961).

SECTION F: METHODS FOR MICROORGANISMS

A Simple and Sensitive Fluorescence Microscopic Method for Detecting Mycoplasma Contamination and DNA Viruses in Cultured Cells

From Russell et al. (1975)

Application: Tissue culture systems.

Designed to show: Mycoplasma contamination and DNA viral infections.

Fixation: None.

Embedding: None.

Preparation of staining solutions:
A. Phosphate-buffered saline (PBS) (see p. 6).
B. DAPI* (4',6-diamidino-2-phenylindole): make small volume of PBS containing DAPI at a concentration of 0.1 μg/ml.

Staining schedule:
1. Cell monolayers grown on coverslips are washed once with PBS.
2. Add small volume of DAPI solution.
3. Incubate cultures at 30°C for 15–30 min.
4. Wash once with PBS.
5. Mount coverslip in PBS on a slide.
6. Examine immediately in fluorescence microscope.

Results: Uninfected cells yield highly fluorescent nuclei and no detectable cytoplasmic fluorescence. In cells contaminated with mycoplasmas, nuclear fluorescence persists and discrete fluorescent foci are seen in the cytoplasm and on cell surfaces. Cells infected with vaccinia virus reveal characteristic "starlike" fluorescent clusters in the cytoplasm which are

---

* *Polysciences Inc.; Aldrich Chemical Co.; Accurate Scientific Co., 28 Tec St., Hicksville, NY 11801.*

suggestive of so-called virus "factories." The early stages of adenovirus infection are also detectable.

NOTE: The DAPI method as described is rapid, simple, and sensitive for detecting DNA-DAPI complexes (Williamson and Fennell, 1975; Russell et al., 1975; Hadjuk, 1976; James and Jope, 1978). DAPI should prove to be a useful tool to tissue culturists who must always be concerned about possible mycoplasma contamination. Although DAPI has been used to detect mitochondrial DNA in yeast cells (Williamson and Fennell, 1975), the fluorochrome does not seem to produce detectable fluorescence in animal cells in the described procedure; if mitochondria did fluoresce, it would be difficult to distinguish fluorescence from that caused by mycoplasmas. The DAPI method is far simpler than previous technics, such as autoradiography or special agar and broth cultivation. The technic is also adaptable to detection of mycoplasmas in cell suspensions and for culture media. As an alternative to DAPI, Russell et al. (1975) mention that the fluorescent DNA-binding benzimidole derivative Hoechst 33258 (Hilwig and Gropp, 1972) gives similar results. Detailed results using Hoechst 33258 were given by Chen (1977), who employed it at a concentration of 0.05 μg/ml in Hanks' BSS on Carnoy's fixed cultures. Chen noted that the fluorochrome is a sensitive indicator of all species of mycoplasma, produces no background fluorescence, and is not quenched during UV exposure like the quinacrines. It is an advantage to use low-density cell cultures in the test since overcrowded cells have their nuclei close together and produce variable results. Micronuclei and nuclear debris fluoresce but are distinguished from mycoplasmas by their larger and variable size as well as more intense fluorescence.

A Fluorescence Brightener Method for Detecting Bacteria in Plant Tissues

From Eng and Cole (1976)

Application: Plant pathology.

Designed to show: Bacteria in plant tissues.

Tissue preparation: Sections of infected plant stems are cut on a freezing microtome, placed in distilled water, fixed in Kirkpatrick's fixative (ethanol:chloroform:formalin, 6:3:1), and mounted on a slide.

Staining solutions:
   A. Erythrosin B counterstain: 0.5% Erythrosin B (C.I. 45430) is dissolved in 5% phenol and filtered.
   B. Tinopal AN*: 0.1% w/v filtered aqueous solution.

Staining schedule:
   1. Stain in Erythrosin solution for 1–2 min.
   2. Rinse in distilled water.
   3. Stain in Tinopal AN for 10 min.

---

* Geigy Co.

4. Rinse in distilled water.
5. Mount in water and view in fluorescence microscope.

Results: Bacteria and infected areas of tissue fluoresce pink, giving clear differentiation from blue, healthy tissue.

NOTE: The technique was applied to lucerne plants (*Medicago sativa*) which were infected with *Corynebacterium insidiosum*. The method is based on the work of Paton and Jones (1973) who described a fluorescence method for observing microorganisms on surfaces of meat, fish, skin, and plants using a fluorescence brightener, Tinopal AN. Presumably, other fluorescent brighteners could be used if this one is not available.

A Dual Fluorochroming Method for Fungi and Mycobacteria in Paraffin-embedded Tissues

From Mote et al. (1975)

Application: Suspected tuberculous tissue and granulomas as a screening procedure for bovine tissues.

Designed to show: Acid-fast organisms and fungi in paraffin sections.

Fixation: 10% neutral formalin.

Embedding: Paraffin.

Preparation of staining solutions:
A. Auramine O solution: Mix 0.3 gm of Auramine O (C.I. 41000) in 90 ml of distilled water, 7 ml of glycerol, and 3.2 ml of liquid phenol.
B. Ferric chloride: Prepare a 10% solution of ferric chloride.
C. Acridine Orange (AO) solution: Dissolve 25 mg of AO (C.I. 46005) in 100 ml of distilled water.

Staining schedule:
1. Deparaffinize sections and bring to water.
2. Stain in Auramine O solution for 10 min.
3. Rinse in running tap water 1 min.
4. Treat with ferric chloride solution for 5 min.
5. Rinse in running tap water 1 min.
6. Stain in Acridine Orange solution 2–5 min.
7. Rinse in running tap water 1 min.
8. Bring to absolute ethanol.
9. Mount from xylene using nonfluorescent mounting medium.
10. Examine in fluorescence microscope.

Results: Mycobacteria fluoresce white to yellow. Fungi fluoresce green to orange according to the species. Normal lymphocytic tissue is green and necrotic tissue is brownish green.

NOTE: The method described is essentially a combination of the fluorescent acid-fast procedure with Auramine O (Fluorescent Microscopy Acid-Fast

Procedure) and Acridine Orange fluorochroming of fungi (Pickett et al., 1960), both of which are described elsewhere in this volume. The combined use of these fluorochromes for acid-fast bacteria was reported by Smithwick and David (1971). The dual technic described here was applied by the authors to a large number of suspected bovine specimens and shown to be 94% accurate in detecting myobacteria and 100% accurate for fungal granulomas. In another series of 100 tissues with typical myobacterial lesions by histopathology, no organisms were detected prior to using the dual fluorescence technic but 2 were observed to be Auramine O positive and 2 contained bacterial colonies.

According to Richards (1941), Auramine O is specific for mycholic acid, the principal acid-fast component of myobacteria. Ferric chloride behaves like acid-alcohol in decolorizing other stained organisms and also quenches tissue fluorescence. In addition, it acts as a mordant and permits the staining of fungi with Acridine Orange. The counterstaining with Acridine Orange had no deleterious effect on the fluorescence of myobacteria and there was no false-positive staining of other bacteria. According to the authors, this AOAO procedure has the advantage over single fluorochroming with Auramine O in that the lesions contrast markedly with normal tissue. Also, examination is rapid, thereby permitting its use as a routine method for microscopical examination of suspected tuberculous bovine submissions.

Anderson and Greiff's Method of Fluorochroming Rickettsiae

From Anderson and Greiff (1964)

Application: Sections and smears of tissues infected with rickettsiae.

Designed to show: Maximum fluorescence of rickettsiae when only a few organisms are present.

Fixation: Tissues are fixed in Carnoy's acetic alcohol (see p. 12) and smears are fixed in absolute ethanol or absolute ethanol-chloroform (1:1).

Embedding: Paraffin.

Preparation of staining solutions:
  A. Alcian blue (C.I. 74220) solution: Glacial acetic acid 3 ml, distilled water 97 ml, Alcian blue 1.0 gm.
  B. Periodic acid: 0.5% H_5IO_6 in H_2O.
  C. Schiff's reagent (see p. 59).
  D. Sulfurous acid: 10% sodium metabisulfite 6 ml, 1 N HCl 5 ml, H_2O 100 ml.
  E. McIlvaine-Lillie buffer (see p. 5).
  F. Acridine orange (C.I. 46005): 0.05% dye in 50 ml buffer diluted to 100 ml with H_2O.

Staining schedule:
  1. Stain with Alcian blue PAS (McManus and Mowery, 1960) modified. Dewax and hydrate sections, stain with Alcian blue 30 min, wash in running water 2 min, rinse in distilled water, oxidize in periodic acid

solution for 10 min. Wash in running water 5 min, rinse in distilled water, place in Schiff's reagent 10 min, rinse in 3 changes of sulfurous acid for 2 min each, rinse in running water for 2 min. Rinse in 3 changes of distilled water.

2. Immerse for 2 min in buffer.
3. Stain for 5 min in Acridine orange solution.
4. Rinse in 3 changes of buffer for 1 min each.
5. Temporary mounts are made in buffer and permanent mounting is done in DPX after drying in air.
6. Examine in fluorescence microscope.

Results: In smears of yolk sac membranes of infected embryonated eggs and scrotal sac tissue of infected guinea pig, rickettsiae are visible as bright orange diplobacillary particulates; granules within individual organisms fluoresce intensely. Similar results are obtained from sections of yolk sac membrane. Noninfected cells do not fluoresce. When acridine orange staining is done without previous staining with Alcian blue-PAS, a non-specific background fluorescence is seen, which reduces contrast and makes it difficult to detect the microorganisms.

NOTE: The advantage of this method over previous methods of fluoro-chroming rickettsiae is that background fluorescence is eliminated by blocking potentially reactive acridine orange sites with Alcian blue-PAS staining.

SECTION G: AUTOFLUORESCENCE TECHNICS

Popper's Method for Vitamin A and Lipofuscin by Autofluorescence

From Popper (1941a, b)

Designed to show: Histologic localization of vitamin A and lipofuscin.

Fixation: Fresh frozen sections mounted in water (avoid formalin fixation according to Hori and Kitamura, 1972).

Preparation of staining solutions and staining schedule: None, view sections rapidly in fluorescence microscope.

Results: Vitamin A and carotene fluoresce green, but vitamin A fades rapidly while the fluorescence of carotene fades slowly. Lipofuscins fluoresce red to brown, according to their location, with the fluorescence persisting during irradiation. Vitamin A is concentrated in liver parenchymal cells but is detectable elsewhere, primarily in endocrines rich in steroid or lipid deposits (adrenal cortex, testis, ovary, adipose). Hepatic lipofuscin gives a brown fluorescence which changes to red after alcohol extraction. Brown to brown-red to red fluorescence is given by lipofuscins in the adrenal cortex, testicular Leydig cells and germinal epithelium, lutein cells of the ovary, and cardiac muscle.

NOTE: Vitamin A is bound to lipids which are themselves demonstrated by common fat stains or by Phosphine 3R, a fluorescent dye. Vitamin A was first discovered in lipid droplets of liver by von Querner (1932). For further information about the distribution of vitamin A, see the original papers of Popper. He gives a thorough description of vitamin A in human tissues under normal and pathologic states and in rat tissues under physiologic conditions. There is disagreement as to whether or not liver Kupffer cells store vitamin A but, according to Hori and Kitamura (1972), these cells are not the main site of vitamin A storage nor are they active in retinol esterification. Little use has been made of the natural fluorescing properties of vitamin A since Popper's classic work. More interest has centered on lipofuscins and ceroid, a related pigment. For a fuller discussion of the native fluorescence of various substances, see page 12 of Lillie and Fullmer (1976). The comparative histochemistry (including autofluorescence) of lipofuscins and other pigments is described on page 1090 of Pearse (1972).

Fluorescence Microscopic Detection of Thiamine (Vitamin B₁) in Nervous Tissue

Modified from von Muralt (1943) and Sjostrand (1946) by Tanaka and Cooper (1968)

Application: Nervous tissue.

Designed to show: Thiamine localization in nerve fibers.

Tissue preparation: Cranial and peripheral nerves are removed rapidly and quickly frozen in isopentane (cooled with liquid nitrogen or an acetone-Dry Ice mixture). Frozen tissues are dried in vacuum at $-35°C$ for 3–5 days. The freeze-dried tissues are embedded in paraffin, sectioned at 5–8 μm, and deparaffinized by the careful addition of one drop of xylene to the slide.

Preparation of solutions:
  A. Cyanogen bromide gas: A 10% solution of cold KCN is added to a beaker of bromine water to decolorize it; the mixture should be used right away.
  B. Ammonia gas: Concentrated ammonium hydroxide solution in a closed beaker.

Staining schedule:
  1. Slide preparation is put into a glass container containing the beakers of cyanogen bromide and ammonia.
  2. Container is sealed and placed in an oven at 40–70°C for 30–60 min.
  3. Slide is removed and tissue is mounted in mineral oil.
  4. Observe in fluorescence microscope using primary filter and/or monochromator to isolate excitation at 362 nm and interference or barrier filters which transmit visible light up to 450 nm but not beyond.

Results: Fluorescence is seen in nerve fibers but not in axoplasm. A faint blue fluorescence may be seen in untreated nerve fibers which is apparently due to autooxidation of the vitamin.

NOTE: The original technics of von Muralt and Sjostrand used an alkaline ferricyanide solution to convert thiamine to its fluorescent derivative, thiochrome. The present technic is said to have an advantage over the ferricyanides which limit diffusion of the thiochrome product. It is essential to use the correct filter-monochromatic system to ensure specificity; otherwise, riboflavin (vitamin B_2) may also fluoresce. Thiamine fluorescence is destroyed by exposing nerve fibers or thiochrome solutions to UV light at 365 nm (only 3 nm higher than the optimal wavelength for excitation of thiamine). It is suggested that thiamine not only serves a role as coenzyme in nerve tissue, but also serves a membrane transport function.

SECTION H: METHODS TO REVEAL METALS

A Fluorescence Method for the Detection of Zinc in Granulocytes

From Smith et al. (1969)

Applications: Detection of zinc in the cytoplasm of granulocytes.

Cell preparation: Human blood smears are made and air-dried for at least an hour.

Preparation of staining solutions:
A. Michaelis universal buffer: 1.94 gm sodium acetate and 2.94 gm of sodium phenobarbital are dissolved in 100 ml of distilled water; to 10 ml of this solution are added 4 ml of physiologic saline (0.85% NaCl), and the pH is adjusted to 8.0 with 0.1 N HCl; approximately 2.0 ml HCl are needed.
B. 8-Hydroxyquinoline (8-quinolinol) solution: 0.1 ml of a 3% solution of 8-hydroxyquinoline in absolute ethanol is added to 25 ml of Michaelis buffer (use Solution A above).

Staining schedule:
1. Stain blood smears for 15 min.
2. Drain excess stain, dip twice in double-distilled water, and air-dry.
3. Control slides are treated with 1% acetic acid for 5 min to extract zinc prior to staining, washed twice with double-distilled water, and air-dried.
4. Mount both preparations in nonfluorescent immersion oil and examine in fluorescent microscope.

Results: Under low-power, cells fluoresce a pale greenish yellow color. With oil-immersion optics, brightly fluorescing, large granules are seen in eosinophils and basophils. Less intense fluorescing granules of smaller sizes occur in neutrophils. Lymphocytes do not fluoresce. Control slides extracted with acetic acid fail to fluoresce.

> NOTE: 8-Hydroxyquinoline forms chelates with many divalent and trivalent metal cations but only calcium, magnesium, and zinc chelates are fluorescent. It is claimed that under the conditions employed here, magnesium and calcium do not interfere with zinc fluorescence. In support of the claimed specificity for zinc detection are identical results obtained by the authors using the dithizone staining method of McNary (1957).

Eggert's Method for Fluorescent Localization of Aluminum in Plant Tissues

From Eggert (1970)

Application: Necrotic plant tissue suggestive of excess deposits of aluminum.

Designed to show: Fluorochromasia of aluminum in woody tissue sections.

Fixation: None, use frozen sections of fresh stem samples cut at approximately 12 μm.

Preparation of staining solutions:
A. 2 N acetic acid: 12 ml concentrated acetic acid added to 88 ml of distilled water.
B. Saturated Morin (C.I. 75660) solution: Saturate a few ml of methyl alcohol with Morin just before dye solution is to be used.

Staining schedule:
1. Place one drop of acetic acid on a glass slide within a previously drawn soft wax circle.
2. Float a single tissue section on the drop for about 30 sec.
3. Allow slide to dry under inverted petri dish.
4. Examine with fluorescence microscope.

Results: Tissue areas which contain high concentrations of aluminum fluoresce green (phloem fiber bundles and xylem). The test is not specific since similar fluorescence occurs in the presence of beryllium, zinc, gallium, and scandium. Lillie and Fullmer (1976) mention that these other Morin-metal complexes are dissolved out in 2 N hydrochloric acid. It is doubtful whether this is a practical procedure for histologic sections.

> NOTE: The method is not used much because of limited application and the lack of specificity. However, Eggert was able to verify the specificity for aluminum localization by first growing apple seeds in a nutrient solution lacking zinc but containing 3 ppm aluminum and examining frozen-dried sections of the stem in an x-ray spectrometer in association with a scanning electron microscope. There was good agreement, in that focal areas exhibiting prominent fluorescence were in the same locations as those showing high concentrations of aluminum.

SECTION I: CYTOCHEMICAL DETECTION OF NUCLEIC ACIDS AND PROTEINS

Acridine Orange Method for Nucleic Acids

Modified from von Bertalanffy and Bickis (1956) and Armstrong (1956)

Designed to show: Differentiation of nucleic acids, viral inclusions, and for detection of malignant cells in exfoliative cytology as a preliminary screening procedure.

Fixation: Acetic alcohol (1:3), methanol-ether (1:1). Fixation time from 10 min to indefinite; fresh frozen sections are used or these may be postfixed. Paraffin sections may give less favorable results. Impression smears and cells on slides or coverslips may be employed (exfoliative cytology, tissue culture).

Preparation of staining solutions:

A. Acridine orange (C.I. 46005) stock solution: 0.1 gm AO in 100 ml of distilled water.

B. Krebs-Ringer solution or phosphate buffered saline (PBS) pH 6.0 (see p. 21).

C. Working AO solution: Just before use, take 1 part stock acridine orange Solution A and dilute with 9 parts of buffer Solution B to make working AO solution.

Staining schedule:

1. Transfer specimen (frozen section, smear, or coverslip preparation from fixative to absolute ethyl alcohol for 2 min (for paraffin sections, carry through normal deparaffinization to water).
2. Transfer to 70% ethyl alcohol, 2 min.
3. Transfer to 50% ethyl alcohol, 2 min.
4. Wash in distilled water for 5 min.
5. Immerse in buffer Solution B for 2 min.
6. Stain in working AO Solution C for 15 min.
7. Drain excess staining solution on absorbent paper.
8. Wash in 2 changes of buffer for 2 min each.
9. Mount in buffer and ring the coverslip with wax or nail polish.
10. Examine immediately with a fluorescence microscope. Preparation is not permanent but slide can frequently be restained.

Results: DNA-nuclei, chromosomes and DNA-containing viral inclusions green to yellow. RNA-cytoplasm, nucleoli, and RNA-containing viral inclusions orange to red.

NOTE: Formalin-containing fixatives may not permit distinct color differences. Fixatives containing heavy metals should not be used. Paraffin-embedded sections do not give the best results. Some workers prefer to stain cells at pH 4.2 in acetate buffer instead of pH 6.0 to achieve sharper color differences between DNA and RNA. Depolymerized DNA fluoresces orange-red like RNA (see review by Kasten (1967a)). False-positive fluorescence reactions are given by mast cell granules, polysaccharides, cartilage matrix (red); vascular elastic fibers (yellow); and keratin (green). Various mucopolysaccharides may be distinguished from each other and from nucleic acids using acridine orange at higher pH and in combination with differential salt extraction and ribonuclease (see "Saunders' Acridine Orange-CTAC Method for Acid Mucopolysaccharides" in Section K).

The binding of AO and related acridine fluorochromes to animal and plant tissues was advocated by Haitinger (1934) and popularized by Strugger (1949). The report by Hicks and Matthaei (1955) gave impetus to AO fluorochroming in histology for people not already familiar with the German literature. The bicolor fluorescence in fixed cells after AO binding was investigated thoroughly by Schümmelfeder and associates in a series of papers (1957). He emphasized the influence of pH in the differential fluorochroming of RNA and DNA in cell structures. He also pointed out the influence of nucleic acid polymerization on the color emission. By use of controlled pH during staining, Schümmelfeder and Stock (1956) found that each tissue has its own characteristic isoelectric point, thus reflecting the influence of proteins on nucleic acid binding. The molecular basis for interaction between aminoacridines and nucleic acids was scrutinized (Morthland et al., 1954; DeBruyn and Smith, 1959) and led to the recognition that the bicolor fluorescence or fluorochromasia represented dye aggregation, as with other metachromatic dyes. Dye aggregation or stacking is influenced by the molar ratio of dye to phosphate on nucleic acids (Bradley and Wolf, 1959) and by the fixative employed (Lamm et al., 1965). Further insight as to the influence of conformational changes in nucleic acids on AO fluorescence was provided by the work of Rigler (1966, 1969), who applied cytofluorometric technics. As further evidence of the sensitivity of AO to molecular alterations, color changes were detected in all cell organelles during cold perchloric acid extraction (Kasten, 1965). The fluorochrome in combination with selective nuclease digestion discriminates as well between single-stranded and double-stranded viruses containing either nucleic acid (Mayor, 1961b). In recent years, the dye has been used to characterize cell populations by flow cytometry, although the mechanism of binding to nucleic acids and other moieties is not always clear (Cambrer et al., 1977; Traganos et al., 1977; Gill et al., 1978). Notwithstanding the difficulties in interpreting the fluorochromasia of AO-stained cells, there has been a wealth of publications focused on the increased content of nucleic acids in neoplasia and the usefulness of AO in detecting such changes. In addition to the work of von Bertalanffy, some others who participated in this practical application include Mellors et al. (1952), Dart and Turner (1959), Kornfield and Werder (1960), Masin and Masin (1960), and Wied and Manglano (1962). For reviews of the theoretical and practical aspects of AO fluorochroming, the reader is referred to the papers of Rigler (1966, 1969) and Kasten (1967a, 1973).

Fluorochroming of DNA with Mithramycin for Rapid Cell Cycle Analysis

From Crissman and Tobey (1974)

Application: High-speed cytofluorometric analysis of normal and tumor cell populations.

Designed to show: DNA patterns and population parameters.

Fixation: 25% ethanol (see below).

Solutions:

A. Mithramycin* in alcohol: 25% ethanol containing 100 μg/ml mithramycin and 15 mM $MgCl_2 \cdot 6H_2O$ (30 mg/ml).

Cell preparation and staining:

1. Cells growing in suspension culture are centrifuged and resuspended directly in the mithramycin solution for 15 min.
2. Cells grown in monolayer culture are trypsinized, centrifuged, and resuspended in mithramycin for 15 min.
3. Cells are examined directly in a flow cytofluorometer with excitation set as close to 395 nm as possible. Fluorescence emission is measured at 535 nm.

Results: Fluorescence is shown to be directly proportional to DNA content.

NOTE: Cells may be fixed in 70% ethanol and stored for at least a week before staining in mithramycin. Aldehyde fixatives should be avoided. The technique is specific for DNA as judged by treatment with DNase (cells become nonfluorescent) and agreement of cellular DNA distributions with cell samples stained by one of Kasten's fluorescent-Feulgen reagents (acriflavine-Schiff). Treatment of fixed cells with RNase has no effect on the fluorescence pattern. It is stated that quantitative DNA distributions are obtained within 20 min after cells are removed from a culture. The technique is derived from earlier biochemical studies of Ward et al. (1965). This antitumor antibiotic binds preferentially to guanine-cytosine base pairs of DNA. Mithramycin binding to DNA has recently been measured in thymocyte nuclei of cells in smears (Johannisson and Thorell, 1977). It was shown that fluorescence intensity could be enhanced by Mg^{2+} with an optimal concentration of 0.01 M $MgCl_2$. Other conditions for the best cytofluorometric measurement of DNA were determined.

Kasten's Fluorescent Feulgen Method for DNA

From Kasten (1959, 1973), and Kasten et al. (1959), also see references cited under Kasten's Fluorescent PAS Method

Application: Tissue sections, smears, monolayer cultures, and suspended cells.

Designed to show: DNA in cells.

---

* *NSC 24559, Division of Cancer Treatment, National Cancer Institute, Bethesda, MD 20014; Sigma Chemical Co.*

Fixative: Carnoy's acetic-alcohol (1:3), 10% neutral formalin, or formol-acetic-alcohol (FAA). Tissues fixed in formalin-containing fluids should be washed thoroughly to remove potentially reactive aldehydes. Avoid Bouin's fixative, which tends to hydrolyze away the DNA and is itself difficult to wash out completely. Also, avoid glutaraldehyde and other dialdehyde fixatives since special treatment is needed to eliminate aldehyde binding by Schiff reagents (Kasten and Lola, 1975).

Embedded in: Paraffin or frozen sections, smears, monolayer cultures, and suspended cells.

Preparation of solutions:
 A. Acriflavine-Schiff reagent: Prepare the same as described for Kasten's Fluorescent PAS Method but use 0.25% dye solution. The dye solution remains yellow in color.
 B. Auramine O-Schiff reagent: This can be used in place of Acriflavine-Schiff. The Auramine reagent has the advantage in producing an especially intense fluorescence as used by the writer. It should be made fresh by preparing a 0.25% aqueous solution and then saturating it with SO_2 gas for 1–2 min (as described in the fluorescent PAS method). The solution remains yellow in color. The reagent is usable for 3 hr after preparation, although it may show signs of precipitation during this period. Do not prepare this reagent by adding to the dye solution thionyl chloride or any combination of HCl and metabisulfite compound.
 C. 1 N Hydrochloric acid: make 8% aqueous HCl solution.
 D. SO_2 water: Saturate 40 ml of tap water with SO_2.

Staining schedule:
 1. Deparaffinize sections and hydrate to water. Smears or monolayer cultures on slides or coverslips are placed directly in water. Cell suspensions are handled in siliconized centrifuge or test tubes, which are centrifuged between treatments.
 2. Place in 1 N HCl at 60°C for 10 min (Carnoy's), 15 min (formalin), or 12 min (FAA). Alternatively, formalin-fixed cells can be hydrolyzed in 6 N HCl at room temperature for 20 min as a more convenient method (Kasten et al., 1959). For room temperature hydrolysis of cells fixed in other ways, the optimal time should be checked first on the material employed.
 3. Place in tap water 1 min.
 4. Stain 60 min in one of Kasten's fluorescent Schiff solutions (Acriflavine-SO_2, or Auramine O-SO_2).
 5. Wash in running water until no further yellow color is extracted.
 6. Place in SO_2 water for 5 min.
 7. Dehydrate in ascending alcohols and mount from xylene in nonflu-

orescent mounting medium, such as Fluormount or DPX. The slide preparation is permanent.

8. Examine in fluorescence microscope or analyze by cytofluorometry using appropriate equipment.

Results: DNA in nuclei fluoresces yellow to orange with the Acriflavine-Schiff, according to the filters employed. With the Auramine O reagent, nuclei fluoresce yellow-green. Intranucleolar DNA is detectable, according to its concentration and the thickness of the section. It is usually seen in whole cell preparations (monolayer cultures, isolated cells). Nonspecific dye binding or autofluorescence from the cytoplasm or other proteinaceous structures does not occur except in some paraffin-embedded material. When present, it is of a weak emission and is actually helpful for histologic recognition. If felt to be undesirable, autofluorescence can be eliminated by treatment of sections with iron alum (4% aqueous solution for 5 min) prior to the acid hydrolysis step. The fluorescent-Feulgen reaction of Kasten is specific for DNA, according to many tests (summarized in references cited above and Kasten (1958; 1960, a and b; 1963; 1964)). Unhydrolyzed control slides can serve as a simple check on the specificity.

NOTE: The fluorescent-Feulgen method has been used by many workers with some modifications, especially for quantitative cytofluorometric applications. See discussion under Kasten's Fluorescent PAS Method (see p. 73) for a brief review of the historical development and applications of these important Schiff analogues and other aldehyde-specific fluorochromes. The mechanism of reaction of Schiff-type reagents has been discussed by several workers (Kasten, 1960a; Stoward, 1967a; Pearse, 1968; Gill and Jotz, 1974, 1976; Jotz et al., 1976; van Ingen et al., 1979).

A Fluorescence Screening Procedure for Determining the Amount of Cytoplasmic Contamination in Nuclear Fractions

From Armstrong and Niven (1957), adapted by Stevens et al. (1969)

Application: Nuclear fractions isolated from subcellular fractions.

Designed to show: Percentage of contamination of nuclear fractions with whole cells and cytoplasm.

Cell preparation: Material from a purified nuclear pellet is suspended in 20 volumes of 0.25 M sucrose made up in 3 mM $CaCl_2$. A drop of cell suspension is put on a slide, air-dried, fixed 5 min in ether-alcohol 1:2, and hydrated to distilled water.

Preparation of staining solutions:

A. Phosphate buffer, pH 6.4: M/15 phosphate buffer.
B. Calcium chloride solution: 0.1 M $CaCl_2$.
C. Acridine Orange (AO) solution (C.I. 46005): Prepare 0.01% solution in M/15 phosphate buffer, pH 6.4.

Staining schedule:
1. Place in 1% acetic acid for 3 min.
2. Stain 3 min in AO solution.
3. Dip in calcium chloride solution.
4. Place in 2 changes of phosphate buffer for 5 min each.
5. Mount in buffer and seal coverslip with Vaseline.
6. Examine in fluorescence microscope.

Results: Nuclei without contamination appear as green fluorescent spheres. Whole cells contain a red halo around each nucleus. Nuclei with cytoplasmic tabs appear with small red clumps attached to the nuclear membrane. Differential cell counts permit one to determine the percent of contamination of nuclei with whole cells and cytoplasmic tabs.

Jensen's Chromomycin A3 Fluorochroming Method for Gynecological Cells

From Jensen (1977)

Application: Human cell samples from vaginal and cervical specimens for detection of dysplasia by flow cytofluorometry and fluorescence microscopy.

Designed to show: Abnormal amounts of DNA in dysplastic cervical cells.

Fixation: 70% ethanol at 5°C for at least 1 hr after scrapings are dispersed (see below).

Cell preparation: Gynecologic scrapings are suspended in 10 ml of cold Polyonic R-148 sterile electrolyte solution* and refrigerated for up to 3 days. Cells are dispersed by vortex mixing and passing through an 18-gauge needle 5–10 times using a 10-ml syringe. The cell suspension in Polyonic solution is added to a 15-ml centrifuge tube containing sufficient 95% ethanol to obtain approximately 70% ethanol. Fix for 1 hr or more at 5°C. Filter through nylon mesh of approximately 75 μm to remove cell clumps.

Preparation of staining solutions:
Chromomycin A3†: Prepare 1.5×10^{-5} in Polyonic solution.

Staining schedule:
1. Centrifuge final cell suspension (see above) at 2500 × g for 5 min.
2. Discard filtrate and add chromomycin A3 staining solution. Store in dark for 30 min.
3. For routine fluorescence microscopic observations or single-cell fluo-

---

* *Cutter Laboratories, Inc., Berkeley, Calif.*

† *Calbiochem, La Jolla, Calif.*

rometric measurements, examine wet mounts on slides using Zeiss excitation filters KP500 and KP490 and barrier filter 530 (or equivalent filters).

4. For flow cytofluorometric analyses, use available cytometer or cell sorter instrumentation. For excitation, the 458 nm line from an argon ion laser excites DNA-bound dye preferentially over dye that is free in solution.

Results: Fluorescence emission of fluorochromed nuclei is in the red region of the spectrum at 590 nm. Using two-parameter flow cytofluorometry, with low angle light scatter as a measure of cell size and chromomycin A fluorescence as a measure of DNA content, histograms are obtained which are typical for the cell types present in the sample. Jensen gives analysis of an isometric projection of such a histogram in which 3 distinct cell types are sorted (single, polymorphonuclear leukocytes; single, normal mononucleate epithelial cells; abnormal cells). Cells from the region of dysplasia are always responsible for high fluorescing signals.

NOTE: As is common with many fluorochromes, chromomycin A3 fades rapidly during extended excitation, but exposures of 1 sec or less produce insignificant fading. For nuclear DNA binding with chromomycin A3, magnesium ions must be present. The dye is not bound to RNA so that RNase pretreatment is unnecessary. This is not the case with the intercalating dyes (ethidium bromide, propidium iodide). Jensen showed that chromomycin A3 stains nuclei more specifically than does ethidium bromide, propidium iodide, and Hoechst 33258. With these other fluorophores, cytoplasmic staining accounts for one-fourth to three-fourths of the total cellular fluorescence in cervical specimens. While the chromomycin A fluorochroming method seems better suited than some other methods, for detecting abnormal cells, it should be kept in mind that "false alarms" can result from measuring normal binucleated cells or small aggregates of normal cells.

Two-Color Fluorochroming of Cytology Specimens

From Cornelisse and Ploem (1976)

Application: Cervical cytology specimens.

Designed to show: Bicolor fluorescence of nuclear DNA and cytoplasmic protein in the same cells.

Fixation: 20% ethanol and postfixed in both formaldehyde and Carnoy's (6: 3:1).

Cell preparation: Cervical cells are obtained with an acryl cotton-tipped applicator and immersed for 24–48 hr in 20% ethanol and 80% PBS. Cells are shaken from the cotton into the ethanol solution, centrifuged for 5 min at 250 × g, resuspended in PBS, and dispersed further by syringing the suspension 10 times through an 18-gauge needle. Cells are centrifuged at 250 × g for 5 min and resuspended in 0.2–0.5 ml of PBS. A drop is placed on a slide, spread by tilting, and fixed for 15 min with formaldehyde vapor at 50°C in a glass jar. Slides are rinsed briefly in distilled water and fixed further in Carnoy's 6:3:1 for 15 min.

Preparation of staining solutions:

A. PBS: See Dulbecco's PBS (p. 22).

B. Carnoy's fixative (6:3:1) (See p. 12).

C. Ethanol-PBS: Add 20 ml of absolute ethanol to 80 ml of PBS (see p. 22).

D. 5 N HCl: Add 45 ml of concentrated HCl reagent grade to sufficient water to make 100 ml.

E. Kasten's acriflavine-Schiff solution: Dissolve 10 mg acriflavine (C.I. 46000) in a solution containing 1 gm $Na_2S_2O_5$, 15 ml of 1 N HCl (9% concentrated HCl in distilled water), and 85 ml of distilled water. Avoid loss of SO_2 gas.

F. SO_2-water: 0.5 gm $Na_2S_2O_5$ in 100 ml of distilled water plus 5 ml of 1 N HCl. Avoid loss of SO_2 gas.

G. Primulin solution (C.I. 49000): 0.01 mM solution (0.48 mg dye in 100 ml of distilled water).

H. SITS* (4-acetamido-isothiocyanatostilbene-2,2-disulfonic acid) dye solution: 0.1 mM solution in distilled water.

Staining schedule:

1. After fixing in Carnoy's, rinse in water.

2. Hydrolyze in 5 N HCl for 30 min at room temperature.

3. Rinse in 3 changes of distilled water.

4. Stain 15 min in acriflavine-Schiff.

5. Rinse briefly for 4 changes each in SO_2-water.

6. Rinse in PBS 1 min.

7. Stain for 5 min in primulin or SITS.

8. Rinse 3 times in distilled water, dehydrate, and mount in Fluormount.

9. Store slides in dark at 4°C.

10. Examine in fluorescence microscope exciting first with UV light (primulin or SITS excitation) and then with blue light (acriflavine).

Results: Protein fluoresces blue and DNA fluoresces yellow. According to the filter combinations used, one can observe bicolor fluorescence of green and yellow (to facilitate recognition of certain cells) or individual color emissions of either blue or yellow.

NOTE: The technique is based on combining Kasten's fluorescent-Feulgen stain with a general protein fluorochrome. The excitation and emission

---

* No. 1010A, Research Organics, Inc., 4353 E. 49th St., Cleveland, OH 44125.

spectra of the two fluorophors are sufficiently separated to permit individual quantitative measurements by two-wavelength measurements or other cytofluorometric instrumentation. The technic is new and has not been applied to practical cytologic use. Rapid automated analysis of cell specimens is possible but will require that cells be dispersed to avoid artifacts due to clumping.

Magun and Kelley's Phenanthrenequinone Fluorescence Method for the Cytochemical Detection of Arginine Residues

From Magun and Kelly (1969)

Application: Tissue sections and blood and sperm smears.

Designed to show: Arginine residues in proteins.

Fixation: 10% neutral formalin overnight.

Cell and tissue preparation: Tissues are embedded in paraffin; smears are air-dried and fixed.

Preparation of solutions:

A. Phenanthrenequinone (PQ) solution: 1 part of 0.5 N NaOH is added to 4 parts of ethanol. To this solution is added 1 part of freshly prepared 1% PQ in dimethylformamide. The solution is red and must be used immediately.

B. Alkaline glycerol: 9:1:1 mixture of glycerol: water: 10 N NaOH; prepare fresh.

Staining schedule:

1. For tissue sections and smears, clear in xylene.
2. Rinse in 3 changes of absolute ethanol, 1 min each.
3. React in PQ solution for 10 min.
4. Rinse in 3 changes of 95% ethanol, 1 min each.
5. Hydrate tissues and mount in alkaline glycerol.
6. Examine in fluorescence microscope.

Results: A green fluorescence is seen in structures containing arginine residues. Fluorescence is most intense in nucleoli, of moderate intensity in nuclei, and high in areas known to have a heavy protein concentration (neuronal soma, cytoplasm of liver, renal proximal and distal tubules, microvillous border of intestinal mucosal cells, collagen, and keratin). Nuclei of chick erythrocytes also fluoresce intensely.

NOTE: The specificity of the cytochemical reaction was tested using a number of agents which block PQ from complexing with the guanidino group of arginine and by performing certain in vitro tests. Magun and Kelly also carried out microfluorometric measurements on chick erythrocyte nuclei. Fluorescence data were presented on the optimal excitation and fluorescence maxima. Parallel experiments were done on tissue stained in the Sakaguchi reaction, which is selective for arginine residues but less chromogenic and sensitive. Recently, the PQ cytochemical method was employed to demonstrate arginine-rich peptide and protein hormones in gastric chief cells, parenchymal cells of the pancreas, GH cells of the adenohypophysis, and glucagon cells of the pancreatic islets (Sundler and Hakanson, 1978). This paper should be consulted for further details of the modifications used.

Ringertz's Method for Demonstrating Histone Proteins by Dansyl Fluorochroming

From Ringertz (1968)

Application: Cell monolayers in tissue culture and protozoa.

Designed to show: Histones and protamines in cytologic preparations.

Fixation: 10% neutral formalin for 24 hr.

Preparation of solutions:
   A. 5% trichloroacetic acid (TCA).
   B. Saturated picric acid.
   C. 0.5% dansyl (dimethylaminonaphthalene sulfonic acid) in 0.035 M borate buffer, pH 8.3 (see p. 8).

Staining procedure:
   1. Rinse formalin-fixed cells for 15 min in distilled water.
   2. Treat with TCA at 90°C for 15 min or by picric acid at 60°C for 4 hr.
   3. Rinse in water.
   4. Stain in dansyl for 2 hr.
   5. Wash for 10 min in water.
   6. Dehydrate in ethanol.
   7. Mount in Fluormount or Entellan.*
   8. Examine in fluorescence microscope.

Results: Cell nuclei of cultured cells emit intense yellow fluorescence. A weak cytoplasmic fluorescence may be seen in protozoa (*Euplotes*). Control slides not treated to remove DNA also exhibit slight cytoplasmic fluorescence.

NOTE: This technic is identical with the alkaline fast green reaction after Alfert and Geschwind (1953) for histones but uses a fluorescent dye. It offers a greater sensitivity and is likely to be useful for cell quantitative studies, although Ringertz did not attempt to demonstrate this point with cell preparations. For a related dansyl technic which does not depend on hot acid extraction of DNA for selective dye binding, see the technic on page 67, "Cytofluorometric Detection of Lysine-rich Proteins with Dansylchloride."

---

* E. Merck, Darmstadt; EM Laboratories Inc., Elmsford, New York.

Cytofluorometric Detection of Lysine-rich Proteins with Dansylchloride

From Rosselet and Ruch (1968)

Application: Isolated nuclei, spermatozoa, tissue culture monolayers, acetic acid-squash preparations of salivary gland chromosomes, and root tips.

Designed to show: Lysine-rich histones and protamines for quantitative cytochemical determinations.

Fixation: Ethanol-acetic acid (9:1) for 3 hr or overnight (tissue pieces).

Embedding: None needed except for tissue pieces which are embedded in glycol methacrylate.

Preparation of solutions: 0.1% solution of dansylchloride in 95% ethanol saturated with sodium bicarbonate.

Staining schedule:
1. Bring fixed cells and sections into 95% ethanol.
2. Stain in dansylchloride for 4 hr.
3. Rinse 30 min in 95% ethanol.
4. Rinse 15 min in 2 changes of absolute ethanol.
5. Mount in Fluormount from xylene.
6. Examine in fluorescence microscope.

Results: There is a selective fluorescence from nuclei and nucleoli which reflects the lysine content of basic proteins according to the authors.

NOTE: The fluorophore decomposes rapidly during UV irradiation, so preliminary observations and positioning of cells must first be made with the aid of phase optics before taking fluorescence measurements. The authors neglected to include other tissues with cytoplasm (besides root tip) to see if cytoplasmic basic proteins fluoresce. Also, they fail to indicate whether or not paraffin-embedded sections are usable with this method. For a different dansyl fluorescent method for histones, see Ringertz's method on page 66.

SECTION J: CHROMOSOME BANDING TECHNICS

Preparation of Chromosomes for Banding Stains by Various Fluorochroming Technics

From Lubs et al. (1973)

Lymphocytes from the buffy coat of blood are grown for approximately 65 hr in GIBCO McCoy's 5A medium with 15% fetal calf serum and phytohemagglutinin (GIBCO lyophilized, 1–2 ml rehydrated per 100 ml of culture medium). Colcemid (0.1 μg/ml of medium) is added 2 hr before chromosome preparations are made. At the end of 2 hr, hypotonic treatment (0.075 M KCl) is carried out for 16–18 min (including 6 min centrifugation). Following fixation in 2 changes of fresh 3:1 methanol:acetic acid (total 25 min) and careful resuspension of the cells, the cell suspension is dropped onto a glass slide held at a 45° angle on which there is an even film of cold water. This is maintained by storing the slides (after cleaning with 7X) in a staining dish filled with cold water.

Spreading of metaphases is carried out first by blowing vigorously at the cell suspension at a right angle to the slide and then by placing the slide directly on a hot plate for 1½ to 2 min. The hot plate should be set to maintain a beaker of water at 65°C. The slide should be hot to the touch but not so hot as to burn. After drying, the preparations are ready for staining by banding technics.

Q-Banding Methods for Chromosome Analyses Using Quinacrine or Hoechst 33258 Fluorochromes

From Caspersson et al. (1968, 1969) and Hilwig and Gropp (1972, 1973); adapted by Lin et al. (1978)

Application: Human chromosome preparations from short-term lymphocyte cultures or other cultured cells.

Designed to show: Differential banding patterns of the Q type using quinacrine or A-T base-specific fluorochromes like Hoechst 33258.

Chromosome preparation: See procedure on page 67.

Preparation of solutions:
  A. Sorenson's phosphate buffer, 0.1 M, pH 6.8 (see p. 6).
  B. Sorenson's phosphate buffer, pH 5.5 (see p. 6).
  C. Acidified water: Adjust 250 ml double-distilled water to pH 4.5 with 0.1 N HCl.
  D. Quinacrine* staining solution: Mix 0.25 gm of quinacrine dihydrochloride in 50 ml of double-distilled water. Adjust pH to 4.5 with 0.1 N HCl (0.9% concentrated HCl in distilled water).
  E. Hoechst 33258† staining solution: Mix 2.5 µg Hoechst 33258 in 50 ml of Hanks' balanced salt solution (see p. 22).
  F. Actinomycin D‡ solution: Prepare at a concentration of 0.128 mM in 0.01 M sodium phosphate buffer at pH 6.8 (see p. 6).

---

* Sigma Chemical Co., P.O. Box 14508, Saint Louis, MO 63178; Polysciences Inc., Paul Valley Industrial Park, Warrington, PA 15976.
† Riedel de Haen Ag, Wunstorfer str. 40, 3016 Seelze, Germany (FR); Polysciences Inc.; Aldrich Chem. Co., 940 W. Saint Paul Ave., Milwaukee, WI 53233.
‡ Calbiochem-Behring Corp., P.O. Box 12087, San Diego, CA 92112.

Staining schedule A (QFQ banding with quinacrine):

1. Dip chromosome preparation in distilled water.
2. Stain in quinacrine solution for 15 min.
3. Rinse in 3 changes of acidified distilled water for a total of 10 min and air-dry.
4. Slides may be stored in refrigerator in a light-tight box or observed immediately by placing coverslip over slide using acidified water.
5. Blot excess water and seal using paraffin wax.
6. Examine in fluorescence microscope using appropriate primary filter for maximum excitation closest to 425 nm.

Staining schedule B (QFH banding with Hoechst 33258):

1. Place chromosome preparation in 95% ethanol for 10 min.
2. Transfer to 70% ethanol for 10 min.
3. Wash in distilled water for 2 min each.
4. Blot off excess water, place 2–3 drops of actinomycin D solution on slide, and put coverslip over coverslip for 20 min.
5. Gently slide off coverslip in phosphate buffer (pH 6.8), rinse in 2 changes of distilled water for a total of 2 min, and air-dry.
6. Place slide in Hoechst staining solution (light-tight box) for 10 min.
7. Rinse twice in distilled water and air-dry.
8. Place coverslip over preparation using 2–3 drops of phosphate buffer (pH 5.5), blot excess water, and seal with paraffin wax.
9. Examine in fluorescence microscope with primary filter chosen to excite close to 360 mm.

Results: Quinacrine and Hoechst 33258 dyes produce similar Q-banding patterns on chromosomes. However, the Hoechst dye also stains the variable bands of human chromosomes nos. 1, 9, and 16. The general fluorescence intensity of bands produced by the Hoechst fluorochrome is brighter and fades slower than quinacrine. For a practical discussion of photography and general methodologic problems of chromosome banding with fluorescent dyes, see Bordelon (1977). Briefly, the following points should be noted: 1) If chromosomes are bright and show no band distinctions, they are either overtreated with colcemid or overstained. 2) If chromosomes are fat and short, reduce the concentration of colcemid or time of exposure. 3) If chromosomes are hazy, there may be excessive mounting media, slides or coverslips may be too thick or uneven in thickness, rinsing may not have been thorough thereby releasing dye into the mounting buffer, or the iris diaphragm needs to be closed down. 4) If there is a background stain of small fluorescent grains with Hoechst 33258, this may be due to mycoplasma infection of the cultures (this dye is also used to detect this organism).

NOTE: Staining with quinacrine mustard has led to many advances to human cytogenetics. Each chromosome pair is clearly depicted with this stain. It possesses high specificity for certain chromosome segments which are rich in deoxyadenylate-deoxythymidilate (A-T) bases. On the other hand, G-C groups have a quenching effect. Other factors, such as protein composition and arrangement plus substructure of chromosomes also affect fluorescent banding patterns (Weisblum and deHaseth, 1972). The Y chromosome exhibits an intense fluorescence at the end of its long arm (Caspersson et al., 1970), which permits its easy recognition and polymorphism, even in interphase cells. The centromeric regions of chromosomes 3, 4, and 13 are specifically stained as well as satellites of certain acrocentric chromosomes. As indicated under results, the bisbenzimidazole dye Hoechst 33258 gives similar Q banding results to quinacrine and has some advantages. Other fluorescent dyes which bind to A-T rich DNA regions include DAPI (4',6-diaminido-2-phenylindole-2HCl) and tert-butylproflavine. Q banding is specific for one type of constitutive heterochromatin (A-T rich) whereas C banding stains all constitutive heterochromatin (Jalal et al., 1974). For general reviews of this fascinating and rapidly developing field of fluorescent chromosome probes, other references are given (Lubs et al., 1973; Dutrillaux, 1977; Latt, 1977; and Comings, 1978). These fluorochromes have been applied as well to the quantitative measurement of DNA and resolution of individual chromosomes by flow cytometry (Jensen et al., 1977; Gray et al., 1979).

Production of D Bands in the Fluorescent Staining of Chromosomes with Daunomycin and Adriamycin

From Lin and van de Sande (1975)

Application: Human chromosome preparations from short-term lymphocyte cultures.

Designated to show: Differential banding patterns on chromosomes similar to those produced by quinacrine technics. Certain advantages are claimed over quinacrine (see Note).

Chromosome preparations: The preparative procedure is given on page 67.

Preparation of staining solutions:
A. Daunomycin* (Cerubidine or daunorubicin HCl): 0.5 mg per ml in 0.1 M sodium phosphate buffer with a pH of 4.3.
B. Adriamycin† (doxorubicin HCl): 0.2 mg per ml in 0.1 M sodium phosphate buffer with a pH of 4.3.

Staining schedule:
1. Stain slides in either daunomycin or adriamycin for 15 min.
2. Wash in 3 changes of phosphate buffer for a total of 6 min.
3. Air-dry and mount in buffer.
4. Examine in fluorescent microscope.

---

* *Sigma Chemical Co.*
† *Adria Laboratories Inc., Dublin, OH 43017.*

Results: Well-defined and reproducible orange-red fluorescent banding patterns are observed on human chromosomes. The banding patterns are similar when using either anthracycline antibiotic and correspond with Q bands. The distal long arm of the Y chromosome fluoresces brightly; the Y-body fluoresces in interphase nuclei. Bright bands are seen on the centromeric area of chromosome no. 3 and on the short arm of chromosome no. 13.

NOTE: The characteristic differential D bands are similar, although not as bright, as those produced by quinacrine technics (Q bands). However, the fading of fluorescence and the disruptive effect of UV light on chromosomes are less pronounced as compared with Q banding. It is claimed that variations in ionic strength during staining do not affect fluorochroming results as in the use of quinacrine. It is possible that D banding may supplant the Q-banding technic for routine chromosome analysis. The D-banding procedure differentiates fluorescent patterns on three specific chromosome regions, which allows the specialized structural and functional properties of these regions to be studied. It is suggested that banding with these anthracycline antibiotics may be due to differential fluorescent quenching by DNA regions with specific base sequences.

SECTION K: CYTOCHEMICAL DETECTION OF POLYSACCHARIDES

Fluorescent Acridine Orange Technic for Acidic Mucins and Fungi

From Hicks and Matthaei (1958) and Pickett et al. (1960)

Designed to show: Acidic mucopolysaccharides and fungi.

Fixation: Formalin or other fixatives; avoid heavy metals.

Embedding: Frozen or paraffin sections.

Preparation of staining solutions:

A. Weigert's iron hematoxylin solution (see p. 108).
B. Acridine Orange (C.I. 46005): 0.1% aqueous solution of AO.

Staining schedule:

1. Bring sections to water.
2. Place in iron hematoxylin for 5 min.
3. Wash 3 min in running tap water.
4. Stain in AO for 1½ min.
5. Wash briefly in tap water.
6. Mount in glycerine.
7. Examine immediately with fluorescence microscope.

Results: Acid mucopolysaccharides, including many mucins, fluoresce reddish orange. Fungi fluoresce red, yellow, or green, according to the species. The background is black. The procedure described by Pickett et al. (1960) is almost identical to that described by Hicks and Matthaei (1958) but emphasizes the fluorescence detection of fungi. Instead of mounting slides in glycerine, they dehydrate in 95% and absolute alcohol, clear in xylol, and mount in a nonfluorescing mounting medium, like XAM.*

NOTE: For the sake of historical accuracy, due recognition should be given to Kuyper (1957) who directed attention initially to the possible identification of acid mucopolysaccharides with the aid of fluorescent basic dyes. He presented test tube evidence for spectral fluorescent shifts of the acridine dye, Coriphosphine (C.I. 46020) in the presence of various acidic mucopolysaccharides and RNA. Hicks and Matthaei (1958) demonstrated that these in vitro interactions occurred as well in tissue sections where histologists and pathologists could make use of them. An important methodologic contribution by Hicks and Matthaei was their use of iron hematoxylin prior to AO staining in order to eliminate tissue autofluorescence and permit a more striking demonstration of mucins. This work and a previous publication of a general nature (Hicks and Matthaei, 1955) helped to generate more interest in fluorescence microscopy among workers not familiar with the pioneering work of the Germans (Hagemann, 1937; Haitinger, 1934; 1938, a and b; Strugger, 1940). Almost hand-in-hand with the 1958 paper of Hicks and Matthaei appeared the work of Toriumi et al. (1959), who claimed to have made the same observations (i.e., acid mucopolysaccharide fluorochroming with acridines) also in 1958 that were reported in Japanese. The method of Saunders (1964), which is summarized elsewhere in this section, also utilizes AO to detect acid mucopolysaccharides. However, by using a selective solvent extraction and RNase, the major acidic mucopolysaccharides can be distinguished from each other. As indicated above, the simplified AO method described here is claimed to detect most fungi, presumably because of their polysaccharide coat. However, Yasaki (1959) stated that by using a relatively concentrated solution of Rheonin (C.I. 46075) preceded by treatment with an aluminum salt, such as AlCl₃, he obtained a selective red-brown fluorescence of *Cryptococcus neoformans* and tissue acidic mucopolysaccharides. Other fungi display weak fluorescence of other colors or no fluorescence at all. This kind of fluorescence discrimination does not seem to occur with AO.

Saunder's Acridine Orange-CTAC Method for Acid Mucopolysaccharides

From Saunders (1964)

Designed to show: Differentiation of various polysaccharides.

Fixation: Fix small pieces of tissue in Newcomer's fixative (see Introduction and General Methods) for 12–24 hr. Unfixed cryostat sections can be used.

Embedding: Process in 1:2 Newcomer/n-butanol for 30 min; 3 changes of n-butanol for 30 min each; transfer to 1:1 n-butanol/paraffin; 3 paraffin changes of 30 min each; cut 5-μm sections and mount.

Preparation of staining solutions:

A. 1% cetyltrimethlammonium chloride (CTAC).
B. Ribonuclease: 1.0 mg/ml in glass-distilled water.

---

* E. Gurr.

C. Acridine orange (C.I. 46005): 0.1% AO in distilled water (pH 7.2).
D. Same as C but use AO in 0.01 M acetic acid at pH 3.2 (0.6 ml/1000 ml distilled water).
E. Sodium chloride: 0.3 M in 0.01 M acetic acid (2.76 gm NaCl/100 ml of 0.01 M acetic acid.
F. Sodium chloride: 0.6 M in 0.01 M acetic acid (5.52 gm NaCl/100 ml of 0.01 M acetic acid).

Staining schedule:
1. Deparaffinize 3 slides (mark 1, 2, and 3) and bring to water.
2. Treat slides in triplicate with Solution A (CTAC) for 10 min.
3. Wash for 10 min in running water.
4. Treat all slides with Solution B (ribonuclease) for 2 hr at 45°C.
5. Treat Slide 1 with Solution A again for 10 min and wash in running water for 10 min.
6. Treat Slide 1 with Solution C for 3 min.
7. Wash Slide 1 in running water for 10 min.
8. Treat Slide 2 with Solution D for 5 min.
9. Wash Slide 2 in running water for 10 min.
10. Wash Slide 2 for 10 min in running water.
11. Treat Slide 3 as Slide 2 but substitute Solution F.
12. Air dry all 3 slides.
13. Mount in Fluormount or other fluorescence-free mounting medium.
14. Examine in fluorescence microscope.

Results:
Slide 1—red fluorescence due to hyaluronic acids.
Slide 2—red fluorescence due to chondroitin sulfates and heparin.
Slide 3—red fluorescence due to heparin.

NOTE: Formaldehyde-containing fixatives should not be used because a diffuse red fluorescence is induced in all tissues. Aqueous fixatives are to be avoided since acid mucopolysaccharides are diffusible. Newcomer's fixative penetrates rapidly and minimizes solution of polyanions of tissues. In stretching tissue sections, float on ethylene glycol monomethyl ether at 35–45°C and pick up on slides without adhesive; water may dissolve out the acid mucopolysaccharides. The staining method is based on the fact that acridine orange precipitates acid mucopolysaccharides, similar to CTAC. The precipitates can be dissolved in solutions of graded sodium chloride concentration. In the technic, CTAC is displaced from mucopolysaccharides by AO. RNA is first removed by ribonuclease to prevent staining of this polyanion.

Kasten's Fluorescent PAS Method

From Kasten (1959, 1964, 1973) and Kasten et al. (1959)

Application: Cells and tissue which contain various carbohydrates. See also page 72.

Designed to show: PAS-positive components, including glycogen, glycoproteins, mucoproteins, neutral mucopolysaccharides, and glycolipids.

Fixative: Various, including 10% neutral formalin, Carnoy, and Zenker.

Embedded in: Paraffin or frozen sections, plastic-embedded media for 0.5-μm sections (Epon, Maraglas, Araldite), smears, monolayer cultures, and suspended cells.

Preparation of solutions:

A. Acriflavine-Schiff reagent: prepare 0.1% Acriflavine (C.I. 46000) aqueous solution. The dye solution is saturated with sulfur dioxide (SO_2); and any of the methods listed below give satisfactory results.

1. Saturate dye solution with SO_2 gas for 1 min and seal with screw-type cap to prevent loss of gas. SO_2 may be obtained from a lecture-size bottle or generated by slowing adding concentrated hydrochloric acid to an excess of sodium or potassium metabisulfite or bisulfite powder in an Ehrlenmeyer flask fitted with a sidearm. The gas generated is led directly into the dye solution with rubber tubing for 1 min and then the dye solution is covered.

2. Add a few drops of thionyl chloride (SO_2) to 40 ml of dye solution and close tightly.

3. Add 0.5 gm of potassium metabisulfite ($K_2S_2O_5$) to 50 ml of the dye solution followed by 10 ml of 1 N HCl (8% concentrated HCl). Seal the jar with a screwcap.

The yellow acriflavine-SO_2 solution does not decolorize as with the conventional Schiff solution. The treated dye solution is ready for use immediately after completion of Methods 1 and 2 for adding SO_2 and, overnight, following Method 3. The concentration of acriflavine may be increased to 0.25% if desired but care should be taken to discard the solution if precipitation develops. Also, with increased dye concentration (as with the 0.5% acriflavine solution of Culling and Vassar (1961), there is a great probability of producing a superimposed fluorochromasia (bicolor fluorescence).

B. Periodic acid solution: 0.4 gm periodic acid (H_5IO_6), 35 ml absolute ethanol, 5 ml 0.2 M sodium acetate, 10 ml distilled water.

Staining schedule:

1. Deparaffinize sections and bring to 70% ethanol. Smears or monolayer cultures on slides or coverslips are brought up to 70% ethanol from water. Cell suspensions are handled in siliconized centrifuge or test tubes. which are centrifuged between treatments.

2. Treat for 5 min in periodic acid.

3. Rinse in 70% ethanol.

4. Rinse in water.

5. Place in acriflavine-Schiff or Flavophosphine N-Schiff (see Note below) solution for 15 min.

6. Wash in running tap water for 1–2 min.
7. Dehydrate and mount from xylene in nonfluorescent mounting medium, like Fluormount.
8. Examine in fluorescence microscope.

Results: PAS-positive substances in various tissues fluoresce yellow to orange (goblet cells, Brunner's glands, salivary mucins, glycogen, basement membranes, striated border of small intestine, thyroid collid, etc.). The same preparations show these reactive sites to appear yellow by brightfield optics when the reactive carbohydrate is present in sufficient concentration. Generally, there is no background autofluorescence but if present, it is slightly green and easy to distinguish from PAS-positive material. The autofluorescence can be eliminated by prestaining the tissue with one of the alum hematoxylins (Ehrlich's, Harris, Delafield's) for 2 min. To check for carbohydrate specificity, control slides should be acetylated (1–24 hr in solution of 16 ml acetic anhydride in 24 ml of pyridine) after periodate oxidation and before staining in fluorescent-Schiff reagent. There should be no fluorescence from sites previously found to be PAS-positive. To check for glycogen, treat control slides with 0.1% diastase in phosphate-buffered saline (see p. 22) for 1 hr before periodic acid treatment.

NOTE: From the historical side, there were initially two reports in which the possibility of substituting other dyes for Basic fuchsin (Pararosaniline) as Schiff reagents were suggested (Ostergren, 1948; Ornstein et al., 1957). The second work by Ornstein and associates considered the usefulness of acriflavine as a fluorochrome in the PAS reaction. Unaware of these papers which were obscure at the time, Kasten independently carried out exhaustive studies to locate suitable alternative dyes and fluorochromes as "Schiff-type" reagents. From over 400 dyes investigated, more than 40 dyes were discovered which could be employed in the PAS and Feulgen reactions. In addition to acriflavine, there were about 15 fluorescent dyes in the group of 40 which were found to replace Schiff's reagent.

One of these fluorochromes is Flavophosphine N or Benzoflavin (C.I. 46065), which produces a more intense fluorescence in the PAS test than any of the others, including Acriflavine. Some of the other fluorescent dyes recommended by Kasten which can be substituted for acriflavine as Schiff-type reagents include Acridine yellow (C.I. 46025), Coriphosphine O (C.I. 46020), Chrysophosphine 2G (C.I. 46040), Phosphine 5G (C.I. 46035), and Phosphine GN (C.I. 46045). Auramine O (C.I. 41000) is an especially useful fluorochrome as a Schiff-type reagent for DNA (see Section I, under "Kasten's Fluorescent Feulgen Method for DNA," page 59). The specificity of fluorescent-Schiff and other Schiff-type dye reactions for tissue polyaldehydes was verified (see papers cited above and Kasten (1958; 1960, a and b; 1964)). Subsequent confirmatory reports of the utility of these fluorescent Schiff-type reagents and the introduction of other aldehyde-binding fluorochromes include papers by Betts (1961), Culling and Vassar (1961), Prenna (1964), Ploem (1967, a and b), Stoward (1967, a and b), Trujillo and Van Dilla (1972), Kraemer et al. (1972), Weinblatt et al. (1975), Cottell and Livingston (1976), and Levinson et al. (1977). It was also shown that nuclear DNA could be measured by cytofluorometry in the Feulgen reaction with Kasten's acriflavine-Schiff reagent (Prenna and Zanotti, 1962; Böhm and Sprenger, 1968; Crissman et al., 1975) and related Schiff-type fluorochromes such as Acridine yellow-SO_2, Coriphosphine O-SO_2 (Böhm and Sprenger, 1968), Auramine O-SO_2 (Bosshard, 1964; Weste and Penington, 1972) and BAO-SO_2 (Ruch, 1964). Indeed, Pararosanilin itself, the classic dye used to prepare Schiff's reagent (Kasten, 1960a), was found to emit a red fluorescence from Feulgen-stained nuclei (Ploem, 1967, a and b) and could be employed as a quantitative tool (Böhm and Sprenger, 1968). Although the PAS cytochemical reaction has not been widely employed for quantitative purposes, there are infrequent reports which claim that glycogen can be measured using the fluorescence from Schiff reagents (Yataganas et al., 1969; Changaris et al., 1977).

SECTION L: METHODS FOR LIPIDS

Popper's Aqueous Phosphine 3R Fluorescence Method for Lipids

From Popper (1941a) and Volk and Popper (1944)

Application: Tissue sections, fecal material, and urine for routine use and for diagnosis of renal and biliary conditions.

Designed to show: Neutral lipids.

Fixation: For tissues, 10% neutral formalin, Baker's formol-calcium, or unfixed frozen blocks. No fixation for fecal material and urinary sediment.

Embedding: After formalin fixation, embed in paraffin; cut 10-μm sections.

Preparation of solutions: Phosphine 3R (C.I. 46045): 0.1 gm in 100 ml distilled water.

Staining schedule:
1. Bring sections to water. To examine feces, suspend a loopful of fecal material in a drop of dye solution. For urine, centrifuge and mix 1 drop of sediment with 1 drop of dye solution on a glass slide.
2. Stain for 3 min.
3. Rinse briefly in water (tissue sections) or cover in dye solution with a coverslip.
4. Mount sections in water or 90% glycerol.
5. Examine with fluorescence microscope.

Results: Lipids produce a silver-white fluorescence, generally in various sizes of droplets within cells (sections) or extracellularly (fecal and urine speciments).

NOTE: The reaction is specific for lipids, as judged by the absence of specific fluorescence in sections previously treated with acetone. Only neutral fats are stained, but not fatty acids, soaps, or cholesterol. The fluorescence does not depend on the presence of vitamin A, which may coincidentally be associated with lipids. Vitamin A autofluorescence is destroyed by continued UV radiation whereas phosphine 3R-induced fluorescence persists. According to Popper, the phosphine 3R method is more sensitive than using alcoholic Sudan III. Of importance here is the fact that the method described is an aqueous staining method and avoids the loss of some fat droplets during staining, as with Sudan and other fat-soluble dyes 3,4-benzpyrene has been used as a sensitive fluorescent detector of lipids but in view of its oil solubility, highly reactive properties, and carcinogenicity, it is not recommended for routine staining purposes.

Neutral Red as a Lipid Fluorochrome and Volutin Stain

From Kirk (1966, 1970)

Application: Studies of plant and animal lipids as well as distribution of oil droplets.

Designed to show: Fluorochroming of oil reserves (guttules) in fungal spores, fat bodies in insects, and simultaneous staining of volutin.

Tissue preparation: Ascospores of aquatic and marine fungi as well as fat bodies of sowbugs are prepared as fresh or formalin-fixed smears.

Preparation of staining solution: Neutral red (C.I. 50040) solution: Prepare 0.1% aqueous solution at pH 6.5 or mixed 1:1 with seawater.

Staining schedule:
1. Add 1 drop of dye solution and 1 drop of artificial or natural seawater to smear preparation and mount.
2. Examine immediately in fluorescence microscope and in white light.

Results: Lipids fluoresce blue-green in some cases and yellow in others, apparently according to their different composition. When the same preparations are examined in white light, volutin appears red and lipid droplets are colorless.

NOTE: The original articles should be read by those interested in more details. Kirk compared results with those obtained by benzpyrene fluorochroming and showed that with the latter the different lipids all fluoresce an intense blue-white color. The strong intensity is said to obscure cytologic detail. Other lipid stains are considered to be ineffective; the solvents used in Sudan staining cause mobility of the lipid droplets and Phosphine 3R was found to lack specificity. The method has had a specialized application and should be applied toward the detection of lipids in higher forms.

Berg's Technic for Masked Lipids Using Benzpyrene

From Berg (1951)

Application: Frozen sections of animal and plant tissues.

Designed to show: Finely dispersed lipids which are not routinely seen with other methods and are referred to as "masked lipids."

Tissue preparation: Frozen sections are fixed in 10% neutral formalin for 6–48 hr. For plant material, the addition of acetic acid to the formalin fixative facilitates diffusion through plant walls.

Preparation of solutions:

A. Benzpyrene-caffein solution (standard preparation): Saturate 110 ml of water with caffein at room temperature (approximately 1.5%) and let stand for 2 days. To 100 ml of filtered solution, add 2 mg of 3,4-benzpyrene (BP) and incubate at 35–37°C for 2 days. Filter, dilute with water by half, and refilter after several hrs. Solution now contains about 0.75 mg/100 ml BP which is usable for months if stored in a closed vessel.

B. Benzpyrene-caffein solution (rapid preparation): To 100 ml of saturated caffein solution, add with brisk stirring, 0.1–0.2 ml of 1% BP in acetone. Dilute by half with water and filter twice.

Staining schedule:

1. Place formalin-fixed frozen sections in BP-caffeine solution for 20 min.
2. Remove and float sections on water for several minutes.
3. Place on nonfluorescent glass slides, allow to dry, press flat as gently as possible without damaging tissue, and mount in water.
4. Examine immediately in fluorescence microscope.
5. Control slides without caffein and without BP should also be examined.

Results: Lipids, especially finely dispersed lipids ("masked"), fluoresce from blue to white to yellow-white when excited with UV light.

NOTE: BP is *carcinogenic* and must be handled with extreme caution. BP exhibits exceptionally strong fluorescence. It is one of the most stable of the carcinogenic hydrocarbons to UV light, and shows little fading with prolonged illumination. Caffeine is employed here as a hydrotropic medium for BP, which is lipid-soluble and insoluble in water or salt solutions. Some interference with lipid fluorescence occurs if sections are left in formalin too long; the natural fluorescence from connective tissue and epidermis increases. Background fluorescence also increases with the use of paraffin embedding or freeze-drying. A conflicting autofluorescence of blue-white color occurs from elastic tissue, expecially in lungs and cartilage (but not skin) and in the internal elastic lamina of arteries. Autofluorescence due to vitamins is not a problem because these compounds are washed out in the BP-caffein solution. Quenching of lipid fluorescence occurs in the presence of melanin, iron pigments, and hemoglobin. Berg compared the BP method with other lipid stains, such as Sudan dyes, and showed that the BP technique is more sensitive. However, the hazards of using BP should be considered in choosing an appropriate method.

SECTION M: NERVOUS SYSTEM METHODS

The Falck-Hillarp Fluorescence Method for Demonstrating Catecholamines

From Falck (1962), Falck et al. (1962), and Falck and Owman (1965); updated and summarized by Moore and Loy (1978)

Application: Nervous system tissues for identification of neurons and their processes according to neurotransmitter content.

Designed to show: Catecholamines (noradrenaline, adrenaline, dopamine) and indoleamines (5-hydroxytryptamine).

Tissue preparation:

A. Stretch preparation—This preparation is relatively simple and allows one to assess certain technical variables in the method. The most common tissue used is the iris of the albino rat. The eye is removed, incising the entire circumference of the globe with iris scissors, removing the lens, and placing the anterior portion of the globe down on a clean glass slide. Using a small forceps, the iris is removed by gently teasing away its attachment to the ciliary body, with the aid of a dissecting microscope. The iris is then moved to another part of the slide where it is gently stretched to its original shape. The remaining tissue is discarded, the excess fluid around the iris is blotted away, and the slide placed overnight in an evacuated desiccator containing phosphorus pentoxide. The same technic can be applied to mesentery or other thin tissue which can be stretched and dried.

B. Smear preparation—This is similar to the stretch preparation except that brain tissue is dissected so that a piece 1 × 2 mm or less is removed and placed on a clean glass slide. A second slide is pulled across it at an acute angle (as in obtaining blood smears) so that a thin, even smear of tissue is obtained. The slide is placed overnight in an evacuated desiccator over phosphorus pentoxide. Tissue organization is disrupted by the smearing, but varicosities are retained in proportion to the original innervation of the area. Thus the catecholamine content (but not 5-hydroxytryptamine) can be assessed by chemical analysis or visualized histochemically in the area being studied. It is useful as a screening procedure for quick checks on pharmacologic interventions or lesions.

C. Freeze-dry method—The animal is sacrificed and the brain removed as rapidly as possible. The brain is dissected into pieces from as small as 2 mm or less to as large as 1 × 2 cm in diameter. The tissue is placed on a card and oriented so that sectioning can be done on the desired surface. A piece of fine-mesh gauze is placed over the tissue and permitted to adhere to it. The tissue is then frozen rapidly in a mixture of liquid propane-propylene or isopentane, cooled with liquid nitrogen, and transferred to the tissue plate of a freeze-dryer where it is stored in liquid nitrogen up to several days until a freeze-drying run is begun. For details of the actual freeze-drying procedure, the reader should refer to instructions for his own equipment and the protocols given by Moore and Loy (1978) and Pearse ((1968) p. 593). The dried tissue can be kept several days in a desiccator prior to formaldehyde vapor treatment without noticeable loss of fluorescence.

Formaldehyde vapor treatment: This treatment is carried out in a 1-liter glass vessel containing about 5 gm of paraformaldehyde in an oven at 80°C for 1–3 hr; the reaction with secondary amines like adrenaline is slower and requires 3 hr. The formaldehyde should be discarded after use. The water content of the paraformaldehyde is critical. Too low a water content yields low fluorescence and few amine-containing structures are visualized. Too high a water content produces excessive and diffuse fluorescence which makes it difficult to visualize morphologic details. In practice, paraformaldehyde is kept in a desiccator with certain concentrations of sulfuric acid and water. Although the optimal humidity must be determined empirically for a given laboratory, good results can be expected between 50 and 70% humidity.

Paraffin embedding and sectioning: After exposure to formaldehyde vapor, tissue should be embedded rapidly in paraffin using a vacuum oven. This is done with the tissue still attached to the paper strips and covered with gauze. After the initial infiltration, the paper is removed, the gauze stripped from the tissue using warm forceps, and the block embedded in a standard way. Sections are cut at 6–15 μm, mounted on clean glass slides in either Entellen, Fluormount, liquid paraffin, or nonfluorescent immersion oil; and then coverslipped. The slide is then warmed gently on a slide warmer at about 60°C to dissolve the paraffin into the mounting medium. Examine in the fluorescence microscope.

Results: Using appropriate primary and secondary filters, catecholamine neurons are seen to emit a green to green-yellow fluorescence. Cell bodies and axon terminals (varicosities) are more readily demonstrated than preterminal axons, except when lesions are employed. The method does not distinguish between noradrenaline, adrenaline, and dopamine fluorophores. However, with the aid of specialized cytofluorometric equipment and procedures, the catecholamines can be distinguished from each other. The 5-hydroxytryptamine fluorophore decomposes rapidly while the catecholamine emitters are relatively more stable.

NOTE: CNS tissue exhibits stable fluorescence in the paraffin blocks for 3–6 months but the deparaffinized sections exhibit pronounced fading in a few days together with increased background fluorescence. Therefore, sections must be analyzed and photographed (Tri-X film) immediately. Although there are some difficulties in interpretation of the material, the Falck-Hillarp method is a valuable one and the only one available for examining large tissue blocks and for demonstrating 5-hydroxytryptamine neurons. The visualization of these particular neurons is improved by pretreating the animal with a monoamine oxidase inhibitor, such as paragyline followed by L-tryptophane, a precursor of 5-hydroxytryptamine (Agahajanian et al., 1973).

Two significant modifications have been made to the Falck-Hillarp method by Hőkfelt and Ljungdahl (1972). They eliminated freeze-drying and introduced the use of formaldehyde-perfused material and the Vibratome. The advantages are that it takes only 2 days to obtain usable sections. The sections are of good histologic quality, and they produce good fluorescence with low diffusion for 1–2 weeks. Newcomers to the field should be aware of other fast moving methodologic developments in this field. Other recent methods for biogenic amines use glyoxylic acid (GA) in combination with the Vibratome or a cryostat. GA methods are more sensitive and simpler than the Falck-Hillarp method. The most rapid procedure involves exposure of cryostat sections to a solution of sucrose-potassium phosphate-glyoxylic acid (SPG method), followed by drying and heating. It is claimed that the SPG method for monoamine transmitters takes as little as 18 min from the time of getting fresh tissue to fluorescence microscopic examination (Torre and Surgeon, 1976; Torre, 1979). The method is sensitive enough to detect preterminal axons. For further details of these exciting developments, one should see the original GA papers by Björkland and associates (cf., Björklund et al., 1972) and the work of Furness and Costa (1975). The Vibratome-GA method of Lindvall and Björklund (1974) also is described in detail in the review by Moore and Loy (1978). The rapid SPG method of Torre has already been cited above. A further increase in sensitivity comes through the addition of aluminum and is referred to as the aluminum-formaldehyde (ALFA) method (Lorén et al., 1980). It is used on freeze-dried paraffin embedded tissue and in Vibratome and cryostat sections of neurons in the central nervous system as well as for demonstrating peripheral stores of catecholamines and indolamines in these same kinds of preparations and whole mount preparations (Ajelis et al., 1979).

Stretton and Kravitz's Method for Visualizing Neurons with Procion Yellow M4RS

From Stretton and Kravitz (1968, 1973)

Application: Living nerve cells.

Designed to show: Fluorescence marking of neuronal geometry and correlation of neuronal morphology with electrical activity.

Fixation: 4% formaldehyde at pH 4.0 overnight. Weigh out 4 gm paraformaldehyde, heat to 90°C in 25 ml distilled water, add 1 ml of 1 N NaOH, 25 ml salts (2% $CaCl_2 \cdot 2H_2O$, 12% NaCl, 12% sucrose), and 50 ml of 0.1 M acetate buffer, pH 4.

Staining solution: 4% Procion Yellow M4RS* (no C.I. number, C.I. Reactive Orange 14): Prepare fresh in distilled water.

Tissue preparation and staining:

1. Expose neurons in ganglia; large invertebrate neurons are easiest to work with. (The original work was done on lobster abdominal ganglia.)
2. Fill electrodes with dye solution; the electrode resistance with dye is 10–20 megohms.
3. Hyperpolarizing current pulses of 1×10^{-8} to 5×10^{-8} amp and 0.5 sec

---

* Polysciences Inc.; Wilson Diagnostics Inc.

long at a frequency of 1 per sec and passed through electrodes until cells are filled with dye for 30–60 min at 10°C.

4. Preparation is kept at 4°C overnight (about 16 hr).
5. Fix with formaldehyde overnight.
6. Dehydrate in graded methanol series.
7. Embed in Epon 812 or Maraglas.
8. Cut 10-μm sections with a steel knife, mount sections on slides with Lustrex.*
9. Examine for cell profile in fluorescence microscope.

Results: The entire cell, including processes, fluoresces orange-yellow.

NOTE: This important technic, although relatively young, has proved to be a powerful tool for solving problems in the invertebrate and vertebrate nervous system. It has permitted the neurophil to be studied systematically and brought about correlations between electrical activity with morphology, particularly in the vertebrate retina. Over 60 different Procion dyes have been analyzed to determine the most suitable ones for iontophoretic injection and filling of neurons plus retention after fixation. Variation in fixative conditions has been found to have a marked influence on dye localization and autofluorescence. Procion yellow is not a true fluorescent probe, as there is no change in fluorescence with membrane excitation. The dye injection technic, when employed properly, even fills fine nerve processes and dendritic spines of 0.2 μm in diameter. Since this is a specialized kind of fluorescence staining, persons interested in more details and applications should consult the references cited, the volume on *Intracellular Staining in Neurobiology* (Kater and Nicholson, 1973), and the chapter by Tweedle in *Neuroanatomical Research Techniques* (Robertson, 1978).

A Simple Fluorescence Method for Serotonin-containing Endocrine (APUD) Cells

From Hoyt et al. (1979)

Application: Plastic-embedded sections of tissue.

Designed to show: Serotonin (5-hydroxytryptophane)-containing endocrine (APUD) cells.

Tissue preparation: Tissue is fixed at room temperature by immersion in 6% v/v phosphate-buffered (0.1 M, pH 7.2) formaldehyde freshly prepared from paraformaldehyde powder. Samples can be left in fixative for weeks before rinsing in water, dehydration in graded alcohols, and embedding in glycol methacrylate (GMA) for cytologic studies. Alternatively, tissues can be embedded in paraffin but with less favorable results. Plastic sections, 2 μm, are cut, floated by water onto glass slides, air-dried, and mounted in Entellen.* Paraffin sections are dewaxed in xylene. Slides are examined with a fluorescence microscope, preferably of the epi type.

---

* *EM Laboratories, Inc., Elmsford, New York.*

Results: Strong yellow fluorescence is seen from cells known to contain serotonin, such as enterochromaffin cells of the gut, rat mast cells, small granule cells of rabbit lungs, and from APUD cells in the thyroid gland and tracheal epithelium of rats.

NOTE: This is a modified formaldehyde-induced monoamine fluorescence technic which avoids losses due to difficulties with frozen sections or lyophilization. Serial sections, 2 μm, are said to be obtained routinely.

SECTION N: INTRAVITAL AND SUPRAVITAL FLUOROCHROMING METHODS

An Intravital Fluorescence Method for the Microvasculature Using Thioflavine S

From Schlegel (1949) and Schlegel and Moses (1950) modified by Dohrmann and Wick (1971)

Application: Microscopic visualization of blood vessels and patterns of blood flow.

Designed to show: Microvasculature of the spinal cord.

Tissue preparation: An adult cat is anesthetized with sodium pentobarbital (22 mg/kg) and a femoral vein exposed. A laminectomy is performed to expose the spinal cord (T10 level) leaving the duramater intact.

Preparation of staining solution: Thioflavine S (C.I. 49010): A 4% aqueous solution is ultracentrifuged at 30,000 rpm for 30 min to remove particulate matter. The supernatant is warmed to 37°C just prior to use.

Staining schedule:

1. Freshly prepared and warmed dye solution is injected rapidly (2–4 sec) into the femoral vein at a dose of 1 ml/kg body weight.
2. Within 10–15 sec after injection, the exposed portion of the spinal cord is excised and trimmed to give a specimen 5–6 mm in length.
3. The specimen is mounted on a tissue carrier and immersed in liquid nitrogen until frozen.
4. Sections of 150 μm thickness are cut in a cryostat and placed on slides.
5. Uncovered sections are examined in a fluorescence microscope with epi-illumination from a high pressure mercury source.

Results: The spinal cord appears a pale blue color while blood vessels emit a bright yellow fluorescence. The microvasculature in both gray and white matter are visualized. The vessels in the spinal cord of noninjected cats do not exhibit autofluorescence.

NOTE: The Thioflavine S perfusion technic was described by Schlegel (1949) and Schlegel and Moses (1950) for fluorescence visualization of blood vessels and lymphatics. Their technic utilized hand-cut sections. However, the write-up given above comes from Dohrmann and Wick

(1971), who adapted the technic for spinal cord material and introduced some improvements; these included the use of microtome-cut sections, a more intense UV light source, and prior removal of undissolved dye particles by ultracentrifugation to prevent occlusion of arterioles. There is no reason why the method could not be used to demonstrate blood vessels in other tissues.

An Intravital Fluorescence Labeling Method with Xylenol Orange for Calcifying Tissues

From Rahn and Perren (1971)

Application: Calcifying tissues.

Designed to show: Growth, turnover, and repair of bone and dentin.

Preparation of staining solution: Xylenol Orange* (no C.I. number): A 3% aqueous solution is employed.

Tissue preparation and staining: Animals to be studied are injected with the dye solution at a dose of 90 mg/kg body weight. In rats, the injection is intraperitoneally; for rabbits, either intraperitoneally, intravenously (0.1 ml/sec), or subcutaneously; for larger animals like dogs and sheep, an intravenous drip of 10–30 ml/min is used. Animals are sacrified from 1 day to a few months after dye administration. Bones and teeth are removed, fixed in 10% neutral formalin (10 days or less), and left undecalcified. Slices of nonembedded material are cut at thickness up to 100 μm with a circular saw under continuous irrigation. Tissue may be embedded in methyl methacrylate. Sections are mounted in Eukitt and examined in fluorescence microscope.

Results: Areas in bone and dentin which were undergoing calcification at the type of dye injection fluoresce orange. The color contrasts with the fluorescence induced by hematoporphyrin (red), tetracycline (yellow), Calcein and Calcein Blue (blue), and DCAF (2,4-bis-N,N-di-(carbomethyl)-aminomethylfluorescein) (green). It is less contrasting from alizarine red S (C.I. 58005).

> **NOTE:** Because of the availability of other fluorescent labels for the mineralization process (see above), it is possible to use the Xylenol Orange technic in time injection sequence with labeling by tetracycline, hematoporphyrin, DCAF, and Calcein or Calcein Blue to demonstrate progressive calcification in the same bone. For single fluorescent labeling, Xylenol Orange is superior to the other fluorochromes because it exhibits little to no toxicity at the recommended dose, has no apparent influence on the calcification process, and fades less from bone sections during UV illuminatione (Rahn and Perren, 1970; Rahn et al., 1970). Also, the final fluorescence is not diminished due to prior bone storage in the frozen state and in

---

* *Sigma Chemical Co.; Eastman Organic Chem.; Aldrich Chemical Co.*

various solvents for many days and weeks (see Rahn and Perren (1971) for details). Tissue sections from Xylenol Orange-injected animals can also be stained with basic fuchsin to permit histologic viewing by bright-field illumination.

For those interested in the classical tetracycline fluorochrome method for mineralizing tissues, further discussions of tetracycline-Ca^{2+} binding, and determination of new bone growth with this technic, see Milch et al. (1957, 1958, 1961) and the papers by Frost (Frost et al., 1960; Frost, 1963, 1966, 1968, 1969). General summaries of tetracycline binding to calcifying tissues are given by Pearse (1972, pp. 1178–1179) and Lillie and Fullmer (1976, pp. 801–802).

Davis and Sauter's Fluorescence Method for Detecting Trypan Blue in Embryonic Tissues after Maternal Injections of Dye

From Davis and Sauter (1977)

Application: Studies of teratogenic action of Trypan blue on embryonic tissues.

Designed to show: Histologic sites of protein-bound and free trypan blue.

Tissue preparation: At approximately 8–9 days of gestation, pregnant rats are injected intraperitoneally with a 1.5% aqueous solution of Trypan blue (C.I. 23850) at a dosage of 75 mg/kg body weight. The particular dye sample must be shown to be free of fluorescent contaminants by UV spot tests. Ninety two hours later, pregnant rats are anesthetized with Nembutal. The 12-day embryos are dissected away from the uterus and placed on a copper disk. Material is frozen rapidly in isopentane cooled to $-160°C$ with liquid nitrogen, transferred on Dry Ice to a freeze-drying apparatus, desiccated, vacuum embedded in paraffin, and sectioned at 10 μm. Sections are mounted on clean, dry glass slides without coverslips, and examined in UV and tungsten light with the aid of a fluorescence microscope.

Results: With UV illumination, intense areas of red fluorescence (protein-bound Trypan blue) are seen within mid- and hind-gut lumina and extending into the allantoic stalk. Less intense red fluorescence is detected within mucosal cells of the mid- and hind-gut. Embryos which still retain visceral yolk sacs reveal red fluorescing vacuoles within the absorptive epithelium (visceral endoderm). Protein-free dye also appears in these cells as nonfluorescing black aggregates. When these same areas are examined with tungsten light, the gut lumen fails to show evidence of dye but the lining epithelium reveals some aggregates of blue dye. The visceral endoderm of yolk sacs contains a few blue vacuoles (corresponding to both red-fluorescing and black nonfluorescing dye seen with UV).

In summary, the method offers a sensitive means of visualizing the teratogenic dye, Trypan blue in bound form (red fluorescence) in maternal blood and yolk sac cavity and in cell vacuoles. These vacuoles probably give rise to intacellular granules, which are nonfluorescing free forms of the dye.

NOTE: It should be pointed out that Trypan blue itself does not fluoresce but when bound to serum albumin (which is also nonfluorescent), a red fluorescence is produced (Hamberger and Hamberger, 1966). This is indirect evidence for the interpretations given by Davis and Sauter (1977) for the histologic results. Binding of the dye to other moieties, such as lipids, is not ruled out. The advantages of the method are derived from the use of fluorescence microscopy to improve sensitivity and the avoidance of solvents (commonly employed in microtechnics) which remove much of the dye.

Similar results are seen with Evans blue (C.I. 23860; Steinwall and Klatzo, 1966), a teratogen and an isomer of Trypan blue. These dyes are aminonaphthalene sulfonic acids that become distorted into a planar configuration and fluoresce when absorbed to protein (Udenfriend, 1962).

Acridine Orange Stain for Lysosomes in Living Cells

From Robbins and Marcus (1963) and Robbins et al. (1964)

Application: Living cells.

Designed to show: Lysosomal granules in viable cells or dispersed lysosomal enzymes in degenerating cells.

Fixation: None.

Preparation of staining solutions:
  A. Stock solution of Acridine Orange (C.I. 46005): Add 1 mg dye to 2 ml of Hanks' balanced salt solution (BSS); adjust to pH 7.3 with minimum volume of 1 N HCl or 1 N NaOH.
  B. Working solution of Acridine Orange: Add 1 ml of stock AO solution to 99 ml of BSS (see p. 22); adjust pH to 7.3 if necessary.

Staining schedule:
  1. Coverslip cultures or suspensions of living cells are drained of medium or centrifuged.
  2. Wash cells with BSS at pH 7.3.
  3. Immerse in working AO dye solution for 5 min.
  4. Rinse a few seconds in BSS.
  5. Invert on glass slide in BSS solution.
  6. Seal edges of coverslip with fingernail polish.
  7. Examine immediately by fluorescence microscopy.

Results: Intact lysosomal particles in viable cells fluoresce orange; dispersed lysosomal enzymes in degenerating or dead cells fluoresce bright red; nuclei fluoresce green. The slide preparation is temporary.

NOTE: The stain deposition within living cells depends on several interacting parameters, including pH, time, temperature, dye concentration, and cell energy expenditure. The Acridine Orange particles (AOP), which fluoresce orange, have been considered by some workers to contain RNA since RNA fluoresces red in fixed cells. Other interpretations are that this fluorescence is due to mitochondria, cellular response to injury, coacervates, or ingested aggregates of dye. These and other interpretations have been reviewed by Wittekind (1958). However, by using controlled experiments, in combination with cytochemistry, electron microscopy and fluorescence microscopy, Robbins and associates demonstrated that the AOP were acid phosphatase-positive multivesicular bodies (MVB). This finding provided the basis for a machine characterization of human leukocytes by AO fluorescence (Adams and Kamentsky, 1971). Apparently, the presence of orange-fluorescing particles is a binding which is compatible with cell viability whereas a general reddening of the cytoplasm results from a prior release of lysosomal enzymes and is indicative of cell death.

The earliest work dealing with the use of Acridine Orange to distinguish dead and living cells by their fluorescent colors dates back to Strugger (1940) who used the technique on yeast cells (see p. 87), "Differential Fluorescence of Live and Dead Yeast Cells with Acridine Orange"). Strugger also employed this fluorochrome to detect live bacteria in soil (1948a) and living trypanosomes in blood (1948b). Schümmelfeder (1950a) applied the technique to animal tissue cells. There is an immense literature dealing with the supravital fluorochroming of cells with AO (cf. Weissmann and Gilgen, 1956; Austin and Bishop, 1959; Wolf and Aronson, 1961; Meisel et al., 1961; Seydel, 1966; West, 1969; Cowden and Curtis, 1974; Nicolini et al., 1979). For a thorough review of the early history of the subject, the monograph by Stockinger (1964) should be consulted. The reviews by Kasten (1967a, 1973) may also be of interest. For a discussion of the interaction of AO with fixed cells, see pages 48 and 57–58.

Differential Fluorescence off Live and Dead Yeast Cells with Acridine Orange

From Strugger (1947), modified by Schwartz et al. (1977)

Application: Yeast cells.

Designed to show: Physiological state of yeast cells.

Cell preparation: Yeast cells are removed from agar slants with an inoculation loop, grown in flask cultures containing broth, incubated at 37°C, and shaken at about 150 rpm.

Preparation of staining solutions:

A. Acridine Orange (C.I. 46005) stock solution: Prepare 0.1% AO aqueous solution and store at 4°C in the dark. Solution is good indefinitely.

B. Acridine Orange working solution: Dilute stock AO solution to 0.01% in 0.005 M phosphate buffer, pH 7.0. Solution is kept at room temperature in the dark and used up to a week.

C. Phosphate buffer, 0.005 M, pH 7.0 (see p. 6).

Staining schedule:

1. For viability staining, a 0.5-ml cell sample is removed from a flask and

mixed on a Vortex stirrer with 0.5 ml of AO working solution in a test tube.

2. After 20 min of staining at room temperature, 4 ml of distilled water are added.

3. Mix on Vortex stirrer.

4. Count approximately 100–200 cells total with Spencer Bright-Line Hemacytometer. It is best to count the orange-red dead cells first.

Fluorescence microscopy: A standard fluorescence microscope is employed and cells are counted under high-dry magnification. Since grid lines of hemacytometer are normally not visible, the problem is overcome by inserting a thin piece of cardboard coated with fluorescent tempera paint under the swing-out lens of the substage condenser. By manipulating the light source and condenser height, the light transmitted upward from the card illuminates the grid lines.

Results: Live yeast cells fluoresce green to yellow-green. Dead cells fluoresce orange to red. Cell lysis is shown by the sudden appearance of a brightly fluorescent halo around the cell and a corresponding loss in cytoplasmic fluorescence. The yeast vacuole or lysosome is normally unstained and appears as a dark area surrounded by green cytoplasm. With aging and loss of physiological activity, the vacuole begins to fluoresce orange, enveloped by green cytoplasm, until eventually the entire cell becomes orange and the cell is dead. The nucleus is a poor indicator of cell viability. It normally appears as a green body near one pole of the cell. With progressive cell degeneration, the nucleus may or may not disappear.

> **NOTE:** AO is employed widely as a vital fluorochrome, especially for protozoa, bacteria, and fungi (see other technics in this section). The application to enumeration of live and dead yeast cells by Strugger (1947) has not been without difficulties, probably because of the complex cytology and varying methods of illumination during counting. The work of Schwartz et al. (1977), which is summarized here should now permit reliable quantitative results to be obtained. The AO method gives results that are in agreement with data obtained with Janus green B or methylene blue as vital stains but is free of certain errors in interpretation. The authors emphasize the need to prevent glare during counting by using a low concentration of AO and cells, employing clean glass surfaces, and noting the effect of cell lysis. When lysis is a problem, the authors recommend the substitution of 10% glycerine in the phosphate buffer in place of water. For references to other literature on the subject, see the original paper by Schwartz et al. (1977). The work of Meisel and associates (cf. 1961) should be mentioned since they studied many fluorochromes to discriminate living and dead yeast cells. As a consequence of radiosensitivity studies, they concluded that a Primulin (C.I. 49000)-Acridine Orange mixture (1:30,000 and 1:100,000 dilutions, respectively) was especially useful and nontoxic. Living cells fluoresce green and dead cells fluoresce yellow.

Love's Supravital Fluorochroming Method with Acridine Orange for Isolated Mast Cells

From Love (1979)

Application: Freshly harvested peritoneal mast cells.

Designed to show: Fluorochromasia of cytoplasmic granules.

Fixation: None.

Cell preparation: Mast cells are obtained from the peritoneal cavity of adult rats, which are sacrificed by CO_2 inhalation. Abdominal skin is reflected and a mid-ventral incision about 4 mm long is made into the peritoneal cavity. Approximately 25 ml of phosphate buffered saline (PBS, see p. 22) is pipetted into the cavity and the incision held closed while the abdomen is gently massaged for about 2 min. The incision is extended to allow a dropping pipette to aspirate the washings. The washings are filtered through a Millipore filter support screen to remove pieces of fatty tissue. The filtered cell suspension is centrifuged lightly for 8 min. Supernatant is discarded and the cell pellet is resuspended in PBS. Mast cells are separated from other cells following the method of Lagunoff (1972) by centrifuging cells through 5 ml albumin, specific gravity 1.100 (Pathocyte 4*) in 20-ml centrifuge tubes at $200 \times g$ for 20 min at 4°C. The nonsedimenting cells at the interface are removed by aspiration and discarded. The residual albumin solution is removed and diluted with 4 ml PBS (see p. 22). Mast cells are collected by centrifugation at $200 \times g$ for 5 min, washed with 5 ml PBS and resuspended in 10 ml PBS. The mast cells comprising 80–90% of the isolated cell population are counted in a hemacytometer and diluted with PBS to give a final stock concentration of 4×10^5 cells per ml. Each animal yields about 1.5×10^6 mast cells.

Preparation of staining solutions:

A. Acridine orange (AO) stock solution: 1.1 mM in distilled water (33.2 mg AO per 100 ml). The dye used by Love was chromatographically purified by the method of Zanker (1952) prior to its use in the stock solution.

B. Working AO solution: Dilute AO solution A with PBS to give a series of six dilutions over the range: 5.5×10^{-7} to 10^{-5} M.

Staining schedule:

1. Centrifuge six 1.0-ml aliquots of the cell stock suspension at $80 \times g$ for 6 min.
2. Decant the supernatant and resuspend the cells in glass test tubes with

---

* *Pentex, Inc.*

2.0 ml of each working AO solution B for 45 min. Seal with Parafilm and shake the tubes every 15 min to keep the cells in suspension.
3. Put a drop of the stained cell suspension on a glass slide.
4. Place coverslip on drop, seal with hot paraffin, and examine with fluorescence microscope.

Results: Fluorescence colors in nucleus, cytoplasm, and cytoplasmic granules vary according to the dye concentration employed. Granules fluoresce yellow-green at a very low dye concentration, red at high dye concentration, and a mixture of green, yellow, and red colors in between, which is most desirable to see all the granules. This ideal dye concentration must first be found by running a series of dilutions as described above for a given batch of dye. Presumably, good results can be achieved with unpurified AO dye samples after the optimal concentration is determined. Nuclei and cytoplasm also fluoresce different colors, generally following the sequence of green to orange to red with increasing dye concentration. The nucleolus is weakly fluorescent, green to yellow.

> **NOTE:** This technic is considered by Love to be a reliable method for visualizing the structures of living mast cells and avoids the variations caused by fixation and handling. The fluorochromasia exhibited by mast cell granules probably results from differences in degree of sulfation of intragranular heparin. With prolonged irradiation, there is a fading of fluorescence. The average time between harvesting of cells and microscopy is 90 min.

Smith-Sonneborn's Rapid Supravital Fluorochroming Method for Cytologic Structures in Paramecia Using Acridine Orange

From Smith-Sonneborn (1974)

Application: Suspensions of paramecia.

Designed to show: Nuclear and cytoplasmic structures in fresh suspensions and autoradiographs.

Cell preparation: Cells are removed from the culture with a micropipet under a dissecting microscope and placed on a slide in a drop of culture fluid.

Preparation of solutions: Acridine Orange solution (C.I. 46005): 0.3 mg per ml of AO is prepared in Dryl's salt solution (see p. 22).

Staining schedule:
1. Add a drop of AO solution to a drop of suspended paramecia.
2. Rotate to mix the solutions.
3. Add coverslip (if desired) lowering gently to avoid rupturing cells.
4. Observe immediately in fluorescence microscope.
5. For staining developed autoradiographs, a drop of AO solution is

placed on the emulsion for 10 sec and rinsed off with distilled water. Dried preparation will retain cytologic differentiation qualities for days.

Results: Nuclei fluoresce greenish-yellow, cytoplasm appears red, and food vacuoles are yellow. Contractile vacuoles do not stain. In cells undergoing nuclear reorganization, the fragmenting old macronucleus is yellow and the developing new nuclei are green, as are the micronuclei. After prolonged observation, the cells rupture. Ruptured cells permit better visualization of internal structures. In autoradiographs, silver grains appear as brown dots over differentially stained structures. However, the emulsion also stains red.

NOTE: This use of AO is meant as a temporary, rapid, and simple method for routine cytology of paramecia. It permits quick and excellent photomicrography.

Vital and Postvital Fluorochroming of Yeast Mitochondrial DNA with DAPI

From Williamson and Fennell (1975)

Application: Yeast cells.

Designed to show: Mitochondrial DNA.

Cell preparation and fixation: For vital staining, growing cells are exposed directly to the fluorochrome dissolved in the medium. For postvital staining, pelleted cells are first suspended in formaldehyde for 30 min and washed in distilled water.

Preparation of solutions:
  A. 3% formaldehyde* in phosphate buffer, pH 7.0: Add 3 ml of concentrated formaldehyde (37–40%) to 97 ml of 0.1 M phosphate buffer, pH 7.0 (see p. 6).
  B. DAPI† solutions (4',6-diamidino-2-phenylindole): For vital staining, add 1.0 μg/ml of medium used to grow yeast cells; for postvital staining, use 0.1–0.5 μg/ml of DAPI in distilled water.

Staining schedule:
  1. For vital staining, treat living cells with fluorochrome-containing medium and examine in fluorescence microscope.
  2. For postvital staining, stain formaldehyde-fixed cells in DAPI overnight at 4°C, wash in distilled water, and examine in fluorescence microscope.

---

* *In the original article, the concentration is said to be 0.3% (w/v). It is assumed that there is an error in printing, and that the decimal point was transposed.*
† *Aldrich Chemical Co., No. 21,708-5; Polysciences Inc.; Accurate Scientific Co., 28 Tec St., Hicksville, NY 11801.*

Results: Individual yeast cells each show single fluorescing nuclei and numerous fluorescing cytoplasmic particles. Evidence to support the idea that all these fluorescing structures contain DNA comes from the fact that the cytoplasmic structures are seen only in yeast strains which contain mitochondrial DNA. Fluorescence of the extranuclear bodies and nuclei is abolished by prior treatment with DNase. RNase and pronase have no effect on fluorescence. Clearer staining is obtained in postvitally stained preparations than in the vital-stained material.

NOTE: According to Williamson and Fennell, the sensitivity of this cytochemical procedure for the detection of DNA seems to be very high. According to their calculations, it is possible to visualize a mitochondrial particle containing as few as two mitochondrial DNA molecules. DAPI has been employed to detect DNA viruses and mycoplasma in cell culture (Russell et al., 1975) and chloroplast DNA (James and Jope, 1978).

A Supravital Staining Method for Cell Cultures with the Fluorescent Dye SITS

From Benjaminson and Katz (1970), modified by Katz (1976)

Application: Single cells or cells in monolayers.

Designed to show: Viable cells.

Fixation: None.

Tissue preparation: Use monolayers of cultured cells.

Preparation of staining solutions: SITS* stain solution: Add 25 mg of SITS (4-acetamido-4-isothiocyano-stilbene-2,2-disulfonic acid) to 10 ml of tissue culture medium to give final concentration of 2.5 mg per ml.

Staining schedule:
1. Incubate cultures in SITS medium for 48 hr.
2. Mount cells in the same SITS medium under a cover slip, seal with paraffin, and view in fluorescence microscope.

Results: Viable cells show violet fluorescence in small cytoplasmic vesicles around the nucleus. Dead or dying cells have fluorescence in nuclear membranes. When cells are fixed in methanol-formalin (9:1), the fluorescent dye is released from vesicles and becomes immediately bound to the nuclei.

NOTE: The dye is activated at about 350 nm and shows an emission peak at 410–420 nm. The activation and emission spectra are constant regardless of pH. The induced violet fluorescence should be distinguished from cellular gray-blue autofluorescence. The earliest reference to the use of SITS is that of Maddy (1964).

---

* *Research Organics Inc., 4353 E. 49th St., Cleveland, OH 44125.*

Rotman and Papermaster's Technic for Fluorochroming Viable Cells Using Fluorescein Diacetate (FDA)

From Rotman and Papermaster (1966)

Application: Isolated cells or organisms and monolayer cultures.

Designed to show: Viable cells.

Fixation: None.

Tissue preparation: Monolayers of cultured cells and cells or protozoa in suspension.

Preparation of staining solutions:

A. Phosphate-buffered saline (PBS) (see p. 22).

B. Fluorescein diacetate* (FDA) stock solution: 1 mg/ml in acetone; store in freezer.

C. Fluorescein diacetate working solution: dilute FDA stock solution with PBS to give a concentration of 6×10^{-7} M.

D. Trypsin-EDTA solution: Prepare 10^{-4} M EDTA and 0.05% trypsin in PBS.

Staining schedule:

1. Remove cells from monolayers by replacing the growth medium with Trypsin-EDTA and incubating for minimum time necessary to remove cells at room temperature (usually 5–15 min).
2. Centrifuge cells and resuspend in PBS to give a concentration of approximately 10^6 per ml.
3. Mix a small volume of cell suspension with working FDA solution to give a final concentration of about 0.1 μg of FDA per ml.
4. Mount on a slide, cover with coverslip; seal edges with silicone grease.
5. Examine with fluorescence microscope or under dark field using tungsten light and an interference filter transmitting between 440 and 480 mμ.

Results: Viable cells immediately begin to display a bright-green fluorescence within each cell, which the authors term fluorochromasia.† According to the concentration of FDA, the fluorescence increases linearly with time until saturation. Injured or dead cells fail to fluoresce. Cells which become mechanically damaged after accumulating fluorescence product immediately release the dye.

NOTE: In an elegant series of experiments, Rotman and Papermaster (1966) gave convincing evidence that cell membranes are permeable to FDA, a substrate for esterases, which is nonpolar and nonfluorescent. In

---

* *Sigma Chemical Co.; Aldrich Chemical Co.; Polysciences Inc.; Eastman Organic Chemicals.*
† *The term fluorochromasia has an additional and special meaning, namely bicolor fluorescence which is similar in mechanism to the metachromasia of thiazine dyes.*

viable cells, the substrate is converted to fluorescein, a polar, fluorescent compound that accumulates within the cell causing fluorescence. Damaged and inviable cells lose their fluorescence as fast as it accumulates because of membrane damage. Positive results were obtained as well in various protozoa. The original paper should be consulted for further details. The FDA method has been applied to the detection of live pollen (see below) and for determining the viability of other cultured plant cells (Widholm, 1972). The reaction kinetics of fluorogenic substrate turnover in living cells was analyzed using an automated flow-through cytofluorometer (Sengbusch et al., 1976.)

A Method for Evaluating Pollen Viability by Induced Fluorochromasia

From Rotman and Papermaster (1966), adapted by Heslop-Harrison and Heslop-Harrison (1970) (slightly modified)

Application: Assessment of pollen quality.

Designed to show: Viability of pollen cells based on integrity of cell membranes.

Preparation of solutions:
  A. Fluorescein diacetate* (FDA) stock solution: FDA is dissolved in acetone at a concentration of 0.4 mg/ml.
  B. Sucrose solution, 0.5 M: Dissolve 17.1 gm sucrose in 100 ml of distilled water.
  C. Working FDA solution, 10^{-6} M in 0.5 M sucrose; Add 0.1 ml of stock FDA solution to 100 ml of 0.5 M sucrose solution. Solution should not be milky or show any precipitation. For more rapid testing of pollen, add FDA stock solution drop by drop to sucrose solution until milkiness just appears; this will occur at a somewhat higher concentration than 10^{-6} FDA.

Staining schedule:
  1. Pollen grains are immersed in a large drop of working FDA solution on a slide for 10 min. Keep under covered petri dish.
  2. Mount with coverslip.
  3. Examine with fluorescence microscope.

Results: Viable pollen grains produce fluorescence, which becomes more intense up to about an hour. There may be a slight background fluorescence in the suspending solution with increasing time. Pollen grains which are nonviable because of prior loss of integrity of their cell membranes fail to fluoresce or lose their fluorescence if they become damaged after immersion in FDA solution. The sucrose concentration was selected to minimize osmotic bursting of pollen grains from 30 flowering plant species tested.

---

* *Polysciences, Inc.; Aldrich Chemical Co.; Sigma Chemical Co.; Eastman Organic Chemicals.*

NOTE: The fluorochromatic reaction (FCR) has been shown by Rotman and Papermaster (1966) to depend, first, upon the rapid entry of the nonpolar, nonfluorescent FDA molecule into the cell, where it is hydrolyzed by esterase to produce the polar product, fluorescein, which fluoresces and is retained by the cell membrane. With progressive entry of new FDA molecules, fluorescence increases. The viability test depends on the integrity of the plasmalemma, which is closely correlated with viability. Cells which lack intact membranes rapidly lose their FDA molecules as fast as they enter and such molecules are not available to be converted to the fluorescent derivative.

Heslop-Harrison and Heslop-Harrison carried out a number of excellent control tests to prove that pollen quality can be accurately assessed by the FCR. They point out that other methods, such as stainability by ordinary dyes, relative sizes of pollen grains, and enzyme reactions (capacity to reduce tetrazolium salts to colored formazans) are inaccurate. The fluorochromasia test is rapid and more accurate than other methods because it is correlated with the viability of the vegetative cell of the pollen grain. The authors recognize that because a male gametophyte gives a positive FCR reaction is no guarantee that it will form a pollen tube and result in fertilization. This same limitation is seen with so-called dye-exclusion staining methods to determine cell viability, such as Trypan Blue (C.I. 23850).

SECTION O: MISCELLANEOUS METHODS

Cytochemical Demonstration of Peroxidase by Fluorescence Microscopy

From Papadimitriou et al. (1978)

Application: Fresh blood smears, peritoneal exudates, and horseradish peroxidase-stimulated lymphocytes.

Designed to show: Peroxidase granules in polymorphonuclear leukocytes and horseradish peroxidase activity in lymphocytes of immunized animals.

Fixation: None.

Cell preparation: Unfixed blood smears are best. Less effective results are obtained after ethanol and glutaraldehyde fixation. Peritoneal exudates are evoked by intraperitoneal injection into guinea pigs of 15 ml of saline solution and cells removed 16 hr later. For details as to the isolation of immunized lymphocytes, the original reference should be consulted.

Staining solutions:
A. Acetate buffer, 0.2 M, pH 6.0: 190 ml of 0.2 N sodium acetate (16.4 gm/liter) plus 10 ml of 0.2 N acetic acid.
B. Incubation medium: 20 mM lead nitrate, 14 mM homovanillic acid, 0.1 mM rhodamine B (C.I. 45170) or rhodamine GG (C.I. 45160) and 0.02% freshly diluted hydrogen peroxide. Medium is prepared by adding the homovanillic acid and rhodamine to the acetate buffer and then adjusting the pH to 6.0 with 0.2 N NaOH. The lead nitrate is then

added slowly to the buffer, stirring until a clear solution results. The incubation medium is prepared fresh and remains stable for an hour.
C. 1% ammonium sulfide solution.

Staining schedule:
1. Place in fresh incubation medium for 10 min.
2. Wash twice for 5 min each in acetic buffer.
3. Wash 3 times in 50% ethanol for 10 min each to remove unbound dye.
4. Dehydrate in ethanol, clear in xylene, and mount in Fluormount.
5. Control slides are incubated in media in which hydrogen peroxide, homovanillic acid, or lead nitrate are omitted.
6. Examine in fluorescence microscope using green light for excitation.

Results: Distinct granules of red-fluorescing material are seen in the cytoplasm of granulocytes. Lymphocytes do not fluoresce. Erythrocytes may fluoresce if incubation is very short but the reaction product is lost. Fluorescence is seen several weeks later when slides are stored at room temperature. Control-treated cells rarely fluoresce. Ethanol and glutaraldehyde fixation induces some autofluorescence, particularly in nuclei, and causes erythrocytes to fluoresce strongly.

NOTE: This technic has been demonstrated to be reliable for peroxidase staining (equivalent to diaminobenzidine technic) and has the added advantage of enzyme localization by its fluorescence properties. It avoids the use of diaminobenzidine, a potential carcinogenic derivative. These workers also demonstrated that horseradish peroxidase-activated lymphocytes could be demonstrated from lymphoid suspensions. The technic is adapted from a biochemical fluorescence method for peroxidases and is based on the dimerization by peroxidase of homovanillic acid (nonfluorescent) into a fluorescent product in the presence of hydrogen peroxide. The product is diffusable and its metal salt does not fluoresce. However, the product with its two adjacent carboxylic acid groups complexes with rhodamine dyes to produce a stable fluorescing structure. The method described above for demonstrating HRP in blood cells has been adapted by Schwerdtfeger (1978) for the identification of injected horseradish peroxidase in the brain.

Fluorescence Cell-Surface Labeling Method with Fluorescamine

From Hawkes and Bartholomew (1977) (slightly modified)

Application: Tissue culture cells for analysis of cell surface by flow microfluorometry.

Designed to show: Fluorescence induced on cell surface proteins and phospholipids of intact cells in culture.

Fixation: 25% ethyl alcohol in Hanks' balanced salt solution (BSS) (see p. 22).

Preparation of solutions:
A. 25% ethyl alcohol in Hanks' BSS: Add 25 ml of absolute ethanol to 75 ml of Hanks' BSS.

B. Hanks' BSS (see p. 22).
C. Borate buffer: pH 9.0 (see p. 8).
D. Fluorescamine*: Dissolve 50 mg fluorescamine in 0.5 ml acetone; add this to 100 ml of borate buffer.

Staining schedule:

1. Wash monolayer cell cultures with warm BSS.
2. Wash cells with warm borate buffer.
3. Stain for 30 sec with fluorescamine dye solution.
4. Wash cells with warm borate buffer and then BSS.
5. Scrape cells from surface with a rubber policeman, centrifuge, and fix cells in Solution A at 4°C for 1 hr.
6. Centrifuge and wash cells in BSS.
7. Analyze cell population in flow microfluorometer.

Results: Fluorescamine-labeled cells yield a blue fluorescence at the cell surface with excitation by the 351.1 and 363.8 nm lines of an argon ion laser. Quantitative results can be obtained on the distribution of fluorescence intensity in the population.

NOTE: This exciting new technic will likely achieve extensive application because it opens up a new cytochemical approach to the study of cell surfaces. Fluorescamine is a novel and sensitive reagent developed for amino acid and protein analysis (Udenfriend et al., 1972). It reacts rapidly with primary amines to form a fluorescent product at an optimum pH of 9. Fluorescamine is shown to label the outer surface of fibroblasts in culture without disruption of the cell monolayer and without labeling internal cell components (Hawkes et al., 1976). Fluorescence results from binding to both surface proteins and phospholipids. Since the fluorescence emission of fluorescamine is in the blue range, it is possible to combine its use on the same cells with a second fluorescent probe, propidium iodide, which intercalates into DNA and emits red light. This dual fluorescence technic was used to show that transformed cells have significantly less surface fluorescence than their normal cell counterparts and that surface fluorescence varies during the cell cycle (Hawkes et al., 1976; Hawkes and Bartholomew, 1977).

A Fluorescence Histochemical Method Using o-Phthaldehyde (OPT) for the Demonstration of Histamine and Glucagon

From Juhlin and Shelley (1966) and Takaya (1970); modified by Hakanson and Owman (1966), Ehinger and Thunberg (1967), Shelley et al. (1968), and Brody et al. (1972); method described is summarized from Brody et al. (1972)

Application: Biological material in tissue sections, smears, washings, and mesentery.

---

* *Hoffman-LaRoche, Inc.; Sigma Chemical Co., Aldrich Chemical Co.*

Designed to show: Histamine and glucagon in cells.

Preparation of tissues: A variety of preparations can be examined, including freeze-dried cryostat or paraffin-embedded sections, smears of rat peritoneal saline washings, and stretch preparations of rat mesentery. Paraffin sections are not deparaffinized with xylene.

Preparation of reacting chambers: A photograph of the equipment is shown in the paper by Brody et al. and should be consulted. Briefly, the unit consists of separate chambers continuing OPT, water and a treatment or reacting chamber. These chambers consist of two 500-ml bottles and a round-bottom flask with three outlets. These outlets in the top of the reacting chamber connect to the bottles containing OPT and water plus a connection to a vacuum manometer and mechanical vacuum pump. The apparatus is kept at 80°C in an oven. A small escape valve is inserted between the water and reacting chambers to allow water vapor to leak out, so as to prevent the build-up of pressure resulting from evaporation during heating. A large piece of filter paper is placed within the water-containing bottle around the sides to increase the evaporation area. One gram of fresh OPT is added to the bottom of the other bottle each day the condensation reaction is to be carried out. The vacuum pump lowers the humidity in the reacting chamber and provides the negative pressure needed to introduce graded amounts of water and OPT.

Reaction procedure:
1. Glass slides with the biological preparations (see above) are placed in a rack and prewarmed in the oven to prevent condensation later on the tissue of either OPT or water. The paraffin melts, leaving the section covered by a thin film of paraffin.
2. The rack of slides is placed in the reacting chamber, which is then closed and evacuated for 2 min to reduce the humidity and allow for a pressure gradient for the introduction of OPT and water.
3. The stopcock leading to the vacuum pump is closed. OPT and water are introduced immediately by opening the connecting stopcocks. The recommended pressure conditions necessary to control the humidity and the proper amounts of reactants vary from 50 to 300 torrs, according to the particular tissue preparation and cell type. This information is given in a table by the authors along with the reaction time, which varies from 1 to 10 min.
4. At the end of the treatment time, the stopcock to the vacuum pump is opened and the chamber is evacuated for 1 min to remove water vapor and OPT gas.
5. The pump is turned off, the chamber is brought to atmospheric pressure, and the slides are removed.
6. The sections are allowed to cool and coverslip is mounted with xylene.
7. Examine in a fluorescence microscope.

Results: Mast cell granules emit an intense yellow fluorescence due to their histamine content. Blue fluorescent histamine-containing epithelial cells are seen in the basal part of rat stomach mucosa (oxyntic gland). In pancreatic A-cells at the periphery of the rat islet of Langerhans, there is a positive blue fluorescence due to their glucagon content. An OPT-induced fluorescence has been observed in rat pinealocytes and in the outer nuclear layer of the retina but this positive reaction is not due to the presence of histamine.

NOTE: The original method has undergone a series of modifications to help overcome diffusion artifacts. The method described here incorporates the prior changes and introduces other modifications of the OPT gas method to make the technic more reliable in terms of reproducibility, morphologic preservation, and stability of the fluorophore. The success achieved depends on the fact that the chemical reaction is done in gaseous form without liquid condensation on the sections.

Staining of Juxtaglomerular Granules with Basic Fluorescent Dyes

From Harada (1969)

Application: Sections of kidney cortex.

Designed to show: Granules of juxtaglomerular cells.

Fixation: Phosphate-buffered 10% formalin for 24 hr.

Embedding: Paraffin.

Preparation of staining solutions:
A. Delafield's Hematoxylin (see p. 105).
B. Coriphosphine O (C.I. 46020) solution: Prepare 0.1% aqueous solution and filter.

Staining schedule:
1. Deparaffinize in xylene and hydrate to water.
2. Stain in Hematoxylin for 30 sec.
3. Wash in tap water.
4. Stain in Coriphosphine O solution for 3 min.
5. Rinse in tap water.
6. Air-dry and mount in Apathy's syrup or synthetic resin.
7. Examine in fluorescence microscope.

Results: JG granules give intense orange-yellow fluorescence in contrast to surrounding tissue elements. Nuclei appear black. JG cells appear at the glomerular root and among arterioles for short distances from glomeruli. Erythrocytes fluoresce weakly. The cytoplasm of some tubular cells fluoresce.

NOTE: If the renal background fluoresces strongly, this may be reduced without affecting fluorescence from JG granules by differentiating for 30–

60 sec in 0.1% acetic acid after Coriphosphine O staining. Hematoxylin staining is interposed in order to quench background cellular fluorescence and permit better visualization of JG granules. Other basic fluorochromes like Acridine Orange (C.I. 46005) and Acridine Yellow (C.I. 46025) give equally good results as does Thioflavine T (Janigan, 1965) but not acid fluorochromes. It should be noted that the technique described above employs sections of paraffin-embedded tissues. In frozen sections, the JG granules can be visualized if the tissue is formol-fixed (Szokol and Gomba, 1971); the granules are destroyed in unfixed frozen and thawed tissue (Gomba et al., 1968).

Thioflavine T Method for Amyloid

From Vassar and Culling (1959), modified by Burns et al. (1967)

Designed to show: Primary and secondary amyloid deposits.

Fixation: Superior results are obtained using unfixed cryostat or frozen sections. For paraffin-embedded material, the choice of fixative is not critical. After prolonged formalin fixation, the staining becomes less intense.

Preparation of staining solutions:
A. Alum hematoxylin (any of the common hematoxylins, Ehrlich's, Harris, Delafield's)
B. Thioflavine T (C.I. 49005): 0.5% dye in 0.1 N hydrochloric acid (0.8 ml. concentrated acid in 100 ml of distilled water). Prepare fresh and filter.
C. 1% acetic acid.

Staining schedule:
1. Deparaffinize and bring sections to water.
2. Stain with alum hematoxylin Solution A for 2 min. Differentiation is unnecessary.
3. Rinse briefly in water.
4. Stain in Solution B for 3 min.
5. Rinse in water
6. Treat with 1% acetic acid for 20 min.
7. Wash well in water, dehydrate, clear and mount in Fluormount.
8. Examine in fluorescence microscope.

Results: Amyloid fluoresces silver blue or yellow, according to whether UV or blue light systems are used. More specificity is achieved with UV light. Paneth and oxyntic cells also give positive fluorescence as do muciphages of the rectal mucosa of man, myelin figures in neurons of amaurotic idiocy and granular cell myoblastoma, acidophil cells of the pituitary gland, zymogen granules in the pancreas of mice, and the juxtaglomerular apparatus in mice (Lehner, 1965). Although acid pH during staining increases the selectivity, one should keep in mind that false-positive fluorescence may be given by mast cells and hyaline substances (autofluorescence).

NOTE: Sections should be thin; with thick sections, autofluorescence becomes more of a problem especially in bone. Thioflavine T solutions do not keep indefinitely; a control section from tissue known to contain amyloid should be used. Alum hematoxylin is used to quench nuclear fluorescence. Thioflavine S, a fluorochrome related to thioflavine T, also is used to demonstrate amyloid. One of the fluorescent fabric brighteners, Phorwhite BBU, is recommended as a more specific fluorochrome for amyloid (Waldrop et al., 1972). It is said that amyloid fluoresces yellow-green in UV and other tissue components appear red (with added red tungsten light).

SECTION P: LIST OF NONFLUORESCENT MOUNTING MEDIA*

1. Apathy's gum syrup (see Lillie and Fullmer (1976), page 119, for recipe).
2. Aqueous buffer solutions.
3. DPX (after Kirkpatrick and Lendrum; distyrene 80, 10 gm; dibutyl phthalate, 5 ml; xylene, 35 ml).
4. Entellen.†
5. Eukitt.
6. Fluorescent-free immersion oil
7. Fluormount.‡
8. Glycerol.
9. Liquid paraffin.
10. Styron 666.§

---

* See de Ment (1950) for a listing of immersion liquids and mounting media for fluorescence microscopy. Other information about mounting media, such as refractive indices and interaction with stained sections, is given in Lillie and Fullmer (1976), between pages 111 and 122.
† E. Merck, Darmstadt; EM Laboratories Inc., Elmsford, New York.
‡ E. Gurr.
§ Dow Chemical Co.

SECTION Q: CHARACTERISTICS OF FLUOROCHROMES USED IN HISTOLOGY AND HISTOCHEMISTRY*

Name	C.I.No.	Dye Group	Mol. Wt.	Fluorescence Color
Acid fuchsin	42685	Arylmethane	585	Red
Acridine orange	46005	Acridine	382	Green, Red
Acridine yellow	46025	Acridine	274	Yellow
Acriflavine	46000	Acridine	296	Yellow
Auramine O	41000	Arylmethane	304	Yellow

* Slightly modified from data presented in Pearse ((1972) p. 1422) and Lillie and Fullmer ((1976) p. 136).

Fluorochromes—*continued*

Name	C.I.No.	Dye Group	Mol. Wt.	Fluores-cence Color
BAO†		Oxdiazole		Blue
Basic fuchsin	42510	Arylmethane	324	Red
Berberine sulphate	75160	Natural dye	509	Yellow
Benzoflavine (Flavophosphine N)	46065	Acridine	350	Yellow-Orange
Congo red	22120	Disazo	697	Red
Coriphosphine O	46020	Acridine	288	Green, Red
Eosin Y	45380	Fluorone	692	Green
Evans blue	23860	Disazo	961	Red
Fluorescein	45350	Fluorone	376	Green
Fluorescein isothio-cyanate		Fluorone	389	Green
Geranine B	14930	Monoazo	498	Red-Orange
Lissamine Rhoda-mine B200 (Sul-forhodamine B)	45100	Rosamine	581	Red
Magdala red	50375b	Quinone-imine	805	Orange
Mercurochrome	None	Fluorone	805	
Methyl green	42585	Arylmethane	458	Red
Morin	75660	Natural dye	302	Green
Neutral red	50040	Quinone-imine	289	Red
Pararosaniline	42500	Arylmethane	324	Red
Phenosafranin	50200	Quinone-imine	323	Red
Phloxine B	45405	Fluorone	761	
Phosphine (Phos-phine 3R)	46045	Acridine	330	Yellow
Phosphine GN	46045	Acridine	330	Orange-Brown
Primulin	49000	Thiazole	476	Blue
Pseudoisocyanin		Quinoline		Yellow
Pyronin Y	45005	Aminoxanthene	303	Red
Quinacrine	None	Acridine	509	Yellow
Quinacrine mustard	None	Acridine	542	Yellow
Rheonin A	46075	Acridine	393	Red
Rhodamine B	45170	Rhodamine	479	Red

† *2,5-Bis(4'-aminophenyl-(1')) or 1,3,4-oxdiazole.*

Fluorochromes—*continued*

Name	C.I.No.	Dye Group	Mol. Wt.	Fluorescence Color
Rhodamine S	45050	Xanthene	375	Yellow
Rhodamine 3G	45210	Rhodamine	437	Orange
Rhodindine	50375a	Quinone-imine	458	Orange
Rivanol	None	Acridine	361	Yellow-Green
Safranin	50240	Quinone-imine	351	Red
Tetracycline	None	Natural dye	444	Yellow
Thiazine red R	14780	Monoazo	600	Violet
Thiazole yellow G	19540	Thiazole	646	Blue
Thioflavine T	49005	Thiazole	319	Blue-White
Thioflavine S	49010	Thiazole		Blue-Yellow

SECTION R: FLUORESCENCE EXCITATION AND EMISSION MAXIMA*

Fluorophore	Excitation Maximum (nm)	Emission Maximum (nm)
FITC conjugates	495	520
DANS conjugates	340	525
Lissamine Rhodamine B (RB200 conjugates)	575	595,710
Acridine orange	430–500	530,650
Coriphosphine O		
Green emission	500	535
Red emission	450	650
Thioflavine T		450
Feulgen (pararosaniline)	570	

* *Derived from Pearse ((1972) p. 1472).*

ANIMAL HISTOTECHNIC

CHAPTER 4

METHODS FOR GENERAL TISSUE
George Clark

INTRODUCTION

In classifying the procedures employed in animal histology, an attempt has been made to separate those intended as general tissue stains from those designated to demonstrate connective tissues or nervous tissue or special purpose stains. In the case of nervous tissue, the separation is quite easy to make, as neurological procedures are quite different from those for general tissue. In the case of connective tissues, however, the grouping followed here must necessarily be more arbitrary; some procedures which bring out connective tissue to good advantage are also fine general tissue stains. In spite of this fact, a rough separation of general purpose and connective tissue procedures proves possible; and that which follows below seems as practical as any.

SECTION A: NUCLEAR STAINING WITH ALUM HEMATOXYLIN

Alum hematoxylin is the most widely used nuclear stain. There is a legion of reported procedures. The first four of the following procedures have been popular for many years; the fifth is suggested for those who only occasionally need an alum hematoxylin nuclear stain. All of these technics can be used following most fixatives and on paraffin, celloidin and frozen sections.

Delafield's Hematoxylin

Emended by Mallory (1938)
Originally described by Prudden (1885)

Preparation of Delafield's hematoxylin: Dissolve 4 gm hematoxylin in 25 ml 95% ethyl alcohol. Add 400 ml saturated aqueous ammonium alum $[(NH_4)_2SO_4 \cdot Al_2(SO_4)_3 \cdot 24H_2O]$, and let stand a week exposed to air and light. Filter, and then add 100 ml glycerin and 100 ml methyl alcohol. Let stand until solution becomes dark (6–8 weeks). (This stock solution may be kept for a considerable period in a tightly stoppered bottle.) Just before use, dilute with an equal volume of distilled water. Stain for 15 min.

Ehrlich's Hematoxylin

From Mallory (1938)
Originally described by Ehrlich (1886)

Preparation of staining solution: Dissolve 2 gm hematoxylin in 100 ml of 95% alcohol. Add 100 ml distilled water, 100 ml glycerin, 3 gm ammonium or potassium alum, and 10 ml glacial acetic acid. It may be ripened at once by addition of 0.1 gm of sodium iodate. Stain for 2–5 min.

Mayer's Acid Hemalum Modified by Lillie

From R. D. Lillie (1965)
Originally described by Mayer (1891)

Preparation of staining solution: Dissolve 5 gm hematoxylin by holding overnight in 700 ml distilled water; add 50 gm ammonium alum $[(NH_4)_2SO_4 \cdot Al_2(SO_4)_3 \cdot 24H_2O]$ and 0.25 gm $NaIO_3$; after these have gone into solution, add 300 ml glycerin C.P. and 20 ml glacial acetic acid. May be used immediately. Stain for 5 min.

Harris' Hematoxylin

From Mallory (1938)
Originally described by Harris (1900)

Preparation of staining solution: Dissolve 1 gm hematoxylin in 10 ml absolute alcohol. Dissolve 20 gm ammonium alum $[(NH_4)_2SO_4 \cdot Al_2(SO_4)_3 \cdot 24H_2O]$ in 200 ml water with heat. Mix 2 solutions, bring to a boil and add 0.5 gm mercuric oxide (HgO). As soon as the solution assumes a dark purple color, remove from flame and quickly cool by plunging flask into cold water. Before use, add 4 ml glacial acetic acid to each 100 ml of staining solution. Stain for 5–15 min.

Alum Hematoxylin for Occasional Use

From Clark (1979)

Stock solutions: 10% hematoxylin in 95% or absolute ethanol, 4% alum (NA, K or NH_4 alums are all satisfactory), 0.1% aqueous $NaIO_3$. All of these are stable.

Preparation of staining solutions: To 40 ml of the alum solution add 1 ml of the hematoxylin and 4.5 ml of the $NaIO_3$ solution and place in paraffin oven overnight. In the AM add distilled water to make 100 ml. The solution may be immediately used, and if kept tightly stoppered in a full bottle, will keep for several months. Stain for 5 min.

Staining schedule for all alum hematoxylins:

1. Dewax and hydrate as usual.
2. Treat with Lugol's and then with 5% $Na_2S_2O_3 \cdot 5H_2O$ each for 1 min if a fixative containing Hg was used.
3. Wash thoroughly.
4. Stain for time given in particular procedure.
5. Wash.
6. Optional differentiation
   a. Immerse in 1% HCl in 70% ethanol until section turns red.
   b. Wash briefly and place in saturated aqueous Li_2CO_3 diluted 1:1 with H_2O until section is blue. If the tap water is sufficiently basic, this will be adequate.
7. Wash thoroughly in H_2O and counterstain as desired.

SECTION B: COUNTERSTAINS FOR HEMATOXYLIN

Eosin Y

Eosin Y is the most commonly used counterstain for hematoxylin. However, the range in individual preferences for the intensity of the eosin stain is so great that only approximate staining and dehydration times can be suggested. Attempted maintenance of a desired staining intensity has led to numerous technics which are more or less successful. The following procedure must be considered tentative and must be adjusted to suit individual preference.

Solution needed: Dissolve 0.5 gm eosin Y in 100 ml distilled water.

Staining schedule:

1. Well washed sections or slides stained with hematoxylin are stained for 1 min in the eosin solution.
2. Rinse in distilled water.
3. 70% and 95% ethyl alcohol; 1 min in each.
4. n-Butyl alcohol (or absolute ethyl alcohol); 2 changes of 2 min each.
5. Clear in xylene and mount.

Results: Nuclei—blue, erythrocytes—light orange, collagen—salmon pink.

Three other counterstains for alum hematoxylin follow: Eosin B, substituted for Eosin Y in above procedure, gives a slightly pinker cytoplasmic stain lacking the yellowish tinge of Eosin Y. Congo red is also a good substitute. It, too, gives a pink cytoplasmic stain but does not stain collagen as heavily as do the eosins. Sirius red F3BA was introduced by Puchtler and Sweat (1964) as a substitute for acid fuchsin in the Van Gieson technic. When dissolved in water (0.05 gm/100 ml), this dye gives a more complete connective tissue stain than any of the other three recommended counterstains. It not only stains collagen but also reticulum and basement membranes.

Iron Hematoxylins

Originals: Heidenhain (1892), Mayer (1899), Weigert (1904),
Verhoeff (1908), Weil (1928)
Emended by Spooner, Read and Clark (1979)

Iron hematoxylins are used primarily for nuclei, elastic tissue and myelin sheaths. A myriad of methods have been reported. Since all of these are special (occasionally used) stains, and since the actual staining solutions are not stable, a method was devised using stable stock solutions and only volumetric measurements.

Stock Solutions:
1. Hematoxylin—10% in alcohol (either 95% or absolute).
2. $FeCl_3 \cdot 6H_2O$—10% in 3% aqueous HCl.
3. Weigert's (1885) Differentiator—modified.

$K_3Fe(CN)_6^-$	12.5 gm
$Na_2B_4O_7 \cdot 10H_2O$	2.5 gm
H_2O to make	1000 ml

Staining Solutions: Nuclear stain (progressive no differentiation needed)

Hematoxylin solution	1 ml
$FeCl_3$ solution	3 ml
H_2O to make	50 ml

Procedure:
1. Can be used immediately. Stain 10 min.
2. Counterstain if desired.
3. Dehydrate, clear and mount.

Myelin sheath and elastic tissue stains (differentiation required).

Hematoxylin solution	4 ml
$FeCl_3$ solution	4 ml
H_2O to make	50 ml

Procedure:
1. Should be used immediately. Stain 10 min.
2. Wash thoroughly; final washing two changes of distilled water.
3. Differentiate in modified Weigert's differentiator.
4. Wash in distilled water, then in tap water.
5. Counterstain, if desired. Dehydrate, clear, and mount.

Results: With the nuclear stain all nuclei should be black but not overstained. Elastic tissue and myelin not stained.

The same staining solution and the same differentiator are used for both myelin sheath and elastic tissue staining. With nervous tissue gray matter becomes golden brown while the myelin remains black. Similarly, when staining elastic fibers these remain black while collagen, muscle and nuclei become golden brown.

Gallocyanine Chrome Alum

Especially with the Van Gieson type stains and those with a phospho-molybdic or phosphotungstic acid step, the various hematoxylin nuclear stains have not been entirely satisfactory. Several hematoxylin substitutes, such as gallocyanine and chromoxane cyanine R, have been suggested as nuclear stains in these technics. These dyes, like hematoxylin, must be chelated with a metal to form the staining solution. In my hands gallocyanine chrome alum is the dye of choice. The blue black staining with this dye is almost impossible to remove and, since it is not an indicator, the color does not change. The Fe^{3+} chelate of this dye was introduced by Proescher and Arkush (1928) and the Cr^{3+} chelate by Einarson (1932). The procedure for preparation of the staining solution suggested by Berube, Powers, Kerkay and Clark (1966) is very satisfactory.

Preparation of staining solution: Dissolve 150 mg gallocyanine and 15 gm chrome alum in 100 ml distilled water and gently boil for 10–20 min. Cool, filter and wash precipitate with sufficient distilled water to restore volume to 100 ml. This staining solution will remain usable for 1–2 weeks, but after this the stain loses intensity and specificity. Berube, Powers, Kerkay and Clark (1966) also give directions for the separation of the active staining compound in the dry state from which the staining solution can be reconstituted.

Staining schedule:

1. Dewaxed and hydrated sections are stained for 16 hr in gallocyanine chrome alum.
2. After thorough washing in water, the slides can be stained by any of the following methods (except those with silver impregnation) for connective tissue. The gallocyanine stained sections are substituted for the hematoxylin stained slides usually specified in these procedures.

Results: Nuclei, Nissl granules and other sites containing cytoplasmic RNA—blue black

Chromoxane Cyanine R

Original Pearse (1957)
Emended by Clark (1979)

This dye, also called Cyanine R, chrome cyanine R, solochrome cyanine R and Eriochrome cyanine R—C.I. 43820, was introduced by Pearse (1957). Used with $FeCl_3$ it gives an excellent nuclear or myelin sheath stain.

Application: Nervous tissue both central and peripheral, general tissue.

Designed to show: Nuclei or myelin sheaths.

Fixation: Formalin.

Embedded in: Paraffin.

Stock Solutions: 0.2% chromoxane cyanine R in 100 ml 0.5% aqueous H_2SO_4, boil 5 min no need to filter. 10% aqueous $FeCl_3 \cdot 6\ H_2O$ in 3% aqueous HCL, 1% HCl in 70% ethanol, 1% aqueous NH_4OH.

Preparation of Staining Solutions: To 40 ml of stock dye solution add 2 ml of $FeCl_3$ solution and dilute to 50 ml.

Staining schedule:
1. Dewaxed and hydrated 6-μm sections are stained for 10 min.
2. Thoroughly wash in tap water.
3. If a nuclear stain is desired differentiate for 1 min in 1% HCl in 70% alcohol, if a myelin sheath stain is desired differentiate for 1 min in 1% aqueous NH_4OH.
4. Wash well in tap water.
5. Counterstain if desired—the counterstains suggested for alum hematoxylins (see p. 107) are all excellent.
6. Dehydrate as usual, clear and mount.

Results: This is an excellent nuclear stain and when counterstained with eosin closely resembles H&E. As a myelin sheath stain the results closely resemble luxol fast blue (see p. 147) but the differentiation is so simple and the total staining time so short that it should supercede the use of luxol fast blue.

Mallory's Phosphotungstic Acid Hematoxylin

Original from Mallory (1938)
Emended by Clark and Powers (1976)

Designed to show: Tissue elements in general, collagen, myofibrils and cross striations in muscle tumors, astrocytes, and neuroglial fibers.

Fixation: Zenker (Mallory) or formalin 1:10 (Peers, 1941; Lillie, 1965)

Embedded in: Paraffin.

Stock solutions: 10% hematoxylin in 95% or absolute ethanol, 0.1% $NaIO_3$, 5% aqueous oxalic acid, 4% iron alum [$Fe_2(SO_4)_3 \cdot (NH_4)_2SO_4 \cdot 24H_2O$].

Preparation of staining solution: To 60 ml distilled water add 1 ml of hematoxylin solution and 4.5 ml of $NaIO_3$ solution. Place in paraffin oven overnight. In A.M. add 1 gm phosphotungstic acid and bring volume to 100 ml. If kept in refrigerator this will keep for several months.

Staining schedule:
1. Dewax as usual.
2. Absolute ethanol, 1 min.
3. Ether-alcohol, 1 min.
4. 1% celloidin, 1 min.
5. Air-dry 1 min.
6. 70% ethanol, 1 min.
7. Rinse in tap and then distilled water.
8. Oxidize in 300 mg $KMnO_4$, 0.5 ml concentrated H_2SO_4 in 100 ml distilled water. This solution is unstable and must be prepared just before use; 15 min.
9. Wash in tap and then in distilled water.
10. 5% oxalic acid until colorless.
11. Wash in tap and then in distilled water.
12. 4% iron alum solution, 2 hr.
13. Rinse briefly in distilled water.
14. Stain in the phosphotungstic acid hematoxylin for 2 hr.
15. Dehydrate in *n*-butyl or *tert*-butyl alcohol.
16. Clear and mount.

Results: Neuroglial fibers—blue, astrocytes—pink to red, myofibrils and cross striations—blue, nuclei—blue, red blood cells—blue, fibrin—blue, collagen—yellowish to brownish red, reticulum—yellowish to brownish red.

NOTE Oxidation with $KMnO_4$ is needed only when blockage of myelin sheath staining is required as in staining for glial fibers. If the oxidative stage is omitted, then it is not necessary to celloidinize the sections. Then 95% ethanol may be substituted for Steps 3, 4 and 5.

Orth's Lithium Carmine

From Mallory (1938, p. 80)

Application: Animal tissue.

Designed to show: Nuclei.

Fixation: Any good fixative may be used.

Embedded in: Paraffin.

Preparation of staining solution: Dissolve 2.5–5 gm carmine in 100 ml saturated aqueous lithium carbonate (approx. 1.25%) and boil 10–15 min. When cool, add 1 gm thymol, and filter.

Staining schedule:
1. Remove paraffin and hydrate sections in usual manner.
2. Stain in above carmine solution for 2–5 min.
3. Transfer directly to acid alcohol (1% conc. HCl in 70% alcohol), 1 or

more changes, for several minutes to fix the dye in nuclei and to differentiate the sections.

4. Wash in tap water.
5. Dehydrate in 95% and 100% alcohol.
6. Clear in xylene.
7. Mount in balsam.

Results: Nuclei—red, cytoplasm—light pink or unstained.

NOTE: A useful application of this staining fluid is in conjunction with orcein in Fraenkel's method for elastic tissue (see p. 124).

CHAPTER 5

METHODS FOR CONNECTIVE TISSUE

George Clark

INTRODUCTION

As stated at the beginning of the preceding chapter, the separation of general tissue stains from connective tissue stains is arbitrary. Connective tissue stains can be divided into five groups. The first of these is the Van Gieson and its variants; second stains which include a phosphotungstic of phosphomolybdic acid step; third, those in which the counterstain is dissolved in Ethyl Cellosolve; fourth, stains devised primarily for elastic tissue; and fifth, those in which silver impregnation is used.

Van Gieson's Stain with Iron Hematoxylin

From Mallory (1938)
Originally described by Van Gieson (1889)

Application: Connective tissue; nervous tissue.

Designed to show: Collagen, muscle, cornified epithelium.

Fixation: Any may be used.

Embedded in: Paraffin or celloidin.

Preparation of solutions:
   A. Any of the alum hematoxylins (p. 105) may be used; or better, iron hematoxylin (p. 108) or, preferably gallocyanine chrome alum (p. 109).
   B. Van Gieson's picro-acid-fuchsin: 1% aqueous acid fuchsin, 5–15 ml; saturated aqueous picric acid (about 1.22%), 100 ml.

   NOTE: Lillie (1965, p. 539) prefers 0.1 gm acid fuchsin to 100 ml picric acid solution, as recommended by Weigert (1904).

Staining schedule:
   1. Dewax, hydrate and apply nuclear stain desired.
   2. Wash in water.
   3. Stain 3–5 min in Van Gieson's solution (Solution B)

4. Wash quickly in water.
5. Dehydrate through ascending alcohols. Differentiation occurs during this step. If the yellow stain is too weak, add a few crystals of picric acid to each of the alcohols and to the clearing solution.
6. Clear and mount. If celloidin sections are used, substitute n-butyl alcohol for the absolute alcohol step and complete dehydration in terpineol-xylene (one part terpineol and four parts xylene).

Results: Collagen—red; smooth and striated muscle—yellowish to brownish. If the muscle fibers are red in color, decrease the amount of acid fuchsin; cornified epithelium—yellow; hyalin—yellow; nuclei—brown, blue or black depending on nuclear stain used.

Fast Green Modification of the Van Gieson Stain

Lillie (1965)

Designed to show: Connective tissue; muscle; cytoplasm; erythrocytes; nuclei.

Fixation: Any fixative.

Embedded in: Paraffin.

Preparation of staining solutions:
  A. Prepare 0.1% fast green FCF in 1% acetic acid, or 3% wool green S for a bluer green.
  B. Dissolve 0.2% acid fuchsin, 0.2% violamine R, or 0.1–0.5% ponceau S in saturated aqueous picric acid.

Staining schedule:
  1. Remove paraffin from sections in the usual manner.
  2. Stain nuclei with an alum hematoxylin (p. 105) or, better, iron hematoxylin (p. 108) or gallocyanine chrome alum (p. 109).
  3. Wash in tap water.
  4. Stain 4 min in Solution A.
  5. Wash in 1% acetic acid.
  6. Stain 10–15 min in Solution B.
  7. Wash 2 min in 1% acetic acid.
  8. Dehydrate, clear and mount.

Results: Connective tissue—red; muscle—gray-green; cytoplasm—gray-green; erythrocytes—green; nuclei—brown or black, depending on stain used; Paneth cell granules—may appear red.

Biebrich Scarlet with Picric Aniline Blue

From Lillie (1965)

Designed to show: Collagen, reticulum, muscle, plasma.

Fixation: Formalin, Orth, etc.

Embedded in: Paraffin.

Preparation of staining solutions:
A. Iron hematoxylin, p. (108).
B. Dissolve 0.2 gm Biebrich scarlet in 100 ml 1% aqueous acetic acid.
C. Dissolve 0.1 gm aniline blue w.s. in 100 ml saturated aqueous picric acid.

Staining schedule:
1. Remove paraffin and hydrate as usual.
2. Stain paraffin sections 5–6 min in hematoxylin Solution A.
2a. If gallocyanine chrome alum is used, stain sections for 16 hr.
3. Wash in tap water.
4. Stain 4 min in Biebrich scarlet solution B.
5. Rinse in tap water.
6. Stain 4–5 min with picro-aniline blue solution C.
7. Transfer directly to 1% acetic acid for 3 min.
8. Dehydrate, clear, and mount in synthetic resin.

Results: Connective tissue—blue, renal glomerular stroma—blue, basement membranes—blue, reticulum—blue, erythrocytes—orange to scarlet, muscle—pink, cytoplasm—pink to gray, nuclei—gray to black, mucus—light blue.

Van Gieson Trypan Blue

Clark (1971)

Application: Connective tissue.

Designed to show: Collagen, including very fine fibrils; muscle; cornified epithelium.

Fixation: Formalin.

Embedded in: Paraffin.

Preparation of solutions:
A. 0.3 gm naphthol yellow S in 100 ml distilled water.
B. 0.2 gm acid fuchsin in 100 ml distilled water.
C. 0.2 gm trypan blue in 100 ml distilled water.

Staining solution: 40 parts Solution A, 2.5 parts Solution B and 2 parts Solution C. For each 40 parts of Solution A add 0.4 ml concentrated HCl.

Staining schedule:
1. Dewax and hydrate paraffin sections.
2. Stain in gallocyanine chrome alum solution (see p. 109) for 16 hr. Iron hematoxylin may be used (p. 108).

3. Wash thoroughly in water.
4. Stain for 10 min in staining solution.
5. Quickly rinse in water and dehydrate through 50%, 70% and 95% ethyl alcohol followed by 2 changes of *n*-butyl alcohol of 2 min each.
6. Clear and mount.

Results: Nuclei—blue black, cytoplasm—yellow, collagen—large fibers red and small fibers blue.

Lillie's Allochrome Connective Tissue Method

Lillie (1951)

Application: Animal tissue.

Designed to show: Collagen, reticulum, basement membranes, sarcolemma, amyloid.

Fixation: Formalin.

Embedded in: Paraffin.

Solutions needed:
A. Schiff reagent (see p. 200).
B. 0.5% $Na_2S_2O_5$ aqueous (make up fresh each time).
C. Iron hematoxylin (see p. 108).
D. Picro methyl blue: 40 mg methyl blue (C.I. 42780) in 100 ml saturated aqueous picric acid.
E. Sweat, Puchtler and Woo (1964) prefer Sirius supra blue GL (C.I. 23160) 70 mg in 100 ml saturated picric acid (aqueous).

Staining schedule:
1. Dewax and hydrate paraffin sections.
2. Oxidize in 1% periodic acid 10 min.
3. Rinse well in running water (5 min).
4. Schiff reagent 10 min.
5. 0.5% sodium metabisulfite, 3 changes, 2 min each.
6. Rinse in running water, 5 min.
7. Stain in iron hematoxylin for 10 min.
8. Rinse well in running water.
9. Stain in picro methyl blue solution, 6 min.
9a. Or stain in picro-Sirius blue solution, 5 min.
10. Rinse quickly in 95% ethanol.
11. Dehydrate in *n*-butyl alcohol, two changes.
12. Clear in xylene and mount.

Results: Collagen and reticulum—blue, basement membranes and sarcolemma—red, amyloid—red to lavender, muscle and cytoplasm—greenish yellow.

SECTION B: CONNECTIVE TISSUE STAINS WITH A POLYACID STEP

Mallory's Aniline Blue Collagen Stain

Mallory (1938)
Originally described by Mallory (1900)

Designed to show: Collagenous and reticulin fibrils, cartilage, bone, amyloid, nuclei. Fibroglia and elastin fibers may remain unstained.

Fixation: Zenker.

Embedded in: Paraffin or celloidin.

Preparation of staining solutions:
A. Acid fuchsin, 0.5 gm; water q.s., 100 ml.
B. Aniline blue, w.s., 0.5 gm; orange G, 2 gm; phosphotungstic acid, 1 gm; water q.s., 100 ml.

Staining schedule:
1. Stain sections, which have been treated with iodine as usual after Zenker's fluid, in Solution A for 1–5 min or longer. If it is desirable to bring out the collagenous fibrils as sharply as possible, omit the staining with acid fuchsin.
2. Transfer directly to staining solution B without washing in water and stain for 20 min to 1 hr or longer.
3. Transfer directly to 95% alcohol, several changes to remove the excess stain.
4. Dehydrate in absolute alcohol, clear and mount.

NOTE: For celloidin sections shorten the staining time, decolorize and dehydrate in 95% alcohol and clear by the blotting-paper-xylene method or in terpineol. Mount in balsam.

Results: Collagenous fibrils—intense blue; ground substances of cartilage, bone, mucus, amyloid—varying shades of blue; nuclei—red; fibroglia—red; myoglia—red; neuroglia fibrils—red; axis cylinders—red; fibrin—red; nucleoli—red; blood corpuscles—yellow; myelin—yellow; elastic fibrils—pale pink or pale yellow or unstained.

Heidenhain's "Azan" Modification

Lillie and Fullmer (1976)
Original from Heidenhain (1915)

Designed to show: Muscle, glia fibrils, collagen, reticulum, glomerular stroma, erythrocytes, nuclei.

Fixation: Zenker, Helly, Bouin or Carnoy.

Embedded in: Paraffin.

Preparation of staining solutions:
A. Dissolve 0.25–1.0 gm of azocarmine B in 100 ml of cold water and add 1 ml glacial acetic acid; or, saturate, by boiling, 1.0 gm azocarmine G in 100 ml distilled water, cool and acidify with 1 ml glacial acetic acid.
B. Stock solution: Aniline blue, w.s., 0.5 gm; orange G, 2 gm; glacial acetic acid, 8 ml; distilled water, 100 ml.

Staining schedule:
1. Stain 30–60 min in Solution A in a covered dish at 50–55°C and then 1–2 hr at 37°C.
2. Wash in distilled water.
3. Differentiate in 0.1% aniline in 95% alcohol.
4. Rinse in 1% acetic acid in 95% alcohol.
5. Mordant 30 min to 3 hr in 5% aqueous phosphotungstic acid.
6. Rinse in distilled water.
7. Stain 1–3 hr in a 1:1 or 1:2 (Mallory recommends 1:3) dilution of Solution B.
8. Rinse in distilled water.
9. Differentiate and dehydrate in 95% alcohol followed by absolute.
10. Clear in xylene and mount in balsam.

Results: Nuclei—red, erythrocytes—red, muscle—orange, glia fibrils—reddish, mucin—blue, reticulum—dark blue, glomerular stroma—dark blue, collagen—dark blue.

Masson's Trichrome Stain

Adapted from Mallory (1938)
Originally described by Masson (1929)

Application: Pituitary gland; epithelium; thyroid gland; nerve (normal and tumors).

Designed to show: Nuclei; argentaffin granules; neuroglia fibrils; collagen; keratin; intercellular fibrils; negative image of Golgi apparatus.

Fixation: Bouin for 3 days, Möller for 24 hr, 10% formalin in 95% alcohol, Zenker, or 10% formalin.

NOTE: Material fixed in formalin should be mordanted in Bouin's for 15 min to 1 hr. This step should be inserted after Step 1.

Embedded in: Paraffin.

Attachment to slides: Masson recommends his own gelatin fixative; but Mayer's albumen fixative (p. 24) seems to be equally satisfactory.

Preparation of solutions:

A. Regaud's hematoxylin:

Hematoxylin	1 gm
95% alcohol	10 ml
Glycerine	10 ml
Water	80 ml

Dissolve hematoxylin in alcohol then add glycerine and water.

B. Picric alcohol:

Saturated solution (about 7%) picric acid in 95% alcohol, 2 vol; 95% alcohol, 1 vol.

C. Ponceau-acid-fuchsin:

Acid fuchsin	0,3 gm
"Ponceau de xylidine"	0.7 gm
Distilled water	100 ml
Glacial acetic acid	1 ml

NOTE: The identity of the "ponceau de xylidine" recommended by Masson is not known. Ponceau 2R gives an acceptable stain and Lillie (1965) lists several other substitutes. Either gallocyanine chrome alum (p. 109) or iron chloride hematoxylin (p. 108) may be substituted for the Regaud's hematoxylin. After either of these, begin with Step. 7.

D. Acetic anilin blue: saturated solution of anilin blue w.s. in 2% aqueous acetic acid.

Staining schedule:

1. Free sections from paraffin by xylene, alcohol and water.
2. Mordant 5 min in 5% $Fe_2(SO_4)_2 \cdot (NH_4)_2SO_4 \cdot 24H_2O$, previously heated to 45°C.
3. Wash in tap water.
4. Stain 5 min in Regaud's hematoxylin (Solution A).
5. Rinse in 95% alcohol.
6. Differentiate in picric alcohol (Solution B).
7. Wash in running top water.
8. Stain 5 min in ponceau-acid-fuchsin (Solution C).
9. Rinse in distilled water.
10. Differentiate 5 min in 1% aqueous phosphomolybdic acid.
11. Without rinsing pour on acetic anilin blue (Solution D), and allow to stand 5 min.
12. Rinse in distilled water.
13. Return to 1% phosphomolybdic acid for 5 min.
14. Place for 5 min in 1% aqueous acetic acid.
15. Dehydrate in 95% alcohol, followed by absolute.
16. Clear in xylene and mount.

Results: Nuclei—black, argentaffin granules—black or red, cytoplasm—vermillion red, neuroglial fibers—vermillion red, collagen—intense blue, mucus—blue, keratin—vermillion red, Golgi apparatus—clear, intercellular fibers—vermillion red.

Lillie Modification of Masson's Trichrome

From Lillie (1940b)

Application: Mammalian tissues.

Designed to show: Cells; cytoplasm; muscle; collagen.

Fixation: Formalin or Orth, but Zenker and Bouin give especially brilliant results; with the last two methods Steps 1 and 2 are omitted and the usual iodine pretreatment is given to Zenker material.

Embedded in: Paraffin.

Preparation of mordant: Phosphomolybdic acid, 5 gm; phosphotungstic acid, 5 gm; distilled water, 200 ml.

Staining schedule:
1. Treat 2 min with saturated alcoholic picric acid.
2. Wash 3 min in running tap water.
3. Stain 5 min in iron chloride hematoxylin (p. 108) or 16 hr in gallocyanine chrome alum (p. 109).
4. Wash in running water.
5. Stain 4 min in 1% Biebrich scarlet in 1% acetic acid.
6. Rinse in tap water.
7. Mordant 1 min in the above phosphomolybdic-phosphotungstic mordant.
8. Stain 4 min in 2.5% fast green FCF in 2.5% acetic acid.

   NOTE: For fast green one may substitute an equal quantity of aniline blue, w.s. or of wool green S.

9. Differentiate 1 min in 1% acetic acid.
10. Carry through alcohol.
11. Dehydrate with acetone or absolute alcohol.
12. Clear and mount in basam saturated with salicylic acid.

Results: Nuclei—black, cytoplasm—brown to pink, muscle—red, erythrocytes—brilliant scarlet, collagen—green (blue or blue-green), myelinated fibers—red, fibrin—red to pink.

SECTION C: DISPLACEMENT CONNECTIVE TISSUE STAINS

A Displacement Connective Tissue Stain

Clark and Barnes (1979)

Application: Various tissues.

Designed to show: Collagen; muscle; cornified epithelium; elastic tissue.

Fixation: Formalin.

Embedded in: Paraffin.

Stock solutions:
- A. Any of the hematoxylins listed (p. 105).
- B. Rose Bengal 1% aqueous.
- C. Bismark brown Y—100 mg dye in 100 ml Ethyl Cellosolve.
- D. Fast green FCF—100 mg dye in 100 ml Ethyl Cellosolve.

Staining schedule:
1. Paraffin sections must be no thicker than 6 μm. Dewax and hydrate two slides.
2. Stain with desired hematoxylin.
3. Stain 10 min or longer in 1% aqueos rose Bengal. Overstaining does not occur.
4. Take one stained slide, rinse in isopropyl or *n*-butyl alcohol, or in Ethyl Cellosolve and stain for 10 min in (displacement solution), 1 part of Fast green FCF (Solution D) and 3 parts Bismark brown Y (Solution C).
5. Rinse briefly in isopropyl or *n*-butyl alcohol or in Ethyl Cellosolve.
6. Complete dehydration in 3 changes of tertiary butyl alcohol, 1 min in each.
7. Clear and mount.
8. Then examine under the microscope and note whether collagen is blue to green and whether muscle is red. If both collagen and muscle are red immerse second slide in displacement staining solution for 12 min. If both muscle and collagen are blue to green, take the second slide and stain in displacement staining solution for 8 min. In either case after removal from displacement staining solution follow Steps 5 to 7.

Results: Collagen and reticular fibers—blue to green, smooth and skeletal—muscle red, elastic tissue—red.

NOTE: It is essential that Ethyl Cellosolve be used. Neither Methyl nor Butyl Cellosolve are satisfactory.

SECTION D: ELASTIC TISSUE STAINS

Weigert's Resorcin Fuchsin

Adapted from Mallory (1938), and Lillie (1965)
Original by Weigert (1898)

Designed to show: Elastic fibers (in wall of blood vessels); nuclei; collagen.

Fixation: Any fixation may be used.

Embedded in: Paraffin or celloidin.

Preparation of staining solution: Basic fuchsin, 2 gm (crystal violet may be used resulting in a deep green stain); resorcinol, 4 gm; distilled water, 200 ml. Bring the solution to a boil and, when briskly boiling, add 25 ml of a 29% aqueous solution of $FeCl_3$. Stir and boil for 2–5 min more. A precipitate forms. Cool and filter. Discard the filtrate. Leave the precipitate on the filter paper until it is thoroughly dry. Then return filter paper and precipitate to the vessel used for the precipitation; this should be dry but still contain whatever part of the precipitate remains adherent to it. Add 200 ml of 95% alcohol and heat carefully on electric hot-plate. Stir constantly and remove the filter paper when the precipitate is dissolved. Cool, filter and add 95% alcohol to make 200 ml. Add 4 ml of conc. HCl. The solution keeps well for months.

> **NOTE:** Puchtler and Sweat (1960) recommend a commercially available resorcin-fuchsin. After removal of paraffin, they bring the sections to 70–80% alcohol and stain for 4–15 hr at room temperature in resorcin-fuchsin (Chroma), 0.2 gm; 70% alcohol, 100 ml and concentrated HCl, 1 ml. They report little difference in staining intensity through the staining times given. After staining, proceed with Step. 3.

Staining schedule

1. Remove paraffin from sections in the usual manner.
2. Stain sections in the above basic fuchsin solution for a period varying from 20–60 min (or longer) to 12 hr according to depth of staining desired.
3. Wash off excess stain in 95% alcohol. If the sections are stained diffusely, differentiate in acid alcohol for several minutes and then wash thoroughly in tap water.
4. The nuclei may be stained with Orth's lithium carmine before the elastic tissue is stained, if no further counterstain is desired, or afterwards with alum hematoxylin followed by either dilute phloxine or Van Gieson's solution. It may be stained in hematoxylin and eosin or any of the usual stains without injuring the elastic tissue stain.
5. Differentiate and dehydrate in 95% alcohol followed by absolute.

6. Clear in xylene and mount in balsam. If celloidin sections are used, clear in oil or origanum or terpineol after 95% alcohol.

Results: Elastic fibers—dark blue to black; nuclei—brilliant red (if carmine is used), or bluish to black (if alum hematoxylin is used); collagen—pink to red; other tissue elements—yellow (if stained with Van Gieson's solution).

> NOTE: A Feulgen reaction (p. 201) can be substituted for Step 5. French (1929) finds that the addition of a trace of dextrin, and the substitution of crystal violet for (or its addition to) the basic fuchsin, makes more certain a satisfactory preparation.

Verhoeff's Elastic Tissue Stain

(See p. 108)

Darrow's Modification with Synthetic Orcein

From Darrow (1952)

Designed to show: Elastic fibers.

Fixation: Zenker, Bouin's, 10% formalin, acetone.

Embedded in: Paraffin.

Preparation of staining solutions:
A. Dissolve 0.4 gm synthetic orcein in 100 ml 70% alcohol containing 1% conc. HCl.
B. Mallory's borax methylene blue. Stock solution: Dissolve 1 gm methylene blue and 1 gm borax in 100 ml distilled water. For use: dilute 1 vol to 9 vol distilled water.

Staining schedule:
1. Remove paraffin from sections in the usual manner.
2. Stain 30 min in Solution A.
3. Rinse birefly with 70% alcohol.
4. Rinse with distilled water.
5. Stain 5 min in diluted Solution B.
6. Rinse with distilled water.
7. Differentiate and dehydrate about 2 min in 95% alcohol containing 0.5% of colophony (rosin). Keep the slide in constant motion and control the results under the microscope.
8. Complete the dehydration quickly with absolute alcohol.
9. Clear in xylene and mount in balsam.

Results: Elastic fibers—dark purple or reddish purple, collagen—practically unstained, nuclei—blue.

Fraenkel's Orcein Method for Elastic Tissue

From Lillie and Fullmer (1976)
Described by Schmorl (1928)

Designed to show: Elastin; collagen; muscle; nuclei.

Fixation: Any fixative.

Embedded in: Paraffin.

Preparation of staining solutions:
A. Prepare a stock solution of orcein, 1.5 gm; 95% alcohol, 120 ml; distilled water, 60 ml; HNO_3, 6 ml. To 70% alcohol containing 3% HCl add enough stock solution to give a dark brown color.
B. Dissolve 0.25% indigo carmine in saturated aqueous picric acid.

Staining schedule.:
1. Remove paraffin from sections in the usual manner.
2. Stain nuclei red for 2–5 min with Orth's lithium carmine (p. 111).
3. Differentiate in 1% alcoholic HCl.
4. Stain in 24 hr Solution A.
5. Differentiate in 80% alcohol.
6. Stain 10–15 min in Solution B.
7. Rinse in 3.5% acetic acid.
8. Dehydrate quickly in 95% and 100% alcohol.
9. Clear in xylene (some suggest that clearing should begin in 50:50 absolute alcohol and zylene).
10. Mount in synthetic resin.

Results: Nuclei—red, elastin—dark brown, collagen—blue-green, muscle—greenish yellow.

Kornhauser's "Quad" Stain

From Kornhauser (1943, 1945)

Designed to show: Most of the elementary structures of complex tissues, elastic, collagenous and muscle fibers, myelinated nerves, blood, with well stained nuclei and cytoplasmic fibrils and granules.

Fixation: Zenker or Helly.

Embedded in: Paraffin or nitrocellulose or Parlodion.

Preparation of solutions:
A. Orcein, synthetic, 0.4k gm; conc. HNO_3, 0.4 ml; 90% alcohol, 100 ml.
B. Dissolve 0.35 gm acid alizarin blue 2B in 100 ml 10% aqueous solution of aluminum sulfate crystals, $Al_2(SO_4)_3 \cdot 18H_2O$. Boil gently for 10 min. Restore evaporated water. Cool and filter. Adjust to pH 2.9 with glacial acetic acid. For 100 ml about 0.5 ml is required.

C. Orange G, 2 gm; fast green FCF, 0.2 gm; glacial acetic acid, 2 ml; distilled water, 100 ml.

D. Phosphotungstic acid crystals, 4 gm; phosphomolybdic acid crystals, 1 gm; distilled water, 100 ml.

Staining schedule:

1a. Paraffin sections are mounted on slides as usual and passed through xylene and absolute down to 85% alcohol.

or

1b. Affix nitrocellulose or Parlodion sections to the slide as follows: Float section from 95% alcohol onto clean slides and straighten out with fine brush or drawn out glass rod. Blot firmly with smooth, hard blotting paper or bibulous paper and immediately flow on several drops of 0.25% Parlodion in ether-alcohol, allowing excess to run down the slide beyond the section. Hold slide at an acute angle to the eye looking across the film and when the surface begins to look dull or slightly wrinkled, place immediately in 85% alcohol to harden the Parlodion. This is accomplished in about 1/2 hr. The smaller the amount of Parlodion the better.

2. Remove any $HgCl_2$ in the sections with iodine in 85% alcohol.

3. Wash out any traces of iodine in 85% alcohol.

4. Stain 1 hr in Solution A or until elastic fibers are deeply stained. (If no elastic fibers are present in the tissues or if one does not wish to stain elastic fibers, this step may be omitted.)

5. Wash out excess of orcein in 2 changes of 85% alcohol and run slides down to water.

6. Stain 5–10 min in solution B.

7. Rinse in distilled water.

8. Decolorize and mordant 10–20 min in Solution D until collagen is destained, examining under microscope from time to time. Fresh solution should be made up when the decolorization of the collagen is unduly slow.

9. Rinse a few seconds only in distilled water. If rinse is too long, the mordanting effect of solution D is lost.

10. Stain 10 min in Solution C.

11. Rinse thoroughly but rather rapidly (about 10 sec) in 50% alcohol and pass through 2 changes of 95% alcohol, allowing a few minutes at least for each change, 2 changes of absolute alcohol, 2 of xylene, and mount in neutral synthetic resin in xylene.

or

11a. With the Parlodion affixed sections, leave in alcohol until the stain comes out of the Parlodion or nitrocellulose surrounding the sections, pass from 95% alcohol through 2 changes of terpineol, 2 of xylene; mount in neutral synthetic resin in xylene.

Results: Elastic fibers—red-brown; nuclei, basophilic granules—blue or purple; cytoplasm and muscle fibers—violet or pink; collagen, reticulum or basement membranes—green; erythrocytes—orange; myelin sheaths— orange; acidophilic granules—orange; plasmosomes—orange.

SECTION E: SILVER METHODS

Bielschowsky's Method

Foot modification
Mallory (1938)
Essentially the same as Foot (1924), Foot and Menard (1927)

Designed to show: Collagenous fibrils; muscle fibers; nuclei; reticulum (in lymphoid tissue, bone marrow, etc.).

Fixation: Zenker.

Embedded in: Paraffin.

Preparation of staining solutions:
A. Ammoniacal silver solution: To 20 ml 10% aqueous $AgNO_3$ add 20 drops 40% aqueous NaOH. Dissolve the brown precipitate by adding about 2 ml strong ammonia water (27–28%) slowly with continual shaking. Do not add an excess of ammonia even at the risk of not dissolving all the precipitate. Make the resulting solution up to 80 ml and filter before use. It must be freshly prepared as needed.
B. Alum hematoxylin, p. 105.

Staining schedule:
1. Remove paraffin from sections in the usual manner.
2. Place 5 min in 0.5% iodine in 95% alcohol.
3. Wash in tap water and place 5 min in 0.5% $Na_2S_2O_3 \cdot 5H_2O$.
4. Wash in tap water.
5. Treat sections 5 min in 0.25% aqueous $KMnO_4$.
6. Rinse in tap water.
7. Place sections 15–20 min in 5% aqueous oxalic acid.
8. Wash thoroughly in tap water.
9. Rinse in distilled water.
10. Leave sections 48 hr in 2% aqueous $AgNO_3$ in subdued light but not in the dark.
11. Wash a short time in distilled water.
12. Place in the ammoniacal silver Solution A 30 min.
13. Wash quickly in distilled water.
14. Reduce 30 min in a 5% neutral formalin solution, changing the solution after the first 10 or 15 min.
15. Rinse in tap water.

16. Tone 1 hr in 1% aqueous "gold chloride."*
17. Rinse in tape water.
18. Remove excess silver by treating sections 2 min with 5% aqueous $Na_2S_2O_3 \cdot 5H_2O$.
19. Wash thoroughly for several hours in running tap water.
20. Stain in alum hematoxylin (p. 105) or in iron hematoxylin (p. 108).
21. Wash in tap water until blue.
22. Counterstain 30 sec in Van Gieson's solution (p. 113).
23. Dehydrate in 95% alcohol followed by absolute.
24. Clear in xylene and mount in balsam.

Results: Coarser and collagenous fibrils—red to rose, finer collagenous fibrils (reticulum)—black to dark violet, nuclei—black, blue or brownish, cytoplasm—grayish yellow, muscle fibers—brighter yellow, elastic fibers—pale yellow.

Lillie Modification

Lillie and Fullmer (1976)

Designed to show: Reticulum; collagen.

Fixation: Formalin or Orth's.

Embedded in: Paraffin.

Preparation of diammine silver hydroxide solution: To 2 ml 28% ammonia water in a small flask, add 35–42 ml 5% aqueous $AgNo_3$, the first 30 or 35 ml fairly rapidly, the rest cautiously, shaking between each addition to clear the brown clouds of silver oxide until a faint permanent opalescence remains. (This solution can be used 1 or 2 days, and is discarded after using once.)

Staining schedule:
1. Remove paraffin through two changes of xylene and two of 100% alcohol.
2. Place for 5–10 min in 1% collodion in ether and 100% alcohol (equal volumes).
3. Drain 1 min.
4. Place for 5 min in 80% alcohol.
5. Rinse in tap water.
6. Treat 5 min in 0.25% aqueous $KMnO_4$.
7. Wash in tap water.
8. Treat 10 min in 5% aqueous oxalic acid.

---

* The "gold chloride" of commerce is ordinarily the yellow crystalline compound, $AuCl_3 \cdot HCl \cdot 3H_2O$. The brown crystals sometimes sold are probably contaminated with metallic gold. For toning purposes, either grade is probably satisfactory.

9. Wash in tap water.
10. Apply one of the following treatments: 5–10 sec in Wilder's 1% aqueous uranyl nitrate; 1 min in Gomori's 2% aqueous iron alum; 2 min in liquor ferri chloridi 1:50 in distilled water; *or* 2 min in 3% aqueous H_2O_2. (With Orth's fixed material, omit this step.)
11. Wash 3 min in running water, and rinse in 2 changes of distilled water. (Wilder omitted this wash, but it does not interfere, even after uranyl nitrate.)
12. Lay slides face up on glass rods over a large pan and deposit on each about 1.5–2 ml of the above diammine silver hydroxide solution for 3 min and decant; or immerse in this solution, followed by draining. Sections should be a golden brown color.
13. Rinse quickly in distilled water.
14. Reduce 2 min in 20% formalin.
15. Wash 3 min in running water.
16. Tone 2 min in 0.2% acid gold chloride ($HAuCl_4$).
17. Rinse in tap water.
18. Fix 2 min in 5% $Na_2S_2O_3 \cdot 5H_2O$.
19. Wash in tap water.
20. Counterstain as desired: acetic alum hematoxylin, 2 min; tap water, 2 min; Van Gieson's picrofuchsin, 1 min; differentiate in 2 or 3 changes of 95% alcohol. *Or* stain by Ziehl-Neelsen technic for acid fast organisms. *Or* stain 5 min in 0.1% safranin, thionin, or toluidine blue O; differentiate 1 min in 5% acetic acid, and wash well in tap water.
21. Dehydrate and decollodionize in 3 changes of acetone.
22. Clear with 1 change of acetone and xylene (50:50) and 2 or more changes of xylene; mount in clarite.

Results: Reticulum—black; coarser collagen fibers—red, if Van Gieson's is used; finer collagen fibers—red to black; nuclei—reddish brown; muscle—yellow.

Wilder Modification

Quoted from Mallory (1938)
Essentially the same as Wilder (1935)

Designed to show: Reticulum; collagen.

Fixation: Formalin 1:10, Zenker or Helly.

Embedded in: Paraffin or celloidin, or frozen sections.

Preparation of solutions:
   A. Ammoniacal silver solution: To 5 ml of 10.2% aqueous $AgNO_3$, add 26 to 28% ammonia water drop by drop until the precipitate which forms is dissolved. Then add 5 ml of 3.1% NaOH and barely dissolve

the resulting precipitate by adding a few drops of ammonia water. Make the solution up to 50 ml with distilled water.

B. Reducing solution: Distilled water, 50 ml; formalin, neutralized with $MgCO_3$, 0.5 ml; 1% aqueous uranium nitrate, 1.5 ml.

Staining Schedule:

1. Remove paraffin or celloidin from sections in the usual manner, and pass them into distilled water.
2. Treat 1 min in 0.25% aqueous $KMnO_4$ or in 10% aqueous phosphomolybdic acid.
3. Rinse in distilled water.
4. Place for 1 min in HBr (Merck's concentrated, 34%, 1 vol; distilled water, 3 vol). This step may be omitted if phosphomolybdic acid is used in Step 2.
5. Wash in tap water, then distilled water.
6. Dip for 5 sec or less in 1% aqueous uranium nitrate (sodium free).
7. Wash 10–20 sec in distilled water.
8. Treat 1 min in the ammoniacal silver Solution A.
9. Dip quickly in 95% alcohol.
10. Reduce 1 min in Solution B.
11. Wash in distilled water.
12. Tone 1 min in 1:500 aqueous "gold chloride" (Merck's reagent).*
13. Rinse in distilled water.
14. Fix 1–2 min in 5% aqueous $Na_2S_2O_3 \cdot 5H_2O$.
15. Wash in tap water.
16. Counterstain, if desired, with alum hematoxylin and Van Gieson's stain or alum hematoxylin and phloxine.
17. Differentiate and dehydrate in 95% alcohol followed by absolute alcohol.
18. Clear in xylene and mount in balsam. Celloidin sections are cleared in oil of origanum following 25% alcohol.

Results: Fine reticulum fibers—black (with great precision), collagen—rose.

---

* *See footnote on p. 127.*

CHAPTER 6

NEUROLOGICAL STAINING METHODS*

George Clark

The nervous system is divided anatomically into two components: central and peripheral. The central nervous system consists of the brain and spinal cord, while the peripheral nervous system includes all cerebrospinal and autonomic nerves. Staining methods suited to the central nervous system may be quite unsuited for peripheral nerves. The reverse is true also, hence in neurological staining, methods devised for specific purposes predominate and generally applicable methods constitute a relatively small minority.

The chief components of brain and spinal cord are nerve cells and their processes, a supporting meshwork formed by neuroglia, fatty material (myelin) and blood vessels. Peripheral nerves consist of bundles of nerve fibers bound together by a matrix of the neuroglial type (endoneurium) and covered by sheaths of ordinary connective tissue (perineurium and epineurium). Nerve cells are abundant in the ganglia of visceral nerves but absent to rare in other peripheral nerves, excepting the ganglia of the dorsal roots. Myelinated nerve fibers are predominant in nerves to voluntary muscles and sense organs, whereas an abundance of nonmyelinated fibers characterize visceral nerves. These structural peculiarities have influenced rather strongly the development of technics.

Methods that have been developed for staining a particular component are usually selective but may partially stain other components. Also, specimens of nervous material, alike anatomically, but obtained from different animals, often vary greatly in their reaction to staining. The resulting inherent difficulties can be minimized only by following a given technic with respect to the kind of fixation, embedding and other processes recommended, as well as the actual staining. Some of the steps in many technics can be

---

* This chapter, which was originally prepared by Dr. H. A. Davenport, is essentially the same as in the second edition. The material has been reorganized, some procedures deleted and some added but much of the text remains unchanged.

varied to assist in securing the results desired. Such variations apply particularly to the concentration of solutions and to timing, and in this respect neurological technics do not differ from others.

A number of technics are influenced by the presence or absence of myelin in the tissue to be stained. Its presence in the central nervous system affects the permeability of the tissue. The permeability in turn affects staining; consequently, some stains function properly only when myelin is present, for example the Golgi method. In other methods (Ranson's and Bielschowsky's for axis cylinders) the myelin has to be extracted with alcohol or pyridine to allow proper permeation of the tissue by the staining reagents. Even though permeability is not involved, some methods require frozen sections with the myelin intact while others are designed for mounted paraffin sections in which the myelin has been removed during dehydration and embedding. Factors other than the presence of myelin are involved in most staining methods, but its presence or absence constitutes a conspicuous feature.

The technician may be consoled by the assurance that failures and mediocre successes are common, but that some successes are often amazing in their clarity of differentiation and high degree of selectivity. When demonstration slides are desired, it is advisable to use a number of specimens from different animals, and then select the best ones from the final products. The familiar cliche that neurological stains are erratic could be expressed more factually by saying that tissues react unpredictably to staining procedures. A good procedure is one that will give a high percentage of successful stains in a large number of specimens from different animals. Tissue from a single individual (man or animal) may range in its reaction to a staining method from excellent to failure, but usually between these extremes. The performance of a method should be judged, therefore, on a statistical basis rather than on a few trials, because methods themselves are not inherently erratic.

The methods given here have been selected as representative of the main staining categories. They are classified according to the tissue components most selectively or characteristically stained.

Cautions Applicable Generally

1. Use C.P. or reagent grade chemicals, high grade distilled water and clean glassware when making metallic stains (silver, gold, etc.)

2. See that fixation is suited to the subsequent staining and considered an integral part of the technic. Examples: myelin sheath stains do not follow fixation in alcohol; most silver stains are not successfully used after Zenker's fluid or other fixatives which contain mercuric salts or bichromate; and the length of time of fixation may be a factor for some methods.

3. As many neurological technics require some experimental adaptation to the specific use to which they are to be put, see that such modifications

are made when necessary. Observe which steps in a technic are critical, that is, whose variation actually affects results, and vary these rather than uncritical steps.

4. Do not allow solutions of ammoniated silver to dry in open dishes. An *explosive compound* may form. Discard solutions after use.

Artifacts

1. *Precipitates.* Many neurological stains are made by reducing a metallic salt so that the metal is left in a state of colloidal dispersion in the tissue elements. During the reduction, coarsely granular precipitates may be formed, and constitute a common artifact. Their extracellular location serves to identify them, but their abundance may spoil a preparation. There is no universally applicable method for eliminating such unwanted precipitation, but some methods contain cautions directed toward minimizing it.

2. *Shrinkage and swelling.* Nervous tissue which is being fixed in formalin weaker than 1:8 first swells and then shrinks. Regardless of what changes take place in the process of fixation the tissue always shrinks during dehydration and embedding. This is true for all fixatives, and most blocks lose 20–40% in lineal shrinkages or about half of their volume. Certain localized shrinkages are particularly annoying, such as the cytoplasm of ganglion cells, formation of large perivascular spaces in the central nervous system, and interfascicular shrinkage of nerve trunks. The worst offenders among fixatives are formalin, ammoniated alcohol and chloral hydrate, especially when the use of these fixatives is followed by paraffin embedding.

3. *Bubbles.* Although not encountered in fixed preparations of peripheral nerves, bubbles which form in the depths of the block are not uncommon in central nervous tissue. These artifacts are recognized easily, when large, by the compression of the tissue around the periphery of the bubble, but when small, they appear as holes 5–20 μm in diameter in the cut sections and may show little compression of adjacent tissue. They may become filled (after aqueous formalin) with myelin derivatives and this material may even take a stain with basic dyes and be mistaken for degenerative changes of glia cells. Such stained material is usually soluble in chloroform.

4. *Fixation.* Since nerve tissue is encased in bone and since large blocks are usually needed, immersion fixation is not satisfactory. It is preferable to fix by perfusion and after removal of the desired tissue to complete fixation and store for a limited period of time in 10% saline formalin. The formalin-acacia perfusion technic of Koenig, Groat and Windle (1945) has had wide acceptance. Under deep anesthesia the thoracic cavity is opened and a cannula is placed into the left ventricle, and the right auricular appendage is removed. Then, for an animal the size of a cat, run 500 ml of Solution A through the cannula and follow with 500 ml of Solution B. It is preferable to wait for several hours before removal of the nerve tissue. After removal place tissue in saline formalin (10% formalin in 0.9% aqueous NaCl).

Gum Acacia Formalin

Solution A		*Solution B*	
NaCl	9 gm	NaCl	9 gm
Formalin	2 ml	Formalin	100 ml
Gum acacia	56 gm	Gum acacia	56 gm
H_2O q.s.	1000 ml	H_2O q.s.	1000 ml

In the preparation of these solutions, the gum acacia should be suspended in a gauze bag in the water to promote solution. If this is not done, the gum acacia forms a gummy mass that takes an interminable time to dissolve. The solutions must be filtered through absorbent cotton before using.

A simpler and equally satisfactory perfusion mixture, derived from Baker (1965), is:

Sugar Formalin

Solution A		*Solution B*	
NaCl	9 gm	NaCl	9 gm
Sucrose	85 gm	Sucrose	75 gm
Formalin	2 ml	Formalin	100 ml
H_2O q.s.	1000 ml	H_2O q.s.	1000 ml

This is used similarly to the acacia-formalin sequence. Routinely we have used granulated sugar purchased from a supermarket. In both solutions (A) a small amount of formalin is used to discourage growth of molds.

SECTION A: METHODS FOR SUPPORTING ELEMENTS

Cajal Gold Sublimate Method for Astrocytes

Slightly modified from Lee (1950, pp. 587–589)
Also from Penfield and Cone, in McClung (1950, pp. 407–413)
Original by Cajal (1913, 1916)

Application: Central nervous system.

Designed to show: Protoplasmic and fibrous astrocytes.

Fixation: About 5 days (limits 2–25 days) in Cajal's FAB (p. 14).

Embedding: None; frozen sections, cut at 15–30 μm.

Preparation of solutions:

Stock Solution A: Gold chloride (brown or yellow crystals),* 1 gm; water, 100 ml.

---

* The "gold chloride" of commerce is ordinarily the yellow crystalline compound, $AuCl_3 \cdot HCl \cdot 3H_2O$. The brown crystals sometimes sold are probably contaminated with metallic gold. For toning purposes, either grade is probably satisfactory. Because of the uncertain chemical composition, it can be denoted here only by its rather indefinite common name.

Stock Solution B: HgCl$_2$, 5 gm; water (warmed to 60°C), 100 ml. Just before use, mix as follows: Solution A, 5 ml; Solution B, 5 ml; water, 40 ml.

Staining schedule:

1. Wash sections through 2 changes of distilled water.
2. Place sections in the above gold-sublimate mixture, flattened out and not lying on top of one another. Allow to stand 3–4 hr; room temperatures (25–27°C) are satisfactory, and ranges of temperature from 18–40°C are permissible. (Cajal recommended staining in the dark, but Penfield states that daylight is not detrimental.) Examine under microscope while still wet; if the reaction is complete, the astrocytes will be dark against a relatively light background. The staining mixture in contact with the tissue accomplishes its own reduction, and the degree of staining can be judged by the intensity of the purple color in the sections.
3. Wash in distilled water.
4. Fix in 5–10% aqueous Na$_2$S$_2$O$_3 \cdot$5H$_2$O (hypo), which completes the process.
5. Wash in tap water, sufficiently to remove hypo.
6. Dehydrate.
7. Clear and mount in balsam or synthetic resin.

Results: Astrocytes—black, background—unstained, or light brownish purple, nerve cells—red, nerve fibers—unstained.

NOTE: The stain is reasonably reliable if the tissue is removed soon after death and fixed for an optimum period. Protoplasmic astrocytes stain better with short fixation, and fibrous astrocytes with long fixation (Penfield and Cone). After about 3 weeks in the fixing fluid, however, stainability is gradually lost. The method of staining can be varied considerably, provided the progress of staining be checked microscopically during the process. Older modifications of the fixing solution have consisted of the substitution of other salts for NH$_4$Br (urea nitrate for young mammals). A newer and more radical departure is the fixing fluid of Lascano (1958), the formula of which is: glycine, 1.05 gm; 1 N HCl, 14.8 ml; formalin (conc.), 15.0 ml; and water to make 100 ml. Modification of the staining solution has been the varying of the proportions of the two salts in the gold-sublimate mixture.

To adapt the method to tissue which had remained in formalin a long time, Globus (1927) recommended treating frozen sections for 24 hr, with dilute ammonia water (NH$_4$OH (conc.), 1 vol; water, 9 vol), rinsing twice in distilled water and then immersing 2–4 hr in dilute hydrobromic acid (HBr, 40%, 1 vol; water, 9 vol). Staining was then carried out as in the regular method after rinsing twice in distilled water containing 0.5% of the above dilute ammonia water.

Del Rio-Hortega's Silver Carbonate Method for Oligodendrocytes

From Penfield and Cone (McClung, 1950, pp. 419–421)
Original by del Rio-Hortega (1917)

Application: Central nervous tissue.

Fixation: Formalin ammonium bromide (FAB) (p. 14) 12 to 48 hr, for frozen sections.

Preparation of ammoniacal silver carbonate: To 5 ml 10% aqueous $AgNO_3$, add 20 ml 5% aqueous Na_2CO_3 (anhyd.). Add conc. (28%) NH_4OH, drop by drop to dissolve the precipitate, taking care not to add an excess. Add water to bring up to 45 ml. Filter and store in a brown bottle.

Staining schedule:
1. Heat the tissue block for 10 min at 45–50°C in fresh fixing fluid; let cool and cut frozen sections 15–20 μm thick.
2. Wash in dilute ammonia water (1 vol conc. reagent to 100 vol water), then in distilled water.
3. Impregnate 1–5 min (exact time to be determined by trial) in the above ammoniacal silver carbonate solution.
4. Wash 15 sec (with gentle agitation) in distilled water.
5. Reduce 30 sec in dilute formalin (1 vol formalin to 100 vol water) without agitation.
6. Wash thoroughly in tap water, tone in 0.2% gold chloride until gray all over.
7. Fix in 5–10% $Na_2S_2O_3 \cdot 5H_2O$.
8. Wash in tap water.
9. Dehydrate.
10. Mount in balsam or synthetic resin.

Results: Processes and cytoplasm of oligodendrocytes—black, background—gray, nuclei—practically unstained.

> **NOTE:** Although this method is probably the most selective for oligodendrocytes, other types of cells may be stained, particularly microglia. To secure staining of the former only, the time of fixation should be kept within 2 days, since long fixation tends to cause staining of microglia, astrocytes and even nerve cells. Penfield (see McClung (1950), p. 420) has modified the method by treating thin blocks of tissue (3 mm) with 95% alcohol for 36–48 hr, washing out the alcohol, cutting frozen sections, and staining in the ammoniacal silver carbonate for from 15 min to 2 hr or until they begin to turn brown; they are then plunged into the dilute formalin and are agitated during the reduction. This modification was found specially valuable for rabbit material.

Penfield's Combined Method

From Penfield and Cone (McClung, 1950, p. 421)
Original by Penfield (1928)

Application: Routine neuropathological use.

Designed to show: Oligodendrocytes; also microglia.

Fixation: About 1 week in formalin (1:10) or in FAB (p. 14) Good results may still be obtained after longer fixation. For frozen sections, cut at 20 μm.

Staining schedule:

1. Cut sections and place in distilled water to which 1% formalin has been added.
2. Treat about 16 hr in a covered dish containing distilled water to which 1% conc. (28%) NH_4OH has been added.
3. Place for an hour at 38°C in 95 ml distilled water to which 5 ml of 40% HBr has been added.
4. Wash through 3 changes of distilled water.
5. Neutralize by placing for 1–6 hr in 5% Na_2CO_3 (anhyd.).
6. Stain in Hortega's ammoniacal silver carbonate solution (p. 136), made up to a final volume of 75 ml instead of 45 ml. Determine the time necessary by leaving individual sections 2, 3, 5 min or longer and examining microscopically after Step 7.
7. Reduce in 1% formalin, agitating during the reduction.
8. Wash in distilled water.
9. Tone in 0.2% gold chloride until gray all over.
10. Fix in 5–10% $Na_2S_2O_3 \cdot 5H_2O$.
11. Dehydrate.
12. Mount in balsam or synthetic resin.

Results: Oligodendrocytes—dark gray to black, microglia—dark gray to black, background—pale gray.

NOTE: McCarter (1940) has further modified this method by adding an equal volume of 5% aqueous ammonium alum to the Na_2CO_3 bath (Step 5) after the sections have been put into it. The precipitate formed does not interfere with subsequent staining. A better differentiation of cell cytoplasm and processes is claimed.

Before detailing the methods for microglia, attention should be called to the similarity between them and those for oligodendroglia (syn. oligodendrocytes). It would be more desirable to have methods which stained these glial elements separately and reliably; but usually both types of cells are stained whether the method is intended for microglia or oligodendroglia. It is generally believed that a short fixation (a few hours to 3 days) predisposes to staining of oligodendrocytes, while longer fixation (4 days to several weeks) tends to favor the staining of microglia. Hortega's method for microglia published by Penfield and Cone (in McClung) is very similar to his (1921) method for oligodendroglia.

Del Rio-Hortega's Method for Microglia

From Penfield and Cone (McClung, 1950, pp. 422–425)

Application: Central nervous tissue.

Fixation: FAB fluid for 2–3 days instead of 12–48 hr. For frozen sections.

Staining schedule:
1. Before sectioning, place 10 min in fresh fixing fluid heated to 50°C.
2. Let cool and cut frozen sections 15–20 μm thick.
3. Wash in distilled water, then in 1% ammonia water, and again in plain distilled water.
4. Stain in Hortega's ammoniacal silver carbonate diluted to 75 ml instead of to 45 ml and apply for a shorter period, removing sections to test after 20 sec, 45 sec, and 2 min. Reduce immediately.
5. Transfer directly to the dilute formalin (1 ml to 100 ml water), and agitate the section by blowing on the surface of the solution.

The remainder of the process is the same as the method for oligodendroglia (p. 137), Steps (6–10).

Results: Microglia and processes—black, background—pale, other glial cells—dark gray to black.

The methods described by Cajal and Hortega for neuroglia require frozen sections to be successful. This has been a source of annoyance to neuropathologists because it requires extra time and because a friable brain tumor often breaks into bits when cut. The desirable system in neuropathological technic is one which permits the diagnosis of all lesions on material fixed in a single fixative and embedded in either paraffin or nitrocellulose. Such a system has not yet been realized but methods have been devised which lead toward it. Selectivity of staining individual types of neuroglial cells have been sacrificed in silver stains used on embedded material. Since tumor cells have as a rule a stronger affinity for silver than have normal cells, the problem of staining them is relatively simple if the neuropathologist relies entirely upon the morphology of the cell type for diagnosis and not upon the selectivity, either real or fancied, of the stain. The stains given below will serve for tumor material or pathological gliosis.

Weil-Davenport Method for Gliomas

From Weil and Davenport (1930)

Application: Central nervous tissue.

Designed to show: Tumor cells and their processes; axis cylinders.

Fixation: Formalin (1:10) for several days or longer.

Embedded in: Paraffin.

Preparation of solutions:
A. Staining solution: Dissolve 8 gm AgNO₃ in 10 ml distilled water and add 90 ml of 95% alcohol.
B. Reducing solution: Dissolve 5 gm pyrogallic acid in 95 ml of 95%

alcohol and add 5 ml commercial formalin. This may be diluted with alcohol to control the intensity of the staining. To keep precipitate from settling on the slide, one may add to about 50 ml of this solution 0.5 ml of diluted corn syrup (1 vol commercial Karo syrup, 3 vol water). Change the reducing solution frequently.

Staining schedule:

1. Cut paraffin sections at 10 μm and mount on slides as usual.
2. Remove paraffin with xylene; and place for about 2 min in each of the following: (1) absolute alcohol; (2) 1:1 mixture of absolute alcohol and anhydrous ethyl ether.
3. Place 2–3 min in a 1:1 alcohol-ether solution of nitrocellulose, a 1.5% solution in the case of Parlodion or celloidin, or 4–5% if the low viscosity type is employed.
4. After partial draining, hold the slide horizontally, upside down, moving it from side to side and tilting from end to end to keep the coating even until the nitrocellulose begins to thicken. Then turn it right side up and lay on a rack until it begins to form a gel. Place the slide in 80% alcohol until ready for next step. (For the method to succeed it is necessary to spread the coating evenly and not to allow it to dry beyond the stage of gelation.)
5. Impregnate for 6–48 hr at 37–40°C in Solution A. The time of application is not very critical, although different tumors respond differently to impregnation.
6. Rinse quickly (in and out) in 95% alcohol. (If subsequent stain is too light, omit the rinsing; if too dark, prolong it.)
7. Reduce in Solution B. The time is usually about 1 min, but timing can be varied to control the staining intensity.
8. Wash the slides thoroughly in running tap water. (Do not put in back to back, as reducing solution may be carried into the gold bath.)
9. Tone 5–10 min in 0.2% gold chloride.
10. Wash in distilled water.
11. Fix 1 min in 10% $Na_2S_2O_3 \cdot 5H_2O$.
12. Dehydrate; and dissolve the coating by including a jar of 1:1 alcohol-ether mixture between the absolute alcohol and the final xylene.
13. Mount in balsam or synthetic resin.

Results: Pathological glia—dark gray to black, background—violet gray, axis cylinders—black or gray, nuclei of cells—less stained than the cytoplasm, sometimes of different hue.

NOTE: This method is very similar to the one for nerve fibers in celloidin sections (p. 151). If 0.5 ml of 7% aqueous nitric acid solution is added to each 100 ml of the alcoholic silver nitrate solution, the staining of normal neuroglia is suppressed. Unmounted celloidin sections can be stained for axis cylinders by using Stender dishes for the fluids.

Stern's Method as Modified by Weil and Davenport

From Weil and Davenport (1933)
Original by Stern (1932)

Application: Brain.

Designed to show: Microglia and oligodendroglia.

Fixation: Formalin (1:10).

Embedded in: Celloidin.

Preparation of ammoniacal silver solution: To 2 ml conc. NH_4OH, add 18–20 ml of 10% $AgNO_3$ (or use 15% $AgNO_3$, if it is desired to favor the staining of oligodendroglia). An exact end point can be obtained if the ammonia is placed in an Erlenmeyer flask and shaken while the $AgNO_3$ is run in from a pipette; the first appearance of permanent opalescence indicates that enough $AgNO_3$ has been added. If too much is added by mistake it is possible to add a drop of ammonia to redissolve the precipitate and then titrate with $AgNO_3$ to the first appearance of permanent opalescence. The proper balance of silver and ammonia is important.

Staining schedule:
1. Cut sections at 15 μm, and wash in distilled water. If desired to favor the staining of oligodendroglia, place for 2–3 min in 0.5% ammonia water.
2. Stain 10–20 sec in the above ammoniacal silver solution; the longer the staining, within the limits given, the more the staining of oligodendroglia is favored.
3. Transfer without washing to: formalin, 3 vol; water, 17 vol. Move the section rapidly in the solution until it is a coffee brown color. Or, to favor the staining of oligodendroglia, use 1:10 instead of 3:20 formalin, and do not begin agitation of the section until the cellodin has blackened and the section has begun to brown. Renew the formalin after each section.
4. Wash in tap water.
5. Dehydrate as usual, except that *n*-butyl alcohol should be substituted for absolute ethyl alcohol to prevent softening of the celloidin. To prevent fading, do not allow sections to remain in xylene longer than necessary.
6. Mount in balsam or synthetic resin.

Results: Microglia—black, background—yellow or brownish yellow. Only pathological microglia stain with any degree of certainty.

NOTE: This method is specially useful when tissue is embedded in nitro-cellulose, and when adjacent sections, stained with dyes, or processed by other technics, are desired.

Nassar-Shanklin Method for Neuroglia in Paraffin Sections

From Nassar and Shanklin (1951)

Application: Neuroglia in general.

Designed to show: All three types of glia cells, but it probably favors astrocytes and microglia. For paraffin sections.

Postfixation treatment and embedding:

1. Soak 10 hr in ammonia water (concentrated NH$_4$OH, 1.5 ml; water 100 ml).
2. Wash through 2 changes of distilled water, 1–2 hr in each.
3. Dehydrate with alcohol, clear and embed in paraffin as usual.
4. Cut 10–15 μm sections and mount on slides with albumen adhesive.

Staining schedule:

1. Dewax and hydrate sections in the customary manner.
2. Transfer from water to 5% Na$_2$SO$_3$ and allow to remain there 2 hr.
3. Pass slides quickly through 3 changes of distilled water.
4. Impregnate at room temperature 2–5 min in an ammino-silver solution prepared as follows: Place 1 ml of concentrated NH$_4$OH (28% NH$_3$) in a small flask and add 7–8 ml of 10% AgNO$_3$ while shaking the mixture. Then add the silver solution drop by drop, shaking after each addition until a faint permanent turbidity is produced. For use, dilute the resulting solution with an equal volume of distilled water. (The method of making this solution is credited to Lillie (1954), p. 335). The optimum time for impregnation should be determined by several trials on the particular material to be stained.
5. Rinse in distilled water 1–2 sec.
6. Reduce 1 min in 2% neutral formalin, with gentle agitation. The formalin solution should be changed frequently.
7. Wash well in water and tone in 0.2% gold chloride for 1 min or less, until the section just becomes gray.
8. Fix in 5% Na$_2$S$_2$O$_3 \cdot$5H$_2$O, wash well, counterstain, if desired, dehydrate, clear and cover.

Results: Neuroglia are dark gray to black against a paler background.

NOTE: Van Gieson's method finds wide acceptance among pathologists. It has been described on p. 103. Mallory's phosphotungstic acid-hematoxylin (p. 110) is a valuable stain also.

Foot's Ammoniated Silver Carbonate Method

From Foot (1924)

Recommended for: Abnormal brain tissue.

Designed to show: Vascular reticulum, tumor cells, and connective tissue around a tumor.

Fixation: Formalin (1:10) recommended; Cajal's FAB and Bouin's fluids also satisfactory.

Embedded in: Paraffin.

Preparation of solutions:
  A. Ammoniacal silver carbonate: To 10 ml of 10.2% $AgNO_3$ add conc. NH_4OH, drop by drop, until the precipitate formed is almost redissolved; then add 10 ml of 3.1% Na_2CO_3, 3 ml; conc. formalin, 1 ml; water, 100 ml.
  B. Reducing solution: 1% Na_2CO_3, 3 ml; concentrated formalin, 1 ml; water, 100 ml.
  C. Intensifying solution: oxalic acid, 2 gm; conc. formalin, 1 ml; water, 100 ml.

Staining schedule:
  1. Remove paraffin from mounted sections and treat for 24 hr at room temperature in a mixture of pyridine and glycerol (2:1, by volume).
  2. Rinse in 95% alcohol, then in distilled water.
  3. Impregnate for 2.5 hr at 40°C in silver Solution A.
  4. Wash in distilled water.
  5. Reduce 5 min in Solution B.
  6. Wash in tap water.
  7. Tone 5 min in 0.2% gold chloride.
  8. Wash in tap water.
  9. Intensify by treating 5 min in Solution C.
  10. Rinse in tap water.
  11. Fix in 5–10% $Na_2S_2O_3 \cdot 5H_2O$.
  12. Wash in tap water.
  13. Dehydrate.
  14. Cover in balsam or synthetic resin.

Results: Tumor cells—reddish to violet-gray, vascular reticulum—black.

> **NOTE:** The stain is not recommended for normal brain. It has considerable value in demonstrating connective tissue reaction around tumors if reticulin fibers are laid down by this connective tissue. If FAB fixation is used, the reticulin staining is said to be suppressed.

SECTION B: METHODS FOR NERVE CELLS, NISSL GRANULES AND CHROMATOLYSIS

Nissl Staining with Cresyl Violet Acetate, Thionin or Toluidine Blue O

Emended from Powers and Clark (1955) and Windle, Rhines and Rankin (1943)

Designed to show: Selective staining of Nissl granules (tigroid bodies, Nissl bodies) and nuclei of neuroglia cells (DNA and RNA).

Fixation: Most fixatives are satisfactory but best results follow perfusion with sugar formalin (see p. 134).

Embedded in: Optional, paraffin, celloidin or frozen sections.

Choice of dyes: The results and methods are similar with cresyl violet acetate, thionine and toluidine blue O. All of these are available as certified dyes. Personally, I prefer cresyl violet acetate as it gives a more intense reddish purple stain while thionin and toluidine blue O give bluer colors. The azures (A, B and C) are also satisfactory but are so readily removed by alcohols that an acetone-xylene sequence must be used for dehydration. The technic of Lillie (1965, p. 163) is quite satisfactory. Neutral red has been used as a Nissl stain but like the azures is too readily removed in dehydration. In conjunction with Luxol Fast Blue MBS the results are brilliant (Lockard and Reers, 1962) but Darrow red appears preferable (see p. 144) as it gives comparable results and the dye is less easily removed.

Preparation of solutions:
A. Stock dye solution: 0.1% in distilled water.
B. Walpole buffer 0.2 M (see p. 3)
C. Staining solution:
   (1) Paraffin sections: Use 10 ml of stock dye solution to each 100 ml of buffer.
   (2) Celloidin and frozen sections: Use 20 ml of stock dye solution to each 100 ml of buffer.
      Routinely use 3 parts of 0.2 M acetic acid to 2 parts 0.2 M sodium acetate solution. This gives a pH of 4.45. The staining solution should be used once and discarded.

Staining schedules:
1. Paraffin sections.
   a. Remove paraffin with xylene, hydrate through graded alcohols and wash in distilled water.
   b. Stain 20 min in Solution C.
   c. Rinse quickly in 95% ethyl alcohol and then place in *n*-butyl alcohol, 2 changes of 2 min each.
   d. Clear in xylene and examine under the microscope. If the nucleoplasm of neurons is not as colorless as the white matter, return briefly to *n*-butyl alcohol and then to 95% alcohol and repeat the 2 changes of *n*-butyl alcohol. If the stain is too weak, return back through the alcohols and restain. If needed, add additional stock dye to the staining solution.
   e. Mount with Permount or similar synthetic resin.
2. Celloidin sections—routinely we place 2–30 sections in a 4-oz jar where they are left for processing. A piece of gauze should be held over the opening when pouring off each solution.
   a. Wash in distilled water.
   b. Stain 20 min in Solution C.

    c. Dehydrate and differentiate through 50%, 70% and 95% ethyl alcohol, and n-butyl alcohol. Check differentiation microscopically. When complete, the nucleoplasm of neurons and the neurophil should be colorless or only faintly blue. Complete dehydration in terpineol-xylene (1 part terpineol to 4 parts xylene) for 12–16 hr.

    d. Clear in xylene and mount with Permount or similar synthetic resin.

3. Frozen sections—mount sections from water on slides lightly coated with albumin (as in mounting paraffin sections) and air dry preferably in a 45°C oven.

    a. Treat sections similar to paraffin sections, removing soluble lipid materials with xylene followed by descending grades of alcohol.

    b. Wash in distilled water.

    c. Stain 20 min in solution C. Dehydrate and differentiate rapidly through 50%, 70% and 95% ethyl alcohol and 2 changes of n-butyl alcohol of 4 min each. If sections are thick (over 20 μm), complete dehydration in terpineol-xylene (see above). Check differentiation as with celloidin sections.

    d. Clear in xylene and mount in Permount or similar synthetic resin.

Results: Nissl granules—purple to violet, nuclei of neuroglia and endothelial cells—slightly bluer than Nissl granules.

Darrow Red

Powers, Clark, Darrow and Emmel (1960)
Emended by Powers and Clark (1963)

Application: Nissl stain, especially recommended as a counter stain for the Weil or Klüver-Barrera myelin sheath stains.

Designed to show: RNA and DNA containing structures.

Fixation: Will follow most fixatives but is usually used on formalin-fixed tissue.

Preparation of solutions: Darrow red, 25 mg; M/5 Walpole buffer pH 3.5, q.s. 200 ml. Wash dye into bottle with buffer solution, cap bottle and place in paraffin oven at about 60°C overnight. Filter before use.

Staining schedule:

1. Stain well washed Weil (p. 146) stained sections or Klüver-Barrera (p. 147) stained sections (with either of these the sections should be no thicker than 20 μm) or unstained sections for 20 min.
2. Pass rapidly through 95% alcohol.
3. Two changes of n-butyl alcohol for a total of 4 min.
4. Terpineol (1 part) and xylene (4 parts) overnight.
5. Clear and mount.

SECTION C: METHODS FOR MYELIN SHEATHS

Large pieces of tissue cannot be stained satisfactorily in osmic fixing fluids because of the poor penetration of the reagent. As a stain following formalin fixation, OsO_4 has been employed to a limited extent. Best results, however, have followed methods employing hematoxylin, and there are a number of these. The original technics of Weigert (1884, 1885, 1891) and the modifications by Pal (1886, 1887) are, to a considerable extent, supplanted by newer modifications. The Weigert and Pal methods require fixation and hardening of the nervous tissues in dichromate solutions and give best results with ripened hematoxylin staining solutions. Newer technics can be employed with formalin-fixed material; the advantages being that time is saved, and alternate sections of a series can be stained by other neurological methods.

Pal-Weigert Method

From Clark and Ward (1934)
Original by Pal (1886); modified from Weigert (1884, 1885)
See also Weigert (1891)

Designed to show: Myelin sheaths in brain and spinal cord; can be used for large peripheral nerves and ganglia.

Fixation: Formalin solutions, e.g. 1:10 formalin, with or without 2% NH_4Br.

Embedded in: Celloidin, low viscosity nitrocellulose or paraffin; or cut frozen sections (fixing fluid with 2% NH_4Br recommended for the latter).

Preparation of staining solution:
A. Saturated aqueous Li_2CO_3, 7 ml; water, 93 ml.
B. Hematoxylin, 1 gm; ethyl alcohol, absolute, 10 ml (95% alcohol probably satisfactory). Does not require ripening. Add 1 vol of Solution A to 9 vol of Solution B.

Staining schedule·
1 Mordant sections for 2–24 hr (as is convenient) in a solution containing 4% aqueous $Fe_2(SO_4)_3 \cdot (NH_4)_2SO_4 \cdot 24H_2O$.
2. Wash in tap water.
3. Place in the staining solution for 1–2 hr.
4. Wash for 2–3 min in tap water.
5. Decolorize partially in the mordanting solution ($Fe_2(SO_4)_3 \cdot (NH_4)_2SO_4 \cdot 24H_2O$). Continue until the gray and white matter are barely distinguishable. (The time varies with amount of staining.)
6. Wash 2–3 min in tap water.
7. Differentiate in 0.4% aqueous $KMnO_4$ until gray and white matter are clearly distinguishable when sections are held up to the light. Sections are colored brown by this solution.

8. Rinse quickly in tap water.
9. Complete the decolorization in equal parts of 1% aqueous oxalic acid and 1% aqueous Na_2SO_3 (or K_2SO_3), mixed *immediately* before using. The gray matter should become completely clear and colorless (except where it contains some myelinated fibers).
10. Wash 2–3 min in tap water.
11. Wash 5 min or more in Solution A to restore blue color lost in decolorizing.
12. Wash thoroughly in tap water.
13. Counterstain if desired.
14. Dehydrate.
15. Clear and mount.

Results: Myelin sheaths—dark blue (not as purplish as with the original Pal-Weigert method). Other structures—unstained unless counterstain has been used.

Weil Stain

Berube, Powers and Clark (1965)
Original from Weil (1928)

Designed to show: Myelin sheaths.

Fixation: 10% formalin.

Embedded in: Celloidin, paraffin or frozen sections. The latter should be mounted on albuminized slides and air dired, preferably in the 45°C oven, and most lipids removed by passing the slides through the xylene-alcohol-water series used for paraffin sections.

Solutions needed:
A. Hematoxylin 1 gm, ethyl alcohol (either 95% or absolute) 100 ml. This solution can be used immediately or kept on the shelf almost indefinitely.
B. Ferric ammonium sulfate ($Fe_2(SO_4)_3 \cdot (NH_4)_2SO_4 \cdot 24H_2O$) 4 gm; distilled water, 100 ml.
C. Potassium ferricyanide ($K_3Fe(CN)_6$) 12.5 gm; Borax ($Na_2B_4O_7 \cdot 10H_2O$) 2.5 gm; distilled water q.s. 1000 ml.

Staining solution: One part Solution A, 1 part Solution B and 2 parts distilled water (see Step 3 below).

Staining schedule:
1. Pass paraffin or air-dried, mounted frozen sections through the usual xylene-alcohol-water series. Celloidin sections (loose) begin with Step 2.
2. Wash in distilled water.
3. Place slides or sections in an empty staining dish. With the total amount

depending on the size of the dish, pour on 2 parts water, 1 part Solution A and 1 part Solution B and immediately mix thoroughly. A precipitate will form but should be disregarded. Staining time 20 min.

4. Wash in water.
5. Differentiate in Solution B until gray and white matter can be distinguished.
6. Wash thoroughly in tap water (several changes) and then in distilled water. Fe^{2+} carried into the final differentiating solution will produce a permanent blue precipitate and ruin the section.
7. Complete differentiation in Solution C.
8. Wash thoroughly in several changes of water. If counterstain is desired, use Darrow red (p. 144).
9. Dehydrate, clear and mount.

Results: Myelin sheaths dark blue to black. Gray matter yellowish to colorless.

NOTE: Iron hematoxylin (see p. 108) is an excellent myelin sheath stain. This procedure has the added advantage that all solutions are stable and all measurements are volumetric.

Luxol Fast Blue MBS

Emended from Klüver and Barrera (1953) method

Application: Myelinated nerve fibers of both central and peripheral nervous systems.

Designed to show: Nerve cells and neuroglia in relation to nerve fiber tracts.

Fixation: Formalin, 10%.

Staining solutions: Luxol fast blue MBS (du Pont), 1 gm; 95% alcohol, 1000 ml. When dissolved, add 10% acetic acid, 5 ml. Filter before using. The solution is stable for a year or more.

Staining schedules:
A. For frozen sections.
1. Soak sections in 70% alcohol 10–15 min.
2. Stain 16–24 hr in Solution A at 40°C. Use the staining solution only once.
3. Rinse off the excess stain with 95% alcohol and wash well in water.
4. Differentiate by dipping the sections singly into 0.05% aqueous Li_2CO_3 solution for a few seconds, then washing through several changes of 70% alcohol, and placing in distilled water. Critical differentiation occurs in the alcohol, and care must be taken not to overdifferentiate. Repeat the process if necessary, observing

under a microscope, until the white matter shows greenish-blue against colorless gray matter.

5. Wash thoroughly with distilled water.
6. (Optional) Counterstain with Darrow red (p. 144).
7. Dehydrate, clear and mount.

An optional procedure which facilitates handling is to mount the sections on slides immediately after cutting. Slides must be clean and a minimum amount of albumen used. The sections are floated with water and dried on a warming plate. When dry, slides are placed in 95% alcohol for 24 hr and staining begun with Step 2 (above).

B. For paraffin sections. This differs from the procedure for frozen sections as follows:

1. Remove paraffin with xylene and pass slides through absolute alcohol and several changes of 95% alcohol.
2. Stain at 57°C instead of 40.
3-7. Same as above.

C. For celloidin sections. Sections are collected in 75% alcohol and transferred directly to the first staining solution. The rest of the process is the same as for paraffin sections except that differentiation of the luxol fast blue may require a stronger lithium carbonate solution (up to 0.5%) for differentation. After staining, the sections can either be mounted or counterstained (see p. 144). A variation is to mount the sections and remove the celloidin before staining.

Results: Fibers—blue.

NOTE: This method will yield excellent preparations but may require some practice to master. Chromoxane cyanine R (see p. 109) stains myelin with a color similar to Luxol fast blue MBS but the staining procedure is much simpler. Actuallly Chromoxane cyanine R will probably supercede Luxol fast blue MBS as a stain for myelin.

SECTION D: METHODS FOR NERVE FIBERS AND NERVE ENDINGS

Most of the methods of this group employ a soluble salt of silver (especially $AgNO_3$) as the means of selectively impregnating nerve fibers. Tissue is allowed to remain in the impregnating solution a suitable length of time and the silver is then reduced by pyrogallol, hydroquinone or other reducing agent. In some methods, the reduced silver in the tissue is replaced by gold to enhance the differentiation of the neural elements (gold toning) but in others, toning may be undesirable. In the Golgi methods, impregnation and reduction occur concommitantly, and no toning is required.

The methods given are representatives of the principal types of technics that include staining in the block and the staining of sections. Most of them have numerous modifications. Theoretical and experimental aspects of staining can be found in articles by Davenport and co-workers (1929, 1930, 1934, 1939), Zon (1936), Bodian (1936, 1937), Silver (1942), Holmes (1943), Foley

(1936, 1943), Nauta and Gygax (1951, 1954), Peters (1955a), Windle (1957), Nauta and Ebbesson (1970) and Robertson (1978).

The Cajal and the Bielschowsky Methods

Both of these methods were introduced by their authors between 1903 and 1904 and have furnished the basic procedures for many subsequent modifications. They are sometimes called reduced silver methods because a chemical reducing agent is applied to the tissue after it has been treated with a silver salt (usually $AgNO_3$). The difference between the Cajal type of staining and the Bielschowsky is in the way that silvering and the subsequent reduction is accomplished. A single silver impregnation with reduction by pyrogallol, hydroquinone or one of the amino phenols characterizes Cajal-type methods, whereas a double impregnation ($AgNO_3$ followed by ammino-silver solution) with reduction by formalin characterizes those of Bielschowsky. Both types of procedure are represented in methods for blocks and methods for sections.

Pyridine-Silver Method

From Ranson (1911), and Davenport, Windle, and Beech (1934)
Modification of original Cajal's formula 3 (Cajal, 1910)

Application: Nervous tissue in general.

Designed to show: Axis cylinders of neurons, especially the smallest fibers; neurofibrils of nerve cells.

Fixation: Fix pieces of spinal cord, brain, ganglia, peripheral nerves or embryos for 2–6 days in the following solution: NH_4OH (conc.), 2 ml; absolute alcohol, 98 ml.

NOTE: To stain small nerves of 2 mm or less, draw them into a piece of spinal cord by means of a needle and thread. After fixation, pare away the excess spinal cord until only about 1 mm thickness remains covering the enclosed nerve. The ends of the nerve should be covered completely by cord. Imperfect staining usually results if this is not done.

Embedded in: Paraffin after staining in bulk.

Staining schedule:
1. After fixation, treat tissues for about 24 hr in 5% pyridine (C.P. or reagent grade). This step is not critical; it can be omitted.
2. Wash embryos 20–60 min in distilled water; adult tissues 2–6 hr.
3. Impregnate for 2 days to 1 wk, depending on size of block, at 37°C in 2% aqueous $AgNO_3$.
4. Wash for 1–15 min in distilled water. *This is a critical step;* long washing results in a light stain; experience will determine the correct time. Some washing is necessary to prevent the periphery from becoming too dark.

5. Reduce for 4–12 hr in a 4% solution of pyrogallol in water (or 1:20 formalin).
6. Dehydrate, embed and section.

Caution: Do not use cork stoppers. Glass stoppered jars must be chemically clean. Use volumes equal to at least 25 times the volume of the tissue.

Results: Axis cylinders—shades of yellow, brown and black, depending on diameter (unmyelinated fibers usually deep black in contrast to brown or yellow axis cylinders of myelinated fibers). Nerve cells—brown or yellow, with black to brown neurofibrils. Connective tissue—yellow.

Chloral Hydrate Method

See p. (156) for the Nonidez method. This is applicable to pieces of brain or spinal cord as well as peripheral nerves and has given splendid results.

Bielschowsky Method

From Davenport, Windle, and Beech (1934), and Beech and Davenport (1933)
Original by Bielschowsky (1904, 1909)

Application: Nervous tissue in general.

Designed to show: Nerve cells and processes; neurofibrils.

Fixation: Formalin 1:10 excepting embryos. Embryos, formalin 1:10 plus 0.25 to 0.5% trichloracetic acid.

Embedded in: Paraffin after staining in bulk.

Preparation of the staining solutions:
A. Dissolve 1.5 gm $AgNO_3$ in 100 ml of water.
B. Mix 5 ml of NH_4OH (conc.) and 40 ml of 2% NaOH. Add from a burette or graduated pipette, while shaking the mixture, enough 8.5% $AgNO_3$ to give a slight permanent opalescence to the solution. Add 3–5 drops of NH_4OH (conc.). The amount of $AgNO_3$ solution used to reach the end point should be 35–45 ml.

Staining schedule:
1. Wash tissue in water for about an hour after fixation.
2. Place in pure pyridine (C.P. or reagent), or for embryos a half and half mixture of pyridine and distilled water, for 1 or 2 days.
3. Wash 2–6 hr in water according to the size of the block.
4. Place in Solution A for about 3 days at 37°C.
5. Wash in water, 20 min for small embryos up to 1 hr for larger specimens.
6. Impregnate 6–24 hr in Solution B.

7. Wash 15 min to several hours in water. (This washing tends to lighten the subsequent stain.)
8. Reduce 1–12 hr in 1:100 formalin.
9. Wash in tap water.
10. Dehydrate.
11. Embed, section and mount on slides.
12. Remove paraffin with xylene and cover.

or

12a. Pass slides through graded alcohols to water and tone with gold, as given in Steps 9–10 of Penfield's combined method (see p. 136) or, if the stain is too light, follow Steps 8–10 of Bodian's protargol method (p. 152). Dehydrate and cover in the usual manner.

Results: (1) Untoned: Nerve fibers and neurofibrils—brown to black, background—yellow to brown. (2) After gold toning: Fibers—gray to black, background—pinkish gray to purplish.

Staining Nerve Fibers in Celloidin Sections

From Davenport (1929), with later emendations by Davenport

Application: This method can be used on sections of tissue embedded in celloidin or other nitrocellulose without removing the embedding material. It is reasonably reliable for myelinated nerve fibers but generally fails to stain the fine, nonmyelinated ones. It may be useful when alternate sections are required to be stained with a dye.

Fixation: 10% formalin or other formalin-containing fixative, without heavy metals.

Embedding medium: Nitrocellulose (any brand).

Staining solutions:
A. An acidified solution of approximately 10% $AgNO_3$ in 85–90% alcohol; made by dissolving 10 gm of the nitrate in 10 ml of water, mixing this solution with 90 ml of either 95% or absolute alcohol, and adding 0.5 ml of 1 N HNO_3.
B. A 2–5% solution of pyrogallol in 90–95% alcohol with 2–3 ml of formalin added to delay air oxidation. Dissolve 5 gm of pyrogallol (pyrogallic acid) in 95 ml of 95% alcohol and add 2–3 ml of formalin.
   If reduction occurs too rapidly, dilute with 1, 2 or more volumes of 95% alcohol to control the staining intensity.

Staining schedule:
1. Impregnate sections singly or well spread out in a shallow dish in Solution A at 37–40°C for 4–24 hr. Transfer them directly from 80 or 95% alcohol to this solution. Keep in darkness or subdued light. Sections become light brown when staining is adequate.

2. Rinse through 2 changes of 95% alcohol of 3–5 sec each.
3. Reduce one to several minutes in Solution B. The reduction can be controlled by observation under a microscope. Deposition of precipitate can be minimized by keeping the sections in motion or by adding to 20–25 ml of Solution B a few drops of 10–20% aqueous dextrin solution (Karo corn syrup diluted 1:3 with water). Change the reducing solution frequently.
4. Wash well in several changes of 95% alcohol.
5. Soak 3–5 min each in 2 changes of *n*-butyl alcohol, clear in xylene and mount in balsam or synthetic resin.

Results: Nerve fibers are light brown to black against a yellow background.

Comments: Connective tissue and osseous structures often stain strongly, as in other silver methods. The acid in the silver bath suppresses connective tissue to some degree and serves no other useful purpose, hence it can be omitted if the staining is found to be well differentiated without it.

Different lots of tissue respond differently, and best results are secured by varying the time of impregnation (Step 1). Generally, maximum differentiation occurs when the minimum time consistent with staining nerve fibers is used. If the stain is too pale, omit one of the rinses of Step 2 and shorten the time. If it is too dark, eliminate more of the silver solution from the section by additional washing.

Gold toning is optional but may effect improvement. The regular gold-toning procedure follows Step 4, but the sections should be soaked 2–5 min in water before placing them in the gold bath.

The method shows peripheral nerves and tracts in the central nervous system, but is unsuited for endings in sense organs, intestinal plexuses or other structures where the fibers are very fine and lacking myelin sheaths.

Bodian's Protargol Method

From Bodian (1936, 1937)
Emended by Clark (1971)

Application: Nervous tissue in general.

Designed to show: Nerve fibers in brain, spinal cord, peripheral nerves and ganglia.

Fixation: Various (see Bodian, 1937) but especially the following: Formalin, 5 ml; acetic acid, glacial, 5 ml; 80% alcohol, 90 ml.

Embedded in: Paraffin.

Preparation of staining and reducing solutions:
  A. Staining solution: 1% Protargol. After weighing the Protargol, dust it from the weighing paper over the surface of the water. *Do not shake*

the solution but allow the Protargol to dissolve from the surface downward, otherwise solution is greatly delayed. Metallic copper is added to this solution just before the slides are put into it. Bright, clean granular copper such as used in chemical laboratories seems to be preferred although wire can be used if the other is not available. Reaction of the copper with the Protargol during the staining process determines the excellence of this stain, hence it is desirable to have enough metallic surface exposed to convert much, if not most of the Protargol to a copper derivative by the end of the staining period, and to do it gradually and progressively. Placing the copper in the bottom of a Coplin jar and allowing the slides to rest on top of it is the usual procedure. About 5 gm of the metal is used. The copper must be perfectly clean. To 7 ml H_2O add 3 ml concentrated HNO_3. Immerse the copper in this until the surface film of oxides is removed. When this occurs the copper is bright in color and all dark areas of oxide removed.

B. Reducing solution: Hydroquinone, 1 gm; Na_2SO_3 (anhyd.), 5 gm; water, 100 ml. This must be made up and immediately used.

Staining schedule:

1. Remove paraffin and run sections down to water.
2. Place in staining Solution A after adding the copper and place in the 60°C oven. The staining time will vary with the particular batch of Protargol-S. Staining is complete when the sections have a golden brown color and the staining time may vary from 4 to 16 hr. If the staining time is too short, only the larger fibers may be faintly stained. However, overstaining produces excessive background which obliterates all fine fibers.
3. Wash very briefly in distilled water. It is only necessary to remove the stain adhering to the surface of the slide and the section. For this only a dip, in and out, is needed. Too long a wash will remove the stain, leaving only the large fibers impregnated.
4. Reduce in Solution B. Original method says 10 min. Can be shortened to 2 or 3 without noticeable change in the stain.
5. Wash thoroughly in water, preferably running. All of the reducing solution should be removed in this washing to prevent its being carried into the gold bath and spoiling the gold solution.
6. Tone 5–10 min in a 1% solution of gold chloride to which 3 drops of glacial acetic acid per 100 ml have been added.
7. Rinse in water.
8. Place 2–5 min in 1–2% oxalic acid solution until sections have a faint purple or blue color. Rinse in water.
9. Place in 5% $Na_2S_2O_3 \cdot 5H_2O$ for 5–10 min.
10. Wash thoroughly in tap water.

11. Dehydrate.

12. Clear and cover.

Results: Axis cylinders of nerve fibers and neurofibrils of cells—black or purplish black.

Variations: For staining of unmounted sections and counterstaining see Foley (1943). For use with insect material: Rogoff (1946).

NOTE: The original method and its modifications were based on the use of German-made Protargol (Winthrop), which is no longer available. Protargol-S, so named to distinguish it from the pharmaceutical product, is certified by the Biological Stain Commission and may be purchased from several sources.

The Bodian stain for nerve fibers is by far the most commonly used. In the second edition of this manual the following methods were also given: Davenport, McArthur and Bruesch (1939), 2-hr Protargol method; Brown and Vogelaar (1956), amino silver method; and Peters (1958), protein-silver method.

When pieces of freshly removed peripheral nerves are fixed in solutions of osmium tetroxide (OsO_4), the myelin of the fiber sheaths becomes blackened. Immersion of the tissues in dilute aqueous solutions (0.5–2.0%) at 20–30°C for 6–24 hr is the usual procedure. Fixation before staining is feasible provided that the fixing fluid is all aqueous, i.e. contains no alcohol or other lipid solvent. Fixation in a mixture consisting of formalin (conc.), 15 ml; formic acid (about 90%), 5 ml; and water, 80 ml; for 24–48 hr gives more rounded and less distorted sheaths than most other fixatives. This is followed by thorough washing in water and staining in a rather dilute (0.1–0.05%) osmic acid solution. In general, better staining of fixed material is obtained with more dilute solutions of OsO_4 than those used for fixation and concommitant staining of fresh material.

Fixation by means of OsO_4 vapor, without immersion of the specimen, can be used on all nerves but is particularly well suited to the optic and olfactory nerves. The following method was used with good success on optic nerves.

Osmium Tetroxide Vapor Method

From Bruesch (1942)
Original osmic acid method for myelin: Schultze and Rudneff (1865)

Designed to show: Myelinated nerve fibers in peripheral nerves.

Embedded in: Paraffin (after staining).

Fixation: Fresh tissue used, which is fixed and stained simultaneously.

Procedure:

1. Suspend pieces of fresh nerves near but not touching the surface of a

2% aq. OsO$_4$ solution. This can be done conveniently by tying the nerve with silk thread between the free ends of a U-shaped frame made of 3 or 4 mm glass rod. The width of the U-frame should be such that it fits inside of a tall Stender dish and stands upright with its ends above the staining fluid. The cover of the dish should be tight. Tie the nerve in the frame, set it in the dish and pour in the osmic acid without its touching the nerve until the proper level of 3–5 mm below the nerve is reached. Allow to stain 12–24 hr according to the size of the nerve.

2. Wash in distilled water for 4–6 hr (several changes).
3. Dehydrate.
4. Embed in paraffin.

Results: Myelin sheaths—black.

The staining of endings of peripheral nerve fibers is rather difficult and results are often disappointing. Two of the most frequently used procedures are those of Ranvier and of Ehrlich. Some of the silver-reduction technics have been used successfully and two will be described. Other methods for staining in bulk, like Ranson's pyridine-silver or staining of sections by the Bielschowsky-Gros technic or modification thereof, can be used.

Gold Chloride Method for Nerve Endings

Löwit (1875) quoted from Fischer (1875)
Variations by Ranvier (1880) Garven (1925) and Combs (1936)

Designed to show: Peripheral nerve fibers and their endings.

Fixation: None.

Staining schedule:
1. Pieces of fresh tissue limited to a 2–3 mm thickness are placed in a strong solution of formic acid (conc. formic acid of 90% strength or over, diluted with 1–3 vol of distilled water). Recent usage indicates that 25% formic acid in the final concentration is adequate. Allow to remain in this fluid 10–15 min. Or, use undiluted lemon juice (Ranvier; Combs) and allow to soak 10–30 min.
2. Remove all excess acid by rinsing quickly in distilled water or (Garven) simply blot thoroughly with a towel.
3. Place for 20–30 min in 1% aq. gold chloride solution.
4. Rinse or blot and place in 25% aqueous formic acid for 24 hr. Combs recommends 10% formic acid, 6–10 hr for rabbit intercostal muscle. (This step is allowed to proceed in the dark.)
5. The older procedures called for rinsing or blotting and teasing the tissue (particularly muscle) in glycerol. A binocular dissecting microscope should be used to isolate fibers bearing motor end plates. Whole

mounts of the portions desired are then made in glycerol or glycerol jelly.

5a. In Combs procedure the tissue is washed after staining and treated 15–20 min in 5% $Na_2S_2O_3 \cdot 5H_2O$, washed thoroughly in running water, and then placed in artificial gastric juice (powdered pepsin, 1 part; HCl, 3 parts; water, 1000 parts) at 37°C for 1–2 hr. Wash by means of several changes of water (not running, tissue is too soft). Drain on clean cloth, select parts for mounting, transfer to a slide and add one or two drops of acacia solution which has about the same consistency as glycerol (rather concentrated). Press the cover glass down upon the tissue to flatten and bring the nerves into view.

5b. (may be used instead of 5 or 5a). Wash, embed, section and mount in the usual manner.

NOTES: Ranvier's original method required placing the fresh tissue directly into a mixture of gold chloride and formic acid and to stain for 20 min to several hours. Recent usages favor a prestaining treatment of the tissue with formic acid or lemon juice. Tartaric, oxalic, pyrogallic, and chromic acids have been used to reduce the gold after staining. Combs warns against over-treatment with $Na_2S_2O_3$ lest the stain be made too light by its action.

Results: Nerve fibers and endings—black, other tissues—purple to red.

Chloral Hydrate Silver Method

From Nonidez (1939)
Similar to the methods of Perez (1931)

Designed to show: Peripheral endings of nerves; nerve fibers of the autonomic system; also neurofibrils and axis cylinders generally.

Fixation: Immerse the tissue 1–3 days in a solution of 25 gm chloral hydrate in 100 ml of 50% alcohol.

Preparation of the reducing solution: Pyrogallol, 2.5–3.0 gm; formalin, 8 ml; water, 100 ml.

Staining schedule:

1. After fixation blot off the excess fixing fluid and place in ammoniated alcohol (95% alcohol, 100 ml; NH_4OH (conc.), 7 drops) for 24 hr. Change at least once if the specimens are large or have much fat on them.
2. Rinse 5 min in distilled water.
3. Place for 5–6 days in 2% aqueous $AgNO_3$ in the dark at 37–40°C. Change the solution after 2 days or when it becomes yellowish brown.
4. Rinse 2–3 min in distilled water.
5. Reduce 24 hr in the reducing solution.
6. Wash for 2–3 hr in several changes of distilled water followed by 50% alcohol for 24 hr changed at least twice.

7. Dehydrate, clear in amyl acetate and embed in paraffin.
8. Section, mount and cover. Counterstaining is usually unnecessary and gold toning does not improve the impregnation.

Results: Nerve endings and fibers—shades of brown to black. Postganglionic sympathetic fibers stain lighter than other autonomics.

SECTION E: METHODS FOR NEUROFIBRILS

NOTE: Nearly all the procedures used for staining axis cylinders of nerve fibers, with the exception of Golgi methods, are suitable for neurofibrils in nerve cells.

SECTION F: METHODS FOR DEGENERATION

Normal myelin has less affinity for fat stains (including OsO_4) than its degeneration products. The Marchi method and its modifications employ means of restraining the staining of normal myelin while allowing the degenerating myelin to stain. For a long time it was believed that mordanting with $K_2Cr_2O_7$ as a preliminary step to staining restrained the normal myelin. Recent evidence indicates that this is not the case but that the method of fixation and composition of the staining fluid itself are the important factors. Trauma of any sort should be avoided in the preparation of tissue for this type of stain, since bruising either before or after fixation generally gives artefacts. The poor penetrating ability of osmic acid makes it necessary to slice brains and spinal cord into 2–4-mm thick pieces. This should be done with a razor or very sharp brain knife with minimum bending of the slices.

The Marchi Method

From Swank and Davenport (1935)
Original by Marchi (1886)

Designed to show: Degenerating myelinated nerve fibers in the central nervous system.

Preparation of solutions:

A. Perfusion fluid: $MgSO_4 \cdot 7H_2O$, 70 gm; $K_2Cr_2O_7$, 25 gm; water, 1000 ml.
B. Staining solution: 1% aqueous $KClO_3$, 60 ml; 1% OsO_4, 20 ml; glacial acetic acid, 1 ml; formalin, 12 ml.

Prestaining schedule:

1. Lesions which will cause degeneration are produced in the living animal by cutting, crushing or electrocoagulation and time allowed for degenerative changes to reach a favorable degree for staining. Human pathological material is suitable if the time between the cause of degeneration and time of securing the material is between 10 days and

about 1 month.* Suitable times for degeneration after injury are 14–20 days for cat and monkey and 10 days for rabbit.

2. Experimental animals should be anesthetized with a barbiturate rather than chloroform or ether. Start the perfusion with Solution A before the heart stops for best results, although it may be done after death. Perfusion through the left ventricle with venae cavae opened allows a free flow and the avoidance of excessive pressure. One meter of hydrostatic pressure should be about the maximum. Physiological saline solution should not be used as a preliminary perfusing fluid, and the use of Müller's fluid is not recommended.

3. Remove the part to be stained without bending or bruising by cutting away all bony and membranous coverings, severing peripheral nerves and other attachments before lifting out.

4. Place the tissue in 1:10 formalin (stock formalin should be neutralized with $CaCo_3$) for at least 48 hr and not over 1 wk.

5. After fixation, cut the material to be stained into slices 2–4 mm thick.

Staining schedule:

1. Place the slices of tissues without washing into about 15 times their volume of staining Solution B.† Exposure of both surfaces of the slices to the staining fluid is very desirable and can be accomplished by suspending them with silk thread or by means of a small glass rod inserted through one edge. If they are not suspended, the use of a small pledget of clean glass wool on the bottom of the staining vessel and turning the slice over every day (or daily turning without the glass wool) will probably give the desired results. Allow 7–10 days for staining. Agitate the fluid daily or oftener.

2. Wash 12–24 hr in running tap water.

3. Dehydrate and embed in celloidin or low viscosity nitrocellulose.

4. Cut sections.

5. Counterstain the nerve cells, if desired, with thionin, safranin, cresyl violet or other basic dye. (See p. 142 for counterstaining.)

6. Dehydrate to 95% ethyl alcohol and use *n*-butyl alcohol instead of absolute ethyl preceding xylene to avoid excessive softening when low viscosity nitrocellulose is used for embedding.

7. Mount in balsam or a synthetic resin.

NOTE: If there is excessive fading of the counterstain in the alcohols during dehydration (Step 6), alkalinize the alcohol by means of a drop of ammonia water added to each jar, or add $NaHCO_3$ (powdered) 0.5 gm per 100 ml, shake well and allow to settle before use. Since the counterstain is a basic dye, any excess acidity in the dehydration system will tend to remove it from the tissue.

---

* *For findings on human material, see Smith (1956, a and b) whose work (with associates) is cited in the last paragraph of "Additional information about this method" below (p. 159).*

† *Diluted with 2–3 vol of distilled water. See Poirier et al. (1954) below.*

Perfusion before fixation is not mandatory for success. Tissues may be removed soon after death and placed directly in neutralized ($CaCo_3$) formalin 1:10. Quality and completeness of staining may be expected to be better after perfusion, however.

Results: Degenerating myelin and fat—black. Other tissues—pale yellow or greenish unless counterstained. Nerve cells—characteristically colored according to the counterstain used.

Additional information about this method: Mara and Yoss (1952) tested the time required for staining by removing 1-mm slices of monkey spinal cord from the staining solution at regular intervals between 1 and 10 days. They found that the 24-hr period was adequate for complete staining of degenerating fibers. In other tests on slices of greater thickness, 48 hr sufficed for such slices up to 24 mm. The time of degeneration was 2 wk.

Poirier, Ayotte and Gauthier (1954) investigated the effects of varying the concentrations of the ingredients of the chlorate-osmic-formalin mixture and found that, although the proportions were about optimum, the concentration of the original fluid was several times greater than necessary. Their revised formula is as follows: 0.5% OsO_4, 11 ml; 1% $KClO_3$, 16 ml; 10% acetic acid, 3 ml; formalin (conc.), 3 ml; and water to make 100 ml. They recommend slicing the tissue to not over 3 mm, using a volume of staining solution equal to at least 20 times that of the tissue, and a minimum staining time of 5 days. Considerable economy in the use of osmic acid could be effected without loss of staining efficiency.

The general belief prevails that survival time after the inception of a lesion should not exceed 2–3 months if successful Marchi stains are to be obtained. However, recent studies by Smith, Strich, Sabrina and Sharp (1956) show that the stain may succeed in human brain and spinal cord even after the extend of the time of degeneration has been 15 months. They found also that long-time storage in formalin-saline (15–100 months) was compatible with successful staining. These observations should encourage studies on human autopsy material that may have been stored and neglected because attempts to stain it for degenerating nerve tracts were considered to be futile.

For a discussion of the histochemical aspects of the Marchi reaction, see C. W. M. Adams (1958).

Lillie's Combined Myelin and Fat Stain

Lillie (1944a)

Designed to show: Intact and degenerating myelin and fat.

Fixation: Formalin, 1:10; or Orth's, Regaud's or other aqueous fixing fluid.

Preparation of solutions:

A. Staining solution: 1% aqueous or alc. hematoxylin 1 to 5 days old, 50 ml; 4% aqueous iron alum, $Fe_2(SO_4)_3(NH_4)_2SO_4 \cdot 24H_2O$, 50 ml. Prepare immediately before use.

B. Staining solution: Sudan II, saturated solution in 93% isopropanol, 30 ml; water, 20 ml. Mix, let stand 10 min, filter and use within 3 or 4 hr.

C. Differentiating solution: iron alum, 0.5 gm; water, 100 ml.

D. Differentiating solution: borax, $Na_2B_4O_7 \cdot 10H_2O$, 1 gm; $K_3Fe(CN)_6$, 2.5 gm; water, 100 ml.

Staining schedule:

1. Cut frozen sections and stain 40 min in Solution A in a covered dish at 56°C (paraffin oven).
2. Rinse in water and differentiate 1 hr in Solution C.
3. Rinse and treat 10 min in Solution D.
4. Stain 10–20 min in Solution B. Float out in water.
5. Mount in Apathy's gum syrup. The gum syrup is made as follows: gum acacia (arabic), 50 gm; cane sugar, 50 gm; water, 100 ml. Place in paraffin oven to facilitate solution. Add 1.5 ml of Merthiolate as a preservative.

Results: Normal myelin—blue-black, nerve cells—gray, nuclei—deeper gray, red corpuscles—yellow to black, fats—orange-yellow.

Polarized Light Method

Setterfield and Baird (1936)

Designed to show: Degenerating myelin in the earliest stages of degeneration.

Fixation: Formalin, 1:10 (aq.).

Procedure:

1. Cut frozen sections at 15–20 μm.
2. Float from water onto slides and mount in a drop of glycerol.
3. Observe with a 4-mm objective in polarized light and between crossed Nicol prisms.

Results: Since normal myelin is birefringent, it will appear alternately light and dark four times in the rotation of the stage of the microscope. Any isotropic material, which includes degeneration products, is not birefringent and will appear dark at all points in the rotation.

NOTE: The preparations cannot be made permanent and photographs are suggested for records.

Nauta and Gygax Method for Frozen Sections

From Nauta and Gygax (1954)
See also Robertson (1978)

This method was designed for use in the study of degeneration in the central nervous system. The novice should bear in mind that there is no known silver stain that is specific for degenerating axons in the same sense that the Marchi stain is specific for the myelin sheaths of degenerating axons. However, both degenerating and regenerating axons have a tendency to show a greater affinity for silver than normal ones, and this is the basis for devising a method to exaggerate their staining differences.

In skilled hands, the Nauta-Gygax method does this quite well for degenerating fibers, and often with startling clarity, but its authors caution against relaying solely on differences of staining as follows: "Although the method is not strictly selective for degenerating axons, a considerable proportion of the normal nerve fibers will remain unstained . . . lack of complete tinctorial selectivity necessitates a thorough familiarity of the observer with the histological characteristics of axon degeneration."

For more information on this type of staining (silver nitrate followed by an ammino-silver solution), of which the Bielschowsky-Gros technic is the prototype, the reader is referred to Garven and Gairns (1952), to Nauta and Ebbeson (1970) and to Robertson (1978).

Application: Central nervous system of mammals.

Fixation: 10% neutral formalin 2 weeks to 6 months.

Embedded in: None, or in 25% gelatin subsequently hardened with formalin.

Staining schedule:

1. Cut frozen sections 15–20 μm thick and soak in 15% alcohol, 0.5 hr.
2. Wash briefly in water and soak in 0.5% phosphomolybdic acid, 0.25–1.0 hr.
3. Without washing, transfer to 0.05% $KMnO_4$, 4–10 min. Turn sections over during this treatment to insure a uniform action, and adjust the time to give the best degree of differentiation between normal and degenerating fibers.
4. Decolorize the sections in an equal-parts mixture of 1% hydroquinone and 1% oxalic acid. This requires 1–2 min.
5. Wash through 3 changes of distilled water and place in 1.5% $AgNO_3$ for 20–30 min. Handle each section individually after this step.
6. Wash briefly in distilled water and transfer to the ammoniated silver solution for 1 min. Formula: $AgNO_3$, 0.9 gm; water, 20 ml. When solution is complete, add pure ethanol, 10 ml; NH_4OH (conc.), 1.8 ml; and 2.5% NaOH, 1.5 ml.
7. Transfer to a reducing fluid consisting of: water, 400 ml; pure ethanol, 45 ml; formalin, 10% not neutralized, 13.5 ml; and 1% citric acid, 13.5 ml. Allow the sections to float evenly on the surface of the fluid. Reduction, indicated by browning, occurs in about 1 min. If the staining tends to be too light, add some NaOH to the ammoniated silver solution; if too dark, add ammonia.

8. Wash, pass sections through 1% $Na_2S_2O_3 \cdot 5H_2O$, and wash thoroughly.
9. Mount on slides, before dehydration, by Albrecht's (1954) method as follows: Soak 5 min or longer in an equal volume mixture of 80% alcohol and 1.5% gelatin. Draw the section onto the slide from this fluid, drain well and wipe around the section. When the edges of the section have just begun to dry, blot gently with a strip of smooth filter paper by bending it down against the section. Immerse the nearly dry section immediately in 95% alcohol. Complete the dehydration, clear, and cover in synthetic resin.

Fink-Heimer Method for Frozen Sections

Original Fink and Heimer (1967)
From Heimer (1970)
See also Giolli and Karamanlidis (1978)

This method is derived from that of Nauta-Gygax and gives better impregnation of degenerating terminals.

Application: Central nervous system of mammals but has also been used on lower vertebrates (see Nauta and Ebbeson, 1970).

Fixation: Perfusion with saline followed by formalin-saline. After careful removal, store blocks in 10% formalin for at least 1–2 wk. Before sectioning, soak blocks in 30% sucrose for 2–3 days.

Embedded in: None or in 25% gelatin subsequently hardened with formalin. Frozen sections are cut at 15–30 μm.

Solutions needed:
  A. Potassium permanganate ($KMnO_4$), 0.1 gm; H_2O $q.s.$, 100 ml
  B. Decolorizing solution: Mix equal volumes of 1% hydroquinone and 1% oxalic acid. This solution must be made up just before use for it is quite unstable.
  C. Uranyl nitrate: $UO_2(NO_3)_2 \cdot 6 H_2O$, 0.5 gm; H_2O q.s., 100 ml.
  D. Silver nitrate: $AgNO_3$, 2.5 gm; H_2O q.s., 100 ml.
  E. Sodium Hydroxide: NaOH, 2.5 gm; H_2O q.s., 100 ml.
  F. Ammonical silver nitrate solution:

Solution D	20 ml
Concentrated ammonia (28%)	1 ml
Solution E	1.8 ml

  G. Nauta-Gygax reducer:

10% formalin	27 ml
1% citric acid	27 ml
95% alcohol	90 ml
H_2O	910 ml

  H. Fixing solution: Sodium thiosulfate ($Na_2S_2O_3 \cdot 5 H_2O$) 0.5 gm; H_2O q.s., 100 ml.

Staining schedule:
1. Rinse sections in H_2O.
2. Solution A, 5–10 min.
3. Rinse in H_2O.
4. Decolorize in Solution B.
5. Rinse thoroughly.
6. Solution C, 10 ml; Solution D, 10 ml; H_2O, 30 ml; 30–60 min.
7. Solution C, 15 ml; Solution D, 35 ml; 30–40 min.
8. Rinse thoroughly in H_2O.
9. Solution F, 1–5 min.
10. Directly to reducer (Solution G) change after 30 sec, total of 1–2 min.
11. Solution H 1 min.
12. Rinse, mount from 70% alcohol, dehydrate, clear and coverslip.

Results: Terminal degeneration as well as degenerating fibers black, background yellow.

Glees' Method for Normal and Degenerating Pericellular Endings

Glees (1946)

The original Glees' method is given first and is followed by a modification by Nauta and Gygax.

Application: Central nervous system; degenerating endings.

Designed to show: Boutons terminaux and free nerve endings about cell bodies.

Fixation: Formalin (1:10) or formalin, 1 part in physiological saline solution, 19 parts.

Embedded in: (a) not Embedded; (b) celloidin.

Preparation of solutions: Mix 20% $AgNO_3$, 3 parts and 95% alcohol, 2 parts. To this mixture add conc. NH_4OH slowly while agitating until the precipitate has just redissolved, then add another 5 drops of the ammonia.

Staining schedules:
(a) For frozen sections.
1. Cut sections 15–20 μm.
2. Place sections in a 0.5% solution of NH_4OH (conc.) in 50% alcohol for 12 hr at 30°C.
3. Wash in water.
4. Place in 10% $AgNO_3$ 12 hr at room temperature.
5. Transfer the sections rapidly through separate dishes containing 10% formalin made up with tap water. Three or 4 changes are needed and the last one should remain free of turbidity.
6. Place in Solution A for 30 sec.

7. Reduce in 10% formalin 30–60 sec.
8. Wash in water.
9. Place in 10% $Na_2S_2O_3 \cdot 5H_2O$ for 10 sec.
10. Wash well in distilled water, dehydrate through alcohols followed by creosote.
11. Mount sections on slides, blot away the creosote and cover. A glass rod must be used for transferring the sections. The type of mounting medium is not stated, hence may be assumed to be uncritical.

(b) For celloidin section. (Gatenby and Beams (1950) state that this may be used for paraffin sections also.)

1. Cut 20-μm sections.
2. Place in distilled water for 0.5 hr.
3. Transfer to an equal parts mixture of 10% $AgNO_3$ and 95% alcohol for 12 hr.

All subsequent steps are the same as the ones for the preceding method excepting that an additional one follows Step 9.

9a. After washing, transfer to 50% alcohol, then to $Na_2S_2O_3 \cdot 5H_2O$ (Step 10).

Results: Nerve fibers and their terminals—black, cells—palely stained.

Variations: Reducing the amount of ammonia that is added in excess of the amount needed to dissolve the precipitate in Solution A intensifies the stain. If the stain is too pale after reducing (Step 7), sections may be restained for a few seconds (Step 6) and reduced a second time.

The modification of the following steps has been proposed by Nauta and Gygax (1951) to avoid the uncertainty of the use of tap water in the formalin solution used in Step 5. (*The steps are numbered as in Glees' method.*)

2. Ammonia concentration 1% instead of 0.5%.
4. $AgNO_3$, 1.5 gm; water, 100 ml; pure pyridine, 5 ml.
5. Omitted. Transfer sections directly from the first to the second silver bath.
6. To 30 ml of a 1.5% $AgNO_3$ solution in 32% alcohol add 2 ml of NH_4OH (conc.) and 2 ml of a 2.5% aqueous solution of NaOH. Allow sections to remain in this solution 2–5 min.
7. Reduce sections until golden brown in 1% formalin in 10% alcohol acidified by the addition of 3 ml of 1% citric acid per 100 ml of the formalin mixture.
8. Omitted.
9. A 2.5% solution of $Na_2S_2O_5 \cdot 5H_2O$ is used instead of 10%.

NOTE: Steps 1, 3, 10 and 11 are the same as in the original method.

Results: Similar to those of Glees' method.

SECTION G: GOLGI METHODS

Golgi stains are made in the block, for there is no section staining counterpart of it as in the Cajal and Bielschowsky methods. All Golgi technics use a dichromate, or in the more recent modifications, a mixture of chromate and dichromate in chemical equilibrium. Either silver (as $AgNO_3$) or mercury (as $HgCl_2$) is the impregnating metal. Methods that use mercury are customarily called Golgi-Cox (or Cox-Golgi).

Differences in opinion are to be found in the literature concerning the use of cover glasses on mounted sections. Fading occurred frequently when Canada balsam was used as a mounting medium. However, with the modern synthetic resins, the component responsible for fading appears to be absent, and covers may be applied when terpene polymers (HSR, Bioloid, Permount) or cumarone resins (Technicon mounting medium) are used.

See Scheibel and Scheibel (1978).

Rapid Golgi Method

From Polyak (1942)
Original by Golgi (1875)

Application: Brain and spinal cord.

Designed to show: Individual nerve cells with their dendrons and axons.

Fixation: For small pieces of nervous tissue: 2–3% $K_2Cr_2O_7$, 4 vol; 1% OsO_4, 1 vol.
1. For larger specimens: 5% $K_2Cr_2O_7$, 2 vol; 1% OsO_4, 1 vol.
2. Prepare the solutions just before using.
3. Fix tissues immediately after killing the animal; preferably perfuse the animal with the fixing fluid. Remove tissues and cut them into slices 2–3 mm thick. Leave these in fixing fluid 24 hr (small pieces) to 2 or 3 days (larger pieces). The higher the temperature the more rapid the fixation. Use of high grade distilled water is recommended.

Embedded in: Celloidin, low viscosity nitrocellulose or soft paraffin.

Staining schedule:
1. Drain tissue blocks to remove excess fixing fluid.
2. Rinse the tissue in 0.75% $AgNO_3$ until the precipitate ceases to form and transfer to fresh solution. Allow to stain for 2 days or longer. Incubating at 37–40°C facilitates staining. It is desirable to have several blocks of tissue and remove one for sectioning at 2-day intervals to determine the optimum staining period for the type of tissue used.

3. Carefully brush the precipitate from the surface of the tissues with a cotton pad.
4. Wash thoroughly in distilled water.
5. Dehydrate in a mixture of acetone, 1 vol; absolute alcohol, 3 vol at 57°C, then wash in several changes of absolute alcohol.
5a. (use instead of 5 if desired). Dehydrate at room temperature with 2 changes of 95% alcohol, one change of absolute alcohol, and one change of ether-absolute alcohol, 1:1, for nitrocellulose; or xylene for paraffin. Allow 1 to 4 hr per change according to the size of the tissue block.
6. Embed.
7. Section at 60–100 μm.
8. Mount on slides and cover with thick Canada balsam. Allow to dry in a dust free location without cover glasses. With the use of a synthetic resin, it is probably safe to apply a cover glass in the usual manner.

Results: Nerve cells and their processes—black (usually only a few neurons stained), neuroglia—astrocytes are sometimes stained black, blood vessels—often stained black or brown, background—dull yellow.

Comments: The duration of fixation influences subsequent staining: 3–5 days is generally considered optimum when the osmic-bichromate mixture is used. If Müller's fluid or bichromate alone is used (regular Golgi process) 2–4-wk fixation is needed.

Emphasis is customarily placed on rapid dehydration and rapid embedding, also completeness of dehydration. Data are not at hand to indicate whether the stain fades or whether this is merely an attempt to avoid excessive hardening. Infiltration of the embedding mass must be complete, however, to secure satisfactory cutting. Low viscosity nitrocellulose (30%) is suggested for the embedding mass. Oven embedding and hardening with chloroform appear to be optional. Paraffin embedding is possible but not recommended for precise work.

Golgi-Cox Method

From Poljak (1942)
Original by Cox (1891)

This method is characterized by fixation and staining occurring simultaneously, with mercury as the impregnating metal.

Combined fixing and staining fluid: $K_2Cr_2O_7$, 10 gm; water, 450 ml; $HgCl_2$, 10 gm; distilled water, 450 ml. Prepare these as two separate solutions and mix when solution is complete. Add slowly and with shaking the following solution: K_2CrO_4, 3 gm; water, 100 ml. A precipitate may form if not mixed in this order.

Staining schedule:

1. Allow the tissues to remain about 2 months in the fixing and staining solution. Small tissue blocks several millimeters thick should be used and are left undisturbed during the entire time of staining. Using several blocks and sampling at 2-week intervals after the first weeks may be done to determine the optimum time.
2. Dehydrate and embed as in the rapid Golgi method.
3. Cut at the desired thickness, usually 60–100 μm.
4. Dip each section in dilute ammonia water (conc. NH$_4$OH, 5 ml; water, 95 ml) to blacken the mercury deposits.
5. Wash, dehydrate and mount as in other Golgi methods.

Results: Similar to rapid Golgi method except that axons may not be so well impregnated, hence following nerve tracts is less feasible. It is generally more reliable, particularly for cell bodies and dentrites.

   NOTE: Formalin fixed material may give satisfactory results with the Cox modification but is generally unsatisfactory for the rapid Golgi method.

Recent Uses and Studies of the Golgi Dichromate-Silver Method and Resulting Modifications

It should be noted that in the original Golgi method, given above, formalin is not used for fixation. The desirability of formalin as a neurological fixative has given rise to a large number of modifications which if given individually and in detail would take considerable space, and probably add more confusion than knowledge to the subject. Hence, a listing of typical variations with brief discussions of them should be more helpful.

I. The standard or "old-fashioned" method. Tissue fixed several days to several years in 10% formalin, either with or without initial perfusion, is chromated in a 3–6% solution of K$_2$Cr$_2$O$_7$, rinsed (or not rinsed) in water and silvered in aqueous AgNO$_3$ of about 1% concentration 2–5 days. That this procedure can give good preparations in some laboratories is shown by its continued use. Anderson (1954) figured excellent preparations so made and stated that about 70% of the trials gave usable sections.

II. The reversed dichromate silver sequence (Porter and Davenport, 1949). Tissue is fixed in 10% formalin containing about 0.5% AgNO$_3$, and with or without pyridine, 0.05–0.1%, for 2 days. Chromating in 2.5% K$_2$Cr$_2$O$_7$ follows, and the activity of this solution appears to be enhanced by the addition of 1–2 ml of 1% osmic acid per 100 ml. This method gives a good return of usable preparations on small rodent brains but may stain too many neurites in some areas. The time in the chromating fluid is uncritical from 3 to 7 days.

III. By adding various chemicals to the fixing fluid, chromating fluid or to both. Ramon-Moliner (1957) listed a large number of such variations, and on the basis of his own experiments, recommended the following procedure:

1. Place pieces of nervous tissue 3–5 mm thick in the following mixture for 3 days with daily changes of fluid: 6% $K_2Cr_2O_7$, 40 ml; 5% $KClO_3$, 20 ml; 20% chloral hydrate, 30 ml; and concentrated formalin, 10 ml.
2. Transfer to 3% $K_2Cr_2O_7$ for 3 days with twice daily changes.
3. Silver 3 days at 20–25°C in 1% $AgNO_3$.
4. Cut frozen sections, dehydrate, clear, and mount in Permount with a cover glass.

The method is recommended for brain of dog, cat and rabbit. Neuroglia, as well as neurites, are stained; the former being shown best in animals 1 mo old.

IV. By the use of an acidified chromate instead of $K_2Cr_2O_7$ for the chromating fluid. Although there is mention of dichromates other than the potassium salt in earlier literature, Fox and associates (1951) appear to have been the first to use an acidified chromate ($ZnCrO_4$ and formic acid) as the chromating fluid. Later, Davenport and Combs (1954) extended the investigation of this type of modification to other chromates and found that the chromates of Cd, Co, K, and Sr behaved much like the zinc salt; also that acetic acid seemed to be preferable to formic as the acidifying agent. Thus it becomes clear than an important feature of the chromating fluid is its pH, with the associated cation playing only a secondary role.

Since other features of the technic are essentially standard procedure, only formulas for chromating fluids will be given.

1. Original (Fox) formula: $ZnCrO_4$, 3–6 gm; formic acid (conc.), 1 ml for each 1.5 gm of chromate used; water, about 45 ml. When the chromate has dissolved, add water to make 100 ml.
2. Davenport and Combs formulas:
   A. Simple mixture recommended for tissues fixed in formalin 2–40 days: 5% K_2CrO_4, 100 ml; glacial acetic acid, 6–8 ml.
   B. General purpose chromating fluid for tissue fixed 10 days or longer; Cd- or Zn-chromate, 1.5 gm; water, 25 ml; glacial acetic acid, 6 ml. When solution is complete, add 65 ml of 5% K_2CrO_4.
   C. Dichromate-acetate mixture: $K_2Cr_2O_7$, 3.0 gm; K-acetate, anhydrous, 1.8 gm; water, 100 ml.
   D. Chromate-nitrate mixture: Cd- or Zn-nitrate (cryst.), 3 gm; water, 25 ml. When solution is complete, add: 5% K_2CrO_4, 65 ml; and glacial acetic acid, 6.0 ml.

In their test series of over 100 blocks of tissue, Davenport and Combs concluded that the various chromating fluids were about equally effective, and any one of the formulas was to be preferred to plain $K_2Cr_2O_7$ solution.

V. Fixation with buffered formalin followed by chromation with $ZnCrO_4$ (Fox formula, slightly modified). On the basis of a rather extensive experimental study, Bertram and Ihrig (1957) concluded that the Golgi method is further improved if pH control be extended to both fixation and chromation. The following steps give the salient features of their procedure (original paper supplemented by a personal communication from Dr. Bertram).

1. Perfuse the brain (pulsating perfusion preferred) with a saline-acacia solution (prepared in lots of 2 liters) with its pH adjusted by adding 1 M NaOH until pH 7.0–7.2 is obtained. When the blood has been washed out, continue the perfusion with formalin-saline-acacia also brought to pH 7.0–7.2 in the same manner and allow fixation to continue 1–2 days, after removal of the brain, in 10% formalin with its pH adjusted similarly.
2. Slice the brain to 2–3 mm slices and allow them to remain in the pH 7.0–7.2 fixing fluids 1–2 days.
3. Chromate 1 day in a chromating fluid at pH 3.1. Formula: $ZnCrO_4$, 60 gm; formic acid (conc.), 35 ml; water to make 1000 ml. Use a large volume of fluid in proportion to the volume of the slices.
4. Silver in 0.75% $AgNO_3$ for 2 days with constant agitation of the slices secured by bubbling air through the fluid.
5. Remove the chromate deposit from the surface of the slices, dehydrate and embed in paraffin.

This method has the advantage of using a short fixation time and has given stained nerve cells in such deep parts of the brain as thalamus and lenticular nucleus. Probably the thin slicing of the fixed brain contributes to the staining of deep structures, but fixations by perfusion, with pH control of both fixation and chromation, may be essential for staining deeply located neurites.

Summary: From the foregoing listing of recent modifications, it is evident that considerable knowledge has been added to the Golgi silver-dicromate procedure. It is evident also that more can be learned because it is still difficult to sort out the critical from the uncritical stages. The value of pH control, of slicing the fixed tissue thin enough (2–4 mm) after fixation to allow penetration of both chromating fluid and silver solution, and of suspending the slices in these solutions or providing constant agitation seem to be well established. Fixation by perfusion, though theoretically desirable, has not been proved to be essential for good results. Some of the reagents used in fixation and chromating have a rather empirical status. The user will probably secure some very good results if a sufficient number of animals be used regardless of the particular method chosen. Considerable variation from animal to animal will be seen.

Golgi-Cox Tungstate Modification

From E. Ramon-Moliner (1958)

The chief modifications of the original method consist of the addition of

Na- or K-tungstate to the impregnating fluid and the blackening of the mercurial deposit in the block.

Fixing and impregnating solution:

A. $K_2Cr_2O_7$, 1 gm; $HgCl_2$, 1 gm; water, 85 ml. Boil 15 min, cool, and if necessary, restore evaporation loss by adding water to bring the volume to 80 ml.

B. K_2CrO_4, 0.8 gm; K- or Na-tungstate, 0.5 gm; water, 20 ml. For use, mix A and B. Use $HgCl_2$ of high purity and prepare the solution just before use.

Alkaline solution: LiOH or NaOH, 1 gm; KNO_3, 15 gm; water, 100 ml.

Staining schedule:

1. Place fresh pieces of brain or spinal cord, between 3 and 6 mm thick in the fixing and impregnating solution for 20–30 days at room temperature and in darkness.
2. Transfer to the alkaline solution, using 50 times the volume of the tissue. Time as follows: frog brain, 4 hr; mouse brain, 10–12 hr; 6-mm slice of cat brain, 24–30 hr.
3. When LiOH is used as the alkali, wash with distilled water; if NaOH, use 1% acetic acid. Wash until no KNO_3 precipitates when the washing fluid is mixed with alcohol.
4. Dehydrate and embed in celloidin. Use celloidin solutions that have been rather freshly made and not exposed to light more than is necessary for the usual handling. Keep the preparations in darkness during infiltration.
5. Cut sections as usual and clear by the following sequence: absolute alcohol-chloroform mixture (3:1), xylene, xylene, iodobenzene, iodobenzene.
6. If more than one section is to be mounted on a slide, handling is facilitated by smearing the slide with a 3% celloidin solution in methyl benzoate, to which the sections stick.
7. Apply a cover glass, using a synthetic mounting medium having a refractive index of about 1.61. (Dr. Moliner used Technicon mounting medium.)

Results: Neurites are darkly stained against a pale background.

CHAPTER 7

MISCELLANEOUS METHODS

George Clark

The preceding chapters have each dealt with fairly distinct groups of procedures. This chapter is divided into sections, most of which contain procedures applicable to some particular tissue or organ. None of these, however, is sufficiently extensive to warrant a separate chapter.

The section on blood and bone marrow is virtually unchanged from the previous edition. This does not reflect lack of research in this field but rather that the standard methods have proved adequate as controls for the popular histochemical and electron microscope studies. These methods have also continued to be adequate for clinical work.

SECTION A: METHODS FOR STAINING BLOOD

Wright Stain

From Conn (1940, p. 172–173)
Original by Wright (1902)

Designed to show: General differentiation of blood corpuscles.

No fixation: Thin smears prepared by streaking a drop of blood over a scrumpulously clean slide by means of a second slide or a cover glass; air-dried quickly at room temperature. Smears must be very very thin, preferably only one cell thick.

or

Fix film prepared as above by flooding with methyl alcohol. This should be done immediately on taking the sample, and is an important step if the smears must be kept any length of time before staining.

Preparation of staining solution: Wright stain can be prepared from eosin and polychromed methylene blue by the method given in the previous edition of this book. This is rarely done now in the laboratory, as Biological Stain Commission certified commercial products in powdered form are available. They should be dissolved as follows:

Dissolve 0.2–0.3 gm Wright stain powder in 100 ml methyl alcohol, absolute, neutral, acetone-free. Allow to stand a day or two before using.

NOTE: This method of preparing the solution seems to be the procedure in most common use in the United States at present, but differs considerably from the original procedure of Wright which was as follows: 0.5 gm dried stain in 100 ml ethyl alcohol, filtered and 25 ml methyl alcohol added.

Staining schedule:
1. Place 1 ml of the above staining solution on the blood film for 1–3 min depending on the behavior of the stain used.
2. Add 2 ml of distilled water; or preferably a phosphate buffer solution adjusted to about pH 6.5.
3. After standing about twice as long as with the undiluted stain, flood off the stain and wash with distilled water (or preferably the above mentioned buffer solution) until the thin portions of the stained film are pink.

NOTE: By staining 10 min in stain diluted with 4 vol of phosphate solution buffered to pH 6.5, results more like those of Giemsa stain may be secured.

4. Dry by blotting carefully. In case of poor results, try altering the reaction of the dilution water or the timing of the staining either before or after dilution.

Results: Erythrocytes—yellowish red; polymorphonuclears—dark purple nucleus, reddish lilac granules, pale pink cytoplasm; eosinophiles—blue nuclei, red to orange-red granules, blue cytoplasm; basophiles—purple to dark blue nucleus, dark purple granules (almost black); lymphocytes—dark purple nuclei, sky blue cytoplasm; platelets—violet to purple granules.

Lillie's Modification of Wright Stain

From Lillie and Fullmer (1976)

Designed to show: General differentiation of blood corpuscles.

Fixation: Thin smears prepared by streaking a drop of blood over a scrupulously clean slide by means of a second slide or a cover glass. Fix films immediately 2 or 3 min in methyl alcohol.

Preparation of staining solution:
A. Stock solution: Dissolve 1 gm Wright stain powder in 50 ml glycerol mixed with 50 ml methyl alcohol.
B. Just before use, mix 4 ml of stock solution, 3 ml acetone, 2 ml M/15 phosphate buffer of pH 6.5, and 31 ml distilled water. This solution may be used for a second group of slides if applied immediately.

Staining schedule:
1. Air-dry the methyl alcohol-treated blood films.
2. Immerse for 5 min in a Coplin jar containing the above diluted stock solution.
3. Wash in distilled water.
4. Air-dry.

Results: Superior to Wright stain in regard to sharpness of nuclear and parasite staining. Similar in general to Giemsa stain.

NOTE: Some samples of Wright stain, which contain an excess of Bernthsen's methylene violet, are not suitable for this procedure.

MacNeal's Tetrachrome Stain

From Conn (1940, p. 173)
Essentially the same as MacNeal (1922)

Designed to show: Differentiation of types of leukocytes.

No fixation: smears prepared as for Wright stain (see above).

Preparation of staining solution: Mix the following dry ingredients: methylene blue chloride, 1.0 gm; azure A, 0.6 gm; methylene violet, Bernthsen, free base, 0.2 gm; eosin Y, 1.0 gm.

NOTE: These dyes mixed in the proper proportions may be purchased from stain companies, either dry or in solution, under the name of tetrachrome stain, MacNeal.

Dissolve 0.15–0.3 gm of the mixed dry ingredients in 100 ml methyl alcohol, neutral, acetone-free, by heating to 50°C. Shake thoroughly and leave 1–2 days at 37°C, with occasional shaking. Filter off any precipitate that forms. If prepared according to directions, the solution should keep fairly permanently; if the above method of dissolving is not carefully followed, however, the solution is likely to deteriorate rapidly.

Staining schedule: Follow directions under Wright stain (see above).

Results: Erythrocytes—yellowish red; polymorphonuclear neutrophiles—dark blue nucleus, reddish lilac granules, pale pink cytoplasm; eosinophilic leukocytes—blue nuclei, red to orange-red granules, blue cytoplasm; basophilic leukocytes—purple or dark blue nucleus, dark purple (almost black) granules; lymphocytes—nuclei dark purple, cytoplasm sky blue; platelets—violet to purple granules.

Giemsa Stain for Thin Films

Lillie and Fullmer (1976)
Original by Giemsa (1904)

Designed to show: Differentiation of types of leukocytes, rickettsiae, bacteria and inclusion bodies.

Fixation: Smears prepared as for Wright stain (see above), or as usual for pus, exudates, etc.

Preparation of staining solution: Dissolve 0.8 gm Giemsa stain in 100 ml of a mixture of equal volumes of glycerol and methanol, by shaking 2 or 3 days in a mechanical shaker or 3–4 days occasionally by hand with glass beads. For use dilute in 50 vol of distilled water buffered to pH 6.0–6.5 with M/100 phosphate.

> **NOTE:** Older procedure called for dissolving 0.5 gm in 33 ml glycerol by allowing to stand 1½–2 hr at 55–60°C, then adding 33 ml methanol and allowing to stand at least 1 day; for use, dilute by adding 1.5–30 ml distilled water.

Staining schedule:
1. Treat dried thin blood film 5–7 min in methyl or ethyl alcohol.
2. Air-dry.
3. Immerse the dried blood films in the above Giemsa solution. (As rapidity of action varies, it is best to test a sample with three or four slides stained at varying periods from 15–50 min.) As long as 2 hr may be needed for undulating membranes of trypanosomes.
4. Wash in distilled water.
5. Air-dry.

Results: Nuclei of leukocytes—reddish purple, rest of leukocytes—similar to Wright's stain, cytoplasm of plasmodia—blue, chromatin—red.

Giemsa Stain for Thick Films

From Lillie (1965)
Essentially the same as Barber and Komp (1929)

Designed to Show: Malarial parasites.

No fixation: Films are prepared by spreading 3–5 drops of blood over a circle of about 15 mm diameter on a scrupulously clean slide; they are air-dried by keeping 18–24 hr at room temperature in a horizontal position, protected from dust and insects.

Preparation of staining solution: Stock Giemsa solution is prepared as above.

Diluted staining solution: Stock solution, 1 ml; pH 7.0–7.2 phosphate buffer, 2 ml; distilled water, 47 ml. (The proportion of stain may be varied from 1:30 to 1:100 according to results desired, although 1:50 is usually satisfactory.)

Staining schedule:

1. Stain 40–120 min in the diluted staining fluid.
2. Wash 5–10 min in distilled water or a phosphate solution of pH 7.0.
3. Dry in the air.

Results: Malarial parasites—clear red chromatin, cytoplasm—clear blue, red corpuscles—not seen on account of laking of the hemoglobin during the drying.

NOTE: For thick and thin films, Lillie recommends the following accelerated Giemsa stain:

Staining solution: Giemsa stain, 4 ml; acetone, 3 ml; phosphate buffer of pH 6.5, 2 ml; distilled water, 31 ml.

Staining schedule: Allow unfixed thick films to dry for 1 hr, then stain 5–10 min; rinse in distilled water; dry and examine.

Results: As above. This brief staining avoids the loss of films which often occurs with short drying periods.

Combined Peroxidase-Wright Stain

From W. D. Stovall (personal communication)
Original by Brice (1933)

Application: Differential diagnosis of chronic leukemias.

Designed to show: Peroxidase granules in leukocytes, especially in neutrophils.

No fixation: smears prepared as for Wright stain (see p. 171).

Preparation of reagent: Dissolve 0.3 gm benzidine base in 100 ml 95% ethyl alcohol; then add 1 ml saturated aqueous sodium nitroferricyanide. (This reagent keeps well for 6–8 months.)

Staining schedule:

1. Flood smear 1 min with the above reagent, counting the number of drops added.
2. Add half as many drops of dilute H_2O_2, and mix thoroughly by blowing or tilting.

NOTE: A successful stain depends on a proper concentration of H_2O_2, which must be determined by trial. One should start with a dilution of 1:100 and vary the strength, keeping the time factor constant, until the best stain is obtained. As the concentration of H_2O_2 is increased, the jet black peroxidase granules in the finished preparation become more minute and discrete, then greenish brown in color, and finally disappear altogether.

3. Allow the combined reagents to act for not more than 2 min.
4. Wash 1 min in tap water and allow to dry.
5. Examine with low power objective. If satisfactory:
6. Stain with Wright stain in the usual procedure (see p. 171).

Results: Red blood cells—same as with Wright stain; myeloblasts, lymphoblasts, and monoblasts—same as with Wright stain; premyelocytes—show a few peroxidase granules; neutrophils, both filamentous and nonfilamentous—show heavy peroxidase granulations; lymphocytes—the same as with Wright stain, no granules; monocytes—the majority as with Wright stain, but a small number show occasional granules.

Brilliant Cresyl Blue for Reticulated Cells and Platelets

From Isaacs (McClung (1937), p. 319)
Essentially the same as Cunningham (1920)

Designed to show: Reticulum of immature red cells; also platelets.

No fixation.

Staining schedule:
1. Polish smooth glass slides or cover glasses.
2. Place a drop of 0.3% alcoholic (ethyl or methyl) brilliant cresyl blue on the glass and allow it to dry.
3. On a clean slide or cover glass place a drop of blood 2–3 mm in diameter, and bring this in contact with the dried stain.
4. Manipulate the slides or covers hinge fashion, up or down, until all the stain is dissolved, and the blood appears blue-black. Then allow the glasses to come in contact so as to spread the drop.
5. Separate the glasses, and allow the films to dry.
6. Counterstain with Wright stain by the standard technic.

> **NOTE:** Record the number of reticulocytes and/or platelets noted in counting 1000 red blood cells (or more). The number of red cells per cubic millimeter should be determined separately in a haemocytometer, and the ratio of reticulocytes or platelets to red cells computed from the stained preparation.

Results: Reticulum of immature red cells—clear-cut blue; background—pale blue (fresh) or eosin-colored; blood platelets—pale blue; or lilac, if counterstained with Wright stain. The brilliant cresyl blue serves to keep the platelets discrete instead of in clumps.

> **NOTE:** Sometimes a precipitate remains in the preparation, and young blood cells may show diffuse basophilia instead of a discrete reticulum. This is ordinarily due to a poor lot of dye and should not occur with the Biological Stain Commission certified product.

Brilliant Cresyl Blue for Counting Reticulocytes

From Conn (1940)
Essentially the same as Robertson (1917)

Designed to show: Reticulate blood corpuscles.

No fixation.

Preparation of staining solution: Add 1.5 gm brilliant cresyl blue to 100 ml of a normal salt solution (0.85% NaCl). Filter. Keep this as a saturated stock solution.

Staining schedule:
1. Add a small quantity of the stock brilliant cresyl blue solution to 80–180 times its volume (varying with the lot of dye used, and determined by a test in the staining technic) of normal salt solution.
2. Mix the solution with blood in a pipette for counting white cells, in the proportion of 1 vol blood to 20 vol of staining fluid.
3. Shake the mixture 5 min in the pipette, and place in a blood counting chamber.
4. Count at once in fresh preparations which are sealed with Vaseline to prevent disturbances due to drying. At least 1000 cells are counted for each test.

Results: Reticulum—stained blue, rest of corpuscle—unstained, leukocytes—nuclei stained blue.

New Methylene Blue as a Reticulocyte Stain

From Brecher (1949)

Application: For identification and enumeration of reticulocytes in peripheral blood.

Preparation of solution: Dissolve 0.5 gm of new methylene blue and 1.6 gm of potassium oxalate in 100 ml of distilled water.

Staining procedure:
1. Draw capillary blood up to the 0.5 mark in a leukocyte diluting pipette, and then draw in an equal volume of the above staining solution by filling until the blood column reaches the 1.0 mark. (Alternatively, approximately equal volumes of blood and staining solution may be mixed on a clean slide and the mixture drawn into a capillary pipette.)
2. Draw the blood and stain on into the bulb of the pipette, mix well and let stand for 10–15 min.
3. Expel small drop of mixture onto slide and smear in usual manner.
4. When smear is dry, examine directly under oil immersion, without fixation or counterstaining.

Results: Red blood cells--light greenish blue, reticulum—deep blue, sharply outlined.

NOTE: Proportions of blood and staining solution have little effect on staining of reticulum, but do determine depth of coloration of the red cells. A slight excess of blood gives optimal color contrast.

Prussian Blue Stain for Non-Hemoglobin Iron

Sundberg and Broman (1955)

Designed to show: Non-hemoglobin cytoplasmic iron.

Preparation of staining solutions:
A. Wright's stain (see p. 171).
B. Prussian blue reagent stock solutions:
  1. Potassium ferrocyanide, 2 gm; H_2O *q.s.*, 100 ml.
  2. Concentrated HCl, 1 ml; H_2O *q.s.*, 100 ml.
C. Prussian blue staining solution: Just before use, take 2 parts of 1 (potassium ferrocyanide solution) and 3 parts of 2 (diluted HCl).

Staining schedule:
1. Rapidly air dry smear or imprint.
2. Flood slide with Wright's stain for 4 min.
3. Dilute stain (on slide) with phosphate buffer pH 5.6. Leave for a period of time suitable to strength of stain and cellularity of film.
4. Immerse slide in Prussian blue staining solution (Solution C) for 10 min.
5. Wash briefly in distilled water (until a pink color appears) and drain dry.
6. If desired dehydrate, clear and mount with synthetic resin.

Results: Slides stained in this way are comparable to the original films stained with Wright's stain but particulate iron is colored a vivid blue or blue-green. The nongranular cytoplasm of many macrophages is also colored a vivid blue or blue-green (hemosiderin?).

Rhodinile Blue for Heinz Bodies

Simpson, Carlisle and Mallard (1970)

Designed to show: Heinz bodies.

Preparation of staining solution: Add 0.5 gm rhodinile blue (E. Gurr, Michrome No. 1156) to 100 ml of a 1% solution of NaCl in water. Shake thoroughly and filter through a No. 1 Whatman filter paper.

Staining schedule: To a few drops of heparinized or EDTA-treated blood in a small vial add the same amount of the staining solution. Swirl gently and let stand 2 min. Prepare thin smears and air dry. The smears can be examined as usual or can be mounted with a synthetic resin.

Results: Heinz bodies—deep purple, erythrocytes—yellow-orange to blue-green, reticulocytes—blue but reticulum not evident unless the staining time is prolonged.

SECTION B: METHODS FOR STAINING BONE MARROW

Smears with a May-Grünwald Stain Preceding Giemsa

From Beck (1938, pp. 233–236)
Original by Pappenheim (1912)

Designed to show: Morphological details of marrow cells.

Preparation of smears: A curetted fragment is held in forceps' tip and drawn lightly across a clean slide in a serpentine streak. It is not allowed to dry completely.

Staining schedule:

1. Fix promptly for 3 min in 97–100% methyl alcohol.
2. Cover for 3 min with stock May-Grüwald, or Jenner stain (*i.e.*, a redissolved precipitate from unpolychromed methylene blue and eosin Y—1 gm in 400 ml methanol).
3. Add an equal amount of distilled water and allow to stand 1 min.
4. Drain without rinsing.
5. Cover for 12 min with dilute Giemsa stain (15 drops to 10 drops distilled water).
6. Differentiate in distilled water, by agitating for about 5 sec and checking under microscope.
7. Blot with fine grain filter paper and mount in clarite.

Results: Similar to Giemsa, but the staining with eosin is more pronounced, giving redder tones throughout.

Custer's Method for Sections

Essentially the same as Kolmer and Boerner (1941, pp. 833–834)
Originally described by Custer (1933)

Application: Bone marrow in biopsy material obtained by means of a trephine.

Designed to show: Morphological details of marrow cells.

Fixation: Formol Zenker (Helly's) solution for 4–6 hr.

Embedding schedule:

1. Wash 1 hr in running water.
2. Decalcify overnight if necessary (until bony button is soft to the prick of a needle) in 5% aqueous formic acid or in the following modification of Wagoner's solution: 85–90% formic acid, 50 ml; mixed with distilled water, 35 ml; and added to a solution of 17 gm sodium citrate in 85 ml warm distilled water. Cut off cortical bone.

3. Wash gently 1 hr in running water.
4. Dehydrate in ascending strengths of alcohol.
5. Clear and pass into paraffin as usual.

Preparation of staining solution:

A. Eosin Y, 0.1% aqueous.
B. Azure II—i.e., equal parts of azure A and methylene blue—0.1% aqueous.

   Just before use mix these stock solutions in glassware reserved for this purpose, in the following proportions: A, 20 ml; B, 10 ml; distilled water, 80 ml. Filter through cotton.

NOTE: If the tissue is very bloody, it is advisable to use only 5 ml of Solution A.

Staining schedule:

1. Cut sections 3–4 μm thick, and mount on slides as usual.
2. Carry through xylene, and absolute alcohol into 95% alcohol.
3. Remove the mercuric precipitate by immersing several minutes in 2% alcohol iodine, then in 80% alcohol, tap water, 5% aqueous $Na_2S_2O_3$, tap water, and 3 changes of distilled water.
4. Place vertically for about 15–16 hr in the staining solution (covered).
5. Pass through 2 changes of 95% alcohol, controlling differentiation by examination with low power of microscope.
6. Wash in 2 changes of absolute alcohol.
7. Clear in 2 changes of xylene.
8. Mount in balsam or gum damar.

Results: Erythrocytes—orange, cytoplasm of lymphocytes and blastocytes—blue, nuclei—deep blue to violet-blue, mast cell granules—violet to reddish purple, cartilage—a more reddish purple, bone matrix—pink.

SECTION C: SUPRAVITAL STAINING

Supravital Staining

Doan and Ralph (1950)

Designed to show: Differentiation of leucocytes in supravital staining.

No fixation.

Preparation of glassware: Place new glassware for 48 hr in cleaning solution (saturated $K_2Cr_2O_7$ in H_2SO_4). Wash 48 hr in running tap water. Rinse 3 times with distilled water and store in alcohol. Wipe dry with clean gauze, avoid lint. Store in dust-proof container.

Preparation of solutions*:
  A. Stock solutions. Must be kept in tightly stoppered containers.
  1. Neutral absolute alcohol (distilled over CaO).
  2. Neutral red, 300 mg; neutral absolute alcohol, 100 ml.
  3. Janus green B, 50 mg; neutral absolute alcohol, 25 ml.
  B. Staining solution for blood. (To be prepared just before using.) To 5
    ml neutral absolute alcohol, add 35–40 drops (Wright's capillary
    pipette) of the neutral red stock solution, and 5 drops (Wright's
    capillary pipette) of Janus green stock solution. Occasionally, a few
    more drops of Janus green are necessary (up to 8–10).

  NOTE: Rapid evaporation changes the concentration. If staining of slides
  is delayed, a new solution must be made up. This 5 ml of staining solution
  is adequate to make 75–100 slides, which then must be kept in dust-proof
  boxes and may be used indefinitely.

  C. For leukemic blood, bone marrow and other cellular tissues, increased
    concentration of dyes may be used, proportional to the number of cells
    present reacting specifically with each dye.

Preparation of slides with stains: Flame slides prepared as above, cool
until only slightly warm, flood with stain rapidly, avoiding overlaying.
Drain excess back into bottle and air-dry. Stain should not be applied
before an open window, or in hot humid atmosphere; 5 ml of staining
solution are adequate for 75–100 slides. They should be stored in a dust-
proof container and may be used indefinitely.

Staining schedule:
  1. Take a small drop of fresh living material on a coverslip (blood, spinal
    fluid, fluid from the serous cavities, bone marrow, tissue or lymph node
    scrapings or suspension).
  2. Invert on stained slide; do not press from above or cells will be injured.
  3. Seal edges of cover slip with Vaseline.
  4. Let preparation stand for at least 5 min before placing on microscope
    for study. (If not to be examined immediately, preparation should be

---

* Neutral red is an indicator with a pH range of 6.8–8.0 (maroon-red to yellow). The dye sensitively
reflects the differing pH of specific intracytoplasmic granules and physiological vacuoles in the
different types of blood and connective tissue cells, thus providing different and sharper differential
criteria for these cells than are available in the fixed film staining technics—with particular
emphasis on the immature and blast forms. Optimum staining of the cytoplasmic organoids may be
accomplished without toxic nuclear reactions or suppression of active cellular motility, and the
differential cell type criteria are dependent upon avoiding any nuclear staining. An acceptable
differential includes: neutral red staining of the eosinophil basophil granules, deep brick red. The
vacuoles of the monocytes are arranged characteristically about the centrosome and stain a uniform
salmon color. The vacuoles of the clasmatocyte or macrophage which segregate all phagocytized
material, reflect the full pH range of the neutral red stain, depending upon the stage of breakdown
of the debris in each individual vacuole. Janus green stains only the mitochrondia in all cells, as
small contrasting dots or rods.

kept in refrigerator.) Study at room temperature (optimum for cellular motility is 37.1°C (98.8°F)).

Results: Basophilic granules—deep brick red, eosinophilic granules—yellow or light orange, neutrophilic granules—pale pink, vacuoles of monocytes—salmon, vacuoles of clasmatocytes or macrophages—reflect the full pH range of neutral red stain (6.8–8.0, red-yellow), mitochrondria-green.

SECTION D: METHODS FOR STAINING BONE AND CALCIUM

Schmorl's Method for Sections

Quoted from Mallory (1938, pp. 172–173)
Original by Schmorl (1899; 1914, p. 136)

Application: Studying bone structure.

Designed to show: Lacunae and canaliculi.

Fixation: Preferably in Müller's or Orth's fluid or 10% formalin. Do not use a $HgCl_2$ solution. Decalcify by one of the slower methods, namely Ebner's fluid (saturated aqueous NaCl, 100 ml; conc. HCl, 4 ml; with 2 ml HCl added each day during fixation); or Müller's fluid, 3% HNO_3.

Embedded in: Celloidin.

Preparation of staining solutions:
   A. Nicolle's carbol thionin solution: Thionin, saturated solution (about 0.25%) in 50% alcohol, 10 ml; 1% aqueous phenol, 100 ml.
   B. Thionin solution: Thionin, saturated solution (about 0.25%) in 50% alcohol, 2 ml; distilled water, 10 ml.

Staining schedule:
   1. Place the sections for at least 10 min in distilled water to remove any trace of alcohol.
   2. Stain 5–10 min or longer in either Solution A or Solution B.
   3. Wash in distilled water.
   4. Place ½–1 min in saturated aqueous picric acid (about 1.22%).
   5. Wash in distilled water.
   6. Place in 70% alcohol for about 5–10 min until no more dense clouds of color are given off.
   7. Dehydrate in 95% alcohol.
   8. Clear in terpineol or in oil of origanum.
   9. Mount in balsam.

Results: Bone substance—yellow to yellowish brown, bone lacunae and canaliculi—dark brown to black, cells—red, fat cells (after Müller's fluid)—reddish violet, osseous tissue—deeper yellow than osteiod tissue.

NOTE: This method is not a true stain but resembles Golgi's method; a precipitation of coloring matter takes place in the lacunae and canaliculi; it also takes place to a considerable extent in other marrow spaces in the tissues. The latter can be eliminated to some extent without injury to the stain by leaving the sections in Step 5 for ½ hr. The canaliculi are then usually brownish red to red, and the bone substance blue to colorless, and it is often best to stain the sections first in alum hematoxylin to bring out the nuclei.

Staining of Canaliculi and Lacunae in Hard Tissues

Powers, Rasmussen and Clark (1951)

Designed to show: Lacunae and canaliculi in bones and teeth.

Fixation: Modified Bouin's—saturated aqueous picric acid, 75 ml; formalin (37%), 25 ml; trichoracetic acid, 1 gm.

Decalcification: Decalcify in 5% trichloracetic acid in 50% alcohol.

Embedded in: Paraffin.

Solution needed: Cupric nitrate $(Cu(NO_3) \cdot 3H_2O)$, 1 gm; H_2O $q.s.$, 100 ml.

Staining schedule:
1. Dewax and hydrate as usual.
2. Mordant overnight in 1% cupric nitrate.
3. Stain by Bodian method (see p. 152).

Results: Canaliculi, lacunae, odontoblasts and dentinal tubules—bluish purple to black.

Alizarin Red S for Small Vertebrates

Quoted from Dawson (1926)
Original by Lundvall (1905)

Application: Small mammals, in whole or in part.

Designed to show: Developing bone.

Fixation: (after complete evisceration): 2–4 days, or even longer, in 95% alcohol.

Staining schedule:
1. Place specimen in 1% aqueous KOH until the bones are clearly visible through the surrounding tissues.
2. Transfer to dilute (0.0025–0.01%) alizarin red S in 1% aqueous KOH, and allow to stand until desired degree of staining is obtained. The smaller the animal, the more dilute the stain may be. Add fresh stain if necessary.

3. Complete clearing by placing in a mixture of 1 vol glycerin to 4 vol of 1.25% aqueous KOH (i.e. glycerin, 20 ml; KOH, 1 gm; water 79 ml). Continue through increasing concentrations of glycerin.

NOTE: Hollister (1934) uses ultraviolet irradiation in this process when applying the method to specimens of fish.

4. Store in glycerin alone.

Results: Bones—red, soft tissue—transparent and unstained.

Alizar in Red S for Fetal Specimens

Quoted from Richmond and Bennett (1938)
Modified from Dawson (1926)

Application: Mammalian embryos.

Designed to show: Minute bones; fetal ossification.

Fixation: Two weeks or more in 95% alcohol or formalin, after eviscerating through a midline abdominal incision.

Preparation of staining solution: Add 6–10 drops 1% aqueous KOH to 0.1% aqueous alizarin red S.

Staining schedule:
1. Rinse in tap water.
2. Place for 4 weeks or longer in 1% aqueous K_2CO_3.
3. Place for 10 days or longer in 1% aqueous KOH till bones are plainly visible through the soft tissues. (Formalin-fixed specimens require about a month in 10% KOH.)

NOTE: If tissues become too soft, harden 12–24 hr in a solution of equal parts of glycerin, 95% alcohol and water, and then return to the clearing solution. The KOH may be reduced to 0.5% the last few days of clearing.

4. Wash 12 hr in running tap water.
5. Stain 30–60 min in the above staining solution freshly prepared.
6. Wash 30 min in running tap water.
7. Decolorize 1–2 wk in aqueous solution of 20% glycerin and 1% KOH. (For small specimens the KOH is reduced to 0.5%.)
8. Mount on a glass frame and dehydrate by passing slowly through alcohol-glycerin-water mixtures, beginning with the proportion 1:2:7 and then in succession, 2:2:6, 3:3:4, 4:4:2, and finally 5:5 without water.
9. Seal in the final glycerin-alcohol mixture.

Modification by Cumley, Crow and Griffin (1939)

Staining solution: 0.01% aqueous alizarin red S, plus a few drops 1% KOH.

Evisceration ommitted for small embryos.

Staining schedule:
Step 2 omitted.
Step 3 same, but time specified is a few hours to a week or more.
Step 4 (washing) omitted; and the remaining steps are as follows:
5. Stain 3–12 hr in the staining fluid.
6. Replace dye solution with 1% KOH containing about 10% glycerin. Gradually add glycerin and alcohol in equal parts until the KOH has been completely replaced. (Decolorizing may be hastened by placing in sunlight).
7. Add 95% alcohol gradually until glycerin has been replaced.
8. Gradully replace 95% with absolute alcohol, changing the latter several
    times.
9. Replace the alcohol gradually with toluene, changing the latter several times.
10. Transfer to toluene saturated with naphthalene.
11. Replace this solution with anise oil saturated with naphthalene and store in this solution.

NOTE: If the specimen shows signs of maceration during the clearing and staining process, gradually add alcohol until the concentration is 80–90%, and allow hardening to go on for a few days before resuming the clearing process.

Results: Bones—red, soft tissues—transparent and unstained.

Alizarin Red S and Toluidine Blue O

Quoted from Williams (1941)
Modified from Dawson (1926)

Application: Mammalian embryos, and mature specimens of Urodela.

Designed to show: Distinction between bone and cartilage; relative amount of ossification.

Fixation: At least 1 week in 10% formalin after eviscerating through an abdominal incision and then washing. (Eviscerating is not necessary for museum specimens.)

Preparation of staining solutions:
A. Toluidine blue O, 0.25 gm; 70% alcohol, 100 ml; 0.5% HCl, 2 ml. Allow solution to stand 24 hr; filter and store in tightly corked container.
B. Alizarin red S, 0.001 gm; 2% aqueous KOH, 100 ml.

Staining schedule:
1. Wash the specimen 24 hr in 250 ml 70% alcohol to which has been added 10 drops of conc. NH_4OH.

2. Stain 1 week in toluidine blue solution (A).
3. Harden and destain 72 hr in 4 changes of 95% alcohol.
4. Macerate 5–7 days (depending upon the size of the animal), in several changes of 2% aqueous KOH.

NOTE: This process is greatly facilitated by exposure to sunlight or ultraviolet rays.

5. Transfer to freshly prepared Solution B. In 24 hr the bones should appear well stained. If the specimen has been insufficiently macerated the soft tissues will appear slightly stained. In this event the specimen may be quickly destained in acid alcohol (1% H_2SO_4 in 95% alcohol).
6. Dehydrate by running the specimen through 3 changes of Cellosolve, of 6 hr each. For small embryos, reduce the time.

NOTE: For this step one may substitute the alcohol series, 50%, 80% and 90%, followed by 3 changes of benzene.

7. Clear by transferring to solutions of 25%, 50% and 75% of methyl salicylate in Cellosolve for 24 hr each. Then transfer to pure methyl salicylate for permanent storage.

NOTE: Modification of technic when glycerin is used in clearing: omit Step 6, transfer directly from KOH-alizarin solution into a series of 50%, 70% and 80% glycerin solution for 24 hr each. May be stored in pure glycerin.

Results: Soft tissues—transparent, osseous tissue—deep red, cartilage—dark blue.

NOTE: The intensity of the stains serves to indicate the relative amounts of ossification and chondrogenesis which may have taken place. The bone and cartilage may be stained separately by omitting Step 2 for bone or Step 5 for cartilage. Excellent permanent preparations may be made of parts of limbs, hands or heads by mounting them in Canada balsam on a deep depression side.

Alizarin Red S for Calcium Deposits

McGee-Russell (1958)

Designed to show: Calcium deposits.

Fixation: Formalin-alcohol (equal parts 40% formalin and absolute alcohol); metallic and acid fixatives should be avoided.

Embedded in: Paraffin, ester wax, celloidin or gelatin.

Preparation of staining solution: With dilute ammonia adjust a 2% alizarin red S solution to pH 4.1–4.3, using a glass electrode pH meter. The solution should be a deep iodine color and is stable.

Staining schedule:
1. Bring sections to 50% alcohol.
2. Rinse rapidly in distilled water.

3. Cover section with the above staining solution, and watch the staining by transmitted light, on stage of staining microscope.
4. In 30 sec to 5 min calcium sites will be covered and surrounded by a vivid orange-red calcium-alizarin lake. The end point of the reaction should be chosen empirically when the deposit is heavy, but not too diffuse.
5. Shake off excess stain and blot section carefully with filter paper (or drain very thoroughly).
6. Plunge immediately into acetone for 10–20 sec.
7. Acetone-xylene (50:50) for 10–20 sec.
8. Clear in xylene and mount in balsam or cedarwood oil.

Results: Sites of calcium—covered and surrounded by a heavy orange-red precipitate which is birefringent between crossed polaroids, background—faint pink.

Silver Nitrate Method for Calcium Deposits

From Lillie and Fullmer (1976)
Slightly modified from Mallory (1938)
Original by Kossa (1901)

Application: Pathological tissue.

Designed to show: Deposits of calcium phosphate.

Fixation: 80–90% alcohol; or neutral formalin 1:10.

Embedded in: Paraffin or frozen sections.

Staining schedule:
1. Bring sections to water.
2. Place for 10–60 min in 5% aqueous $AgNO_3$ and expose to strong light.
3. Wash in distilled water.
4. Expose for 10–60 min to bright but diffuse daylight.
5. Wash in distilled water.
6. Remove excess silver by treating 2–3 min in 5% aqueous $Na_2S_2O_3 \cdot 5H_2O$.
7. Wash in distilled water.
8. Counterstain 20–30 sec in 0.2–0.5% aqueous safranin O.
9. Dehydrate in 95%, then absolute alcohol.
10. Pass through: absolute alcohol + xylene; xylene; balsam or clarite.

NOTE: If paraffin sections are ragged and friable, immerse 10 min in 1–2% celloidin in alcohol-ether, before or after removing the paraffin. Drain off the excess, wipe back of the slide, harden (and deparaffinize) in chloroform, pass through 95–80% alcohol to water; stain as above, doubling all time intervals; dehydrate in 2 changes of isopropyl alcohol and xylene (1:1); clear in xylene. By this procedure the celloidin is not dissolved; but it may be removed by using acetone instead of isopropyl alcohol in the dehydration.

Results: Calcium (in masses)—deep black, calcium (dispersed)—gray, nuclei—red.

Glyoxal Bis(2-Hydroxyanil) (GBHA) Method for Calcium

Kashiwa (1970)

The preceding procedures all have definite limitations of sensitivity and reliability and, furthermore, are only useful for deposits of insoluble calcium salts. The ionic calcium or calcium present as soluble salt is not localized by these methods. The GBHA method localizes both the soluble and insoluble calcium salts in the same tissue section.

Application: Unfixed, fresh tissue.

Designed to show: Deposits of soluble calcium salts or ionic calcium and insoluble calcium salts.

Preparation of tissue: Hand slice fresh tissue about 1 mm thick with a razor blade under a dissecting microscope. If the tissue is firm, such as cartilage or bone from a newborn animal, the Smith-Farquhar chopper (Ivan Servall, Inc.) may be used. The fresh tissue block or section is immersed directly in the GBHA staining solution. Fixed tissues are not recommended because the fixative may elude the soluble calcium. Freeze-dry or freeze-substitution methods may be used if hand sectioning of fresh tissues is not possible.

Solutions needed:
  A. Prepare concentrated and dilute stock solvents and add GBHA powder into the solvent immediately before use:

Stock Solvents: NaOH in 100 ml of 95% EtOH	Staining Solution: GBHA per 2 ml of Solvent	Concentration of GBHA Solution
3.4 gm	0.2 gm	10%
1.7 gm	0.1 gm	5%
0.08 gm	0.05 gm	2.5%

The staining solution darkens and is degraded within 3 hr.
  B. Counterstain: 0.08% methylene blue (aqueous).

Staining schedule:
  1. Immerse fresh tissue blocks in the GBHA staining solution for 1–3 hr. Use dilute 2.5% or 5% GBHA staining solution to localize the soluble calcium salts or 10% GBHA staining solution to localize both soluble and insoluble salts.
  2. Rinse excess GBHA in 100% ethanol.
  3. For specificity, immerse stained tissue for 15 min in 90% ethanol

saturated with KCN and Na_2CO_3. This decolorizes all chelates other than those of calcium.

4. Dehydrate in 100% ethanol.

5. Clear in xylene. (If the tissue is thin and can be whole mounted, mount the tissue on slide with preservaslide, Matheson, Coleman and Bell, Inc., Cincinnati.)

6. Embed in paraffin.

7. Section at about 7 μm and float section on water bath (about 40°C) alkalinized with NaOH solution to prevent decolorization of the red Ca-GBHA complexes.

8. Affix to albuminized slide and air dry.*

9. Deparaffinize in xylene, bring down to water and counterstain with 0.08% methylene blue if desired.

10. Dehydrate in ethanol and clear in xylene.

11. Mount in preservaslide. Other mounting media such as permount or balsam decolorizes the red Ca-GBHA complex.

Results: Calcium deposits—bright red, nuclei—blue.

SECTION E: METHODS FOR STAINING FAT

Herxheimer's Technic

Essentially the same as Herxheimer (1901, 1903)

Fixation: Formalin.

Frozen sections.

Preparation of staining solution: Dissolve 0.1 gm Sudan IV† in a mixture of 50 ml 70% alcohol and 50 ml acetone C.P.

NOTE: This is Herxheimer's second formula (1903). His earlier formula (1901) employed a saturated solution in 1–2% NaOH in 70% alcohol, and has been used chiefly in neurohistology; it is unstable and must be made fresh every 2–3 days. Globus (1937, p. 256) recommended a saturated solution in 3.3% NaOH in 50% alcohol.

Staining schedule:

1. Dip for an instant in 70% alcohol.

2. Stain for 2–5 min in the above staining solution.

3. Wash quickly in 70% alcohol and transfer to distilled water.

4. Counterstain in alum hematoxylin.

---

* If the insoluble calcium salts are not stained, place the slide (after Step 8) horizontally on a staining rack and pour the 10% or 5% GBHA solution on the paraffin section. Allow to stain for 1–3 min, decant the GBHA solution, rinse excess GBHA in 100% ethanol and continue with Steps 9–11.

† The name Biebrich scarlet T, medicinal, has been applied to Sudan IV, and this has been corrupted in Scarlet R or Scarlet Red. These are erroneous terms, because Scarlet R is recognized more correctly as the name of two or three other, unrelated dyes.

NOTE: If it is desired to give the nuclei a clearer blue, to contrast more sharply with the red of the fat, follow the counterstain with 2–5 min in 1% acetic acid.

5. Wash thoroughly in tap water.
6. Mount in glycerin or glycerin jelly.

Results: Nuclei—blue, fat—orange to red, cholesterol—less brilliant red, normal myelin—unstained, fatty acids—unstained.

Nile Blue A

From Lillie (1956a, 1956b)

Designed to show: Differentiation of melanins and lipofuscins.

Fixation: Formalin or other appropriate fixatives.

Preparation of staining solution: Dissolve 0.05% Nile blue A in 1% H_2SO_4 (H_2SO_4 98.5%, 1 ml; water, 99 ml).

Staining schedule:
1. Remove paraffin from sections in the usual manner.
2. Stain 20 min in the above staining solution.
3. Wash 10–20 min in running water.
4. Mount in glycerol gelatin.

Results: Lipofuscins—dark blue or green-blue, melanins—dark green. Cytoplasms: muscle—pale green, red corpuscles—greenish yellow to greenish blue, myelin—green to deep blue. Nuclei—poorly or not at all.

Variant 1: Step 3: Do not wash in water, but rinse quickly in 1% H_2SO_4 to remove excess dye and dehydrate at once in 4 changes of acetone of about 15-sec duration each, rinse in equal volumes of acetone and xylene, clear in 2 changes of xylene and mount in xylene cellulose caprate.

Results: Lipofuscins—unstained but appear in their native yellow to brown color, melanins—dark green, mast cells—purple-red, cytoplasms (muscle, myelin and red cells)—unstained, nuclei—sometimes green but often remain unstained.

NOTE: If the pH level of the Nile blue solution is set at 2.0 with a buffer of sulfuric acid and sodium acid phosphate, nuclei usually stain greenish blue in Variant 1, but the picture is otherwise quite similar to that obtained in the 1% H_2SO_4 technic.

Smith and Mair's Stain; Dietrich's Modification

From Mallory (1938)
Derived from Smith and Mair (1908), Dietrich (1910)

Designed to show: Lipoids.

Fixation: Formalin 1:10.

For frozen sections.

Preparation of solutions:

A. Staining fluid: 10% alcoholic hematoxylin 10 ml; 2% aqueous acetic acid, 90 ml. Allow to ripen for at least 6 weeks, or add 0.2 gm $NaIO_3$.

B. Weigert's borax-potassium-ferricyanide: $K_3Fe(CN)_6$, 2.5 gm; borax, 2 gm; distilled water, 100 ml.

C. Mounting medium: Levulose, 30 gm; distilled water, 20 ml. Mix and let thicken 24 hr at 37°C.

Staining schedule:

1. Mordant sections 24–48 hr in 5% aqueous $K_2Cr_2O_7$ at 37°C.
2. Wash in distilled water.
3. Stain 4–5 hr in Solution A at 37°C.
4. Wash in distilled water.
5. Differentiate overnight in Solution B.
6. Wash thoroughly in distilled water.
7. Mount in levulose syrup (Solution C).

Results: Lipoid substances—blue-black.

Baker's Acid Hematein Test

Baker (1946)
Lillie and Fullmer (1976)

Designed to show: Phospholipids.

Preparation of solutions:

A. Formaldehyde-calcium: Formalin (40% HCHO), 10 ml; 10% $CaCl_2$, 10 ml; distilled water, 80 ml. Keep powdered chalk in the solution. The solution is stable.

B. Dichromate-calcium: $K_2Cr_2O_7$, 5 gm; $CaCl_2$, 1 gm; distilled water, 100 ml. The solution is stable. (A small precipitate may be neglected.)

C. Gelatin for embedding: 0.25% aqueous solution of cresol, 10 ml; gelatin 25 gm. Soak gelatin in cresol solution for 1 hr, then warm until gelatin dissolves and strain through muslin while still warm. The gel is stable.

D. Acid hematein: To 50 mg of hematoxylin add 48 ml distilled water and 1.0 ml 1% sodium iodate; bring to a boil, then cool and add 1 ml glacial acetic acid. Prepare fresh daily.

E. Borax-ferricyanide: 250 mg each of borax ($Na_2B_4O_7 \cdot 10H_2O$) and potassium ferricyanide ($K_3Fe(CN)_6$) in 100 ml distilled water. Solution is stable if kept in refrigerator.

Staining schedule.

1. Fix in Solution A for 6 hr.
2. Transfer directly to Solution B at room temperature for 18 hr.
3. Transfer to a second bath of Solution B at 60°C for 24 hr.
4. Wash in running water 6 hr.
5. Transfer to melted gelatin (Solution C) at 37°C for overnight.
6. Solidify gelatin in refrigerator, then cut out rectangular block containing tissue and harden overnight in Solution A.
7. Wash block in running water 30 min.
8. Cut frozen sections at 10 μm.

NOTE: Washing at Step 4 can be continued overnight, and Steps 5–7 omitted, the material then being sectioned directly without imbedding in gelatin.

9. Put loose sections into Solution B at 60°C for 1 hr.
10. Wash in several changes of distilled water (5 min total).
11. Put sections in Solution D at 37°C for 5 hr.
12. Rinse sections in distilled water and transfer to Solution E at 37°C for 18 hr.
13. Wash in several changes of distilled water (10 min total).
14. Mount in Farrant's medium or glycerin jelly; or dehydrate, clear and mount in balsam.

Results: Phospholipids and certain proteins—blue, blue-black, or gray; background—pale yellow, (Feeble reactions, such as very pale dirty blue, are regarded as negative.)

Baker's Pyridine Extraction Test

Baker (1946)

Designed to show: Distinction between phospholipids and certain proteins which give a positive reaction in the acid-hematein test.

Preparation of solutions: "Weak Bouin's Fluid"—Saturated aqueous picric acid, 50 ml; formalin (40% HCHO), 10 ml; glacial acetic acid, 5 ml; distilled water, 35 ml. Other solutions as for the acid hematein test (Solutions A–E).

Staining schedule:

1. Fix small (3 mm) pieces of tissue in weak Bouin's fluid for 20 hr.
2. 70% alcohol, 1 hr.
3. 50% alcohol, 30 min.
4. Wash in running water for 30 min.
5. Pyridine at room temperature, 2 changes of 1 hr each.
6. Pyridine at 60°C, 24 hr.
7. Wash in running water for 2 hr.

8. Transfer to dichromate-calcium (Solution B) at room temperature and proceed with schedule for the acid hematein test starting at Step 2.

Results: Phospholipids remain unstained after pyridine extraction. Nuclei stain after extraction, but not before. Mitochondria and myelin are blue or blue-black without extraction, negative after extraction. Erythrocytes stain both with and without pyridine extraction.

Proescher's Oil Red with Pyridine

From Conn (1940)
Essentially the same as Proescher (1927)

Application: Central nervous system.

Designed to show: Lipoid degeneration.

Fixation: Formalin, Müller-formalin, or picro-formalin.

For frozen sections.

Preparation of staining solution: Oil red O, 3–5 gm; 70% pyridine (i.e. 70 vol to 30 vol distilled water), 100 ml. Let stand 1 hr at room temperature with occasional stirring; it is then saturated and ready for use. Keep the solution in a glass-stoppered bottle, protected from light; filter before use.

Staining schedule:
1. Immerse sections in 50% pyridine for 3–5 min.
2. Transfer to the above staining solution for 3–5 min—(in case of the central nervous system, 20–30 min).

   NOTE: The staining dishes should be well covered to prevent undue evaporation of the pyridine.

3. Differentiate in 60% pyridine for several minutes.
4. Transfer to water.
5. Counterstain 2–3 min with Delafield's hematoxylin (see p. 105). In case of the central nervous system, an acidified solution is recommended, to prevent overstaining of cytoplasm; this is prepared by adding 2 ml of glacial acetic acid to 16 ml of the staining fluid.
6. Mount in gum acacia or Apathy's gum syrup (p. 194).

Results: Fats and lipoids—brilliant deep orange.

Sudan Black B Stain

From Lillie and Fullmer (1976)
Original from Chifelle and Putt (1951)

Designed to show: Fat in animal tissue.

Fixation: Formalin.

Frozen sections.

Preparation of solution: Dissolve 0.7 gm Sudan black B in 100 ml propylene glycol at 100–110°C. Do not exceed 110°C. Filter hot through Whatman No. 2 filter paper. Cool and filter with vacuum through a medium-porosity fritted glass filter.

Staining schedule:
1. Cut frozen sections, wash 2–5 min in distilled water to remove formaldehyde.
2. Dehydrate 3–5 min in pure propylene glycol, moving sections at intervals.
3. Transfer to the above staining solution for 5–7 min, agitate occasionally.
4. Differentiate 2–3 min in 85% propylene glycol.
5. Wash 3–5 min in distilled water.
6. Use nuclear counterstain if desired.
7. Float onto slides, drain and mount in glycerol gelatin.

Results: Neutral fats—greenish black, myelin—greenish black, mitochondria—greenish black, other lipids—greenish black, cytoplasm—unstained.

NOTE: Sudan IV may be substituted for Sudan black B in this technic.

Fat Stains in Supersaturated Isopropanol

Quoted from Lillie and Fullmer (1976)

Designed to show: Fat in animal tissue.

Fixation: Formalin.

Preparation of solutions:
A. Staining fluid: Stock saturated solution of Sudan IV, oil red O, or Sudan III, in 99% isopropanol. Dilute 6 ml of stock solution with 4 ml water. Let stand 5–10 min and then filter. The filtrate can be used for several hours.
B. Apathy's gum syrup: Dissolve 50 gm acacia, 50 gm sucrose in 100 ml distilled water at 55–60°C with frequent shaking; restore the volume with distilled water; 1.5 ml 1% Merthiolate or 100 mg thymol is added as preservative; place in vacuum chamber for a few minutes to remove air bubbles.

Staining schedule:
1. Stain thin frozen sections 10 min with the diluted stock solution.
2. Wash in tap water.
3. Stain 5 min in an acid alum hematoxylin of about 0.1% strength (*e.g.* Mayer's undiluted, Lillie's diluted 1–4 in 2% acetic acid, or Ehrlich's diluted 1–5 in 2% acetic acid).

4. Place in 1% aqueous Na_2HPO_4 or in tap water, until blue.
5. Float out in water, and take up on a slide.
6. Mount in a suitable aqueous medium such as Apathy's syrup, Zwemer's glychrogel or Maiser's mounting medium.

Results: Fat–deep orange red with oil red O, lighter with Sudan IV, orange-yellow with Sudan II; nuclei—blue; erythrocytes—sometimes green; cytoplasm—lighter green.

NOTES: Oil red O, oil red 4B or EGN and Sudan red 4B give a deeper orange-red or red fat stain and are more stable in dilute isopropanol solutions than Sudan IV. Sudan II gives a bright yellow-orange; it is especially good for demonstrating degenerating myelin (orange-yellow) in contrast to a Weigert myelin stain on frozen section. It is more stable in dilute isopropanol solutions than Sudan III.

Sudan brown, Sudan brown 5B and oil brown D give a satisfactory brownish red stain. As to the intensity and stability of their dilute isopropanol solution, see Lillie (1944b).

Marchi Method for Fatty Degeneration

Details for this procedure are given in Chapter 6 and need not be repeated here.

SECTION F: METHODS FOR STAINING GLYCOGEN

Best's Method

Slightly modified from Bensley (1930)

Fixation: Alcoholic formalin or 5% acetic acid plus 10% formalin alcohol. Good results may often be attained with neutral 10% formalin in water.

Embedded in: Paraffin preferably. Celloidin, or low viscosity nitrocellulose may be used.

Preparation of staining solution: Boil 2 gm carmine, 1 gm K_2CO_3, and 5 gm KCl in 60 ml distilled water for several minutes until color darkens. Cool and add 20 ml 28% NH_3-water. Ripen 24 hr and store at 0–5°C. This stock solution keeps only a few weeks. Mix 8 ml of the above stock solution with 12 ml 28% NH_3 and 24 ml methanol. Do not filter. Use within 24 hr and use only once.

Staining schedule:
1. Remove paraffin from sections through 2 changes each of xylene and 100% alcohol and place in 1% celloidin in 50:50 ether alcohol mixture for 5–10 min.
2. Drain 1 min with slides vertical.
3. Harden 5 min in 80% alcohol.
4. Transfer to tap water.

5. Stain 5 min in alum hematoxylin (p. 105).
6. Wash briefly in tap water.
7. Stain 20 min in the above diluted staining solution.
8. Rinse paraffin sections with 3 changes of fresh methanol; dehydrate and decollodionize with 2 or 3 changes of acetone; clear with 50:50 acetone-xylene mixture followed by 2 changes of xylene and mount in synthetic resin.

or

8b. Wash celloidin or nitrocellulose sections in several changes of 10 ml water, either tap or distilled, 8 ml ethanol and 4 ml methanol; dehydrate in 80% and 95% alcohol; clear in origanum oil; mount in balsam.

Results: Glycogen—red, nuclei—blue.

Bauer's Method

Slightly modified from Bensley (1930)

Fixation: Alcoholic formalin or 5% acetic acid plus 10% formalin alcohol. Good results may often be attained with neutral 10% formalin in water.

Embedded in: Paraffin.

Preparation of staining solution: Schiff reagent. Dissolve 1 gm basic fuchsin in 100 ml hot (80–100°C) distilled water. Filter at 60–80°C. Cool and add 2 gm $NaHSO_3$ and 10 ml N HCl. Stopper tightly and let stand 18–24 hr in the dark. Add 300 mg finely ground charcoal, shake 1 min, filter, and store at 0–5°C. The reagent should be clear straw yellow and should be discarded if it turns pink. It may be used several weeks.

Staining schedule:
1. Remove paraffin from sections through 2 changes each of xylene and 100% alcohol to 1% celloidin in 50:50 ether alcohol for 5–10 min. Drain 1 min on edge. Harden 5 min or as much longer as convenient in 80% alcohol.
2. Wash in tap water and immerse 1 hr in 5% CrO_3, pouring in and out of Coplin jar each 10 min.
3. Wash 5 min in running water.
4. Immerse 15 min in the above staining solution in covered Coplin jar, agitating every 3–5 min. Do not return the Schiff reagent to stock bottle.
5. Wash 90 sec each in 3 changes of M/20 $NaHSO_3$ (10.4% $NaHSO_3$, 6 ml; tap water, 114 ml; making 3 portions of 40 ml each).
6. Wash 10 min in running water.
7. Counterstain 2 min in acid hemalum.
8. Blue in tap water containing 2–3 drops of saturated aqueous Na_2CO_3.
9. Dehydrate.

10. Clear in xylene-alcohol mixture and 2 changes of xylene and mount in synthetic resin.

Results: Glycogen—purplish red, cartilage matrix—deep purple, chitin—red-purple, mucin—purplish pink, cellulose—red-purple, nuclei—blue.

Other Methods

See periodic acid leucofuchsin, (p. 200).

SECTION G: CYTOLOGY

Orcein and Feulgen Technics

Gay and Kaufmann (1950)

Application: Corneal epithelium.

Designed to show: Mammalian somatic mitosis.

Fixation: Fix entire eye for 30 min at room temperature in acetic acid-alcohol (3 parts glacial acetic acid and 11 parts ethyl alcohol). If fixation is prolonged or material is to be stored, refrigerate at 4°C.

Staining schedules:
A. Orcein staining.
  1. Remove eye from fixative and place for 30 min in a saturated solution of orcein (2 gm per 100 ml) in 60% acetic acid (prepared by refluxing 2 hr, cooling and filtering).
  2. Destain 5 min in 95% alcohol. During destaining, detach cornea from posterior portion by cutting circumferentially close to sclerocorneal junction. Make 4 incisions extending toward center from equidistant points on the circumference.
  3. Mount with epithelial surface uppermost in gum sandarac medium or euparal or diaphane.
B. Feulgen method (formulas for Solutions A and B given on p. 196).
  1. After fixation, rinse eye in distilled water for 5 min.
  2. Stain 20 min in leucobasic fuchsin (Solution A).
  3. Sulfite rinse (Solution B)—2 rinses 5 min each.
  4. Rinse in water.
  5. Counterstain in 0.25% fast green FCF or 0.25% orange G in 95% alcohol.
  6. Dissect (see above), while destaining in 95% alcohol.
  7. Mount in gum sandarac medium.

Simultaneous fixation and staining:
  1. Combine 11 parts of saturated solution of orcein in 95% alcohol with 3 parts saturated solution of orcein in glacial acetic acid and 2.4 parts of N HCl.
  2. Place unfixed eye in this orcein solution in stain for 45–60 min.

3. Destain 7 min in mixture of 3 parts of glacial acetic and 11 parts of alcohol. During destaining, detach cornea from rest of eye and prepare for flattening.
4. Wash 3 min in alcohol.
5. Mount in gum sandarac.

Squash preparations:
1. After enucleation, eye is held cornea downward, ½-inch above surface of a layer of glacial acetic acid in small beaker for 1–2 min.
2. Immerse in Ringer's solution (0.14 M or 0.85% NaCl) for a few seconds.
3. Detach epithelium from underlying tissue with sharp scalpel. With blade of scalpel held perpendicular to eyeball, scrape epithelium from edge toward center of cornea.
4. Immerse in Ringer's solution. Layer of scraped cells is floated free from eyeball onto slide in saline solution.
5. Remove slide from saline and cover the adhering epithelium with few drops of aceto-orcein and stain 15 min.
6. Place coverslip over tissue. With slide between several layers of filter paper or paper toweling, squeeze out stain by gentle pressure.

Results: Chromatin—dark translucent red (appearing black if green filter is used); cytoplasm—uncolored or light red.

Synthetic Orcein

Personal correspondence from B. P. Kaufmann
See Demeric and Kaufmann (1957, p. 27)

Application: Salivary gland cells of *Drosophilia* (and other Diptera) and meristematic cells of root tips.

Designed to show: Chromosomes.

Preparation of staining solution: Boil 1 gm orcein in 50 ml 45% acetic acid, refluxing for 2 hr. Cool and filter. Use as stock solution.

Staining schedules:
A. Salivary-gland chromosomes:
1. Dissect glands from well fed, third-instar larvae, preferably at low temperature (e.g. using a slide resting on a piece of ice).
2. Transfer glands into drop of acetic-orcein on albuminized slide. Allow to stain 3–6 min. Half-strength acetic-orcein is usually satisfactory (1 part stock solution: 1 part 45% acetic acid).
3. Add coverslip and flatten glands by pressing on cover with fingers (paper toweling may be used to absorb extruded dye and protect fingers) and then tapping or stroking cover with a blunt instrument.

4. Transfer slides to 95% alcohol-vapor chamber (a covered dish lined with paper toweling saturated with alcohol and containing rack in which slides have been placed). Allow to remain few hours to overnight.

5. Transfer slides to alcohol and allow to remain there until the covers drop off or can be removed easily. The squashed cellular material should adhere to the albuminized slide.

6. Mount in euparal.

B. Root-tip chromosomes:

1. Fix root tips in acetic-alcohol (1 part glacial acetic acid: 3 parts 100% alcohol).

2. Transfer tips to mixture of equal parts alcohol and conc. HCl for 5 min to facilitate separation of cells, and then rinse thoroughly in alcohol alone.

3–6. As given above, except that full-strength (stock solution) acetic-orcein may be required, the time of staining may need to be extended to 6–9 min, and greater pressure may be required to dissociate cells in squashing.

Results: Chromosomes—purplish to brownish red, cytoplasm and nucleoli—little coloration.

SECTION H: MISCELLANY

Methyl Green with Pyronin

From Taft (1951), and personal communication (1958)

Application: Animal tissue.

Designed to show: Nucleoli and cytoplasmic granules of liver cells (*i.e.* to differentiate cytoplasmic and nucleolar basophilia from chromatin).

Method of fixation: Carnoy (absolute alcohol, chloroform, glacial acetic, 6:3:1).

Preparation of staining solution: To 100 ml hot distilled water add an amount of methyl green representing 0.5 gm of pure dye (*e.g.* with methyl green NG-27, dye content 80%, add 0.625 gm). When cool, extract the solution in separatory funnel with successive 20–30 ml aliquots chloroform until the latter remains colorless or is only slightly tinged with green. Add to the above aqueous solution an amount of pyronin Y representing 0.05 gm of pure dye, and shake to dissolve. Store in amber, glass-stoppered bottle. Solution need not be filtered prior to use and may be reused.

NOTE: If pyronin B is to be used (although pyronin Y is considered superior) the initial methyl green solution should be prepared in 100 ml M/10 acetate buffer at pH 4.4, and the amount of pyronin B should be equivalent to 0.06 gm of pure dye.

Staining schedule:
1. Remove paraffin from sections and hydrate in usual manner.
2. Stain 3–5 min in the above staining solution.
3. Rinse in distilled water and blot with smooth filter paper.
4. Before completely dry, differentiate at least 2 minutes in tertiary butyl alcohol-absolute alcohol, 3:1.
5. Clear in 2 changes of xylene (5 min each), and mount in synthetic resin.

Results: Chromatin—blue green; nucleoli—rose; cytoplasmic granules—dark rose; cytoplasm of plasma cells—dark rose, occasionally almost purple; cartilage matrix and mast cell granules—refractile, orange-red.

    NOTE: Pyronin Y and B give slight differences in hue.

Periodic Acid Leucofuchsin (PAS) Method

From Lillie and Fullmer (1976)

Designed to show: Glycogen, mucin, reticulum, collagen, basement membranes, fibrin.

Fixation: Zenker or 10% formalin (although any may be used).

Embedded in: Paraffin. Cut sections at 5–7 μm.

Preparation of solution: Schiff's reagent: Dissolve 1 gm basic fuchsin in 80 ml distilled water and add 2 gm $NaHSO_3$ (or 1.9 gm $Na_2S_2O_5$) and 20 ml N HCl. Stopper tightly and shake at intervals for 2 hr, by which time the solution should be clear and light yellow. Add 500 mg finely powdered fresh charcoal, shake 1 min and filter. The solution should be clear and colorless. Store at 5°C. When kept at this temperature it remains active for some months. A white precipitate may appear after a time, and should be filtered out. When a pink tint appears, the solution should be discarded. Make periodic tests for potency on known material.

Staining schedule:
1. Remove paraffin from sections and hydrate through alcohols in the usual manner.
2. Rinse in tap water and oxidize 10 min in 1% periodic acid (H_5IO_6).
3. Wash 5 min in running tap water.
4. Immerse 10 min in Schiff's reagent.
5. Pass directly to 3 successive baths of 2 min each in M/20 (0.52%) $NaHSO_3$.
6. Wash 10 min in running tap water.
7. Stain 1–2 min in Weigert's acid iron chloride hematoxylin (p. 108), or acid hemalum (see p. 106) or 6 min if a cytoplasmic stain is to be used.
8. Wash in tap water.

9. If desired, counterstain 1 min in a saturated aqueous picric acid or 1% aqueous orange G. (Other counterstains may be used or a Gram-Weigert staining procedure may be substituted for Steps 9–11 of this method.)
10. Dehydrate in 2 changes each of 95% and 100% alcohol.
11. Clear through 100% alcohol-xylene (1:1) and 2 changes of xylene; mount in clarite.

Results: Nuclei—black or blue (if hematoxylin used); collagen—pink, red (or orange, if picric acid is used as counterstain); reticulum—purplish red (or slightly orange-red, if picric acid is used in Step 9); glycogen—dark purplish red; epithelial mucin—red-purple to violet; fibrin—pink (pink to violet if the Weigert fibrin variant was used as a counterstain); cytoplasm—gray, yellow or orange (depending on counterstain).

Feulgen Stain for DNA

Original Feulgen and Rossenbeck (1924)
Emended by Itikawa and Ogura (1954) and Lillie and Fullmer (1976)

Application: Most tissues.

Designed to show: DNA.

Fixative: Formalin but most fixatives are satisfactory.

Embedded in: Paraffin.

Solutions required:
1. Schiff reagent (see p. 200).
2. 5 N HCl.
3. 0.05 M (about 0.5%) $Na_2S_2O_3$.

Staining schedule:
1. Dewax and hydrate followed by the usual iodine thiosulfate sequence if a fixative containing Hg was used.
2. Hydrolyze in 5 N HCl for 40 min at room temperature.
3. Rinse in distilled water.
4. Stain for 10 min in Schiff reagent.
5. Three rinses of 2 min each in $Na_2S_2O_3$ solution.
6. Wash in tap water for 10 min.
7. Dehydrate, clear and mount.

Results: DNA is stained deep red-purple, other tissues elements unstained to very lightly stained.

Alcian Blue 8 GX for Carbohydrate Polyanions

Mowry (1975)

Application: Various tissues.

Designed to show: Complex carbohydrates with free acidic groups.

Fixation: Preferably neutral buffered formalin.

Solutions needed: 0.5 N HCl, 0.25% alcian blue 8 GX in 0.5 N HCl (add a crystal of thymol), 3% acetic acid, 1% alcian blue 8 GX in 3% acetic acid.

Staining schedule:
1. Dewax and hydrate.
2. Two changes of 0.5 N HCl 3 min each.
3. Stain 1 hr in 0.25% alcian blue 8 GX in 0.5 N HCl.
4. Wash in running water for 1 min.
5. Two changes of 3% acetic acid 3 min each.
6. Stain 1 hr in 1% alcian blue 8 GX in 3% acetic acid.
7. Two changes of 3% acetic acid.
8. Wash in running water.
9. Rinse in three changes of acid alcohol (1 ml HCl in 100 ml of 70% ethanol) 3 min in each.
10. Rinse in 2 changes of 70% ethanol, each for 3 min.
11. Dehydrate in 95% and absolute ethanol, clear and mount.

Results: Carbohydrate polyanions (mast cell granules, ground substance of connective tissue, cartilage cells and matrix, and most epithelial mucous secretions are colored turquoise blue.

NOTE: If Steps 4–9 are replaced by 3 changes of 0.5 N HCl of 2 min each followed by rinsing in running water the stain is limited to sulfated polyanions. After above note or after Step 9 either PAS (see p. 200) or aldehyde fuchsin (see p. 204) may also be used.

Prussian Blue Method for Hemosiderin

From Highman (1942) and Lillie and Fullmer (1976)
Original by Perls (1867)

Designed to show: Deposits of hemosiderin.

Fixation: 10% formalin, buffered at pH 7.0. Dissolve 4 gm $NaH_2PO_4 \cdot H_2O$ and 6.5 gm Na_2HPO_4 in 100 ml 37–40% formaldehyde solution and 900 ml distilled water.

Embedded in: Paraffin.

Preparation of staining solution: Dissolve 1 gm potassium ferrocyanide in 50 ml distilled water and 50 ml 2% HCl, C.P. (or 5% acetic acid).

NOTE: The ferrocyanide solution should be freshly made. Freshly formed deposits of iron pigment react well with the acetic variant and are less likely to be dissolved out. Older deposits may require the stronger acid for adequate reaction.

Staining schedule:
1. Remove paraffin and hydrate sections in usual manner.
2. Place sections in above staining solution for 30 min at 60°C or 1 hr at room temperature.
3. Rinse in distilled water.
4. Counterstain for 2 min in 0.2% safranin O in 1% acetic acid.
5. Wash in 1% acetic acid.
6. Dehydrate with 95% and 100% alcohol.
7. Clear in xylene.
8. Mount in synthetic resin.

Results: Hemosiderin—blue or green, nuclei—red, background—pink.

Ferric Ferricyanide Method for Reducing Groups

From Lillie and Fullmer (1976)
Original by Goldetz and Unna (1909)

Application: Demonstration of: chromaffin cells of adrenal medulla; enterochromaffin substance of basal granular cells of gastrointestinal mucosa; certain lipofuscin and melanin pigments; thyroid colloid; sites of keratinization.

Designed to show: Chemical groups which reduce ferri- to ferrocyanide—thus producing Prussian blue at site of reaction.

Preparation of staining solution: Combine 30 ml of 1% ferric chloride with 10 ml of freshly prepared 0.4% potassium ferricyanide.

Staining schedule:
1. Fix, embed in paraffin, section and deparaffinize.

   NOTE: Fixation requirements differ according to application. For chromaffin, $K_2Cr_2O_7$-formaldehyde mixtures without added acid. For sulfhydryl groups (in sites of keratinization), brief formaldehyde fixation. For thyroid colloid, alcoholic fixatives without dichromate. Otherwise, routine fixation is usually satisfactory. See Lillie (1965) for details.

2. Take to water. If necessary, certain blockade methods may be employed to distinguish the various substances which give positive reactions.
3. Wash in water and immerse for 10 min in freshly prepared ferricyanide reagent.
4. Wash in 1% acetic acid.
5. If desired, counterstain 10 min in 1:5000 fuchsin or new fuchsin in 1% acetic acid.
6. Rinse in 1% acetic acid.
7. Dehydrate, clear and mount in a synthetic resin.

Results: Reducing sites—blue, background—pale pink or faint green, nuclei—pink to red.

NOTE: Various similar methods are used for demonstrating iron and iron-containing pigments. See Lillie and Fullmer (1976) for details.

Toluidine Blue O for Metachromatic Substances

J. Lowell Orbison, personal communication (1959)
(See also Kramer and Windrum, 1955)

Designed to show: Metachromatic staining of acidic substances.

Fixation: 10% neutral formalin, cold absolute alcohol, or Carnoy (absolute alcohol-glacial acetic, 3:1).

Embedded in: Paraffin. Sectioned at 6 μm.

Staining solution: Dissolve 0.1 gm of toluidine blue O in 100 ml of distilled water.

Staining procedure.
1. Remove paraffin and bring sections to water in usual manner.
2. Stain for 1–2 min in the above 0.1% solution of toluidine blue (or for 5–10 min in a 0.01% solution).
3. Rinse in distilled water.
4. Apply coverslip directly from distilled water.
5. Blot around edges of coverslip and seal with Vaseline or fingernail polish.

Results: Certain acidic carbohydrates—metachromatic (pink to red or violet). Nuclei and cytoplasm—orthochromatic (blue). (Connective tissue mucins, ground substances of cartilage, mast cell granules and many epithelial mucins are metachromatic.)

NOTE: Aqueous preparations should be examined promptly. If a permanent preparation is desired, sections at Step 4 can be passed rapidly through 2 changes of 95% and 2 changes of 100% alcohol, 1 min each, then cleared in xylene and mounted in Permount. Sections so treated will show general diminution of metachromasia and may show loss of metachromasia in some sites.

Aldehyde Fuchsin Method

From Gomori (1950); Lillie and Fullmer (1976)

Designed to show: Elastic fibers, mast cells, granules of pancreatic islet β cells, and certain other cell granules.

Fixation: Formalin or Bouin preferred. Mercury fixatives result in pale lilac background. Avoid chromates.

Embedded in: Paraffin.

Preparation of staining solution: Dissolve 0.5 gm basic fuchsin in 100 ml 70% alcohol. Add 1 ml conc. HCl and 1 ml U.S.P. paraldehyde. When mixture becomes a deep violet (about 24 hr) it is ready for use. Store in refrigerator.

Staining schedule:
1. Remove paraffin and hydrate sections in usual manner.
2. Treat 10–60 min with 0.5% iodine.
3. Decolorize for 30 sec with 0.5% sodium busulfite.
4. Wash for 2 min in water.
5. Transfer to 70% alcohol.
6. Stain in above aldehyde fuchsin solution 5–10 min for elastic fibers, 15–30 min for islet cells, 30 min to 2 hr for hypophysis. Rinse in 70% alcohol and inspect microscopically from time to time.
7. Wash in several changes of 70% alcohol.
8. Stain 2–5 min in Ehrlich's hematoxylin (p. 106)
9. Blue in tap water.
10. Stain 1–2 min in 1% aqueous orange G.
11. Rinse in tap water.
12. Dehydrate, clear and mount.

Results: Elastic tissue—violet to purple, granules in mast cells—violet to purple, gastric chief cells—violet to purple, pancreatic beta cells—violet to purple, certain hypophyseal β cells—violet to purple.

NOTE: At Step 8 various other counterstaining procedures may be substituted, e.g. Masson trichrome, Van Bieson, or Mallory-Heidenhain Azan. Aniline blue should be replaced by light green SF yellowish or fast green FCF in these methods.

Methenamine Silver Method for Argentaffin Cells

From Gomori (1948); Burtner and Lillie (1949)

Designed to show: Argentaffin granules in enterochromaffin cells.

Fixation: Formalin, Bouin, or other formalin-containing fixatives. Dichromates and mercury salts impair contrast between granules and background.

Embedded in: Paraffin.

Preparation of solutions:
A. Weigert's iodine: Dissolve 1 gm iodine, 2 gm KI in 100 ml distilled water.
B. Dissolve 5 gm sodium thiosulfate ($Na_2S_2O_3 \cdot 10H_2O$) in 100 ml distilled water.
C. Stock methenamine solution: Dissolve 3 gm methenamine (hexamethylene tetramine) in 100 ml distilled water and add 5 ml of 5%

aqueous $AgNO_3$. Shake until initial heavy white precipitate dissolves. Solution can be stored in cool dark place for months.

D. Borate buffer, pH 7.8: Dissolve 1.7 gm H_3BO_3, 0.76 gm borax ($Na_2B_4O_7 \cdot$ $10H_2O$) and make up to 200 ml with distilled water.

E. *Working solution:* 30 ml stock methenamine solution (C) and 8 ml borate buffer solution (D).

NOTE: Coplin jars should be chemically clean. Previous silver mirror deposits should be removed with conc. nitric acid.

Staining schedule:

1. Remove paraffin and hydrate sections in usual manner.
2. Treat 10 min with Solution A.
3. Bleach for 10 min with Solution B.
4. Wash for 10 min in running water.
5. Rinse in 2 changes of distilled water.
6. Place sections in Coplin jars containing the buffered methenamine silver solution (E) at room temperature and put in a 60°C (paraffin) oven for 3–3½ hr. (Preheating the solution to 50°C reduces the impregnation time by ½–1 hr.)
7. Rinse in distilled water.
8. Tone in 10 min in 0.1% gold chloride.
9. Rinse in distilled water.
10. Fix for 2 min in Solution B.
11. Wash for 5 min in running water.
12. Counterstain for 5 min with 0.1% safranin O in 0.1% acetic acid.
13. Dehydrate with acetone, clear in xylene, and mount.

Results: Argentaffin cells—black (optimal at 3–3½ hr; impregnation partial and cells in reduced numbers at 2–2½ hr). Coarse connective tissue of the submucosa sometimes shows a variable amount of blackening by 3–3½ hr. Granules of eosinophil leucocytes, nuclei, smooth muscle and surface epithelium show additional blackening after incubation beyond 3½ hr. By 4½ hr a silver mirror begins to appear on slides and sides of Coplin jar. Granules of mast cells remain brilliantly red after nuclei and reticulin are blackened.

Relative Basophilia

Original Dempsey, Bunting, Singer and Wislocki (1947)
Emended by Clark (1971)

Designed to show: Basophilia of various tissues and tissue elements.

Fixation: Formaldehyde solutions.

Embedded in: Paraffin.

Preparation of staining solutions:

A. Stock solution of cresyl violet acetate:

Cresyl violet acetate	0.1 gm
H_2O q.s.	100 ml

B. Buffer solutions: Use HCl acetate or HCO_2CH_3 acetate buffers of pH 1.5, 2.0, 3.0, 4.0 and 5.0.

C. Staining solutions: Add 5 ml Solution A to 40 ml buffer.

Staining schedule:

1. Dewax and hydrate 5 slides.
2. Stain for 40 min in each of the prepared staining solutions.
3. Place the 5 stained slides in single staining rack.
4. 95% alcohol for 30 sec.
5. n-Butyl alcohol, 2 changes of 2 min each.
6. Clear and mount.

Results: The least basophilic structures stained in the pH 5 staining solution are much less strongly stained at lower pH's. Only the most basophilic structures are stained at pH 1.5.

Biebrich Scarlet at an Alkaline pH

Clark and Spicer (1979), Spicer and Lillie (1961)

Application: Various tissues.

Designed to show: Differential staining with changes in pH.

Fixation: Buffered $HgCl_2$ (see p. 19), formalin may be used but results are not as clear.

Embedded in: Paraffin.

Solutions needed: M/10 $Na_2B_4O7 \cdot 10 \ H_2O$, M/10 HCl and M/10 NaOH.

Preparation of staining solutions: Using above solutions prepare 100 ml each of solutions at pH 6, 8, 9.5 and 10.5. In each of these dissolve 40 mg of Biebrich scarlet.

Staining schedule:

1. Take four slides, dewax, hydrate, iodine-thiosulfate sequence for removal of Hg precipitates.
2. Wash thoroughly.
3. Stain on slide for 50 min in each of the above solutions (pH 6, 8, 9.5 and 10.5).
4. Drain briefly and immerse in absolute ethanol for 1 min.
5. Two changes of tertiary butanol for 3 min each.
6. Clear in xylene and mount.

Results: At pH 6 all structures are stained. At pH 8 collagen is not stained and muscle is only lightly stained. At pH 9.5 collagen, muscle and red blood cells are not stained. At pH 10.5 spermatazoa heads, elastic fibers, Paneth cell granules, eosinophilic granules and at least most "acidophilic inclusion bodies" are the only elements stained.

Displacement Stain for Acidophilic Structures

Clark (1979)

Application: Various tissues.

Designed to show: Differential staining with time. This is the stain of choice for acidophilic inclusion bodies and is far superior to Lendrum's phloxine-tartrazine (1947).

Fixation: Formalin, buffered $HgCl_2$.

Embedded in: Paraffin.

Solutions needed: 1% aqueous rose Bengal, 50 mg Bismark brown Y in 100 ml ethyl Cellosolve.

Staining schedule:
1. Take four slides, dewax, hydrate, iodine $Na_2S_2O_3 \cdot 5H_2O$ sequence if needed.
2. Wash well.
3. Stain in 1% aqueous rose Bengal for 10 min (overstaining does not occur).
4. At intervals of 10 min rinse slide briefly in isopropyl or n-butyl alcohol or in ethyl Cellosolve and place in Bismark brown Y solution.
5. After the fourth slide has been in Bismark brown Y displacing solution for 20 min remove all slides and rinse in isopropyl or n-butyl alcohol or in ethyl Cellosolve.
6. Two changes of tertiary butyl alcohol for 2 min each, clear in xylene and mount.

Results: After 20 min in Bismark brown Y collagen and muscle will be only lightly stained. Other structures will still be heavily stained. After 50 min only sperm heads, Paneth cell granules, and acidophilic inclusion bodies (*e.g.* Russell bodies) remain heavily stained with the rose Bengal. Usually erythrocytes, too, remain red colored. Elastic fibers survive 40 min displacement but lose at least a portion of the red stain at 50 min.

Papanicolaou Stain

Papanicolaou (1941)
From Lillie and Fullmer (1976)

Designed to show: Vaginal smears for detection of vaginal, cervical and uterine cancer. It is also valuable for smears and imprints from a variety of sources. There are a considerable number of variants, most of which will give good results.

Fixation: Smears are not allowed to dry but are fixed immediately in ether-alcohol (diethyl ether, one part; absolute alcohol, one part).

Preparation of staining solutions:

A. Harris hematoxylin (see p. 106).
B. 0.03 N HCl (approximately a 1/400 dilution of the concentrated acid.
C. Orange G (OG-6).

Orange G	0.5 gm
Phosphotungstic acid	0.015 gm
Alcohol (95%) *q.s.*	100 ml

D. Counterstain (EA 50).

Light green SF yellowish (0.5% alcoholic solution)	4.5 ml
Bismark brown (0.5% alcoholic solution)	10 ml
Eosin Y (0.5% alcoholic solution)	45 ml
Phosphotungstic acid	200 mg
Li_2CO_3 (saturated aqueous solution)	1 drop

Staining schedule:

1. 80%, 70% and 50% alcohols and distilled water, 30 sec each.
2. Harris hematoxylin (without acetic acid), 5 min.
3. Distilled water, 6 dips.

NOTE: Each dip should take about 1 sec. The actual movement of the slide is needed to promote even staining. Slow dips should take about 5 sec.

4. 0.03 N HCl, 8 dips.
5. Running water, 6 min.
6. Distilled water, 50%, 70%, 80% and 95%, 30 sec each.
7. Solution C, 1½ min.
8. 95% alcohol, 2 changes each, 2 slow dips.
9. Solution D, 1½ min.
10. 95% alcohol, 3 changes each, 2 slow dips.
11. Absolute alcohol, 30 sec.
12. Equal parts absolute alcohol and xylene, 4 min.
13. Xylene, 2 min.
14. Xylene, 3 min.
15. Mount in synthetic resin.

Buffered Azure Eosin Method

Modified Nocht's method
From Lille and Fullmer (1976)

Designed to show: Cell granules, microorganisms, nuclei.

Fixation: Zenker or formalin.

Preparation of staining solution:

	Coplin jar	*Technicon*
Azure A	4 ml 0.1%	6 ml 1%
Eosin B	4 ml 0.1%	6 ml 1%
M/5 Acetic acid	1.7 ml	34 ml
M/5 Sodium acetate	0.3 ml	6 ml
C.P. acetone	5 ml	90 ml
Distilled water	25 ml	585 ml

Staining schedule:
1. Bring paraffin sections to water, using 0.5% iodine, 5% sodium thiosulfate sequence for mercuric chloride fixed material.
2. Stain 1 hr in the above staining solution.
3. Dehydrate in 2–3 changes of acetone.
4. Clear in a 50:50 acetone xylene mixture.
5. Two changes in xylene.
6. Mount in synthetic resin. Permount, HSR, polystyrene, Groat's copolymer and ester gum are satisfactory.

   NOTE: The buffer mixture (pH 4.1) may be altered, lower pH levels giving redder effects, higher levels more blue. Zenker fixed materials require pH 4.5 to 5.0; formalin fixed, pH 3.75 to 4.3. One may substitute 0.5 ml 1% azure eosinate in glycerol and methanol for the separate aqueous solutions of eosin (10 ml for Technicon schedule).

   Nitrocellulose sections and collodionized sections require 2–2½ hr staining, or 2–3 times as much dye. If necessary to preserve the collodion coat, substitute isopropanol or *n*-butyl alcohol for acetone in the dehydration and clearing schedule.

Results: Nuclei, tigroid, bacteria, rickettsia—blue; mast cell and basophile leukocyte granules—blue-violet; cartilage matrix—reddish violet; calcium deposits—dark blue; cytoplasm of surviving cells—light blue to violet or lavender; cytoplasm of necrosing and necrotic cells, muscle fibers—bright pink; secretion granules in pancreatic and salivary gland acini and Paneth cells—pink; cytoplasm of gastric gland parietal cells—pink; cytoplasm of chief cells—blue; chromaffin after chromate fixation only—yellowish green to green; eosinophil and pseudoeosinophil granules—pink; keratin, amyloid, fibrin, muscle cytoplasm, thyroid colloid, nuclear and cytoplasmic oxyphil inclusion bodies and bone matrix—pink; erythrocytes and hemoglobin—orange-pink; various hyalin degeneration products—pink; mucin varies from unstained, pale greenish blue to fairly deep blue-violet.

   NOTE: The larger quantities prescribed for the Technicon schedule are allowed to stand 1 hr after mixing, and then filtered. The mixture is used repeatedly, discarding at the end of each week. The smaller, Coplin-jar quantities are used at once and discarded after use.

Bismark Brown with Methyl Green

Quoted from Conn (1940, p. 60)
Essentially the same as List (1885)

Application: Embryonic tissue; trachea; intestine.

Designed to show: Mucin; cartilage; goblet cells.

Fixation: Bouin or Zenker.

Embedded in: Paraffin.

Staining schedule:
1. Remove paraffin from the sections in the usual manner.
2. Stain in a 1% aqueous bismark brown for 5–10 min.
3. Rinse in 95% alcohol.
4. Transfer to 0.5% aqueous methyl green until the slide appears dark green.
5. Dehydrate.
6. Clear and mount in synthetic resin.

Results: Cartilage—dark brown, mucin—light brown, nuclei of all the cells—green.

Thionin for Frozen Sections

From Mallory (1938, p. 35)

Application: Biopsy material.

Designed to show: Malignant cells.

No fixation: Frozen sections.

Staining schedule:
1. Stain 30 sec to 1 min in 0.5% thionin in 20% alcohol.
2. Wash in tap water.
3. Mount in tap water.

Results: Nuclei—blue to purple, collagen—red, elastin—light green.

NOTE: A more intense nuclear stain can be obtained by using in the same way a 1% solution of toluidine blue in water or in 20% alcohol.

Thionin or Azure A for Fixed Material

From Lillie (1965)

Designed to show: Mast cells; mucin; cartilage.

Fixation: Any formalin or sublimate method.

Staining schedule:
1. Stain 30–60 sec in 1:1000 aqueous solution of thionin buffered at pH 5.0 or 4.0 with acetate in M/200 final level; or stain 4–5 min at pH 3.
2. Wash briefly with tap water.
3. Dehydrate in acetone.
4. Clear in xylene and mount.

Results: Nuclei—deep blue, cytoplasm—light blue, erythrocytes—green, mucus—red-purple to violet, cartilage matrix—red-purple to violet, mast cell granules—deep violet.

> **NOTE:** At pH 1 (1% dilution of conc. HCl) and 1.5000 thionin, only mast cells, cysteic acid, and some cartilages stain. At about pH 1.5 nucleic acid staining is obtained.
> With staining at pH 3, erythrocytes are refractile, pale yellow, cytoplasm unstained, some mucin failing to stain. Nuclei and mast cell granules still stain at pH 2.
> At pH 4 muscle and cytoplasm are pale green, erythrocytes deeper green.
> At pH 7 mast cells stain red, nuclei lose sharp definition.
> The stain can be further diluted (1:10,000 or 1:100,000) and the time prolonged to overnight. Overstaining does not occur.
> In low pH ranges, if HCl-KH$_2$PO$_4$ buffers are used, substitute Azure A, since phosphates precipitate thionin.

Mucicarmine with Hematoxylin and Metanil Yellow

Mayer (1896)
Adapted from Masson (1923, p. 697–698)

Designed to show: Mucin.

Fixation: Formalin 1:10; any fixative will do.

Embedded in: Paraffin.

Preparation of staining solutions:
A. Mucicarmine. Dissolve 1 gm carmine and 0.5 gm anhydrous AlCl$_3$ in 2 ml distilled water by aid of gentle heat for 2 min, stirring constantly until the mixture becomes dark. Continue stirring while adding gradually 100 ml 50% alcohol. Let it stand for 24 hr, then filter. Keep this as a stock solution. Dilute with 10 vol of distilled water or preferably 50–70% alcohol just before use. Lillie (1948) prefers a freshly prepared solution. This is essentially the same staining solution as Mayer (1896).

> **NOTE:** Southgate (1927) recommends the following method of preparing this solution, for it gives a more constant reaction.
> To 1 gm carmine and 1 gm Al (OH)$_3$ in a flask, add 100 ml 50% alcohol; add 0.5 gm freshly powdered AlCl$_3$ and place on a boiling water bath, with

frequent agitation. After just 2½ min of boiling and shaking, cool under a stream of water and filter. Dilute 1 vol of this stock solution with 9 vol of water a couple of days or less before using.

NOTE: If it is desired only to stain the mucin, omit Steps 3–5.

B. Metanil yellow solution: Place 0.25 gm metanil yellow on filter paper in a funnel and pour over it 1000 ml 0.25% acetic acid.

Staining schedule:

1. Remove paraffin from sections in usual manner.
2. Wash sections in distilled water.
3. Stain sections 1 min in Weigert's iron hemotoxylin
4. Wash in distilled water.
5. Stain 30 sec in metanil yellow (Solution B).
6. Wash in distilled water.
7. Stain 45 min in mucicarmine (Solution A).
8. Wash quickly in distilled water, then in 95% alcohol, and finally in absolute.
9. Clear in xylene and mount.

Results: Nuclei—black, connective tissue—yellow, mucin—red.

Modification of Mayer's Mucihematein

Original Mayer (1896)
Quoted from Laskey (1950), Lillie (1965)

Application: Glandular tissue—particularly mucins derived from epithelial cells.

Designed to show: Mucin.

Fixation: Absolute alcohol for 5–8 hr or in 5% sublimate solution for 5 hr.

Embedded in: Paraffin or celloidin.

Preparation of staining solution: Dissolve 1.0 gm hematoxylin in 100 ml 70% ethyl alcohol, add 0.5 gm aluminum chloride and 40–100 mg $NaIO_3$ (4–10 ml of 1% aqueous $NaIO_3$) and make up to 500 ml with 70% ethyl alcohol. (The smaller amount of $NaIO_3$ is preferred as the solution keeps better. More may be added if a staining test shows insufficient ripening.)

Staining schedule:

1. Deparaffinize sections and bring to water as usual.
2. Place on staining rack and rinse several times with distilled water.
3. Place 2 ml of the staining solution on each slide and allow to remain 5–10 min.
4. Drain off stain and wash well with distilled water 3 times for 5 min each.

5. Dehydrate in 2 changes of 95% alcohol, 2 changes of absolute alcohol, absolute alcohol + xylene (50:50), and 2 changes of xylene.
6. Mount in polystyrene, Permount, or gum damar.

Results: Mucin—deep violet, cell nuclei—pale gray-blue, connective tissue—pale gray to colorless.

Crystal Violet Stain for Amyloid

Procedure derived by R. D. Lillie from statements in Conn (1940) concerning methyl, gentian and crystal violets

Application: Pathological human tissue.

Designed to show: Amyloid.

Fixation: Absolute alcohol or 10% formalin.

Embedded in: Paraffin.

Preparation of modified Apathy's gum syrup: Dissolve 50 gm acacia (gum arabic) and 50 gm cane sugar in 100 ml distilled water by frequent shaking at 500–60°C. Restore volume with distilled water, add 15 mg Merthiolate (sodium ethylmercurithiosalicylate) or 100 mg thymol, as a preservative. Place in vacuum chamber for a few minutes, while warm, to remove air bubbles. Highman (1942) adds 50 gm potassium acetate or, preferable 10 gm NaCl to this formula, to prevent bleeding of crystal violet amyloid stains.

Preparation of staining solution: Crystal violet, 1.0 gm; methyl violet 2B, 0.5 gm; approximately 10% alcohol, 100 ml.

Staining schedule:
1. Stain sections of unfixed or fixed tissue in the above staining solution for 3–5 min.
2. Wash in a 1% aqueous solution of acetic acid.
3. Wash thoroughly in water to remove all traces of acid.
4. Examine in water or in glycerin. The stain will keep for some time if the sections are mounted in a saturated solution of potassium acetate (about 200 gm to 100 ml water) or in levulose.
5. Mount in a modified Apathy's gum syrup or Lieb's Abopon (Glyco Products Co., New York).

Results: Amyloid—violet-red, tissue—blue.

Benhold's Congo Red for Amyloid

From Puchtler and Sweat (1962) and Lillie (1965)
Originally described by Bennhold (1922)

Application: Pathological human tissue.

Designed to show: Amyloid.

Fixation: Formalin or alcohol.

Embedded in: Paraffin or celloidin, or frozen sections.

Staining schedule:
1. Dewax and hydrate through graded alcohols, removing formalin pigment and mercury precipitates as usual.
2. Stain 10 min in Mayer's acid hemalum (p. 106).
3. Wash in 3 changes of distilled water.
4. Transfer to alkaline NaCl alcohol for 20 min. To 40 ml saturated sodium chloride solution in 80% alcohol (stable stock solution) add 0.4 ml 1% NaOH just before using.
5. Stain 20 min in freshly alkalinized Congo red solution. To 40 ml of the stable stock saturated solution of Congo red in 80% alcohol saturated with NaCl, add 0.4 ml 1% NaOH, filter at once and use alkalinized solution within 15 min.
6. Dehydrate quickly in 3 changes of absolute alcohol, clear in xylene and mount in Permount.

Results: Amyloid—red to pink, elastic tissue—lighter red, nuclei—blue, other structures—largely unstained.

CHAPTER 8

STAINING METHODS FOR CELL TYPES IN ENDOCRINE GLANDS

Robert E. Coalson

The selective staining of specific cell types within the various endocrine glands depends upon the presence of granules which represent either stored secretory products or their precursors. The ease with which individual cell types can be identified will be influenced, therefore, by the functional state of the cell as reflected by the degree of granularity and by the efficacy of the fixing solution.

The empirical stains presented in this section represent only a small sampling of the technics currently available for the selective demonstration of cell types. It should be noted, however, that many of the methods which have not been included yield preparations which are equivalent or superior to those presented. A high priority has been given to simple procedures which are adaptable to formalin-fixed material. The latter requirement was considered because frequently only formalin-fixed material is available for study.

The histochemical procedures which have been included in this section operate on mechanisms which are at least partially understood and which demonstrate chemical groups or configurations which are known to be associated with the specific hormones elaborated by these cells. The chemical components demonstrated by these methods are not peculiar to the secretory products of endocrine cells, but the selectivity of staining and geographical location, as well as granule size and morphology permit identification of individual cell types. Furthermore, since the chemical composition of many hormones is similar in different vertebrate groups, these methods can be applied to comparative studies.

As with all cytological technics, the successful application of these empirical and histochemical procedures requires well preserved material and thin sections.

ISLETS OF LANGERHANS

There is general agreement that the islets of most vertebrates contain three types of granulated cells but it should be noted that additional cell types (granular and nongranular) have been reported. The terminology used in this section is based on the three classically recognized granular cells.

Alpha cell = A_2 alpha, non-argyrophil, silver negative, glucagon producing cell

beta cell = insulin producing cell

D-cell = A_1 alpha, argyrophil, silver positive, metachromatic, carboxyl-rich polypeptide producing cell, pancreatic gastrin cell.

Mallory-Heidenhain Azan—Gomori's Modification for Islet Cells

From Gomori (1939)
Original from Heidenhain (1915)

Designed to show: Alpha, beta and D-cells simultaneously in the islets of Langerhans.

Tissue preservation: Bouin's, Helly's.

Sections: 3–5 μm paraffin sections.

Preparation of reagents:

A. Azocarmine: Add 0.1 gm azocarmine G to 100 ml of distilled water and boil for 5 min. Restore to original volume with distilled water and add 2 ml of glacial acetic acid. Before use, heat to 60°C and filter into warm staining dish. This stain will keep for months at room temperature.

B. Differentiation solution: 99 ml of 90% alcohol with 1.0 aniline oil.

C. Mordant: 5% aqueous solution of iron alum (w/v).

D. Stock solution of aniline blue-orange G: 0.5 gm aniline blue and 2.0 gm orange G in 100 ml of distilled water. Do not add acid.

Staining procedure:

1. Deparaffinize and take through graded alcohols to water.

2. Wash in running water to remove all traces of picric acid.

3. Stain for 45–60 min in covered staining dish at 56°C in azocarmine stain (Solution A).

4. Rinse briefly and blot.

5. Under microscopic control, differentiate in aniline-alcohol (Solution B) until acinar parenchyma becomes almost completely decolorized and beta cells stand out red against the pinkish alpha cells (see Results).

6. Rinse briefly in distilled water.
7. Transfer to 5% iron alum solution for 5 min or more (Solution C)
8. Rinse briefly in distilled water.
9. Stain for 2–20 min under microscopic control in aniline blue-orange G (Solution D) diluted 1:3 with distilled water.
10. Rinse briefly and blot.
11. Differentiate and dehydrate in absolute alcohol. (To intensify orange colors which may be extracted during this step, dehydrate in absolute alcohol saturated with orange G and use a second bath of absolute to remove unbound dye.)
12. Clear in xylene and mount in synthetic resin.

Results: The selectivity of the azocarmine and orange G for alpha and/or beta cell granules varies with the type of fixation employed and with the species used (Gomori, 1939, 1941). In human islets after Bouin fixation, the alpha granules are bright red, beta granules are dingy orange-brown and D-cell granules are dark blue. In guinea pig pancreas after Bouin fixation, the alpha granules are orange-tan, beta granules are fiery red and the D-cell granules are deep blue.

Comments: This procedure is somewhat long and tedious to perform but when all conditions are optimal, it produces very striking preparations. This procedure is applicable to cell types in the adenohypophysis.

Mallory-Heidenhain Azan after Permanganate Oxidation

From Gomori (1939)
Original from Heidenhain (1915)

Designed to show: Alpha cells (A$_2$) only in all animals.

Tissue preservation: Alcohol fixation gives poor results but any other type may be used. Strong overfixation in mercuric fixatives will abolish the staining property of alpha cells.

Sections: 3–5 μm paraffin sections.

Preparation of reagents:
A–D. Same reagents as in preceding technique.
E. Oxidizing solution: 0.25 gm $KMnO_4$; 100 ml distilled water. Add 0.5 ml concentrated sulfuric acid just before use.
F. Bleaching solution: 1–5% aqueous potassium metabisulfite or 5% oxalic acid (w/v).

Staining procedure:
1. Deparaffinize and take through graded alcohols to water.
2. Wash in running water to remove all traces of picric acid or treat with

Lugol's and sodium thiosulfate to remove mercury precipitates if necessary.

3. Oxidize sections for 30–60 sec with freshly prepared permanganate (Solution E).
4. Bleach sections in potassium metabisulfite or oxalic acid for 1 min or until sections are colorless (solution F).
5. Wash in running water for 5 min.
6. Stain in azocarmine for 45–60 min at 56°C in a covered staining dish (Solution A).
7. Wash in running water for about 1 min. No other differentiation may be required. If section is too dark, complete differentiation in aniline-alcohol as with Mallory-Heidenhain procedure above using Steps 8 through 12.

Results: In all species examined and with almost all fixatives the following results are obtained: alpha cells—coarse bright red granules; beta and D-cells cannot be distinguished.

Comments: This procedure can also be applied to the adenohypophysis. Oxidation increases the affinity of pituitary acidophils for azocarmine, but the effect is less pronounced than that obtained with the pancreatic islet cell.

Phosphotungstic Acid Hematoxylin (PTAH)

From Levene and Feng (1964)
Original from Mallory (1938)

Designed to show: Alpha cell (A_2) granules in man and all common laboratory mammals.

Tissue preservation: Formalin-acetic or Bouin's recommended. Formalin-fixed material may be used but may require overnight pyridine extraction at 37°C to remove stainable mitochondrial remnants. Wash thoroughly to remove pyridine.

Sections: 3–5μm paraffin sections.

Preparation of reagents:
A. Oxidizing solution: 0.3 gm of $KMnO_4$ in 100 ml distilled water. Add 0.32 ml concentrated sulfuric acid just before use.
B. Bleaching solution: 4% aqueous potassium metabisulfite (w/v).
C. Mordant: 4% aqueous iron alum (w/v).
D. PTAH staining solution: 0.1 gm hematoxylin in 100 ml of warm distilled water, cool and add 2.0 gm phosphotungstic acid and shake to dissolve. Ripen by adding 2.5 ml of 1% aqueous $KMnO_4$ and shake

vigorously; let stand for 48 hr, filter and store. Refilter before each use.

Staining procedure:

1. Deparaffinize and take sections through graded alcohols to water.
2. Oxidize in freshly prepared acidified-permanganate for 5–40 sec (Solution A).
3. Rinse in water.
4. Decolorize in potassium metabisulfite for 5–10 sec or until sections are colorless (Solution B).
5. Wash in running water for 5 min.
6. Rinse in distilled water.
7. Mordant in iron alum for 30 min to 2 hr (Solution C).
8. Rinse in distilled water.
9. Stain in PTAH for 16–48 hr (Solution D).
10. Pass slides quickly through graded alcohols to 95% alcohol or blot and take directly to 95%.
11. Rinse in 95% alcohol until no more stain runs from the tissue—about 10 sec to 2 min.
12. Dehydrate in absolute alcohol, clear in xylene and mount in synthetic resin.

Results: Alpha cell granules are clearly stained a deep blue. Nuclei of all cells are sharply but unobtrusively stained. Red blood cells are deep blue-black. Background staining is suppressed and may appear pale rose-pink or almost colorless.

Comments: This method produces consistent reproducible staining, permits wide variation in time at each stage and does not require differentiation under microscopic control. The staining is progressive but does not tend to overstain. Sections may be loosened or detached during longer staining times. Since mitochondrial remnants are stained in formalin fixed material, this technique might be useful for demonstrating cells known to contain large numbers of mitochondria, e.g., oxyphil cells in the parathyroid gland.

Tryptophan: Post-Coupled p-Dimethylaminobenzaldehyde (DMAB) Reaction for Indoles

From Barka and Anderson (1963)
Original from Glenner and Lillie (1957a)

Designed to show: Indoles (tryptophan). Glucagon producing alpha cells (A_2) are demonstrated in islets because of their high tryptophan content (approximately 7%).

Tissue preservation: 3–6 hr in 10% neutral buffered formalin. Alcohol hardening overnight or longer after brief formalin fixation is helpful for paraffin sections.

Sections: 4–8 μm paraffin sections.

Preparation of reagents:
  A. DMAB reagent: 1.0 gm *p*-dimethylaminobenzaldehyde in 10 ml concentrated HCl and 30 ml glacial acetic acid.
  B. Coupling reagent: The diazotate of S-acid (8-amino-1-naphthol-5-sulfonic acid) should be prepared just before use. 240 mg of S-acid in 3.0 ml of 1 N HCl and 6 ml distilled water. After cooling in an ice bath to 4°C add 1.0 ml of 1 N $NaNO_2$. Solution should be stirred frequently for 15 min at 4°C or in an ice-water bath.

Staining procedure:
  1. Deparaffinize sections and rinse in absolute alcohol (2 changes) to remove most of the xylene. Air-dry for 30 sec.
  2. Treat sections for 5 min at room temperature with DMAB reagent (Solution A).
  3. Rinse in glacial acetic acid for 30 sec.
  4. Wash in 2 changes of glacial acetic acid, 1 min each.
  5. Transfer sections into a Coplin jar containing 40 ml of glacial acetic acid and 1.0 ml of the coupling reagent (Solution B) for 5 min at room temperature.
  6. Wash in glacial acetic acid for 30 sec.
  7. Wash in glacial acetic acid, 2 changes, for 1 min each.
  8. Clear in 50:50 mixture of glacial acetic acid-xylene.
  9. Final clearing in 3 or 4 changes of pure xylene.
  10. Mount in synthetic resin.

Results: Structures containing tryptophan or other indoles are stained varying intensities of blue depending on the concentration. Alpha cells (A_2) are the only islet cells stained.

Comments: The reaction is easy to perform and produces very good color if formalin fixation has not been prolonged. The zymogen granules of the exocrine pancreas are well stained but are easily distinguished from alpha cells granules by their size and location. Glenner and Lillie (1957a) have reported strong indole reactions with this procedure in one type of pituitary basophil (beta) and in pituitary colloid.

Aldehyde-Fuchsin

From Gomori (1950)

Designed to show: Beta cells of all vertebrates examined and those of some functional beta cell tumors (those storing granules).

Tissue preservation: Formalin, Bouin's. Avoid dichromate and mercury containing fixatives if possible to prevent excessive background staining.

Sections: 3–6 μm paraffin sections.

Preparation of reagents:
  A. Lugol's iodine solution: Iodine, 2 gm; KI, 4 gm; 100 ml distilled water.
  B. Bleaching solution: 5% aqueous sodium thiosulfate (w/v).
  C. Aldehyde-fuchsin stain: Add 1.0 ml concentrated HCl and 1.0 ml paraldehyde U.S.P. to 100 ml of a 0.5% solution of basic fuchsin in 60–70% alcohol. Let stand 24 hr at room temperature before use. If refrigerated, the stain is good for 6 mo.

Staining procedure:
  1. Remove paraffin and take slides to water.
  2. Remove mercury precipitate with Lugol's (10 min to 1 hr) (Solution A). Lugol's is recommended for all slides; it shortens the staining time and makes the shades darker.
  3. Remove iodine with sodium thiosulfate (Solution B).
  4. Wash in running water for 5 min.
  5. Rinse in distilled water.
  6. Place slides for 3 min in either 60 or 70% alcohol depending on alcoholic concentration of the aldehyde-fuchsin stain.
  7. Stain in aldehyde-fuchsin (Solution C) for 5 min to 2 hr or longer. The stain is very selective and does not overstain. The stain can be rinsed from the slide with alcohol and checked under the microscope any number of times. When sufficiently dark, rinse with several changes of alcohol. Do not expose the slide to dilute alcohols or water until excess stain has been removed or precipitation will occur. After staining and alcoholic washes are complete, the slide can be exposed to aqueous solutions without danger since the stain is insoluble in water. The stain can be extracted slowly from sections with acid alcohol only if the stain has not aged for more than 3–4 days before use. After this time, the stain is almost impossible to remove.
  8. If desired, counterstain with hematoxylin and eosin or orange G.
  9. Dehydrate in absolute alcohol.
  10. Clear in xylene and mount in synthetic resin.

Results: Beta cell granules are sharply stained in deep purple. Elastic fibers, mast cell granules and mucus are also stained. The background is almost colorless after formalin fixation, but is somewhat darker after Bouin fixation. Background staining may be objectionable after dichromate or mercuric fixatives or if other oxidants are substituted for Lugol's.

Comments: Aldehyde-fuchsin is probably the best method for demonstrating the granules in beta cells. This method has also been used to demonstrate some type of pituitary basophil (thyrotrophs and gonadotrophs have both been reported to stain).

Chrome Hematoxylin-Phloxin

From Gomori (1941)

Designed to show: Beta cells.

Tissue preservation: Bouin's, formalin or chromate fixatives.

Sections: 3–5-μm paraffin sections.

Preparation of reagents:
A. Bouin's fluid: 75 ml of saturated aqueous picric acid; 25 ml formalin; 5 ml glacial acetic acid.
B. Oxidizing solution: 0.3 gm $KMnO_4$ in 100 ml distilled wter. Add 0.32 ml concentrated sulfuric acid just before use.
C. Bleaching solution: 2–5% aqueous solution of sodium metabisulfite (w/v).
D. Hematoxylin: Mix equal parts of 1% aqueous solution of hematoxylin (w/v) and 3% aqueous chrome alum, $CrK(SO_4)_2 \cdot 12H_2O$ (w/v). Add to each 100 ml of this mixture, 2.0 ml of 5% potassium dichromate and 2.0 ml of 0.5 N sulfuric acid. The mixture is ripe after 48 hr and can be used as long as a film with a metallic luster will form on the surface after standing 1 day in a Coplin jar (about 4–8 weeks). Filter before each use.
E. Acid alcohol: 100 ml of 95% alcohol with 1.0 ml concentrated HCl.
F. Phloxin stain: 0.5 gm phloxin in 100 ml distilled water.
G. 5% phosphotungstic acid, aqueous (w/v).

Staining procedure:
1. Deparaffinize sections and take through graded alcohols to water.
2. Treat Zenker and Helly fixed material to remove precipitate using Lugol's followed by sodium thiosulfate bleach. Wash 5 min in running water.
3. Refix all sections in Bouin's for 12–24 hr (Solution A).
4. Wash thoroughly in running water to remove picric acid.
5. Oxidize in freshly prepared acidified permanganate for 1 min (Solution B).
6. Decolorize in sodium metabisulfite (Solution C).
7. Stain in chrome hematoxylin under microscopic control until beta cells are deep blue (10–15 min) (Solution D).
8. Differentiate in acid alcohol for about 1 min (Solution E).
9. Counterstain with aqueous phloxin for 5 min (Solution F).
10. Immerse in 5% phosphotungstic acid solution for 1 min (Solution G).

11. Wash in running tap water for 5 min. Sections should regain red color.
12. Differentiate in 95% alcohol. If sections are too red, rinse for 15–20 sec in 80% alcohol.
13. Dehydrate in absolute alcohol.
14. Clear in xylene and mount in synthetic resin.

Results: Beta cells are stained steel blue to black. Alpha cells and D-cells are stained pink to red but are not differentially stained.

Comments: This method may give results when all others fail due to postmortem change or to less than ideal fixation. Basophils and acidophils in the adenohypophysis are also well shown by this technic. However, the stain is so precise and intense that many poorly granulated cells (chromophobes) are also stained and the overall number of chromophobes will appear to be reduced.

Alcoholic Silver Nitrate

From Hellman and Hellerström (1960)
Original from Davenport (1930)

Designed to show: Argyrophilic cell (A_1, D-cell or metachromatic islet cell).

Tissue preservation: Bouin's, Romeis or 10% formalin.

Sections: 3–5 μm paraffin sections.

Preparation of reagents:
  A. Bouin's fluid: 75 ml of saturated aqueous solution of picric acid; 25 ml formalin; 5 ml glacial acetic acid.
  B. Alcoholic silver nitrate: 10 gm silver nitrate; 10 ml distilled water; 90 ml of 95% alcohol; 0.1 ml of 1 N nitric acid. Before use, the solution is diluted 1:6 with distilled water and the pH adjusted to 5.

  NOTE: The silver nitrate solution should not be used with a fiber junction electrode; a special sleeve-type double junction reference electrode must be used if a pH meter is utilized in this step. However, since true pH values cannot be determined for alcoholic solutions with a pH meter, indicator paper (pHydrion paper in the pH 3.0–3.5 range) can be used for the final pH adjustment. Titrate carefully using dilute ammonium hydroxide (1:10) and 1 N nitric acid.

  C. Developing solution: 5 gm pyrogallic acid; 100 ml 95% alcohol; 5 ml formalin.

Staining procedure:
  1. After careful deparaffinization, the sections are refixed in Bouin's solution at 37°C for 2 hr (Solution A).
  2. Wash in running tap water for 1 hr.

3. Pass slides through graded alcohols to 95% alcohol.
   CAUTION: *From this point, use plastic or paraffin coated forceps and avoid contact with other metal objects, e.g., staining racks, Coplin jar lid-liners, etc.*
4. Stain for 12–18 hr at 37°C in the dilute alcoholic silver nitrate solution, pH 5 (Solution B); use a tightly sealed Coplin jar and avoid exposure to light.
5. Wash with constant agitation in 95% alcohol for no more than 10 sec.
6. Develop for 60 sec (Solution C).
7. Rinse in 95% alcohol 3 times for 1 min each.
8. Dehydrate in absolute alcohol.
9. Clear in xylene and mount in synthetic resin.

Results: The granules of the argyrophilic alpha cell (A_1 or metachromatic D-cell) are impregnated with silver and appear black.

Comments: The silver impregnation with this procedure appears to operate on a carboxyl mechanism since the staining is abolished by methylation blockade or by staining at pH 4. Demethylation (saponification) restores the staining at pH 5. This method can also be used to stain at least some of the argyrophilic cells found within the gastrointestinal mucosa. The degree of staining overlap between this alcoholic argyrophilic method and other silver methods (argyrophilic, e.g., Grimelius and argentaffin, e.g., methenamine silver) used for enterochromaffin cell identification is not known.

Toluidine Blue Metachromasia

Original from Manocchio (1964)

Designed to show: Metachromatic islet cell (D-cell, argyrophilic, A_1 cell).

Tissue preservation: Bouin's, formalin.

Sections: 3–5 μm paraffin sections.

Preparation of reagents:
   A. Methylating reagent: 0.1 N HCl in absolute methyl alcohol (0.8 ml concentrated HCl in 99.2 ml absolute methanol).
   B. Saponification solution: 1.0 gm KOH in 100 ml of 70% ethanol.
   C. Staining solution: 0.05 gm toluidine blue O in 100 ml of 0.1 M acetate buffer at pH 5.

Staining procedure:
   1. Prewarm slides to melt paraffin and place heated slides in 3 changes of xylene for 5 min each or use prewarmed xylene baths for the same length of time.
   2. Wash slides thoroughly in 2 changes of absolute ethanol for 5 min each.

3. Wash slides in 2 changes of absolute methanol for 2–3 min each.
4. Incubate in methylating reagent (Solution A) at 37°C in a tightly sealed Coplin jar for 24–48 hr or until all tissue basophilia is suppressed.
5. Transfer slides to absolute ethanol and take to water through graded alcohols (30 sec to 1 min per change).
6. Wash in running water for 5 min.
7. Blot slides and place sections in 80% ethanol for 3 min.
8. Place slides in saponification solution (Solution B) in a tightly sealed Coplin jar at room temperature for 25–30 min.

NOTE: Tissues will be very fragile from this point and may be partially detached from the slide.

9. Gently lower slide into 80% ethanol for 3 min.
10. Carefully transfer slides through graded alcohols (70, 60, 50, 35, 10%) for 2 min each. It may be necessary to blot slides gently during this procedure to prevent section loss.
11. Wash in gently flowing water for 5 min.
12. Rinse in distilled water.
13. Stain in pH 5 toluidine blue (Solution C) for 10–20 min.
14. Rinse slides in distilled water to remove unbound dye.
15. Mount sections in water and examine at once.

Results: The D-cells are stained metachromatically; all other islet cells and tissue components will be unstained. If the demethylation (saponification) is shortened, the alpha cells (A_2) may also be stained but will appear in the orthochromatic blue color.

Comments: Permanent preparations can be made by dehydrating through 2 changes of absolute *tert*-butyl alcohol, clearing in xylene and mounting in resin. The metachromasia is alcohol labile and only blue colors will appear on the finished slide; this is not objectionable if the microscopic check indicates that only the metachromatic D-cells have stained. Slides prepared by this method can be restained with alcoholic silver nitrate. The mechanism of this staining reaction, as with the alcoholic silver nitrate, appears to depend upon the presence of carboxyl groups. The selectivity depends upon the suppression of other anionic groups by methanolysis (sulfate esters) or by extraction (nucleic acids). This procedure or the acid hydrolysis variant of Solcia et al. (1968) for enterochromaffin cells can be used to demonstrate a wide variety of different endocrine cell granules, i.e., enterochromaffin and other basophil cells of the gastrointestinal mucosa (G-cells), alpha and D-cells of pancreatic islets, adrenaline cells of the adrenal medulla, C-cells of the thyroid and some type of pituitary basophil (thyrotrophs?).

ADRENAL GLANDS

The adrenals of man and higher vertebrates are composite organs consisting of cortex and medulla which differ developmentally, structurally and functionally. In lower vertebrates, the cortical and medullary components retain their identity as separate structures and are referred to as interrenal and chromaffin tissues respectively.

ADRENAL CORTEX OR INTERRENAL TISSUE

The interrenal tissue of all placental mammals is arranged in three relatively distinct layers or zones which are named according to the configuration of their cellular cords. Since all cortical cells produce steroid hormones and all are very similar morphologically, they are distinguished primarily on the basis of their location within a particular layer of zone. In humans, the cortical cell types and the major class of steroid hormones produced by each are:

Glomerulosa cells—mineralocorticoids (aldosterone)
Fasciculata cells—glucocorticoids (cortisol)
Reticularis cells—sex hormones (androgens) and glucocorticoids

Specific steroid hormones appear to be both synthesized and released as needed from a common cholesterol-rich lipid precursor which is stored in the form of lipid droplets. The methods used for visualizing these droplets of hormone precursor are almost without exception general lipid solubility methods.

Oil Red O Method

From Lillie and Fullmer (1976)
Original from Lillie and Ashburn (1943)

Designed to show: Droplets of hydrophobic lipids (cholesterol, fatty acids and neutral fats) which serve as the precursors for biosynthesis of adrenal hormones.

Tissue preservation: 10% neutral buffered formalin, calcium acetate formalin, Müller's, Orth's. Avoid fixatives with acetic and formic acids and those containing organic solvents, e.g., Carnoy's.

Sections: 5–10-μm frozen sections, free-floating or mounted.

Preparation of reagents:
A. Staining solution: Stock saturated solution of oil red O in 99% isopropanol. Before use dilute 24 ml of the saturated stock solution of oil red O with 16 ml distilled water. Allow mixture to stand (10–15 min), filter before use. The working solution is stable for several hours.
B. Harris's hematoxylin: Dissolve 1 gm hematoxylin in 10 ml absolute

ethanol. Dissolve 20 gm ammonium alum $(NH_4Al(SO_4)_2 \cdot 12H_2O)$ in 200 ml distilled water with the aid of heat. Mix the two solutions, bring to a boil and add 0.5 gm mercuric oxide (HgO). As soon as the solution becomes dark purple, remove from heat and cool quickly by plunging flask into cold water. Before use, add 4 ml glacial acetic acid to each 100 ml of staining solution.

Staining procedure:

1. Wash frozen sections for 2–5 min in several changes of distilled water.
2. Transfer sections to 60% isopropanol for 1–2 min.
3. Stain sections in filtered working solution of oil red O (Solution A) for 10–20 min at room temperature.
4. Wash sections in running water to remove excess staining solution.
5. If desired, sections may be counterstained with Harris' (or other alum) hematoxylin. Differentiate nuclei with dilute acid water (1% HCl v/v). (A 10–30-sec dip in undiluted hematoxylin followed by a short bluing wash will permit visualization of nuclei without acid differentiation.)
6. Wash sections in running water to blue nuclei or use 1% aqueous sodium acetate (w/v).
7. Drain sections and mount in gum syrup or other aqueous mounting medium.

Results: Lipid droplets are stained bright red, nuclei are blue. In normal animals, lipid droplets are largest and most abundant in the zona fasciculata. Droplets are much smaller and much less abundant in the glomerulosa and reticularis and in some cases discrete droplets cannot be visualized.

Comments: Oil red O and the red Sudan dyes are absorbed or solubilized by hydrophobic unsaturated triglycerides, cholesterol esters and fatty acids. The hydrophilic phospholipids and glycolipids are at best faintly stained. Staining of hydrophobic lipids by lipid soluble dyes occurs only when the lipid is near or above its melting point, consequently the depth of staining increases with higher staining temperatures. Although free cholesterol (m.p. 148°C) is probably not demonstrated by this procedure, the associated unsaturated triglycerides, fatty acids and fatty acid chains of cholesterol esters are probably visualized. The steroid producing endocrine cells of the testis and ovary do not store significant amounts of lipid precursor and are rarely demonstrable by routine lipid soluble dye methods.

Sudan Black B Stain

From Lillie (1965)
Original from Chiffelle and Putt (1951)

Designed to show: Lipid droplets and lipid-rich structural components of tissues.

Tissue preservation: 10% neutral buffered formalin, calcium acetate formalin, Müller's, Orth's. Avoid fixatives with acetic and formic acids or those containing organic solvents.

Sections: 5–10-μm frozen sections, free-floating or mounted.

Preparation of reagents:
  A. Dissolve 0.7 gm Sudan black B in 100 ml propylene glycol at 100–110°C. Do not heat above 110°C. Filter the hot solution through Whatman No. 2 filter paper. Cool and filter with vacuum through a medium-porosity fritted glass filter. (Particulate matter can also be removed by filtering through disposable Millipore filters (0.2 μm) or by centrifugation.)
  B. Kernechtrot counterstain: Dissolve 5 gm aluminum sulfate in 100 ml distilled water and add 0.1 gm Kernechtrot (nuclear fast red).

Staining procedure:
  1. Wash mounted or free-floating frozen sections in 3 changes of distilled water for 2 min each change.
  2. Place sections or slides in two changes of pure propylene glycol for 3 min each—agitate solutions gently.
  3. Stain sections or slides in Sudan black B (Solution A) for 5–10 min—agitate solutions gently.
  4. Wash and differentiate in 85% propylene glycol for 2–3 min.
  5. Wash in 2 changes of distilled water for 3 min each.
  6. Counterstain nuclei with Kernechtrot (Solution B) for 10 min.
  7. Rinse in distilled water.
  8. Mount and coverslip sections or slides with glycerol gelatin or other aqueous mounting media.

Results: Lipid droplets of the adrenal cortex are stained greenish-black. The structural lipids of myelin and mitochondria are also stained. Cholesterol, cholesterol esters and other high melting point lipids are not stained at room temperature. Nuclei are pink to red after Kernechtrot staining but some nuclear staining (nonlipid?) may be evident after staining with Sudan black B alone. The use of acetylated Sudan black B prevents this electrostatic binding.

Comments: The Sudan black B method appears to be much more sensitive than the oil red O technic since many structural lipids, e.g., remnants of myelin and mitochondria are stained even in paraffin sections. Since the amount of stainable lipid in paraffin sections is increased by postfixation chromation, the two procedures can sometimes be combined to demonstrate cells possessing an abundance of bound lipid in membrane associated organelles (endoplasmic reticulum, mitochondria). The steroid producing endocrine cells of the gonads possess a well developed smooth endoplasmic reticulum and such cells are frequently well demonstrated by Sudan black B staining in paraffin sections of chromated tissues.

Perchloric Acid Naphthoquinone Method for Cholesterol

From Adams (1965)

Designed to show: Cholesterol and related 3-hydroxy-$\Delta^{5,7}$-sterols.

Tissue preservation: 10% neutral buffered formalin, calcium acetate formalin.

Sections: 5–8-μm unmounted frozen sections. Collect sections and leave free-floating in calcium formol for at least one week (3–4 weeks is better) to undergo preliminary atmospheric oxidation. (Atmospheric oxidation may be accelerated by mounting sections immediately and storing in a loosely covered staining dish over a thin layer of formalin.)

Preparation of reagents:
A. Naphthoquinone reagent: Mix 50 ml absolute ethanol; 25 ml of 60% perchloric acid; 22.5 ml distilled water; 2.5 ml formalin (37–40%). To the above solvent, add 100 mg 1,2-naphthoquinone-4-sulfonic acid.
B. Mounting media: 60% aqueous solution of perchloric acid.

Staining procedure:
1. Oxidized free-floating sections are mounted on slides, drained and allowed to air dry (oxidized mounted sections are now allowed to air-dry).
2. Using a small camel hair brush, paint the air-dried sections lightly with the naphthoquinone reagent (Solution A) and heat to 60–70°C for 5–10 min or until the initial red color turns dark blue. Use bottom of oven or hot plate. DO NOT OVERHEAT.
3. Coverslip using a drop or two of the 60% perchloric acid (Solution B) as mountant.
4. Examine immediately, photograph if permanent records are required.

Results: The cholesterol-rich lipid droplets of the adrenal cortex are stained dark blue.

Comments: The color is stable in perchloric acid for a few hours and completed preparations should be examined and photographed as soon as possible. The color is not stable to water, glycerin-jelly or other aqueous mounting media. Histologically, this procedure appears to be less destructive than the Shultze or other Liebermann-Burchardt methods. Although the three preceding lipid technics are used to demonstrate the precursor lipid directly, it should be emphasized that the vacuoles seen in routine histological preparations are negative images of the same precursor lipid.

ADRENAL MEDULLA OR CHROMAFFIN TISSUE

In man, two types of catecholamines are produced by the cells of the adrenal medulla: epinephrine and norepinephrine. Separate cells are responsible for the production of each catechol; they are referred to as epinephrine and norepinephrine cells respectively and both cell types exhibit chromaffin reactions. In animals known to produce a single catecholamine, only one type of medullary cell can be demonstrated.

Chromaffin Reaction

From Lillie and Fullmer (1976)
Original from Henle (1865)

Designed to show: Chromaffin cells of the adrenal medulla and sympathetic chain and collateral paraganglia.

Tissue preservation: The chromaffin reaction occurs during fixation. Excellent results may be obtained with either Müller's or Orth's solutions.

Sections: 5–8 μm paraffin sections.

Preparation of reagents:
  A. Müller's fluid: Dissolve 2.5 gm of potassium dichromate ($K_2Cr_2O_7$) in 100 ml distilled water and add 1.0 gm sodium sulfate ($Na_2SO_4 \cdot 10H_2O$).

or

  B. Orth's fluid: Immediately before use, add 10 ml formalin (37–40%) to 100 ml of Müller's fluid.

Staining procedure:
  1. Fix tissues for 1–3 days in either Müller's or Orth's fluid, change solution daily.
  2. Wash tissues thoroughly (overnight) in running water.
  3. Washed tissues may be: (a) processed routinely for paraffin sections *or* (b) stored in 10% formalin for frozen sections.
  4. Deparaffinize and wash sections in xylene before mounting in synthetic resin. (Sections may be overstained with hematoxylin and eosin or other suitable stains before permanent mounting.) Frozen sections may be stained by the oil red O or Sudan black B procedures before mounting in a suitable aqueous mountant.

Results: Chromaffin cells of the adrenal, synpathetic chain and collateral paraganglia are stained brown. Epinephrine cells are dark brown, norepinephrine cells are light brown to golden yellow.

Comments: The chromaffin reaction occurs during fixation and is inhibited by formaldehyde. Tissues may be exposed to formaldehyde during but not before the initial exposure to dichromate solutions. A similar procedure with longer reaction times (14+ days) can also be used to demonstrate the endocrine cells (enterochromaffin) of the gastrointestinal mucosa.

Ferric Ferricyanide Reduction Test

From Lillie and Fullmer (1976)
Original from Golodetz and Unna (1909)

Designed to show: Reducing substances including the catecholamines in chromaffin and other cells.

Tissue preservation: Alcoholic formalin, Carnoy's, Mercuric chloride-formalin mixtures and chromate mixtures.

Sections: 5–10-μm paraffin sections.

Preparation of reagents:
A. Ferric ferricyanide solution: Mix 30 ml of 1% ferric chloride ($FeCl_3 \cdot 6H_2O$), 4 ml of 1% potassium ferricyanide ($K_3Fe(CN)_6$) and 6 ml distilled water. Prepare fresh just before use.
B. Washing solution: 1% aqueous solution of acetic acid (v/v).

Staining procedure:
1. Deparaffinize sections of appropriately fixed material and take to water.
2. Place sections in freshly prepared ferric ferricyanide reagent (Solution A) for 10 min at room temperature (25°C).
3. Wash sections in 1% acetic acid (Solution B) for 1–2 min.
4. Counterstains are not recommended but if necessary, stain for 10 min in 0.02% new fuchsin in 1% acetic acid.
5. Dehydrate through graded alcohols, clear and mount in synthetic resin.

Results: Reducing sites (catechols) in the chromaffin cells of the adrenal and paraganglia are stained dark blue.

Comments: The reaction is easy to perform and very sensitive. Unfortunately the reagent is easily reduced by substances other than the catecholamines, epinephrine and norepinephrine, of the chromaffin cells. With various modifications, this procedure can be used to demonstrate a wide variety of reducing compounds including melanin, lipofuscin and sulfhydryl groups, as well as the reducing substance in gastrointestinal enterochromaffin cells. It may be helpful when only formalin fixed material is available and the chromaffin reaction cannot be utilized.

Formaldehyde Induced Fluorescence Method (FIF)

From Jarolim (1975)
Original from Falk and Owman (1965)

Designed to show: Catecholamines and related compounds including indolamines.

Tissue preservation: Freeze dry (see details below).

Sections: 5 μm paraffin sections (thinner if possible).

Preparation of reagents: Equilibrated paraformaldehyde (50%): A relative humidity of approximately 50% can be obtained by placing 12.5 N sulfuric acid in the bottom compartment of a desiccator. Equilibration is effected by exposing a thin layer of paraformaldehyde powder to this atmosphere for about 10 days at room temperature.

Staining procedure:
1. Pieces of tissue 1–6 mm in thickness are flash frozen in isopentane cooled to the temperature of liquid nitrogen. Frozen tissues are transferred to prechilled flasks and attached to the lyophilizer manifold. To prevent thawing, lyophilization flasks are immersed in liquid nitrogen until adequate vacuum (2.3 μm) and temperature (−58°C) are attained by the lyophilizer. Drying time will vary from a few hours to several days depending upon specimen size, density and equipment efficiency. When lyophilization is complete, flasks are immersed in warm water to heat specimen above ambient temperature (this prevents condensation of atmospheric moisture on the dried specimen). Dried specimens are transferred to screwcap vials and stored in a desiccator over Drierite at −4°C until needed.
2. Bring specimen vial to ambient temperature, unseal and place on a small watchglass or other shallow receptacle. Place specimen and watchglass in a screwcap specimen jar containing 5–10 gm of equilibrated paraformaldehyde reagent and seal tightly. Heat for 1 hr at 80°C or for 3 hr at 60°C to effect formaldehyde condensation.
3. After condensation is complete, the dried specimen is vacuum embedded in paraffin and blocked in the usual manner.
4. Cut paraffin sections at desired thickness and mount directly on albuminized slides without water or other floatation media. Heat mounted sections on a warming tray at 50°C until sections are as flat as possible.
5. Without removing paraffin, coverslip using a few drops of nonfluorescent paraffin oil and examine with a UV-microscope equipped with a 410–420 nm excitor filter and a barrier filter passing above 460 nm.

Results: Catecholamines are identified by their apple-green fluorescence. Indolamines (serotonin) exhibit yellow fluorescence.

Comments: The FIF procedure itself is simple, sensitive and possesses a high degree of specificity. Tissue preparation is, however, technically difficult and requires good freeze-drying equipment. Technically, the most difficult step is that of obtaining flattened sections without water floatation (the originators advise strict anhydrous conditions). A comparison of dry and water mounted sections from the same blocks indicated that the intensity of fluorescence was enhanced in water mounted sections and that differences in localization could not be detected. Paraffin sections reexamined after immersion in distilled water for 72 hr at room temperature exhibited a slight reduction in intensity and localization of fluorescence. For most studies, brief water floatation followed by draining, blotting and rapid drying at 50°C would probably be acceptable and would produce much better preparations. For critical work with unfamiliar tissues or with very low levels of fluorescence, the anhydrous conditions should probably be maintained. A recent modification for indolamines in enterochromaffin cells utilizes perfusion or immersion fixation with regular processing in paraffin (see "Enterochromaffin Cell System").

ENTEROCHROMAFFIN CELL SYSTEM

In 1870, Heidenhain described a type of gastrointestinal epithelial cell which stained deep yellow after exposure to potassium dichromate solutions. These cells were later noted to possess basal granules (Kultschitzky, 1897) and were subsequently termed enterochromaffin cells (Ciaccio, 1906).

Gastrointestinal epithelial cells with basally situated granules can also be demonstrated by a wide variety of nonchromaffin techniques, and although the term enterochromaffin is commonly used to describe any basally granulated cell demonstrated by any nonchromaffin technique, it should be emphasized that the degree of overlap between different staining methods is not known. The enterochromaffin cell system appears to consist of a large population of gut associated epithelial cells which are almost identical morphologically (basal granules) but which are quite different functionally and biochemically. It is possible that different staining procedures may demonstrate complete or partial subpopulations within the total enterochromaffin cell system.

To avoid confusion in terminology, cells demonstrated by a specific method will be described as positive for that technique.

Enterochromaffin Reaction

From Lillie and Fullmer (1976)
Original from Heidenhain (1870)

Designed to show: Enterochromaffin (Heidenhain) cells of the gut and gut associated structures.

Tissue preservation: Müller's fluid or 2.5–5.0% aqueous solution of potassium dichromate for 1–4 weeks. Change solution every other day.

Sections: 5–8 μm paraffin sections.

Preparation of reagents: Dissolve 2.5–5.0 gm of potassium dichromate ($K_2Cr_2O_7$) in 100 ml distilled water. For Müller's fluid, add 1 gm sodium sulfate for each 100 ml.

Staining procedure:
1. Fix tissues for 7–30 days in aqueous dichromate solutions, change solutions every other day.
2. Wash tissues thoroughly (overnight) in running water.
3. Washed tissues may be: (a) processed routinely for paraffin sections *or* (b) stored in 10% formalin for frozen sections.
4. Sections (paraffin or frozen) can be mounted for direct examination or counterstained with appropriate nuclear and cytoplasmic overstains.

Results: Enterochromaffin cells of the gut and gut associated structures are stained deep yellow to brown.

Comments: Better histological preparations are obtained if chromation times are limited to about 10 days. Nuclear preservation and staining is improved by adding formalin after the second or third day of fixation. Use 10 ml formalin per 100 ml dichromate solution and change daily. The reaction between dichromate and enterochromaffin granules appears to require longer exposure times than the catecholamines of chromaffin cells. Counterstains should be light and transparent for optimum demonstrability.

The Grimelius Argyrophil Reaction

Original from Grimelius (1968)

Designed to show: Argyrophilic cells in the gastrointestinal mucosa and other gut associated structures.

Tissue preservation: Bouin's, formalin, glutaraldehyde-picric acid mixtures.

Sections: 3–6-μm paraffin sections.

Preparation of reagents:
   A. Buffered silvering solution: Mix 10 ml 0.2 M acetate buffer, pH 5.6, with 87 ml distilled water and 3.0 ml 1% aqueous silver nitrate (w/v). Prepare fresh just before use.

B. Reducing solution: Dissolve 5.0 gm sodium sulfite in 100 ml distilled water and then add 1.0 gm hydroquinone. Prepare fresh just before use and warm to 40–45°C.

C. Nuclear fast red (Kernechtrot) solution: Dissolve 5.0 gm aluminum sulfate in 100 ml distilled water then add 0.1 gm of nuclear fast red (Kernechtrot). Use heat to effect solution, cool, filter and add thymol crystal to preserve.

Staining procedure:

1. Deparaffinize sections and take to water. If necessary, wash thoroughly to remove all traces of picric acid.
2. Rinse sections in three changes of distilled water 1–2 min each.
3. Place sections in Coplin jar containing buffered silvering reagent (Solution A), seal, and incubate for 24 hr at 37°C or for 3 hr at 60°C.
4. Drain slides without allowing sections to dry and place in reducer (Solution B) at 40–45°C for 1 min.
5. Rinse sections thoroughly in distilled water 3–4 times for 2 min each.
6. Counterstain nuclei with nuclear fast red (Solution C) for 1–10 min.
7. Wash in distilled water 2 times for 2 min each.
8. Dehydrate through graded alcohols, clear in xylene and mount in synthetic resin.

Results: Granules of argyrophil cells are dark brown to black, nuclei are dirty red to pink according to depth of background staining.

Comments: This procedure is similar to the Davenport alcoholic silver nitrate method which is used to demonstrate argyrophilic (A_1) cells in the pancreatic islets. This aqueous version (Grimelius) stains A_2 and some A_1 cells. Both procedures will stain some argyrophilic cells in the gut but identical populations of argyrophil cells are probably not revealed by the alcoholic and aqueous methods.

Methenamine Silver Argentaffin Reaction

From Burtner and Lillie (1949)
Original from Gomori (1948)

Designed to show: Substances which reduce silver salts in the dark without the addition of reducers. Enterochromaffin cells are demonstrated because of the argentaffin character of their granules. With modifications, melanin and lipofuscin can also be demonstrated.

Tissue preservation: Formalin containing fixatives, Bouin's. Avoid chromate and mercury containing fixatives.

Sections: 4–6-μm paraffin sections.

Preparation of reagents:

A. Stock methenamine silver solution: Dissolve 3.0 gm methenamine (hexamethylene-tetramine) in 100 ml distilled water and add 5.0 ml of a 5% aqueous silver nitrate solution (w/v). Shake to dissolve the white precipitate. Solution can be used for months if refrigerated or for about 2 weeks if unrefrigerated.

B. Holmes' borate buffer solution, pH 7.8:
   (1) Dissolve 1.24 gm boric acid in 100 ml distilled water to make 0.2 M solution.
   (2) Dissolve 1.90 gm sodium tetraborate (borax) ($Na_2B_4O_7 \cdot 10\ H_2O$) in 100 ml distilled water to make 0.05 M solution.

   To prepare pH 7.8 buffer, mix 16 ml of boric acid solution (Solution B-1) and 4.0 ml of the borax (Solution B-2). Check pH before adding silver solutions.

C. Silvering solution: Mix 30 ml of stock methenamine silver (Solution A) and 8.0 ml Holmes' borate buffer, pH 7.8 (Solution B).

All glassware should be chemically clean. Treat with concentrated nitric acid to remove old silver deposits.

D. Weigert's iodine solution: Dissolve 2 gm potassium iodide (KI) in 5 ml of distilled water and add 1.0 gm iodine crystals. After iodine crystals are in solution, add distilled water to make final volume of 100 ml.

E. Toning solution: Dissolve 0.1 gm gold chloride ($HAuCl_4$) in 100 ml distilled water.

F. Fixing solution: Dissolve 5.0 gm sodium thiosulfate in 100 ml distilled water.

G. Nuclear counterstain: Dissolve 0.1 gm safranin 0 in 100 ml distilled water and acidify by adding 0.1 ml glacial acetic acid.

Staining procedure:

1. Deparaffinize and take sections to water.
2. Treat with Weigert's iodine (Solution D) for 10 min.
3. Bleach in sodium thiosulfate (Solution F) for 2 min.
4. Wash in running water for 10 min.
5. Wash 3 times in distilled water, 2–3 min each.

CAUTION: *From this point, use plastic or paraffin coated forceps and avoid contact with other metal objects, e.g., staining racks and Coplin jar lid-liners, etc.*

6. Place slides in preheated buffered methenamine silver (Solution C) at 60°C for 3 hr. Preheating reduces impregnation time by 30–60 min.
7. Rinse in distilled water.
8. Tone in gold chloride (Solution E) for 10 min.
9. Rinse in distilled water.
10. Fix in sodium thiosulfate (Solution F) for 2 min.

11. Wash in running water 5 min.
12. Counterstain nuclei with acidified safranin (Solution G) 1–5 min.
13. Dehydrate with acetone, clear in xylene and mount in synthetic resin.

Results: The maximum number of argentaffin cells is blackened at about 3 hr.

Comments: Other argentaffin reactions can be used, but identical cell populations may not be demonstrated. Argyrophil reactions will demonstrate cells in areas where none are demonstrable by argentaffin methods.

Alkaline Diazonium Method for Enterochromaffin Cell Granules

From Lillie *et al.* (1961)
Original from Gomori (1952)

Designed to show: Diazo positive cells of the enterochromaffin cell system.

Tissue preservation: Formalin, Bouin's. Avoid fixatives containing mercury and alcohol.

Sections: 4–6 μm paraffin sections.

Preparation of reagents:
A. Veronal acetate buffer, pH 8.0:
   (1) Dissolve 1.17 gm sodium acetate (anhydrous) and 2.94 gm Veronal (sodium diethylbarbiturate) in 100 ml distilled water.
   (2) Add 0.85 ml concentrated HCl to 99.0 ml distilled water.
   For pH of approximately 8.0, mix 25 ml of Veronal-acetate (Solution A-1) with 8.5 ml of 0.1 N HCl (Solution A-2) and dilute to 100 ml with distilled water.
C. Fast Garnet GBC or Fast Red Salt B (use 1.0 mg/ml of buffer solution).
D. Mayer's hemalum: Dissolve 5.0 gm ammonium or potassium alum in 100 ml distilled water and add 0.1 gm hematoxylin. Add 0.02 gm sodium iodate, 0.1 gm citric acid, 5.0 gm chloral hydrate and shake until all are in complete solution. Solution keeps for months.

Staining procedure:
1. Deparaffinize and take sections to water.
2. Place slides in screwtop Coplin jar with 40 ml, pH 8.0 buffer (Solution A) and add quickly 40 mg of stable diazotate. Seal and mix by gentle shaking for 1–2 min. (As an alternative, mix diazotate and buffer and shake vigorously, filter rapidly through several thicknesses of tissue (Kleenex or Chemwipe) into an empty Coplin jar containing slides.) Speed is essential because diazotates decompose rapidly in alkaline solutions.

3. Wash thoroughly in running water.
4. Stain nuclei with Mayer's hemalum (Solution C) for 6–10 min.
5. Wash in running water for 15–30 min.
6. Dehydrate in graded alcohols, clear in xylene and mount in synthetic resin.

Results: Granules in the enterochromaffin cells are fiery orange-red, nuclei are dark blue, cytoplasmic structures are yellow.

Comments: Other stabilized diazotates can be substituted for the Fast Garnet GBC and Fast Red B. Efficiency of coupling, i.e., number of cells demonstrated and final color varies with diazotate. See Lillie et al. (1961).

Acid Diazonium Method for Aromatic Amines and Phenols

From Pearse (1968)

Designed to show: Diazo positive cells of the enterochromaffin system.

Tissue preservation: 10% neutral buffered formalin, Bouin's. Avoid mercury and alcohol containing fixatives.

Sections: 4–6-μm paraffin sections.

Preparation of reagents:
A. 0.2 M acetate buffer pH 5.2:
  (1) Dissolve 1.64 gm sodium acetate in 100 ml distilled water.
  (2) Add 1.2 ml glacial acetic acid to 98.8 ml distilled water.
For pH 5.2, use 39.5 ml sodium acetate (Solution A-1) and 10.5 ml acetic acid (Solution A-2). Check pH with meter or indicator paper.
B. Fast red GG (use 1.0 mg/ml of buffer solution).
C. Mayer's hemalum: See preceding technic.

Staining procedure:
1. Deparaffinize sections and bring to water.
2. Immerse in acetate buffer, pH 5.2 (Solution A) containing 1 mg Fast Red GG per ml of buffer for 20–40 min. Mix diazotate and buffer then filter. Diazotates are relatively stable in acid solution but couple slowly.
3. Wash in running water for 5 min.
4. Counterstain nuclei with Mayer's hemalum (Solution C) for 3–5 min.
5. Wash in running water 10–30 min to differentiate nuclei—check with microscope.
6. Dehydrate through alcohols, clear and mount in synthetic resin.

Results: Granules of enterochromaffin cells and carcinoid tumor cells are orange red. Cytoplasm is very pale yellow, nuclei are blue. Tissue components containing aromatic amines (intestinal pseudomelanins) are colored to a lesser extent.

Comments: Although the acid coupling reaction is slower than the alkaline diazonium procedure, the enterochromaffin cells are very sharply delineated against the pale yellow background staining. Preparations are excellent for photographic purposes and for studies involving cell enumeration.

Clara's Dilute Hematoxylin

Original from Clara (1935)

Designed to show: Granules of enterochromaffin cells.

Tissue preservation: 10% neutral buffered formalin, Bouin's, picro-glutaraldehyde mixtures.

Sections: 4–6-μm paraffin sections.

Preparation of reagents:
  A. Stock hematoxylin solution: Dissolve 1.0 gm hematoxylin in 100 ml absolute ethanol.
  B. Phosphate buffer, pH 6.5.
    (1) Dissolve 0.138 gm monobasic sodium phosphate (NaH_2PO_4) in 100 ml distilled water.
    (2) Dissolve 0.142 gm dibasic sodium phosphate (Na_2HPO_4) in 100 ml distilled water.
  For pH 6.5, mix 36 ml monobasic sodium phosphate (Solution B-1) and 14 ml dibasic sodium phosphate (Solution B-2).
  C. Dilute hematoxylin, pH 6.5: To 50 ml of the 0.01 M phosphate buffer, pH 6.5, add 0.5 ml alcoholic hematoxylin (Solution A). Prepare fresh and use once.

Staining procedure:
  1. Deparaffinize sections and take to water.
  2. Wash thoroughly to remove all traces of picric acid if necessary.
  3. Stain in dilute, pH 6.5 hematoxylin (Solution C) for 24–36 hr. Slides can be removed and checked for staining progress under microscopic control.
  4. Dehydrate through graded alcohols, clear in xylene and mount in synthetic resin.

Results: Enterochromaffin cells and basal cells in some carcinoid tumors show blue to black granules. Hemosiderin, enterosiderin, granules of eosinophilic leukocytes, cutaneous eleidin, keratohyaline are also stained.

Comments: Despite the length of time required for staining, the method is very simple and enterochromaffin cells are beautifully demonstrated. The mechanism for dilute hematoxylin staining is not known. Granules are thought to stain via their catechol groupings.

Simple Fluorescence Method for Serotonin-containing Endocrine Cells

Quoted from Hoyt, Sorokin and Bartlett (1979)

Designed to show: Serotonin-containing endocrine cells in the gut and some gut associated structures, e.g., lung, thyroid.

Tissue preservation: By perfusion or immersion in freshly prepared 0.1 M phosphate buffered formaldehyde, pH 7.2.

Sections: 4–6-μm paraffin or 2 μm plastic (glycol methacrylate) sections.

Preparation of reagents:
  A. Stock 0.2 M phosphate buffer solutions:
   (1) Dissolve 27.58 gm monobasic sodium phosphate (NaH_2PO_4) in 1000 distilled water.
   (2) Dissolve 28.38 gm dibasic sodium phosphate (Na_2HPO_4) in 1000 distilled water.
  B. For each 100 ml of 0.1 M phosphate buffer at pH 7.2, mix 14.0 ml monobasic sodium phosphate (Solution A-1) and 36.0 ml dibasic sodium phosphate (Solution A-2) and dilute to 100 ml.
  C. Phosphate buffered 6% formaldehyde (w/v): Dissolve 6.0 gm para-formaldehyde powder in 100 ml 0.1 M phosphate buffer (pH 7.2) preheated to 90°C. Filter, cool and readjust pH to 7.2. Fixative must be freshly prepared just before use.

Staining procedure:
  1. Tissues are fixed by perfusion or immersion in freshly prepared fixative for 24 hr to 1 month. After fixation, tissues are rinsed in water, dehydrated and embedded in paraffin or in plastic (glycol methacrylate) for sectioning.
  2. Paraffin sections are dewaxed in xylene and mounted directly for UV examination. (The authors suggest Entellen for mounting but other nonfluorescent mountants (nonaqueous, e.g., paraffin oil or aqueous, e.g., phosphate buffer, glycerine, etc.,) could probably be used.)

Results: Strong yellow fluorescence from cells known to contain serotonin, e.g., enterochromaffin cells of the gut, mast cells of rat, small granule cells of rabbit lung.

Comments: This method does not demonstrate catecholamines in the adrenal medulla or dopamine cells in the thyroid gland or tracheal epithelium; these amines can, however, be demonstrated in all of these sites by the standard freeze-dry procedure of Falck and Owman. Thyroid C-cells and certain tracheal epithelial cells become fluorescent if serotonin precursor (5-hydroxytryptophan) is administered before sacrifice (APUD or *A*mine *P*recursor *U*ptake and *D*ecarboxylating cells). This procedure can be applied to cell marking experiments since the fluorescent cells can be marked (photography or camera lucida drawing) and subjected to other staining methods for enterochromaffin cells.

Acid Hydrolysis Basophilia

From Solcia et al. (1968)

Designed to show: Enterochromaffin, basophil cells of gastrointestinal mucosa (gastrin or G-cell), alpha and D-cells of pancreas, adrenaline cells of adrenal medulla, thyroid parafollicular (C-cells) and pituitary thyrotrophs.

Tissue preservation: Bouin's with 1% acetic acid (v/v) or picro-glutaraldehyde (1 part 25% glutaraldehyde, 3 parts saturated aqueous picric acid, 1 part of 1% sodium acetate).

Sections: 4–6 μm paraffin sections.

Preparation of reagents:

A. Hydrolyzing solution: (a) 0.5% (v/v) solution of hydrochloric acid in water *or* (b) 0.2 N HCl (1.7 ml concentrated HCl in 98.3 ml water).

B. Acetate buffer, pH 5.0:
   (1) Dissolve 1.64 gm sodium acetate in 100 ml distilled water.
   (2) Mix 1.2 ml glacial acetic acid in 98.8 ml distilled water.
For pH 5.0, mix 35.2 ml sodium acetate (Solution B-1) and 14.8 ml acetic acid (Solution B-2). Check pH with meter or indicator paper.

C. Staining solution: 0.01 gm toluidine blue O (or Azure A) in 100 ml acetate buffer, pH 5.0.

Staining procedure:

1. Deparaffinize sections and take to water.
2. Wash thoroughly in running water to remove all traces of picric acid.
3. Hydrolyze in dilute acid (Solution A) at 60°C for 3 to 4 hr in a sealed screwcap Coplin jar. Hydrolysis time varies slightly with different fixatives but 3 hr (4 for thyroid C-cells) is about optimum. (Hydrolysis progress may be checked by rinsing slides in water and staining briefly in toluidine blue; if incomplete return slides to hydrolyzing solution.)
4. Wash slides thoroughly in distilled water (several changes for 1–2 min each).
5. Stain in dilute toluidine blue O or Azure A at pH 5.0 (Solution C) for 20 min.
6. Mount in a drop of staining solution for microscopic examination. Photograph if permanent records of metachromasia are needed or rinse quickly in distilled water, blot, dehydrate in tertiary butanol, clear in xylene and mount in synthetic resin.

Results: Aqueous mounted sections stained with toluidine blue will show metachromatic staining of pancreatic D-cells, enterochromaffin and other basophilic cells of gastrointestinal mucosa and thyroid C-cells. Pancreatic A_2 cells, adrenaline cells of adrenal medulla and pituitary basophils (thyrotrophs ?) are usually stained orthochromatically. After dehydrating, clearing and mounting in resin, all of the above cells appear in the orthochromatic blue.

Comments: The acid hydrolysis eliminates basophilia of nucleic acids (DNA and RNA) and mucopolysaccharides. The hydrolysis either induces or unmasks basophilia at pH 5. Argentaffin, argyrophil, azocoupling and FIF procedures for enterochromaffin can precede this technic. Although this method is similar to the methylation-demethylation procedure used for pancreatic D-cells, this aqueous version is easier to perform. Unfortunately, this aqueous version appears to be less sensitive than the alcoholic procedure.

THYROID GLAND

The thyroid gland contains two types of endocrine cells: follicular and parafollicular (clear or C-cells). The follicular cell is the more abundant and is responsible not only for the synthesis and storage of thyroglobulin (colloid) but also for its subsequent phagocytosis and degradation to release the active hormone thyroxine. Although droplets of ingested colloid are sometimes demonstrable, the epithelial follicular cells are nongranular and cannot be demonstrated selectively by staining methods alone. After the administration of labelled iodine and the application of autoradiographic technics, follicular cells can be identified with great precision. In routine preparations from normal glands, these cells can be identified by their location and arrangement with respect to the follicles and colloid. Parafollicular (clear or C-cells) cells which produce the hormone calcitonin (thyrocalcitonin) possess cytoplasmic granules and can be identified by several of the staining methods which have been presented for other endocrine glands. The methods which can be used without modification are:

1. Alcoholic silver nitrate (see "Islets of Langerhans").
2. Toluidine blue metachromasia (see "Islets of Langerhans").
3. Formaldehyde induced fluorescence for catecholamines—reaction intensified following administration of the catechol precursor, L-dopa (see "Adrenal Medulla or Chromaffin").
4. Simple fluorescence method for serotonin-containing cells—reaction only after administration of serotonin precursor, 5-hydroxytryptophan (see "Enterochromaffin Cell System").
5. Acid hydrolysis basophilia (see "Enterochromaffin Cell System").

PARATHYROID GLAND

The parathyroid gland of adult humans contains two major cell types: chief (or principal) cells and oxyphil cells. As its name implies, the chief cell is the most abundant and important cell throughout life and is responsible for the production of parathyroid hormone. In young children and in almost all laboratory animals, it is the only cell type found in the parathyroid. Chief cells are known to occur in dark and light varieties. The dark chief cell is considered to be active in hormone production; the light chief cell (water clear cell) is thought to be inactive or resting and is clear or more lightly stained because of its glycogen content. Since parathyroid hormone appears to be synthesized and released as needed, chief cells are nongranulated and cannot be demonstrated by selective staining methods. The less abundant oxyphil cells are rich in mitochondria and seem to originate from chief cells. Although the number of oxyphil cells increases with age, these cells do not appear in children before the onset of puberty. The significance of oxyphil cells in parathyroid function is not known but they are easily recognized and distinguuished from chief cells by their larger size, marked acidophilia and deeply stained nuclei.

PINEAL GLAND

Although an endocrine function for the pineal gland is widely accepted, neither a distinct functional role nor a secretory product has been identified. Despite the absence of an identifiable hormone or secretory product, a parenchymal cell type (the pinealocyte) is described and can be demonstrated by selective staining procedures. The following method is one of the very few applicable to routinely processed material.

Achúcarro-Hortega Stain for Pinealocytes

Quoted from DeGirolami and Zvaigzne (1973)

Original from del Rio-Hortega (1962) and Achúcarro (1911)

Designed to show: Pinealocytes in normal and abnormal (neoplastic) pineal glands.

Tissue preservation: Fix specimens for 2 days or longer in 10% formalin neutralized with excess $CaCO_3$.

Sections: 6–10-μm paraffin sections.

Preparation of reagents: Special silver solutions:
A. To 100 ml of 2% aqueous silver nitrate, add 1.2 ml pyridine.
B. Add 30 ml of 5% sodium carbonate to 10 ml of 10% aqueous silver nitrate. Then add concentrated NH_4OH drop by drop with constant

shaking until the precipitate is just dissolved. (Avoid adding an excess of ammonia. The finished solution should have no smell of ammonia.) Add 110 ml distilled water, filter and add 2.0–2.2 ml pyridine to give about 150 ml of the finished solution.

C. To 30 ml 10% aqueous silver nitrate, add concentrated NH_4OH drop by drop until the precipitate is almost dissolved, filter.

D. Reducing mixture: 1% neutral formalin containing 0.2% (v/v) pyridine. Three changes of this solution are required.

E. Toning solution: 0.2% gold chloride.

F. Fixing solution: 5% aqueous sodium thiosulfate ($Na_2S_2O_3$).

Staining procedure:

1. Deparaffinize and take sections to distilled water.

    NOTE: Avoid metal forceps, staining racks and lid liners.

2. Impregnate in Solution A for 2 hr at 56–58°C.

3. Without washing, transfer sections to Solution B for 16–24 hr at room temperature (about 25°C) and in darkness.

4. Reinforce the impregnation in Solution C for 2 min.

5. Transfer directly to reducing mixture (Solution D) for 2–5 min. The reducing solution should be changed 3 times. Make the first change after 3–5 sec.

6. Wash in running water for 5 min. Rinse once with distilled water.

7. Tone in gold chloride (Solution E) for 10–15 min.

8. Rinse twice in distilled water.

9. Fix in thiosulfate (Solution F) for 1 min.

10. Wash in running water for 5 min followed by a rinse in distilled water.

11. Dehydrate in 95% and absolute ethanol, clear in 2 changes of xylene and mount in synthetic resin.

Results: Pinealocytes, including the diagnostic club-shaped processes, are impregnated. Glial and other cell types found in the pineal region are not impregnated.

Comments: The contributor has had no experience with this procedure; it has been included for the sake of completeness.

GONADS

The steroid producing endocrine cells of the testis (Leydig or interstitial cells) and ovary (thecal and lutein cells) are not characterized by the presence of lipid droplets. Because of this, gonadal steroid producing cells are difficult or impossible to demonstrate by the general lipid solubility methods used for adrenal cortical cells. As a general rule, steroid producing cells possess numerous mitochondria and a particularly prominent endoplasmic reticulum, both of which contain large amounts of structural lipid. When these structural lipids are preserved and stabilized by appropriate treatment, these cells (especially the Leydig) can sometimes be demonstrated in paraffin sections stained with Regaud's method for mitochondria.

Regaud's Hematoxylin Method for Mitochondria

From Gurr (1956)

Original from Regaud (1910)

Designed to show: Mitochondria and other lipid containing membranous structures.

Tissue preservation: Fix thin slices of tissue for 3 days in modified Orth's (4 parts Müller's, 1 part formalin), change fixative every day. Following fixation, immerse tissues in 3% aqueous solutions of potassium dichromate for 6–8 days. Wash in running water for 24 hr before processing in paraffin. (Tunica albuginea should be removed from testis if possible to allow penetration.)

Sections: 4–6-μm paraffin sections.

Preparation of reagents:
  A. Mordant: Dissolve 10 gm iron alum in 100 ml distilled water.
  B. Regaud's hematoxylin solution: Dissolve 1.0 gm hematoxylin in 10 ml absolute ethanol and then add 10 ml glycerine and 80 ml distilled water.
  C. Differentiating solution: Dissolve 5 gm iron alum in 100 ml distilled water.

Staining procedure:
  1. Deparaffinize and take sections to distilled water.
  2. Mordant sections for 1–3 days in 10% iron alum (Solution A) at 37°C. Use a tightly sealed staining jar.
  3. Rinse sections for 5 min in running water.
  4. Stain in Regaud's hematoxylin (Solution B) for 24 hr.
  5. Rinse sections in distilled water and differentiate in 5% iron alum (Solution C) under microscopic control until Leydig cells are visible against more lightly stained cells.
  6. Rinse sections in running water for 20–30 min.
  7. Blot section and rinse with 95% alcohol.
  8. Dehydrate with absolute alcohol, clear in xylene and mount in synthetic resin.

Results: Cells containing large amounts of bound lipid in membrane form (mitochondria and endoplasmic reticulum) are stained gray-blue to intense black. Nuclei and background are stained in various shades of brown.

Comments: The Regaud method for mitochondria is an empirical version of the more sophisticated controlled chromation-hematoxylin methods for phospholipids. Chromated lipids retain their sudanophilia and the Sudan black B procedure (see "Adrenal Cortex") can be used instead of the hematoxylin staining.

PITUITARY

The pituitary (hypophysis) is composed of an anterior and a posterior lobe. In some vertebrates, the two lobes may remain completely separated while in others, including man, partial fusion may occur. The anterior lobe (adenohypophysis) develops from oral ectoderm via Rathke's pouch and subsequently differentiates into the pars distalis, tuberalis and intermedia. The cavity of Rathke's pouch may persist either as a cleft or series of small cavities (follicles) separating the pars distalis and pars intermedia. The posterior lobe (neurohypophysis) develops from the infundibular process of the diencephalic floor plate.

ADENOHYPOPHYSIS

In mammals, seven hormones are known to be associated with the adenohypophysis. Immunofluorescent studies have indicated that each hormone is produced by a different cell and that the adenohypophysis contains at least six and possibly seven functional cell types. Only three basic categories of cells (acidophils, basophils, chromophobes) can be demonstrated in the adenohyophysis by routine histological staining procedures, but additional cell types can be demonstrated within these basic categories by combining one or more histochemical procedures with one or more empirical staining methods. Unfortunately, there is not universal agreement as to which functional cell type is being demonstrated. Extreme caution should be used in applying staining results obtained in one animal group to another even when the groups are closely related. Most of the methods selected for inclusion are simple, reproducible and conservative (noncontroversial). Functional cell types are assigned to basic staining categories according to the following schedule.

Acidophils— somatotrophs and lactotrophs (mammotrophs).

Basophils— thyrotrophs, gonadotrophs, cortico/lipotrophs.

Chromophobes— poorly or nongranular acidophils and basophils (resting).

Methyl Blue-Eosin B Technique

From Glenner and Lillie (1957b)

Original from Mann (1894)

Designed to show: Acidophils, basophils and chromophobes.

Tissue preservation: 10% neutral buffered formalin, Helly's (Zenker formol).

Sections: 4–6-μm paraffin sections.

Preparation of reagents:
  A. Eosin B stock: 1% aqueous eosin B (w/v).
  B. Methyl blue stock: 1% aqueous methyl blue (verify C.I. 42780).
  C. Acetate buffer, 0.01 M, pH 4.5–4.6.
    (1) Dissolve 0.82 gm sodium acetate in 1000 ml distilled water.
    (2) Mix 0.6 ml glacial acetic acid and 1000 ml distilled water.
    For pH 4.5–4.6, mix 16 ml sodium acetate stock (Solution C-1) and 24 ml acetic acid stock (Solution C-2). Check with pH meter or indicator paper and adjust pH if necessary.
  D. Buffered eosin-methyl blue staining solution: Mix 8.0 ml stock eosin B (Solution A), 2.0 ml methyl blue (Solution B) and 30 ml acetate buffer, pH 4.5–4.6. Prepare fresh each time, just before use.
  E. Iodine solution: Dissolve 0.5 gm iodine in 100 ml 70% ethanol.
  F. Bleach: Dissolve 5.0 gm sodium thiosulfate in 100 ml distilled water.

Staining procedure:
  1. Deparaffinize and take sections to distilled water. Treat mercury containing sections with iodine alcohol (Solution E) for 10 min, rinse briefly in water and bleach for 2 min in sodium thiosulfate (Solution F). Wash in running water for 10 min.
  2. Place slides in screwtop Coplin jar containing 40 ml of buffered staining solution, pH 4.5–4.6 (Solution D) at room temperature. Seal tightly and stain in 60°C oven for 60 min.
  3. Wash 5 min in running water.
  4. Dehydrate in 50, 80, 100% acetone, clear in xylene/acetone (1:1) followed by two changes of pure xylene. Mount in synthetic resin.

Results: Granules in acidophils are stained dark red, basophil granules are dark blue, chromophobes are gray to pink. Colloid red to lavender, red blood cells are orange red and collagen blue.

Comments: This staining technic demonstrates that the so-called pituitary basophil is not truly basophilic. In this case, the staining of basophils is accomplished by the anionic dye, methyl blue. The terms alpha and beta, which are more accurate, are frequently substituted for the classical terms acidophil and basophil. In well fixed material, satisfactory results may be obtained with many different alum hematoxylins and a wide variety of acid counterstains, e.g., eosin, phloxine, chromotroph 2R, orange G. Gomori's chrome hematoxylin-phloxin method (see "Islets of Langerhans") gives excellent results.

PFAAB, PAS, Orange G Method for Human Hypophysis

From Adams (1956)

Designed to show: Acidophils, several types of basophils and chromophobes.

Tissue preservation: Formol-mercury, 10% neutral buffered formalin.

Sections: 4–6-μm paraffin sections.

Preparation of reagents:

A. Oxidizing solution: Performic Acid—Mix 8.0 ml formic acid (90%) and 31 ml hydrogen peroxide (30%) and add 0.22 ml concentrated sulfuric acid. Keep temperature below 25°C. Maximum concentration (4.7%) is formed in about 2 hr. Decomposition begins in a few hours. Prepare fresh daily.

or

Peracetic Acid—Mix 95.5 ml glacial acetic acid and 259.0 ml hydrogen peroxide (30%) and add 2.2 ml concentrated sulfuric acid. Let stand for 3 days and add 40 mg disodium phosphate (Na_2HPO_4) to stabilize. Store at 0–5°C. Stabilized peracetic acid can be stored for several months. Maximum concentration (8.6%) is formed in 80–96 hr.

B. Alcian blue, pH 1.2: Dissolve 1.0 gm alcian blue 8GX in 100 ml distilled water, adjust pH to 1.2 with 1 N HCl.

C. Periodic acid: Dissolve 0.5 gm periodic acid in 100 ml distilled water.

D. Schiff's reagent: Dissolve 1 gm basic fuchsin in 200 ml of distilled water, heat to boiling and shake for 5 min. Cool to 50°C, filter and add to filtrate 20 ml 1 N HCl. Cool to 25°C and add 1 gm sodium or potassium metabisulfite. Stand this solution in the dark for 14–24 hr. Add 2 gm activated charcoal and shake for 1 min. Filter. Keep filtrate in dark at 4°C. Allow to reach room temperature before use.

E. Orange G stain: Saturate 100 ml absolute ethanol with orange G. Filter or decant carefully before use.

Staining procedure:

1. Deparaffinize and bring sections to distilled water. If necessary, remove mercury precipitates with Lugol's and thiosulfate bleach followed by 5 min wash in running water.

2. Oxidize sections in performic or peracetic acid (Solution A) for 10 min. (Peracetic acid must be prewarmed to room temperature before use. Sections may be loosened during oxidation—handle carefully in subsequent steps.)

3. Rinse carefully in gently running water for 5 min.

4. Rinse gently in 2 changes of distilled water for 1 min each.

5. Stain in alcian blue, pH 1.2 (Solution B) for 1 hr.

6. Rinse carefully in gently running water for 5 min.

7. Rinse in distilled water, blot and transfer to periodic acid (Solution C) for 5 min.
8. Rinse carefully in gently running water for 2 min followed by 2 rinses in distilled water for 1 min each.
9. Place in Schiff's reagent (Solution D) at 25°C for 15 min.
10. Wash carefully in gently running water for 10 min.
11. Rinse in 2 changes of distilled water for 1 min each.
12. Counterstain nuclei lightly with hematoxylin if desired.
13. Blot sections and transfer to 70% ethanol for 5 min.
14. Stain in orange G (Solution E) for 5 min—sections will not overstain.
15. Rinse in 2 changes of 100% ethanol for 2–3 min to complete dehydration and remove unbound dye.
16. Clear in xylene and mount in synthetic resin.

Results: Acidophils are stained deep orange. Mucoid granules of the R-type stain magenta-red; mucoid granules of the S-type are purple-blue to deep blue. Nuclei appear as negative images when hematoxylin counterstin is omitted.

Comments: The significance of R-type and S-type granules is not completely understood. The magenta-red (R-type) granules are thought by some to belong to corticotrophs, by others thyrotrophs. The deep blue to purple (S-type) granules are thought to represent gonadotrophs.

NEUROHYPOPHYSIS

The two hormones (oxytocin and vasopressin) associated with the posterior pituitary are synthesized in hypothalamic neurons and transported via their axons to the neurohypophysis for storage and release.

Performic Acid Alcian Blue Method for Neurosecretory Substance

From Lillie and Fullmer (1976)

Original from Adams and Sloper (1956)

Designed to show: Neurosecretory substances.

Tissue preservation: Picroformalin or picroglutaraldehyde.

Sections: 10–15-μm paraffin sections.

Preparation of reagents:
A. Performic or peracetic acid: Prepare as in preceding technique.
B. Alcian blue, pH 0.2: Add 5.4 ml sulfuric acid (98%) to 94.6 ml distilled water. Add 3.0 gm alcian blue 8GS and dissolve with the aid of gentle heat. Cool and filter.

Staining procedure:
1. Deparaffinize sections and take to water (increase times slightly to compensate for thicker sections).
2. Oxidize either in performic or in peracetic acid (Solution A) for 5 min.
3. Wash carefully in several changes of distilled water—loose sections may be reattached and bubbles removed by gently blotting and partial drying.
4. Stain in alcian blue, pH 0.2 for 1 hr.
5. Wash in gently running water for 5 min.
6. Blot sections gently and dehydrate rapidly in 95% and 100% ethanol.
7. Clear in absolute/xylene mixture (1:1) followed by final clearing in two changes of pure xylene. Mount in synthetic resin.

Results: Neurosecretory substances in hypthalamic nuclei, pituitary stalk and neurohypophysis are stained bright blue. Cerebral corpora amylacea and some connective tissue elements may also be stained. Nuclei are usually unstained.

Comments: Other basic dyes can be substituted for alcian blue. Dilute toluidine blue O or Azure A (0.02% in 0.1 N HCl, pH about 1) solutions are satisfactory.

Bargmann's Chrome Hematoxylin for Neurosecretory Substance

From Bargmann (1950)
Original from Gomori (1939)

Designed to show: Neurosecretory substance in hypothalamus, stalk and neurohypophysis.

Tissue preservation: Bouin's picro-formol, picro-glutaraldehyde.
Sections: 5–10-μm paraffin sections.

Preparation of reagents:
A. Bouin-chrome alum mordant: Mix 25 ml formalin (37–40%) and 75 ml saturated aqueous picric acid solution with 5 ml glacial acetic acid. Dissolve 3 gm chrome alum per 100 ml Bouin's.
B. Oxidizing solution: Dilute 5.0 ml 2.5% aqueous potassium permanganate (w/v) with 35 ml distilled water and add 5 ml 5% sulfuric acid (v/v). Prepare fresh just before use.
C. Bleach: Dissolve 1.0 gm oxalic acid in 100 ml distilled water.
D. Chrome hematoxylin: Mix 50 ml of 1% aqueous hematoxylin, 50 ml 3% aqueous chrome alum, 2.0 ml 5% aqueous potassium dichromate and 1.0 ml 5% aqueous sulfuric acid (v/v). Allow to ripen for 48 hr before use. Stain can be used for several weeks if stored at 0–4°C. Filter before use.

E. Acid alcohol: Add 1.0 ml concentrated HCl to 100 ml 95% ethanol.

F. Phloxine counterstain: 0.5% aqueous solution of phloxine.

G. Phosphotungstic acid solution: 5% aqueous solution of phosphotungstic acid.

Staining procedure:

1. Deparaffinize and bring sections to water.
2. Mordant in Bouin-chrome alum (Solution A) at 37°C for 12-24 hr.
3. Wash in running water to remove all traces of picric acid.
4. Oxidize in acidified permanganate (Solution B) for 2–3 min.
5. Rinse sections in running water to remove excess oxidant.
6. Bleach in oxalic acid (Solution C) for 1 min.
7. Wash in running water for 5 min.
8. Stain in freshly filtered chrome hematoxylin (Solution D) for 10–15 min. Stain should be allowed to warm to room temperature before use.
9. Rinse briefly in water and differentiate in acid alcohol (Solution E) for about 30 sec.
10. Wash in running water for 1–2 min.
11. Stain in phloxine (Solution F) for 2–3 min.
12. Immerse sections directly in phosphotungstic acid (Solution G) for 5 min.
13. Wash in running water for 5 min to restore red color.
14. Differentiate in 95% ethanol (if sections are too red, rinse for 15–20 sec in 80% ethanol).
15. Dehydrate in absolute ethanol, clear in xylene and mount in synthetic resin.

Results: Neurosecretory substances in hypothalamic neurons, axons and neurohypophysis as well as nuclei are stained deep purple; other structures are stained in shades of pink. The neurosecretory substance in axons of the pituitary stalk (Herring bodies) are very well shown.

Comments: This method is almost identical to the Gomori chrome hematoxylin presented in the section on pancreatic islets and recommended for pituitary basophils. When applied to the pituitary, the Gomori method will occasionally demonstrate the neurosecretory substance within the pituitary stalk axon very clearly but for reasons which are not readily apparent it is much less consistent in its staining of neurosecretory substance than the Bargmann technic.

CHAPTER 9

METHODS FOR INVERTEBRATES

George E. Cantwell

The number of species of invertebrate animals has been estimated at well over one million, the vast majority of which are insects. When one considers that each species has at least three distinct stages in its life cycle, *i.e.* egg, immature (larva, nymph, pupa), and adult, and that these species and/or stages may differ greatly in ecology, anatomy and physiology, it is easily understood why, in fact, little is known of their normal histology.

Fortunately, many of the standard histological procedures that apply to the tissues of higher animals may be used intact or with slight modifications to study these invertebrates. Various fixatives are used, namely Bouin's (or Duboscq-Brazil's modifications of it), Carnoy's, Zenker's or Kahle's. Fixation and infiltration, however, do present difficult problems, particularly in those life stages such as the egg with its tough shell or an insect larva or pupa whose outer covering seems, at times, to be impervious. Methods are available for overcoming these problems, however.

Stains frequently used for invertebrates are a simple hematoxylin such as Delafield's and Ehrlich's or Mayer's hemalum followed by an eosin counterstain. Popular also are Heidenhain's iron hematoxylin, Giemsa's and Feulgen's and sometimes a triple stain such as Mallory's. Other staining technics frequently used include basic fuchsin, methyl green-pyronine, Pappenheim's, and toluidine blue. Many of the stains employed to detect infectious organisms in higher animals have been adapted for use in invertebrates, for example, the use of Giemsa for detecting insect viruses, Giemsa and Macchiavello for rickettsiae, Gram and Giemsa for bacteria and Grocott's stain for fungi.

Despite the fact that invertebrate tissues and diseases have been the subject of study since the work of Bassi, Pasteur and Metchnikoff over one hundred years ago, today, very little is known of the normal anatomy and histology of most of the animal species that inhabit this earth. This presents a vast area for investigation to the future generations of histologists.

255

Oyster Tissue

Pauley (1967)

Designed to show: Collagen fibers.

Fixation. Zenker's solution.

Embedded in: Paraffin.

Preparation of solutions:

A. Ammonium hydroxide   3 drops
   Water   1000 ml
B. Acid fuchsin   0.5 gm
   Distilled water   100.0 ml
C. Aniline blue, water soluble   0.5 gm
   Orange G   2.0 gm
   Phosphotungstic acid   1.0 gm
   Distilled water   100.0 ml

Procedure:
1. Deparaffinize with xylene.
2. Pass through absolute and 95% ethanol to water.
3. Place in 70% alcoholic iodine solution for 10 min.
4. Wash in running water for 5 min.
5. Clear in 5% sodium thiosulfate (hypo/solution).
6. Wash in running water for 10 min.
7. Stain in Harris' or Mayer's hematoxylin for 20 min (see pp. 105–107).
8. Rinse in tap water for 5 min.
9. Differentiate in 1% acid alcohol (3–5 quick dips) until nuclei are distinct.
10. Rinse in running water for 5 min.
11. Dip in Solution A until sections are deep blue.
12. Rinse in running water for 5 min.
13. Stain in Solution B for 7 min.
14. Rinse in tap water (2 quick dips).
15. Blot to remove excess stain without touching the tissue.
16. Stain in Solution C for 15 min.
17. Rinse in tap water (1 or 2 dips).
18. Pass through 3 changes of 95% ethanol for 1 min each.
19. Pass through 2 changes of 100% ethanol and then 3 changes of xylene for 3 min each.

Results: Nuclei—dark reddish-blue; collagen fibers—blue; cartilage—blue; epithelium—orange-red; mucous cells—bluish-purple; pigment cells—brown; Leydig cells—pink; leukocytes—pink to orange.

Modified Barbeito-Lopez Trichrome Stain for Oyster Pathology

Heaton and Pauley (1969)

Application: Has the benefit of being a trichrome and bacterial stain simultaneously.

Fixation: Ice cold Solution A.

Preparation of solutions:

A.	Formaldehyde	150 ml
	1.3% calcium chloride (anhydrous, granular, tech grade)	850 ml
B.	Formaldehyde	4.0 ml
	Glacial acetic acid	0.2 ml
	Distilled water	100.0 ml
C.	Basic fuchsin	1.0 gm
	Phenol	5.0 gm
	90% ethanol	10.0 ml
	Distilled water	100.0 ml

Grind the fuchsin with phenol in a mortar, add the alcohol in 10 successive lots and continue grinding. Wash out the mortar with 10 successive lots of water. Filter and age for 30 days.

D.	Solution C	10.0 ml
	Glacial acetic acid	0.2 ml
	Distilled water	100.0 ml
E.	Aniline blue (water soluble)	0.5 gm
	Orange G	2.0 gm
	Phosphotungstic acid	1.0 gm
	Distilled water	100.0 ml

Procedure:
1. Clear in xylene, 2 changes for 3 min each.
2. Rinse in 100, 95, 70 and 50% ethanol for 1 min each.
3. Rinse in running tap water for 5 min.
4. Transfer to Solution B for 1 min.
5. Rinse in tap water.
6. Stain in Solution D for 1–5 min.
7. Transfer to Solution E for 30–60 min.
8. Rinse in 95% ethanol for 45 sec.
9. Rinse in 100% ethanol, 2 changes for 1½ min each.
10. Clear in xylene, 2 changes—3 min each.

Results: Mucous glands—orange; Leydig cell cytoplasm—pale orange; cell nuclei—violet; collagen—blue, cartilage—brilliant blue; epithelial structures—violet; leukocytes—violet; bacteria—red.

Insect Chromosomes

Crozier (1968)

Application: For species with small, numerous chromosomes and in situations where only a small amount of material is available.

Preparation of solutions:
  A. Sodium chloride   14.0 gm
      Calcium chloride   0.4 gm
      Potassium chloride   0.2 gm
      Sodium bicarbonate   0.2 gm
      Water   1000 ml
  B. Colcemid-Ringer 0.05% Colcemid (Ciba) in Solution A.
  C. Orcein   1 gm
      Lactic acid 85%   28 ml
      Glacial acetic acid   22 ml

Procedure:
  1. Place specimen in Solution B at 20–25°C for at least 5 hr.
  2. Transfer to 1% sodium citrate for 10–20 min.
  3. Transfer to acetic-methanol (1:3) for 30 min.
  4. Transfer to a drop of 60% acetic acid (aqueous) on a clean warmed slide and macerate if necessary.
  5. Transfer to a drop of acetic-methanol and gently warm.
  6. Place the slides in acetic-ethanol (1:3) for 4 hr.
  7. Rinse in 70% ethanol.
  8. Stain in Solution C, apply coverslip and place in heat for 12 hr at 50°C.
  9. Loosen coverslip in acetic-ethanol.
  10. Dehydrate with 95% and absolute ethanol and clear in xylene.

Stain for Trematode Whole Mounts

Gower (1939)

Preparation of solutions:
  A. Acidified carmine
      Acetic acid 45%   100 ml
      Carmine   10 gm
      Bring to a boil, then cool and filter. The residue is acidified carmine.
  B. Acidified carmine   1 gm
      Alum   10 gm
      Distilled water   200 ml
  C. Potassium chlorate crystals   100 gm

| Hydrochloric acid (concentrated) | 0.1 ml |
| Ethanol 70% | 100 ml |

Add the HCl to the potassium chlorate in a covered dish. As chlorine is given off, add the ethanol.

Procedure:

1. Bring specimen to water.
2. Stain in Solution B for 12–36 hr depending on size of the worm.
3. Wash in water.
4. Dehydrate through 20, 35 and 40% ethanol.
5. Destain in Solution C.
6. Transfer to 80% ethanol.
7. Dehydrate in absolute alcohol.
8. Add cedarwood oil until mixture is up to 50%.
9. Clear in cedarwood oil.

Results: Nuclei or principal organs appears a deep rose red; cytoplasm, parenchyma and muscle are unstained.

Trematodes and Cestodes

Smyth (1951)

Designed to show: Egg-shell material.

Fixation: 0.5% formol-saline.

Preparation of solutions:

A. Acidified carmine (see p. 258)	1 gm
Alum	10 gm
Distilled water	200 ml
Dissolved by heating-cool-filter	
B. Malachite green	0.5 gm
Distilled water	100 ml
C. Orange G	1 gm
Absolute alcohol	99 ml

Procedure:

1. Bring sections down to water.
2. Stain in Solution A for 2 hr.
3. Rinse in water.
4. Stain in Solution B for 2 min.
5. Upgrade through 50, 70 and 90% ethanol.
6. Differentiate in absolute alcohol.
7. Counterstain in Solution C for 1 sec.
8. Rinse in absolute alcohol, clear in xylene.

Results: Nuclei—red; cytoplasm—pink and yellow; egg-shell material—green or greenish blue.

Insect Musculature

Kramer (1948)

Designed to show: Dipteran larval musculature *in toto.*

Fixation: Modified Bouin. Water, 190 ml; picric acid, 2.4 gm; formalin, 60 ml; and glacial acetic acid, 12.5 ml.

Procedure:
1. Transfer to 50% ethanol for 10 min.
2. Transfer to 70% ethanol for 1 hr.
3. Transfer to 95% ethanol for 10 min.
4. Stain in eosin (0.5% eosin in 95% ethanol) for 6–8 hr.
5. Transfer to 95% ethanol for 30 min.
6. Immerse in 95% ethanol and add 4–6 drops of oil of wintergreen (methyl salicylate) hourly and swirl until missible.
7. Continue adding oil of wintergreen until specimen turns orange red.
8. Transfer to oil of wintergreen.

Results: Under a binocular scope the muscles appear pink; cuticle is transparent green.

Rae's Modified Masson Trichrome for Insect Tissue

Rae (1955)

Preparation of solutions:
  A. Water   50 ml
      Ferric ammonium sulfate   0.8 gm
      Sulfuric acid, conc   50 ml
      Ethanol, 95%   50 ml
      Hematoxylin   0.5 gm
  B. Xylidene Ponceau (G. T. Gurr), 0.25% in 1% aqueous acetic acid.
  C. Phosphomolybdic acid, 1% aqueous.
  D. Light green, SF, 2% in 1% aqueous acetic acid.

Procedure:
1. Transfer to Solution A for 12 min.
2. Transfer to Solution B for 3.5 min.
3. Rinse in distilled water.
4. Transfer to Solution C for 4 min.
5. Rinse in distilled water.
6. Transfer to Solution D for 3 min.

7. Rinse in distilled water.
8. Dehydrate quickly in absolute ethanol.
9. Clear and mount.

Results: Nuclei—brownish purple; connective tissue—green; muscle—green striations on reddish pink or green background; nerve tissue—pink and greenish gray.

Mosquitoes

Matta and Lowe (1969)

Designed to show: Cells infected with the mosquito iridescent virus.

Fixation: Carnoy's.

Embedded in: Paraplast.

Procedure:
1. Xylene via alcohols to water.
2. Stain in Delafield's hematoxylin for 5 min (see p. 105).
3. Rinse in distilled water for 10 sec.
4. Decolorize in 1.0% hydrochloric acid for 30 sec.
5. Rinse in distilled water for 5 sec.
6. Transfer to 0.035% ammonium hydroxide for 2 min.
7. Rinse in distilled water for 5 sec.
8. Counterstain in 5% aqueous eosin for 3 min.
9. Transfer to 70, 95 and 100% ethanol—10 sec each.
10. Transfer to xylene, 2 changes—3 min each.
11. Mount in neutral, synthetic mounting medium.

Results: Under dark field, virus—infected cells—dark purplish-brown; cuticle—green; epidermal cells—light reddish brown; eye facets—bright red; imaginal disks—light reddish-brown with green interior; muscle-yellow; brain and ganglia—yellowish-green with light brown nuclei; nerve cord—green; Malpighian tubules—yellowish green with light brown nuclei; fat body—green with brown nuclei, globules—yellow; gut epithelium—yellow with green nuclei.

A Modified Azan Staining Technic for Insects

Hamm (1966)

Designed to show: Polyhedrosis and granulosis inclusion body viruses in lepidopteran insect larval tissues.

Fixation: Alcoholic Bouin's. Picric acid—1 gm, glacial acetic acid—15 ml, formalin—60 ml, 80% ethyl alcohol—150 ml.

Embedded in: Paraffin.

Preparation of solutions:
  A. Dissolve 0.1 gm of azocarmine G in 100 ml of distilled water and boil the solution for 5 min. Allow to cool and add 2 ml glacial acetic acid. Filter before use.
  B. Dissolve in 100 ml of distilled water, 1.0 gm phosphotungstic acid, 0.1 gm aniline blue (water soluble), 0.5 gm orange G, and 0.2 gm fast green FCF.

Procedure:
  1. Toluene via alcohols to water.
  2. Transfer to 50% acetic acid for 5 min.
  3. Rinse in distilled water for 2 min.
  4. Stain to Solution A for 15 min.
  5. Rinse in distilled water for 5 sec.
  6. Transfer to aniline (1% in 95% ethanol) for 30 sec.
  7. Rinse in distilled water for 5 sec.
  8. Counterstain in Solution B for 15 min.
  9. Transfer to 50% ethanol for 10 sec.
  10. Transfer to absolute alcohol, 2 changes, 30 sec each.
  11. Transfer to toluene, 2 changes.
  12. Mount in neutral, synthetic mounting medium.

Results: Virus inclusion bodies—red; epicuticle—red; endocuticle—blue; muscle—light blue to blue green; epidermal cells—yellowish-green; fat body cells—yellowish green with dark green nuclei; nerve tissue—light blue; silk gland—green, contents red or blue; midgut epithelium—green and blue.

Silver Stain for the Central Nervous System

Blest (1961)

Application: Recommended for the display of the CNS especially of the locust, honey bee, and Lepidoptera.

Fixation: Alcoholic Bouin's for 24 hr (see p. 15).

Embedded in: Paraffin.

Preparation of solutions:
  A. Dissolve 20 gm silver nitrate in water up to 100 ml.
  B. Boric acid, 0.2 M   27.5 ml
     Borax, 0.05 M   22.5 ml
     Silver nitrate, 1% aqueous   10.0 ml

Pyridine	6.0 ml
Distilled water	250.0 ml
C. Hydroquinone	3 gm
Sodium sulfite	30 gm
Distilled water	300 ml

Procedure:
1. Dewax and hydrate quickly.
2. Transfer to Solution A for 2–3 hr.
3. Rinse in distilled water for 5 min.
4. Transfer to Solution B at 33°C for 16–20 hr.
5. Transfer to Solution C at 60°C for 3 min.
6. Wash in running water for 3 min.
7. Tone in 0.2% gold chloride solution for 1–3 min.
8. Rinse in distilled water.
9. Transfer to 2% oxalic acid for 5 min.
10. Wash in distilled water for 3 min.
11. Fix in 5% sodium thiosulfate for 1 min.
12. Wash thoroughly in distilled water, dehydrate, and mount.

Results: Fibers—reddish purple.

Silver Staining of Insect Central Nervous System's

Modified from Gregory (1970)

Fixation: Alcoholic Bouin's (see p. 15).

Embedded in: Paraffin-cut sections at 10 μm.

Preparation of solutions:
A. Dissolve 2 gm Protargol-S on the surface of 100 ml water, add to Coplin jar containing 5 gm clean copper (5–10 μm lengths, 19 AWG).

B. Hydroquinone	1 ml
Sodium sulfite hydrate	5 gm
Distilled water	99 ml
Make up shortly before use.	
C. Gold chloride	1 gm
Distilled water	100 ml
D. Oxalic acid	2 ml
Distilled water	98 ml
E. Sodium thiosulfate	5 gm
Distilled water	100 ml

Procedure:
1. Bring sections to water.
2. Impregnate in Solution A for 18–24 hr at 37°C.
3. Rinse briefly in distilled water.
4. Develop in Solution B for 10–15 min.
5. Wash in distilled water 3 times, 2 min each.
6. Repeat Steps 1–4 using fresh Solution A.
7. Intensify in Solution C for 15 min.
8. Rinse briefly in distilled water.
9. Remove residual silver salts in Solution E for 10–15 min.
10. Wash in distilled water 3 times, 3 min each.
11. Dehydrate, clear and mount.

Results: Nuclei-purplish or bluish black; fibers and cell bodies—red-purple.

Cresyl-fast Violet Stain for Oyster Tissue

Heaton and Pauley (1969)

Application: Facilitates the observation of both bacteria and fungi in the same oyster tissue.

Fixation: Ice cold Solution B.

Embedded in: Paraffin.

Preparation of solutions:
A. Cresyl-fast violet   5.0 gm
    Distilled water   500.0 ml
    Glacial acetic acid   0.5 ml
    pH should be 3.7. Allow to stand for 16–20 hr before use. Do not filter.
B. Formaldehyde   150 ml
    1.3% calcium chloride (anhy- 850 ml
    drous, granular, tech. grade)

Procedure:
1. Clear in xylene, 2 changes, for 3 min each.
2. Rinse in 100, 95, 70 and 50% ethanol for 1 min each.
3. Rinse in tap water for 2 min.
4. Stain with Solution A for 10–20 sec.
5. Wash in running tap water for 5 min.
6. Drain well and remove excess moisture.
7. Rinse in 50, 70 and 95% ethanol for 1 min each.
8. Rinse in 100 ethanol, 2 changes, for 10–15 sec each.
9. Clear in xylene, 2 changes, for 3 min each.
10. Mount in Technicon mounting medium.

Results: Fungi-blue or purple, cell nuclei—dark blue, bacteria—dark blue, cartilage—pink to red, mucous glands—pink to red.

NOTE: The 4 rinses in ethanol after staining provide differentiation by reducing the overstaining that occurs in oyster tissue, even with the short staining time. Cresyl violet acetate dye stains in a similar manner, but is not as satisfactory as cresyl fast violet.

Wismar's Quadrachrome

Wismar (1966)

Designed to show: Chiton and nucleic acids in invertebrates.

Fixation: Formalin-sublimate-acetic acid (no more than 24 hr).

Embedded in: Paraffin.

Preparation of solutions:

A. Fixative—formalin, 20 ml; mercuric chloride, 4 gm; glacial acetic acid, 5 ml; distilled water, 80 ml.

B. Lugol's iodine

Potassium iodide	6 gm
Iodine	4 ml
Distilled water	100 ml

C. Alcian blue—dissolve 1 gm Alcian blue 8GS in 1% acetic acid.

D. Verhoeff's Hematoxylin

Hematoxylin (5% in absolute alcohol)	50 ml
Ferric chloride, 10%	20 ml
Lugol's iodine (solution B)	20 ml

E. Woodstain scarlet-acid fuchsin—2 parts of 0.1% woodstain scarlet in 0.5% aqueous acetic acid and 1 part of 0.1% acid fuchsin in 0.1% aqueous acetic acid.

F. Extract 6 gm saffron in 100 ml of 100% ethanol at 58–60°C for 48 hr, decant and store in a tightly sealed bottle.

Procedure:

1. Take sections through alcohol changes to 80%.
2. Place in a 15% solution of Lugol's iodine (Solution B) in 70% ethanol for 5 min.
3. Decolorize in a 5% aqueous sodium thiosulfate for 5 min.
4. Wash in running water for 5 min.
5. Stain in Solution C for 30 min.
6. Wash in running water for 1 min.
7. Stain in Solution D for 4 to 6 hr.
8. Place in 95% ethanol for 3 min.
9. Stain in Solution E for 3 min.

10. Place in 1% aqueous acetic acid for 1 min.
11. Differentiate in 5% aqueous phosphotungstic acid for 20 min.
12. Place in 1% aqueous acetic acid for 1 min.
13. Differentiate in 10% aqueous iron chloride.
14. Dehydrate in acetone and 2 changes of absolute alcohol.
15. Stain in solution F for 5–8 min.
16. Rinse in 2 changes of absolute alcohol.
17. Clear in xylene and mount.

Results: Nucleic acids—purple to black; chitin and cytoplasm—bright red to lavender.

Acetic-Orcein for Insect Chromosomes

LaCour (1941)

Application: Recommended for quick chromosome counts and chromosome structure.

Preparation of solutions: Dissolve 1 gm of orcein in 45 ml of hot (near boiling) glacial acetic acid. Allow to cool; then add 55 ml of distilled water, shake and filter.

Procedure:
1. Tease out cells in a drop of stain for 30–60 sec.
2. Smear the coverslip with a thin layer of Mayer's (see p. 106) albumen and place over low flame until a gray smoke appears.
3. Cool and place over tissues.
4. Apply pressure under blotting paper to coverslip.
5. Pass slide over low flame 2 or 3 times.
6. Ring for temporary mount.

For Permanent Mounts:
1. Remove coverslip in 10% acetic acid for 3–10 min.
2. Transfer through 80% absolute alcohol for 2 min each.
3. Pass through 2 changes of cedarwood oil, 5 min each.
4. Add balsam and recombine slide and coverslip.
5. Blot and remove excess cedarwood and dry on hot plate.

For Drosophila Salivary Glands:
1. Use modified staining solution by making it 2% orcein and increasing the acetic acid content to 70%.
2. Dissect out tissues in staining solution.
3. Let stand in stain for 5–10 min.
4. Apply coverslip.
5. Apply gentle pressure for flattening.

NOTE: Staining is good after acetic-alcohol fixation. Prefixed tissues should be rinsed in 45% acetic acid before staining. Material previously stored in 70% alcohol should be transferred to 45% acetic acid for 1 hr or more.

Differential Stain for Chitinoid Material

Gier (1949)

Application: Recommended for detecting early chitin formation, chitinous linings, egg capsules, and insect exoskeleton fragments in water.

Preparation of solutions:

A. *Stock*—dissolve 1 gm of powdered Giemsa in 66 ml of glycerin at 60°C for 24 hr, then add 66 ml of acetone-free methyl alcohol.

B. *Working solution*—mix 1 ml of the stock solution with 50 ml distilled water; then add 0.5 ml of a 0.1% $NaHCO_3$ solution to make the mixture slightly alkaline.

Procedure:

1. Take the slides down to tap water.
2. Stain in working solution for 1 hr.
3. Rinse in water.
4. Differentiate in acetone—about 30 sec.
5. Dip in xylol.
6. Mount.

Results: Chitinous material—light green, nuclei—blue to black, cytoplasm—pink to red.

NOTE: Insect fragments in aqueous suspension may be stained by adding stain to the suspension at the rate of 1 drop of stock solution of stain to 1 ml of suspension. Staining time is approximately 1 hr at room temperature.

Aldehyde Fuchsin for Neurosecretory Products in Insects

Ewen (1962)

Preparation of solutions:

A. Basic fuchsin   1 gm
   Water   200 ml
   HCL (concentrated)   2 ml
   Paraldehyde   2 ml

Dissolve fuchsin in boiling water for 1 min. Cool, filter and then add HCl and paraldehyde. Store in a stoppered jar for 4 days at 20°C (red color should be gone). Filter and dry the precipitate on filter paper. Store the crystals in a stoppered bottle.

B. Staining solution, stock
   Dry dye crystals   0.75 gm
   Ethanol (70%)   100 ml

C. Working stain solution
   Solution B   25 ml
   Ethanol 70%   75 ml
   Glacial acetic acid   1 ml

D.	Potassium permangenate	0.15 gm
	Concentrated sulfuric acid	0.1 ml
	Distilled water	50.0 ml
E.	Sodium bisulfite	2.5 gm
	Distilled water	100 ml
F.	Ethanol, absolute	100 ml
	HCL, concentrated	0.5 ml
G.	Phosphotungstic acid	4.0 gm
	Phosphomolybdic acid	1.0 gm
	Distilled water	100.0 ml
H.	Distilled water	100.0 ml
	Light green SF yellowish	0.4 gm
	Orange G	1.0 gm
	Chromotype 2R	0.5 gm
	Glacial acetic acid	1.0 ml

Procedure:
1. Fix material in Bouin's, Helly's or Susa's.
2. Hydrate sections as usual.
3. Transfer to Solution D for 1 min.
4. Rinse in distilled water.
5. Decolorize in Solution E.
6. Dehydrate through several water rinses to 30 and 70% ethanol.
7. Stain in Solution C 2–10 min.
8. Wash in 95% ethanol.
9. Differentiate in Solution F for 10–30 sec.
10. Pass through rinses of 70 and 30% ethanol to distilled water.
11. Mordant in Solution G for 10 min.
12. Rinse in distilled water.
13. Counterstain in Solution H for 1 hr.
14. Rinse in 0.2% acetic acid in 95% ethanol.
15. Dehydrate rapidly through absolute ethanol, clean in xylol.

Results: Cytoplasm—light green, Nuclei—orange, neurosecretion—dark purple, neuropil mass—light green, corpus paracardiacum—green, corpus allatum—pale green.

Insect Tissue

Levy (1943)

Application: Recommended for insect tissues that have been fixed and preserved in 70% ethanol for long periods.

Embedded in: Paraffin.

Preparation of solutions:

A. Methylene blue—saturated solution in 95% dioxane.
B. Erythrosin—saturated solution in 95% dioxane.

Procedure:

1. Transfer to xylene for 2–3 min.
2. Transfer to xylene for 1 min.
3. Transfer to dioxane 100% for 2–3 min.
4. Transfer to dioxane 100% for 1 min.
5. Stain in solution A until tissues turn blue green.
6. Wash in dioxane 100% for 5 sec.
7. Transfer in Solution B until desired contrast is achieved.
8. Wash in 100% dioxane for 15 sec.
9. Transfer to 100% dioxane for 2–3 min.
10. Transfer to xylene for 2–3 min.
11. Transfer to xylene for 1 min.
12. Mount in gum damar.

Results: Nuclei of epithelial, fatty and glandular tissue—blue; cytoplasm, muscle fibers, nerve cord, hypodermis—reddish pink; epicuticle—black; exocuticle—yellow to green.

A Modification of Gomori's for Crustaceans

Hubschman (1962)

Application: Recommended for both larval and adult crustacean tissue, especially marine and freshwater shrimp.

Fixation: Bouin's picro-formalin or any alcoholic modification is recommended as a fixative (see p. 15).

Embedded in: Paraffin.

Preparation of solutions:

A. Dissolve 0.1 gm of azocarmine in 100 ml of distilled water and boil the solution for 5 min. Allow to cool and add 2 ml of glacial acetic acid. Filter before use.
B. Prepare a solution of distilled water, 1000 ml; phosphotungstic acid, 10.0 gm; aniline blue, 1.2 gm; and orange G, 4.4 gm.

Procedure:

1. Transfer from xylene through series of alcohol solutions to water.
2. Transfer to Solution A for 15 min.
3. Rinse in distilled water.
4. Transfer to aniline (1% in 95% alcohol) for 30 sec.
5. Transfer to distilled water.
6. Counterstain in Solution B for 15 min.

7. Transfer through 3 changes of absolute alcohol for 1, 1 and 10 min.
8. Transfer through 2 changes of xylene.
9. Mount in Piccolyte or Permount.

Results: Epicuticle—red; endocuticle—two shades of blue; epidermal cells, cytoplasm—yellowish pink; nuclei—red orange; muscle—blue purple; nerve tissue, ganglionic nuclei—red, neuropil—pale lavender; hepatopancreas—pale orange to gray cytoplasm, brushborders bluish gray; oocytes—gray with brilliant red nucleoli; sperm—orange.

Wolbach's Giemsa Colophonium Stain

Killick-Kendrick and Canning (personal communication)

Giemsa Colophonium is the best general method of demonstrating exo-erythrocytic schizonts and sporogonic stages of Haemosporidia (malaria parasites and related organisms) and endogenous stages of coccidia and gregarines. It is also excellent for trypanosomatids in sections of tissues of invertebrate hosts. It most nearly displays the colors obtained in smears by Romanovsky staining.

Fixation: Carnoy's fluid is the fixative of choice. Pieces of tissues, not more than 5 mm thick, are fixed for 3–4 hr. Insects, after removal of wings and legs, are fixed for 30 min or longer according to size. Wash out fixative with absolute ethanol before clearing and embedding.

Embedded in: Tissues are routinely embedded in paraffin wax. Insects may require double embedding in celloidin and wax. Cut sections 3–5 μm (or thicker for trypanosomatids).

Double embedding technic:

Solution A.	Celloidin flakes	1 gm
	Methyl benzoate	100 ml
Solution B.	Paraffin	100 gm
	Benzene	10 ml

1. Add celloidin to methyl benzoate, shake and let stand, repeat until completely dissolved.
2. Transfer fixed and washed specimen through 50%, 70%, 90% and absolute ethanol for 2 hr each.
3. Transfer to Solution A for 24 hr.
4. Transfer to fresh Solution A for 48 hr.
5. If necessary, transfer to third change of Solution A.
6. Transfer through 3 changes of benzene for 4, 8, and 12 hr.
7. Transfer to Solution B in embedding oven for 1 hr.
8. Transfer through 2 changes of paraffin for 1 hr each.
9. Embed in paraffin.

Preparation of solutions:

A. Giemsa's stain, 10 ml; acetone, 10 ml; methanol, 10 ml; and phosphate buffered distilled water, 100 ml.

NOTE: Not all brands of Giemsa stain are suitable; Hopkin and William's "Revector" is recommended.

B. 15% colophonium resin in acetone.
C. 30% acetone in xylene.
D. Xylene.

Procedure:

1. Dewax and hydrate as usual.
2. Stain in the Giemsa's Solution A; Carnoy-fixed material for 1 hr.
3. Rinse in tap water. Shake off excess water and place slide on a staining platform over a sink.
4. Put 2 drops of water on the sections and flood slide with colophonium acetone (B), rocking the slide during addition.

NOTE: Speed of differentiation is increased if more water is present initially and is halted by addition of pure colophonium acetone.

5. Wash in acetone xylene mixture (C).
6. Wash in several changes of xylene to ensure that acetone is removed.
7. Mount in Permount.

Results: Polychromatic as of Romanovsky stains: generally cytoplasm—blue, depth of color according to nature of material; nuclei—purple; erythrocytes—pink.

NOTE: Overstained sections: slide off coverslip, wash off Euparal vert with Colophonium acetone mixture and repeat from Step 4. Understained sections: remove coverslip and Euparal vert, wash in acetone and repeat from Step 2.

Iron Hematoxylin Staining

Modified after Heidenhain by various persons
Victor Sprague, (personal communication), University of Maryland

Application: Iron hematoxylin staining is primarily a cytological, rather than a histological, method. Being extremely versatile, it is, perhaps, the most generally useful of all cytological staining methods for both plant and animal cells. It is indispensible for studying protozoan cells. This modification (being simple, reliable, very flexible and not excessively time consuming) is highly recommended to students of Microsporida who have generally avoided using iron hematoxylin because older modifications have the reputation of being time consuming and difficult.

Designed to show: DNA and many other cell components.

Fixation: Bouin or any common fixing solution containing acetic acid.

Embedded in: Paraffin.

Preparation of solutions:
   A. Lang's mordant
          4% ferric-ammonium sulfate
          (C.P., purple crystals)   300.0 ml
          Acetic acid (C.P.)   5.0
          Sulfuric acid (C.P.)   0.6 ml

   B. 0.5% aqueous hematoxylin. Prepare stock solutions as follows: Dissolve 10 gm hematoxylin crystals in 100 ml of 95 or 100% ethanol (or other amounts in the same proportion). The stock improves with age. Prepare the staining solution by adding 1 part of the stock to 19 parts of distilled water. (The stain improves with use for several weeks and then deteriorates.)

   C. Saturated aqueous picric acid. Prepare a stock solution by adding an excess of crystals so that some remain on the bottom of the bottle.

Procedure:
1. Bring smears or sections down to distilled water.
2. Mordant in Lang's solution (A) about 10–30 min or longer. (Time not critical.)
3. Wash 5 min in running tap water.
4. Rinse in distilled water.
5. Stain in 0.5% aqueous hematoxylin (B) about 1½ times as long as time in mordant. (Time not critical.)
6. Destain in saturated aqueous picric acid solution (C) for 10–60 min (depending on amount of differentiation wanted). (Time not critical.)
7. Wash for at least 1 hr in running tap water. (Thorough washing is critically important.)
8. Dehydrate, clear and mount.
9. A light counterstain (eosin, lightgreen, orange G, etc.) may be used if desired.

Results: Where short time intervals are used, DNA is black and other elements are black or various shades of gray. When long time intervals are used, DNA is black and most other elements are essentially colorless (unless counterstained).

Varient: Excellent for microsporidian spores following the PAS treatment. Polarcap (sometimes also polar filament) is red and nuclei black. Contrasts good, probably because of a beneficial oxidizing effect of the periodic acid on some cell components.

Wright's Stain for Earthworm Coelomic Cells

Ramy R. Avtalion, Elizabeth Stein and Edwin L. Cooper
University of California, Los Angeles, California

Preparation of cells: Spread a drop of coelomic fluid on a glass microscope slide in the same manner as when making a blood smear. Allow to air dry.

Preparation of solutions:

A. Wright's stain—Add 100 mg Wright's powder to 20 ml absolute methanol. Warm gently in water bath for 20–30 min or until powder is dissolved. Filter before using. Stain should be made fresh or used for no more than 2 days.

B. Buffer solution, pH 6.3 to 6.4 (see "notes")—The buffer should be of low ionic strength (approximately 0.02 M) and preferably made up fresh each time. It may be made from phosphate buffer or from standard Beckman buffers. 10 ml of the standard pH 6.86 buffer is added to distilled water to make 200 ml, and the pH adjusted to 6.4 by adding a few drops of pH 4 buffer.

Procedure:

1. Cover smear with a *thin* film of stain and allow to nearly dry (approximately 1 min).
2. Rinse slide briefly in buffer and blot dry.
3. Add Wright's stain again, this time flooding slide. Leave on 1 min and then drain off excess stain.
4. Rinse again in buffer and leave immersed in buffer for 5 min.
5. Blot dry.
6. Slides may be mounted or examined directly.

Results: Epithelial-like cells—medium blue and dark blue-violet nuclei. They frequently contain small basophilic inclusions. Macrophage-like cells—vary pale blue to lavender; cytoplasm free of inclusions. Nuclei are large, rose-violet, and occasionally contain a single blue nucleolus. Eosinophilic cells are of two types and contain either small bright pink granules or larger vesicles. Another kind of granular cell contains coarser inclusions that may be pinkish-red, blue, light violet or a mixture of the three. Chlorogogen cells exhibit two types of granulation: one is bright blue and ovoid to round and the other is granular, dark blue-violet and irregular in shape.

NOTES: Standard methods of Wright's staining, as used for mammalian blood, do not produce the desired color range in coelomic cells. In this two-step staining procedure, the first step is designed to promote red coloration. Soaking in buffer also helps in this regard and is particularly useful in removing some of the dried coelomic fluid that surrounds the cells and would otherwise partially obscure them. Since the pH of the buffer affects the red-blue balance of the stain, a slightly redder coloration results at pH 6.3 than at 6.4.

Acid Phosphatase Determination for Earthworm Coelomic Cells

Barka *et al.* (1963) modified by Stein, Avtalion and Cooper
University of California, Los Angeles, California

Application: Demonstration of the presence of acid phosphatase activity in earthworm coelomocytes.

Preparation of cells: Prepare a thin smear by spreading a drop of coelomic fluid on a glass slide in the same manner as when making blood smears. Allow to air-dry; then fix in formalin vapor for 5 min.

Preparation of solutions:
  A. Stock solutions.
   (1) Veronal acetate buffer—9.714 gm sodium acetate trihydrate and 14.714 gm sodium barbituate dissolved in CO_2-free distilled water to make 500 ml.
   (2) Substrate—100 mg naphthol AS-BI phosphate dissolved in 10 ml *N,N*-dimethyl formamide. Keep refrigerated.
   (3) Dye—2 gm pararosanalin-HCl added to 50 ml 2 N HCl and gently heated. Filter when cool.
   (4) 4% sodium nitrate—4 gm sodium nitrate dissolved in 100 ml distilled water.
  B. Working solutions—Should be made up prior to use. Mix 10 ml of buffer stock solution with 24 ml distilled water and add 2 ml substrate stock. Mix 1.6 ml pararosanalin HCl stock solution with 1.6 ml sodium nitrate. Allow to stand for a few minutes; then add to solution. Adjust pH to 5.0 with N NaOH. This makes nearly 40 ml of incubating medium, enough to fill a Coplin jar.

Procedure:
  1. Incubate smears in medium for 60–90 min at room temperature.
  2. Rinse briefly in distilled water.
  3. Counterstain if desired with hematoxylin or other nuclear stain.
   (a) Hematoxylin method—Stain in Harris' hematoxylin for (see p. 106) 10–15 min; then rinse in tap water. Differentiate *briefly* (1–2 dips) in acid alcohol (1 ml concentrated HCl in 99 ml of 70% ethanol). Rinse again. Dip briefly in ammonia water (2 drops concentrated ammonium hydroxide in 50 ml tap water) to blue nuclei. Give final rinse.
  4. Air-dry and mount if desired.

Results: Acid phosphatase activity is indicated by the presence of the rose-red color of the azo dye. Cells may show different levels of activity ranging from negative (no activity) to very strongly positive. In the latter case the entire cell, except for the nucleus, stains an intense red. Use of a nuclear counterstain is not mandatory but is helpful when working with cells that tend to clump, as do earthworm coelomocytes.

NOTES: Slides should be processed as soon as possible after preparation of smears since acid phosphatase activity will decrease with time. The purpose of mixing pararosanalin HCl with sodium nitrate is to diazotize the dye to form hexazonium pararosanalin. This then acts as a coupler to the substituted naphthol that is released upon hydrolysis of the substrate by acid phosphatase. The resultant rose-red azo dye is quite stable and will not fade after mounting.

Vital Staining of Active Mucous Secretion of a Marine Coelomate

Bang and Bang (1972)

The free urn cell of the marine coelomate *Sipunculus nudus* produces two distinct types of secretion, each with a separate secretory apparatus. Each type of secretion fails to stain with vital dyes that stain the other. One is secreted as mucous droplets or granules that accumulate in membrane-bound sacs and that stain metachromatically with methylene blue, brilliant cresyl blue and neutral red. The other type of secretion is activated by specific stimuli; it is produced in four or five loci each of which synthesizes a free-flowing stream of secretion. Each locus concentrates the rose-violet form of Janus green and secretes the dye, either in its violet or its blue-green form, depending on the stimulus. The two systems can be stained either differentially or simultaneously. The average urn cell is about 25 μm in diameter.

Preparation of solutions:

A. Methylene blue and brilliant cresyl blue are prepared as 1:5000 solutions in boiled, cooled, filtered seawater; neutral red is prepared as a 1:3000 solution. One drop is added to one drop of cultured urn fluid.

B. Janus green or Janus green B (British Drug House) preferred; other stocks may not yield the same results) is prepared as a 1:5000 solution in boiled filtered seawater and filtered; the filtrate is prepared as a final 1:5000 solution. One drop is added to 1 drop of cultured urn fluid.

Results: Within a few minutes the basic aniline dyes metachromatically stain both the actively secreting mucous droplets and the coalescent granules that have accumulated in the dependent central sacs.

Janus green is concentrated by the secretory loci within about 5 min. In unstimulated urns these appear as discrete rose-violet spots peripheral to the central sacs. The same cells may then be stimulated to produce free-flowing secretion: when certain bacteria, or human saliva or nasal fluid are added, the resulting secretions emerge as violet-colored streams. When the cells are stimulated by human serum, the emerging secretory streams are a clear blue-green. The newly emergent streams always push the original violet secretory loci distalward.

Thus the secretory apparatus presumably concentrates the leuco violet form of the dye; then when the cells are stimulated, the dye is bound to a protein component of the emerging secretion either in its reduced (violet) form or in its oxidized (blue-green) form, depending on the stimulus.

Coelenterates with Calcified Exoskeleton

Cheney, University of Hawaii at Hilo (personal communication)

Application: Fixation and embedding procedure for corals and other coelenterates with calcified exoskeletons. Many coelenterates secrete a calcified exoskeleton composed of fine aragonite crystals of $CaCO_3$. The morphology, density and thickness of this skeleton is species-specific within a range of environmentally induced variation. In some forms the skeleton and/or tissue is pigmented. If only small samples are desired, tissue from species with larger polyps can be teased away from the skeleton before or after fixation. It is usually necessary, however, to decalcify the tissue prior to embedding. Generally, the decalcification and embedding procedures are analogous to those used for bone and yield good results at the light microscope level. The procedures have been infrequently applied at the electron microscope level. Best results for general histology have been obtained with corals from the families Pocilloporidae and Acroporidae.

Designed to show: General microscopic anatomy of tissue from coelenterates with calcified exoskeletons.

Fixation: Buffered formalin (10% in seawater of 0.4 M Sorensen's, pH 7.8) for 24 hr; Bouin's for 12 hr.

Postfixation treatment: Decalcification. Formalin-fixed specimens are readily decalcified in a 3% solution of hydrochloric acid in 70% alcohol. The material is hardened in 70% alcohol and then transferred to a large volume of acid-alcohol in a loosely covered container. If the solution is changed or replenished daily, decalcification is usually completed in 3–4 days. In some cases, it is desirable to use chelating agents for decalcification as they seem to cause less disruption of the soft tissues. A 10% aqueous solution of EDTA (disodium salt of ethylenediaminetetraacetic acid) will decalcify most tissues in 3–7 days if the specimens are held in a vacuum. Bouin's fixed material, if not already decalcified by the fixative, may be treated with either decalcifying solution. After decalcification, the specimens are rinsed in 3 changes of buffer (Sorensen's, 0.4 M, pH 7.8) for 1 hr each.

Embedded in: Paraffin. Infiltrate and embed according to usual procedures. Small sections can be processed rapidly, with 15–30 min per infiltration step. A vacuum bath should be used in the last paraffin step to eliminate bubbles in blocks. Cut at 5–7 μm.

Procedure: No special procedures are needed for general histology. Excellent results are obtained with Harris hematoxylin and eosin, Masson's trichrome, and PAS stains for paraffin section (see pp. 106, 118, 201).

Results: Nematocysts and goblets cells will stain a deep pink with hematoxylin and eosin. Zooxanthellae exhibit intense nuclear staining and usually some residual yellow to brown cytoplasmic pigmentation. Some corals (ex. *Porites*) have a heavy non-zooxanthellae pigmentation that usually remains visible throughout the processing steps.

NOTES: Decalcification is best done on as small a piece as possible. This can be facilitated by first breaking up the fresh or fixed specimens with a hammer and chisel. However, the tissues are fragile and it is difficult to break up massive colonies; therefore, it is far better to section fixed material with a circular diamond saw. Excess calcified material can be easily removed by this method and the resulting thin blocks are rapidly decalcified. Decalcification proceeds from the outside of the block to the dense inner skeleton and it is often possible to reduce the time to decalcification by stripping the more rapidly decalcified outer tissue mass from the remaining core. If desired, complete decalcification can be verified by carefully inserting a needle into the section or by adding 1 ml of 5% aqueous ammonium or sodium oxalate to 5 ml of the decalcifying solution containing the specimen. If the solution is clear after 5 mm, decalcification is complete. If a precipitate forms, decalcification must be continued. When DNA-sensitive techniques such as the Feulgen reaction or DNA autoradiography are employed, decalcify with EDTA. EDTA decalcification of formalin-fixed material from some sorals such as *Porites* results in very soft and easily damaged preparations. In these cases, much better results are obtained with glutaraldehyde fixation.

Demonstration of Invertebrate Nervous Systems in Whole Mounts or Sections

Jennings and LeFlore (1972)

This method demonstrates details of invertebrate nervous systems down to the finer ramifications of single neurones. It is especially applicable to adult and larval platyhelminthes, smaller annelids and copepod and ostracod Crustacea.

Preparation of solutions:

A. Fixative—Formalin, 10 ml; distilled water, 90 ml; sodium phosphate monobasic, 0.4 gm; sodium phosphate dibasic, 0.6 gm. Final pH 7.0. This gives 10% buffered formalin, pH 7.0.

B. Incubation medium.
   (1) Stock solution:

Cupric sulfate	0.3 gm
Glycine	0.38 gm
Magnesium chloride	1.0 gm
Maleic acid	1.75 gm
4% aqueous sodium hydroxide	30.0 ml
40% aqueous (saturated) sodium sulfate	170.0 ml

     This stock keeps indefinitely.
   (2) Working solution: 20 mg acetylthiocholine iodide dissolved in 1 ml distilled water and added to 10 ml of stock solution. Make up as required.

C. 40% aqueous sodium sulfate.
D. Dilute yellow ammonium sulfide.

Procedure:
1. Fix in 10% buffered formalin, pH 7.0 at 1°C, for 1–12 hr depending on size of specimens. Cercariae fix adequately in 1 hr; polychaetes up to 3 cm long require 12 hr.
2. Immerse in incubation medium at room temperature for 2–12 hr.
3. Wash well in three changes of 40% sodium sulfate.
4. Follow last sulfate wash with dilute (pale yellow) ammonium sulfide, 2 min.
5. Wash well in distilled water.
6. Counterstain if desired in 0.5% aqueous eosin, Nuclear fast red or other appropriate stain.
7. Wash in distilled water.
8. Mount in glycerol jelly or other aqueous medium.

Results: Sites of cholinesterase activity (cholinergic nerves) appear in shades of brown.

Comment: If sections are required, specimens can be gently dehydrated in graded ethanols to 95% and then impregnated in polyester wax, blocked and sectioned.

    In controls, the enzymic and thermolabile basis of the reaction is demonstrated by preincubation of specimens in 10^{-4}M eserine, heating before incubation to 90°C for 2 min, or omitting the specific substrate from the incubation medium. All three treatments prevent development of the final, visible reaction product.

Demonstration of Alimentary Systems in Small Invertebrates

Jennings and LeFlore (1972)

This method demonstrates acid phosphatase activity in the alimentary canals of small invertebrates and their larvae and in whole mounts, it allows clear differentiation of the gut and its associated glands from surrounding structures. Specially suitable for platyhelminth larvae.

Preparation of solutions:
A. Fixative—see method for 10% buffered formalin (see p. 13)
B. Incubation medium:
  4 mg naphthol AS-BI (or TR) phosphate dissolved in 0.25 ml dimethyl formamide
  24 ml 0.2 M acetate buffer pH 5.2
  34 mg red violet LB salt
  2 drops 10% manganese chloride

Procedure:
1. Fix in 10% buffered formalin, pH 7.0, at 1°C for 1–12 hr.
2. Rinse in distilled water.
3. Remove lipid by immersion in graded acetones (the Red violet LB salt is soluble in lipid and stained lipid droplets may be confused with a true positive reaction.)
4. Return to distilled water down graded acetones.
5. Incubate for 1–12 hr in incubation medium at room temperature.
6. Rinse in distilled water.
7. Counterstain, if desired, in Mayer's hemalum (see p. 106).
8. Wash, mount in glycerol jelly or other aqueous medium.

Results: Sites of acid phosphatase activity (concentrated in gut wall) red.

Comments: In controls, the enzymic and thermolabile basis of the reaction may be demonstrated by omitting the specific substrate or the manganese ions from the incubation medium or by holding the specimen at 90°C for 2 min prior to incubation.

Wright-Nicotine-Oxalic Acid for Invertebrate Blood

Yeager (1938)

Application: Heat-fixed blood insects, lobsters, crayfish, earthworms, etc.

Designed to show: The invertebrate blood cell types in their passive form. It will prevent form changes due to passive-active transformation during the process of cell coagulation.

Preparation of solutions:
A. Wright's blood stain—dissolve 0.5 gm Wright stain powder in 100 ml methy alcohol (acetone free). Reflux for 6–7 hr, pour into a clean bottle and allow to cool.
B. Nicotine solution—dissolve 2.5 ml of the pure redistilled alkaloid in 97.5 ml of distilled water.

   C. Oxalic acid   0.6 mg
       Distilled water   100 ml

Procedure:
1. Immerse insect in water at 55°C for 5 min.
2. Remove insect and dry.
3. Sever appendage; collect and smear a large drop of blood on a clean slide.
4. Flood dried smear with undiluted stain (Solution A) for 1.5 min and then diluted with an equal volume an equal volume of solution B for 3.5 min.
5. Pour off stain and wash with distilled water.
6. Blot and dry carefully over a small flame.
7. Dip quickly 2–4 times in 70% ethanol.
8. Blot and dry carefully over a flame.
9. Flood with Solution C for 1–2 min.
10. Blot and dry over a flame.
11. Mount in balsam under a coverslip.

CHAPTER 10

METHODS FOR PROTOZOA

John L. Mohr

UNSTAINED MATERIAL AND VITAL STAINS

Except in sections, most protozoans have sufficient difference in the refractive index of their structures that study in a medium permitting change of orientation of individuals is very useful. Alternatively, Sharp's (1914) practice of mounting between coverslips and examining specimens from both sides yields fuller observations than do ordinary preparations. Exploitation of ambient fluids of different refractive index or of special optical systems (polarizing, phase contrast, interference microscopes), or combinations of these reveal some structures more satisfactorily than stains. For examples of water miscible fluids with a useful range of refractive indices, refer to Amann's media (1896) or to Table 13 p. 111 of Lillie (1954).

SECTION A: GENERAL METHODS

Janus Green B

Michaelis (1900)
From Wenrich and Diller (Jones 1950, pp. 438–439)

Designed to show: Mitochondria.

No fixation.

Preparation of staining solution: Stock solution; 0.5% aqueous or 1% janus green B in neutral absolute alcohol.

Staining schedule:
Aqueous stain—
1. Add 1 drop of stock solution to 20 of water, sea water, or physiological saline solution (for parasites).
2. Mix 1 drop of diluted stain with a drop containing protozoans, cover with coverslip and observe over 2–10 min.

or

2a. Put coverslip on a drop with protozoans, and while observing, place a drop of diluted stain at one edge. With care draw off fluid by touching a strip of filter paper or towel to opposite side until stain encloses most of the protozoans. Observe over 2–10 min.

NOTE: Try a 1:10,000 dilution first. For many protozoans this is too high; dilution as low as 1:180,000 has been found effective for some kinds.

Alcoholic Stain—
1. Spread thin film of alcoholic solution on a clean slide; allow to dry.
2. Add drop of fluid with protozoans and observe as above.

Results: After a few minutes, the period varying with kind and physiological state: Mitochondria—deep blue green or green.

NOTE: As protozoans die, other organelles stain unspecifically, revealing nothing not demonstrable in unstained material. As oxygen in the preparation is exhausted, janus green is reduced, sometimes in characteristic patterns, with the pink of the diethylsafranin moiety of the dye appearing, sometimes fading as the leucobase is produced.

Neutral Red

From Wenrich and Diller (Jones, 1950, p. 438)

Designed to show: Progression of pH in food vacuoles of ciliates; some cytoplasmic parts of various protozoans.

NOTE: The concept of "vacuome," the cell-total of vacuoles stainable with both neutral red and osmium tetroxide, as the undistorted form of Golgi's internal net apparatus, is based in part on neutral red stained protozoans (see Parat (1928)).

No fixation.

Preparation of staining solution: Stock solution (1 gm neutral red in 100 ml neutral absolute alcohol); dilute stock solution about 1 drop to 10 ml of fluid approximating that in which protozoans are living (filtered pond water, physiological saline solution, etc.).

Staining schedule: On microscope slide place 1 drop of diluted stain to 4 of fluid with protozoans. With heterotrophs observe through the formation of several food vacuoles and change of color in vacuoles. With other protozoans observe until color appears in cytoplasm, usually within 5 min.

Results: Acid food vacuoles—cherry red, basic vacuoles—yellow, general cytoplasm—pink. Some cytoplasmic areas, as sap-filled vacuoles of dinoflagellates—reddish.

NOTE: If nuclei stain, they are dead. Lower concentration of stain; avoid overheating slide.

Inert Particulates

Any very fine, inert particulate material may be used to observe effects of single flagella or of fields of flagella (as of hypermastigotes from termites) or cilia. For short study ordinary commercial carbon inks may serve, but adjuvants may cause abnormal responses, particularly in extended studies. Very fine synthetic organic pigments give particularly striking patterns with epi-illumination.

Designed to show:
   A. Patterns of currents set in motion by cilia or flagella.
   B. Distinction between suspension feeders and others (Fenchel, 1965).

Preparation of particulate suspension: Shake 1 cc of amorphous carbon, graphite, well rubbed cochineal, or very fine pigment (such as Höchst Colloisol series) in 5 ml distilled water in test tube until most or all is wetted. Scoop off any that does not settle below the water surface.

Application: Squirt suspension with fine-drawn pipette into drop containing protozoans.

   NOTE: For detecting suspension feeders it is enough to observe the preparation briefly, as the particles are characteristically ingested as soon as they come within the feeding range. To observe the effects of ciliary or flagellar fields, vary heaviness of suspension, direction of jet, and vigor of jet. To get reproducible results may require use of an appropriate cell and a device to control position and angle of pipette (see Jones (1950), p. 492 *ff.*, or Langeron (1942), p. 269 *ff.*).

HANDLING OF PROTOZOANS FREE-LIVING OR SEPARABLE FROM HOST CAVITIES OR TISSUES

   Because of their size (larger than bacteria, but rarely large enough to treat individually) and numbers (often great), protozoans require their own methods for concentration and transfer. Free-living and enteric (especially rumen and cecum) protozoans may be so abundant they may be fixed *en masse* and be carried from step to step by centrifuging out of solution. Critical washings are necessarily more numerous than with smears or sections as it is important to remove interstitial fluids. Protozoans too delicate to withstand needed centrifugal forces may often be handled by settling and decanting or drawing off fluid with capillary siphon or filter paper. With settling, the process may be very slow, concentration and washing time requiring major adjustments as compared with processes for smears and sections. Reactions that cannot be carried out with diluted reagents and that cannot be prolonged must be done some other way. Final dehydration for resinous mountants requires special care.
   For technics that cannot be handled well by centrifuging or settling or where there are too few individuals for one to bear the losses inherent in those methods (unless the protozoans are large enough to manipulate one by one under a dissecting microscope), one affixes individuals on a coverslip or

microscope slide by coagulating the ambient serum or gut solute around them or pipettes them onto coagulable material.

Affixation in Albumen Smears

From Wenrich and Diller (Jones (1950), p. 442)

Designed to show: Attach loose protozoans to microscope slide or 22-mm coverslip.

Fixation: Appropriate to method or methods to be employed subsequently.

Preparation: Fresh egg albumen or Mayer's albumen fixative.

Schedule:
1. By settling method bring material to 85% alcohol in tapered centrifuge tube.
2. Spread thin smear of albumen or Mayer's fixative on microscope slide or 22 mm No. 1 coverslip.
3. Pipette a small drop of alcohol with protozoans onto smeared surface. The alcohol spreads with the protozoans which should be trapped by coagulating albumen.
4. Allow fluid to evaporate a little, then immerse in 95–96% alcohol 5–10 min for hardening of the albumen.

   NOTE: Steps 3 and 4 are critical and should not be tried for a first time with material in short supply. Too large a drop can overdilute the albumen. Too short evaporation will result in loss of much material; at best there is usually some loss. Too much evaporation desiccates the specimens which are then useless.

5. Treat subsequently by any technic appropriate for a wet smear.

Individual or Mass Section

Protozoans large enough to be handled individually can be oriented precisely by placing them in warmed formol-agar, cooling and manipulating the resultant transparent agar-cum-protozoan like a bit of tissue (Balamuth, 1941). Forms attached firmly to host tissues (as some cephaline gregarines, Licnophora, etc.) or which have been fused to a very small slice of fresh frog liver (or similar tissue base) by dropping protozoan(s) in albumen solution on the surface, oriented under a dissecting scope with a fine needle, then coagulated by gentle addition of fixative (Goodrich, 1937) can also be processed as bits of tissue. Otherwise, protozoans may be carried in a section of glass tubing of appropriate diameter with an end annealed and capped with nylon netting or bolting cloth of No. 20–24 mesh, or by pipette from reagent to reagent (dehydrating, clearing, infiltrating), or kept in one dish with reagents pipetted in and out. There is some advantage in clearing with a somewhat viscous fluid like cedar oil (Sharp, 1914) or terpineol when one carries a mixed batch of large protozoans to paraffin. These one can pick up seriatim individually with a tuberculin syringe by a controlled twist of the plunger, picking up negligible amounts of ambient fluid at the same time. Sharp eliminated solvent from cattle ciliates by pipetting them into a succession of gelatin capsules with melted paraffin. Handling with net-covered tubing until the last paraffin, then removing the cap, is somewhat simpler.

To get a paraffin block large enough to work with easily, the portion containing the protozoan or protozoans is included in a larger mass of paraffin lightly tinged with fat-soluble dye (for example, Sudan IV) so that the position of the protozoan in the unstained portion is apparent. Protozoans prestained are sectioned, placed on slides and flattened, dried, decerated in a fluid that dissolves both paraffin and mountant, drained, then covered with mountant and coverslip. Unstained sections are carried through appropriate reagents.

Marine protozoans require special care. Alcoholic fixatives such as Brasil's or Schaudinn's fluids may cause precipitation of gypsum, making material worthless or nearly so. The alcoholic fixatives can be used if the protozoans are fixed in so little sea water and so much fixer (for example 1 vol to more than 10 of fixative) that the sulfates do not precipitate or the precipitate is negligible or acid can be added (as to Schaudinn's) or increased. Alternatively other fixatives of somewhat similar composition, such as Heidenhain's SUSA or Hollande's cupric picro-formal-acetic, but without alcohol may give desired results.

Dissolution of skeletons in fixative or preservative occurs particularly with foraminiferans and acantharians (celestite spicules). Accordingly samples (especially sediment samples) of foraminiferans may be dried or may be fixed and preserved in alcohol. If material is to be stained unsectioned, fixation should be brief and material should be worked up promptly. For most sectioning methods, tests will be decalcified. In buffered formalin dissolution of acantharian spicules is moderate (Botazzi, Screiber, and Bower, 1971), but prolonged storage in it is presumably unfavorable for staining purposes. For critical work on acantharians and on whole radiolarians, it is essential to pipette individuals from plankton hauls as soon as they come on deck, fixing them singly or a very few together in separate small vials, avoiding shattering, irreversible clumping and covering with fine debris. Fix for minimal periods.

PROCEDURES WITH KILLED MATERIAL

Rose Bengal

Walton (1952)

Designed to show: Cytoplasm of foraminiferans alive at time of fixation of marine sediments "without incidental staining of organic or inorganic debris."

Fixation: Not specified; usually alcohol.

Procedures:
1. Wash sediment repeatedly with fresh water in stainless steel sorting sieve.

> NOTE: 0.075 opening in mesh suggested.

2. Stain 10 min in 0.1% rose bengal.

Results: Protoplasm—deep red through wet arenaceous tests, and particularly plainly through walls of wet or dry calcareous tests permitting enumeration of foraminiferans that had living protoplasm at the time of collection as contrasted with empty tests with no stained material.

> NOTE: Concentration of dye is not critical, the live-fixed protoplasm being recognizable with stains of different intensity. Counts made by this method are "within the accuracy of known sampling and counting techniques."

Cupric Picro-Formol-Acetic Fluid as Fixative and Stain

Modified from Hollande (1918)

Designed to show: Chromatoplastids in contrast with other organelles of plant flagellates, diatoms and desmids.

Preparation of solution: Add 4 gm picric acid little by little to 100 ml 2.5% aqueous neutral cupric acetate, then 10 ml formalin and 1.5 ml glacial acetic acid. For this purpose Hollande's fluid keeps indefinitely.

Fixing-Staining schedule:
1. Add material from bucket of plankton net to an equal amount of Hollande's fluid.
2. When organisms have settled, decant supernatant fluid.
3. Transfer a portion by pipette to 10% glycerin.
4. Mount in glycerin jelly or glychrogel.

or

3. Dehydrate with centrifuging or settling (if fragile forms are present).
4. Clear in terpineol or xylene, mount in balsam or synthetic resin.

> NOTE: Hollande fixing-and-staining is particularly applicable to blooms of marine dinoflagellates, silicoflagellates, chrysomonads and diatoms or freshwater flagellates (volvocales, euglenoids, dinoflagellates), diatoms or desmids. For marine armored forms such as Ceratium, Gonyaulax, or Dinophysis or diatoms and for freshwater forms, the formula may be made up with distilled water; but in a mixture of marine species with naked forms such as Gymnodinium and Noctiluca, the cupric acetate should be dissolved in filtered seawater.

Results: Chromatoplastids—green modified by whatever pigments are dominant in life, chloroplastids—green. Background—paler green with stigma, nucleus, flagella marked by differences in refractive index.

NOTE: Flagellates preserved and stained by this method, while not of living colors, have markedly contrasting appearances. Forms like Chilomonas with leucoplastids are pale, while the chloroplastids of Euglena, Volvox or desmids are green, and chromatoplastids of Ceratium, Gonyaulax or diatoms are brownish green, olive green, yellowish green, a little too green, but having a useful resemblance to color in life.

If part of the material is to be stained, as with carmine, iron-hematoxylin, or activated protein silver (all of which work well after Hollande's fluid), the fixative should be fresh and its proportion to water and organisms should not be reduced below one to one, and the staining should be carried out within a few weeks; otherwise they may be stored indefinitely in 10% formalin or in 5% glycerin in 70% alcohol.

Relief or Negative Staining

Night Blue-Phloxine Relief Stain

Modified from Mackinnon and Hawes (1961, p. 423)
Original Bresslau (1921)

Designed to show: Cilia, flagella and cortical patterns.

No fixation.

Preparation of staining solution:
  A. Night blue or Victoria blue B stock solution: dye 3 gm; distilled water, 100 ml.
  B. Phloxine B stock solution: dye, 6.5 gm; distilled water, 100 ml.
  C. Staining fluid: Add 4–6 drops of Solution B to Solution A just before use.

Staining schedule:
  1. Place small drop of protozoan culture or of body fluid with parasites on slide.
  2. Put drop of staining fluid next to protozoan drop and mix with glass rod.
  3. Observe under low power: With ciliates the stain is ready when contractile vacuoles have bluish fluid; for other kinds, 1–2 min.
  4. Air-dry, clear in xylene, mount in resinous medium.

  NOTE: Bresslau used a hot-air hair drier.

Results: Locomotor organelles—blue in relief against lighter background; depressions in pellicle—blue as with stain filled puddles; nucleus—vaguely pink; food vacuoles and sometimes contractile vacuoles—pinkish and bluish.

NOTE: Bresslau, modifying a bacteriological stain using Chinablau (anilin blue, w.s.), stated his Opalblau was *probably* a "disulfosäure des triphenyl-pararosanilin." Opalblau is listed in some older compendia as a basic dye similar to anilin blue, w.s. Because Bresslau intended that the protozoans (ciliates and flagellates) should be living when he started the drying out and because basic dyes tend to be less toxic, it is suggested that night blue or Victoria blue B be used in place of the anilin blue, w.s. used by Mackinnon and Hawes, Opalblau being unavailable. The particulate suspensions (above), 10% nigrosin, and 0.25% anilin blue, w.s. in 0.75% nigrosin have all been used similarly.

Osmium-Toluidine Blue Temporary Stain

From Kirby (1947)
Modified from Gelei (1927)

Designed to show: Cilia and pellicle of ciliates and flagella.

Fixation: 1% OsO_4.

Preparation of staining solutions:
A. Dissolve 0.1 gm toluidin blue O in 100 ml of distilled water.
B. Dilute 1 ml of Solution A with 9 ml of distilled water (= 0.01%).

Staining schedule:
1. Place drop with living protozoans on clean slide.
2. Invert slide placing drop over shallow cell with drops of 1% OsO_4, for 10–20 sec.
3. Turn drop upward, add a drop of one dilution (A or B) and observe.

NOTE: The weaker stain is effective with protozoans with stout structures such as the membranelle zone of Stentor; the strong solution is needed for such more delicate structures as the flagella of Chilomonas. Suitable action on forms of intermediate stoutness may be obtained by starting with the weaker stain and progressively reinforcing it with the stronger.

Results: Cilia, flagella, cirri, membranelles—deep blue, nuclei and plastids—blue, cytoplasm generally—transparent.

Iodine-Potassium Iodide Stain

From Lillie (1954, p. 92)
Original by Caventou (1826)

Designed to show: Starch, glycogen and other polysaccharides.

Fixation: None for temporary mounts; ice-cold Rossman's fluid or Gendre or Carnoy for permanent preparations.

Preparation of solutions: Dissolve 10 gm KI in 20 ml of distilled water; add 5 gm I_2 and allow to dissolve; add 80 ml of distilled water.

Staining procedure:
1. To drop of protozoans or to Rossman- or Gendre-fixed sections, add staining solution for a few min.
2. Pour off staining solution and rinse with distilled water.
3. Add glycerin and cover.

Results: Starch—deep blue to nearly black, glycogen—brown, nuclei—brown, skeletal plates of entodiniomorphs—brown, chloroplastids—brown or blue, cellulose (cell walls)—yellowish, some proteins—brown (not eliminated by diastase digestion).

NOTE: Iodine-KI has wide use in protozoology from a quick fixation and general stain or, particularly with eosin Y (see Microbiology section), for intestinal protozoans of man and a demonstration of skeletal plates in entodiniomorphs to a presumptive reagent for polysaccharides in plant flagellates and many other protozoans.

Methods for Glycogen

Best's method and Bauer's method as given in detail on pages 195–196 are directly applicable to whole protozoans or to protozoans in sections of tissues.

SECTION B: NUCLEAR STAINS FOR TEMPORARY PREPARATIONS

Iron Aceto-Carmine

From Conn and Darrow (1943)
Original by Belling (1926)

Designed to show: Chromatin in temporary preparations.

No fixation.

Preparation of staining solution: Add 1 gm carmine to boiling 45% glacial acetic acid and boil about 2 min. Cool and filter. To half of this add a few drops of a solution of ferric hydroxide in 45% glacial acetic acid until the liquid becomes bluish red, but without visible precipitate; then add the rest of the untreated aceto-carmine.

Staining procedure:
1. Place protozoans, in a drop of fluid or teased from host tissues, on to slide.
2. Cover with a drop of staining fluid and put on a coverslip.
3. Heat, short of boiling, to intensify stain.
4. Examine immediately, preferably with light that has passed through a green filter.

NOTE: To prolong usefulness of the temporary mount, add glycerin at one side of the preparation, drawing the staining fluid away with towelling or filter paper at the opposite side until glycerin occupies the whole area under the coverslip.

For methods of making iron aceto-carmine preparations permanent, see Chapter 8.

Results: Cytoplasm—unstained, chromatin—dark, translucent red (appearing black in green light).

> **NOTE:** Iron aceto-carmine and acidulated methyl green (next below) act similarly and well with stains of good quality, but trials with both are required to see which works better with particular protozoans.

Acidulated Methyl Green

Essentially as in Lee (1890)

Designed to show: Chromatin in temporary preparations.

No fixation.

Preparation of staining solution: Dissolve 1 gm methyl green in 1% acetic acid.

Staining schedule:
1. Place protozoans, in drop of culture or taken from host, on to slide.
2. Cover with a drop of staining fluid.
3. When protozoans stop deepening in color, rinse with several changes of 1% acetic acid, leaving a little of the rinsing fluid on them.
4. Add a drop of glycerin. When the fluids are in equilibrium, observe.

Results: Chromatin—bright green, nucleoli—unstained. Extranuclear structures—may be blue or violet (particularly if the dry stain or the solution is old), or transparent.

> **NOTE:** Solutions must all be acidulated. Staining is nearly instantaneous; "overstaining never occurs."

SECTION C: PERMANENT NUCLEAR STAINS

Precipitated Borax Carmine

Essentially as in Galigher (1934, p. 193)
Original by Lynch (1929)

Designed to show: Nuclei and cytoplasmic organelles in whole organisms.

Fixation recommended: Bouin, Hollande, Schaudinn and others not containing chrome salts.

Preparation of staining solution: Add 3 gm carmine and 4 gm borax to 100 ml of distilled water; boil until dissolved (about ½ hr); add 100 ml of 70% methanol and let stand 2 days; filter.

Staining schedule:
1. Transfer material from 35 or 50% to stain for 3–24 hr.

2. Add concentrated HCl dropwise, agitating container vigorously until all the carmine is precipitated as a brick red floc. Let stand 6 hr to overnight.

NOTE: With the small volume of material usually stained in protozoal work, it is easily possible to pass from basic to a strongly acid solution with the dye again soluble, the floc being dissolved before one is aware that the process is well under way. In such very acid solutions the protozoans may be consumed. After each drop the container should be shaken or tipped until no more action (precipitation) is apparent. End point is reached when there is little or no more of the original deep red translucent solution. If, with a drop of concentration HCl, the floc begins to dissolve again, add a small drop of borax carmine staining solution.

3. Add about an equal volume of 3% alcoholic HCl (either in 50% or 70% alcohol) and agitate gently to mix thoroughly. Let stand until stained material settles. Decant or pipette off stain suspension, repeating process several times, as needed to remove most of the stain.

NOTE: It is this stage which limits the convenience of this stain for protozoans. Individuals smaller than large Stentors, if they are not attached to tissues (as Licnophora on respiratory tree wall) or in smears (as termite flagellates or blood parasites) should be affixed to coverslips (see p. 284).

4. Cover material about 3 mm deep in fresh 3% HCl in 70% alcohol in a petri dish and observe under microscope until nuclei, zones of membranelles and other organells retaining stain are a deep pink.

NOTE: If decolorization appears to be happening in a few minutes, put material in 70% alcohol until the process is stopped; examine some in glycerin under the microscope. If the general cytoplasm is still stained, continue differentiation in acid-alcohol, but with more dilute, 1% or even 0.5%, HCl-alcohol.

5. When cytoplasm is transparent (nuclei and fibrillar structures should still be deep pink), remove acid alcohol.
6. Wash with two 5-min changes of 80% alcohol; hold in a third change for 60 min.

NOTE: Galigher counterstained ciliates and flagellates with indulin, but most batches do not work well in this technic; indigo carmine (0.1% in 80% alcohol briefly) contrasts well with the red of the carmine in motile structures, but these may be seen without counterstain in properly differentiated material.

7. Dehydrate, clear, mount in resinous medium.

NOTE: Lynch's carmine gives much more transparent stains than hematoxylins on the same subjects; it gives useful stains of Opalina and Nyctotherus, or of small flagellates and trichonymphas in the same termite gut smear or small and large rumen ciliates in the same batch; this is not usually possible with hematoxylins.

Feulgen Nucleal Reaction in Heat-Coagulated Albumen

From McArdle (1959)

Designed to show: DNA of chromatin, cortical structure and tri-dimensionality of ciliates.

Method of fixation required: 2% OsO_4 on slide.

Preparation of reagents:

A. Dilute Mayer's albumen 1 part to 4 parts of distilled water.

B. Schiff reagent: Dissolve 0.5 gm basic fuchsin in 100 ml of hot distilled water; filter still warm (about 50–60°C). cool and add 1 gm $NaHSO_4$ and 10 ml N HCl. Stopper tightly and let stand 24 hr in the dark. Add about 0.25 gm of activated charcoal, shake about 1 min, filter and store in a tightly stoppered bottle in refrigerator. The solution should be clear straw-yellow to brownish; discard it if it turns pink.

C. Dilute sulfurous acid: Add together 1 gm $NaHSO_3$, 10 ml N HCl, and 200 ml of distilled water.

Staining schedule:

1. On a clean microscope slide mix a drop of 2% OsO_4 into a drop containing protozoans.
2. After a few sec add 1–2 drops of diluted Mayer's albumen (A); mix vigorously with a plastic toothpick.
3. Heat slide in 60°C oven until drop browns and congeals (at least 1 hr).
4. Hydrolyze in 1 N HCl; time of hydrolysis must be determined by experiment; try 12 min in first run.
5. Rinse in 1 N HCl at room temperature, then in distilled water.
6. Stain in Schiff reagent (B) 1 hr or longer.
7. Rinse in 3 changes of solution C, each for 1 min.
8. Wash in distilled water.
9. Dehydrate, clear and mount.

Results: Chromatin—reddish violet, cortical cytoplasm—brownish, cilia and parts of infraciliature—brownish.

NOTES: Osmic fixation preserves body shape and heating colors cortex so that the 3-dimensional character of the specimens is apparent as it is not in other Feulgen methods. Loss of specimens is minimal. Cytoplasm shrinks slightly, there are sometimes cracks, and resolution at highest powers is reduced.

Oxidative Schiff-Aniline Blue-Orange G

Slightly modified from Wagner and Foerster (1964)
After Hotchkiss (1948)

Designed to show: Coccidians in tissues.

Method of fixation recommended: Stieve ($HgCl_2$, saturated aqueous, 76 ml; formol, 20 ml; glacial acetic acid, 4 ml).

Embedded in: Paraffin

Preparation of solutions:
A. Oxidizing solution: Mix 0.4 gm periodic acid; in 45 ml of distilled water and 5 ml of M/5 sodium acetate.
B. Reducing rinse: Dissolve 1 gm KI; 1 gm $Na_2S_2O_3$ in 45 ml distilled water; with stirring add 30 ml absolute alcohol and 0.5 ml 2 N HCl.
C. Schiff reagent: As B in Feulgen Nucleal Reaction above.
D. Dilute sulfurous acid: As C in Feulgen Nucleal Reaction above.
E. Counterstain solution: Stock solution—0.5 gm aniline blue, w.s., 2 gm orange G, 100 ml of distilled water, 8 ml glacial acetic acid. Just before using, dilute 1 part of stock solution with 2 parts of distilled water.
F. Differentiating solution: 1 gm phosphotungstic (or phosphomolybdic) acid, 100 ml of 90% isopropyl alcohol.

NOTE: Ethyl alcohol must *not* be used as it extracts aniline blue, w.s. vigorously.

Staining schedule:
1. Iodize 2–3-μm sections to remove mercury; wash in 70% alcohol.
2. Immerse in oxidizing Solution A 5 min.
3. Rinse in 70% alcohol.
4. Immerse in reducing Solution B 5 min.
5. Rinse in 70% alcohol.
6. Stain in Schiff reagent C 15–45 min.
7. Wash in several changes of sulfurous acid solution D.
8. Rinse in water, then counterstain in aniline blue-orange G (Solution E) 1 min.
9. Wash in distilled water.
10. Immerse in 1% phosphotungstic acid 1 min.
11. Immerse in phosphotungstic acid in isopropyl alcohol (solution F) 1 min.
12. Wash-dehydrate in 3 changes of absolute isopropyl alcohol each for 5 min.
13. Clear in xylene and mount in resinous medium.

Results: Nuclei—blue green; nuclei of red corpuscles, nucleoli, "hyaline bodies," and peripheral bodies of macrogametocytes—orange; glycogen and other carbohydrates—red; collagen—blue; elastic fibers—violet; muscle and cytoplasm—blue-gray.

SECTION D: SILVER STAINS

Various protozoologists experimented with silver staining methods based on those of Cajal and Golgi, but until Klein (1926) reported his "dry" method for ciliates and the silver-line system it sometimes revealed, there were no useful methods specifically for protozoans. The Klein dry method darkened parts of the exposed surface of the pellicle with deposits of specific pattern as well as portions of the infraciliature in unpredictable combinations. Subsequently, Chatton and Lwoff (1936) and other workers, particularly Kirby and his students, adapting Bodian's (1937) activated protein silver impregnation, provided two consistently effective methods which, with relatively few modifications, serve to bring out motor structures of most flagellates and ciliates and their relations to nuclei. Wessenberg (1972) provides a current review of variants and special cases.

Silver-Gelatin Stain

From Corliss (1953)
Original by Chatton and Lwoff (1930, 1936)

Specifically recommended for: Whole ciliate protozoans.

Designed to show: Cilia and infraciliature.

Method of fixation required: Champy followed by DaFano.

Preparation of solutions:
   A. Champy fixative: Just before using, mix 3% $K_2Cr_2O_7$, 7 parts; 1% CrO_3, 7 parts; 2% OsO_4, 4 parts.
   B. DaFano fixative: Mix 1 gm, $Co(NO_3)_2$, 1 gm NaCl, 10 ml of formol, and 90 ml of distilled water. Keep in stoppered bottle.
   C. Saline gelatin: Dissolve 10 gm powdered gelatin (Nutritional Biochemicals and Grayslake gelatins have been found to work; others may be satisfactory) and 0.05 gm NaCl in 100 ml of distilled water. Store in test tubes in refrigerator.

Staining procedure:
   1. Fix concentrated ciliates in an embryological watch glass in Champy fluid (Solution A) 1–5 min.

   NOTE: Use a watch glass or similar scrupulously clean glass container permitting microscopical check on process.

   2. Replace Solution A with daFano fluid (Solution B), changing the fluid twice and letting the third change act at least several hours.

   NOTE: Cilates may remain in solution B "for weeks."

   3. Wash out solution B with distilled water.
   4. Place a small drop of concentrated ciliates on a clean, warmed slide and one warmed (40–45°C) drop of saline gelatin (Solution C); mix thoroughly with warmed needle, and *quickly* withdraw excess fluid so that gelatin just covers the largest ciliates present.

NOTE: These operations are listed as one step to emphasize that they must be done consecutively with dispatch; temperature of the gelatin (Solution C) must not be allowed to drop much below 40°C.

5. Place slide immediately in a chilled (5°C) moist chamber for 2 min to coagulate the gelatin.
6. Immerse in cold (5–10°C) 3% $AgNO_3$ in the dark for 10–20 min.

NOTE: If final results are too light, increase time in $AgNO_3$; if too dark, decrease it.

7. Flush slide gently but copiously with cold distilled water; transfer to a bath of cold (not over 8–10°C) water in a white-bottomed dish.
8. Expose in *cold* to sunlight or UV lamp until basal granules of cilia are dark brown or black.

NOTE: It may be useful to rest dish on cracked ice. Keep "conventional 2537 Å lamp" 20–30 cm from gelatin-covered ciliates. If many preparations are being made, replace cold water from time to time.

9. Dehydrate, transferring from water to 70% alcohol, then to higher alcohols, keeping slides a few min in each stage.
10. Clear and mount in resinous medium under No. 0 or 1 coverslip.

Results: Argyrophil structures—black, background—clear or slightly yellow.

Activated Protein Silver

Modified from Kirby (1947, pp. 107–108)
Original by Bodian (1937)

Designed to show: Motor organelles and their relationship to nuclei in whole flagellates and cilates.

Fixation recommended: Hollande or Bouin.

NOTE: In order to have whole protozoans within the focal length of highest power objectives, smear or affix them with albumen on No. 1, or with thick forms, on No. 0 coverslips.

Preparation of solutions:

A. Activated protein silver: Sprinkle Protargol-S in proportion of 1 gm to 100 ml of distilled water gently on to water containing coverslips or slides with protozoal smears or settled protozoans; when Protargol has dissolved, add clean coiled copper wire or sheet copper in proportion of 5 gm per 100 ml of solution.
B. Reducer: Dissolve 1 gm hydroquinone crystals in 100 ml of 5% Na_2SO_3.

Staining procedure:

1. Transfer slips or slides to staining dish.

2. Add silver protein solution and copper as detailed in A; let react 1–2 days.

NOTE: Length of reaction and possibly temperature (raised for stronger reactions) must be adjusted to particular protozoans. Some workers change protargol and copper for a second day's reaction.

3. Wash in distilled water.
4. Reduce in solution B 5–10 min.
5. Wash several times in distilled water.
6. Dehydrate, clear and mount in resinous medium.

Results: Nuclei, cilia, flagella, basal bodies, trichocysts, fibrils, axostyle, costa, and other organelles—dark brown to brownish black; background— transparent golden brown.

NOTE: Settled or centrifuged materials can be stained by this method, but it is not possible to control the process closely. Stain only a few milliliters of protozoans in one batch, using a container that will hold much more (200 ml for the settling process). Use an amount of solution A about 10 times the volume of the protozoans. Shale the protozoans very gently into suspension several times during the first day. At the end of the first day bring the volume to fill the container with distilled water. As soon as all the protozoans have settled, decant. Repeat. Add Solution B diluted to fifth strength, decanting in not less than a half hour, or when the protozoans have settled (this may be nearly a day). Wash by adding water, waiting for settling, and decanting. Add glycerin and allow to penetrate. Pipette a small drop in a slightly larger drop of warmed glychrogel and cover with No. 1 or 0 coverslip.

Pretreatment with Bleach

Note: Some workers bleach protozoan materials *before* treating them with activated protein silver and have darkened. Bleaching may be desirable with materials that have been stored in alcohol. Proceed as follows:

Bleaching procedure:

1. Wash in fresh 70% and 50% alcohol and distilled water.
2. Immerse in 0.5% $KMnO_4$ 5 min.
3. Wash in distilled water.
4. Immerse in 5% oxalic acid 5 min.
5. Wash several times in distilled water, then proceed with stain as above.

NOTE: See comment after section on Toning below.

Toning

NOTE: After reducing and washing in the Bodian technic, many workers "tone" before dehydrating, clearing and mounting. To do this, after 5 min in the staining procedure, continue as follows:

6. Immerse in 1% gold chloride (made up with either yellow or brown crystals), 4–5 min.

7. Wash in distilled water.
8. Immerse in 2% oxalic acid until material turns faintly purple (usually in about 3 min).
9. Wash well several times.
10. Treat with 5% $Na_2S_2O_3$ 5–10 min.
11. Wash in distilled water several times during 5 min.
12. Dehydrate, clear, and mount.

Results: Nuclei and other argyrophil structures—purplish black, background—transparent lavender.

> **NOTE:** Hollande-fixed termite flagellates stained with some batches of protargol submitted in 1969 for certification were bleached as prescribed above; all batches stained less well following bleaching than without. Batches of Protargol-S certified between 1969 and 1972 stain fresh Hollande-fixed Zootermopsis flagellates well in 2 days at room temperature. Elephant ciliates fixed in Hollande 5 years before staining have responded well to the settling method.

SECTION E: HEMATOXYLIN

For general use hematoxylin stains are most widely applied and of these the iron hematoxylins give the most precise cytological results. Principal significant variations on the basic Heidenhain method include a) use of different fixatives; b) substitution of alcoholic for aqueous solutions; c) substitution of hematein for hematoxylin in making up the stain (Dobell); d) control of pH of staining solution by addition of acid or of buffers; e) differentiation with various agents; and f) short-cuts particularly by heating reagents. Kirby (1967, pp. 79–81) gives a useful short account of these. Heidenhain's iron hematoxylin is indispensable and can be modified to the requirements of special materials.

Heidenhain's Hematoxylin

Modified from Conn (1940), derived from Mayer (1899)
Original by Heidenhain (1892)

Designed to show: Nuclei, karyophores, locomotor organelles, plastids and other structures.

Fixation recommended: Schaudinn (for smears); Zenker, Heidenhain's SUSA and Bouin (for protozoans in tissues), and Hollande (for smears, tissues and marine planktonic forms).

Preparation of solutions:
A. Mordant: Dissolve 1.5–4 gm amethystine crystals of $NH_4Fe(SO_4)_2 \cdot 12H_2O$ in 100 ml of distilled water. Ferric alum solutions should ordinarily not be kept for more than a few weeks, should not be reused,

and should be discarded on appearance of cloudiness or discoloration.

B. Staining solution: Dissolve 0.5 gm hematoxylin in 100 ml of distilled water; let stand 4–5 weeks to ripen before use.

Staining schedule:

1. Mordant in solution A 30 min to 12 hr.
2. Rinse in tap water, then in distilled water.

NOTE: With loose material (settled or centrifuged protozoans), make at least two changes of water.

3. Stain in Solution B 1–12 hr.
4. Rinse in distilled water.
5. Differentiate in a fresh quantity of Solution A, observing process under microscope.
6. Wash in running tap water at least 1 hr.

NOTE: A. If the laboratory tap water is unreliable (turbid, far from neutral, etc.), subject to contamination, make 10 or more changes of deionized or distilled water during at least an hour, using 0.1% Li_2CO_3 for the penultimate wash. Life of the stain is directly dependent on the elimination of all unbound ferric alum.

B. With settled or centrifuged protozoans, except very large ones which can be handled individually, washing enough to prevent early degrading of stain by remaining ferric alum is very difficult. It is a principal reason for using albumen affixation to coverslips which is also time-consuming. Use half strength or weaker differentiating solution and decant *before* nuclear and fibrillar detail is visible under the microscope. Wash by replacement as in A, but slowly as determined by settling rate of the protozoans. Several days to a week of replacements may be required to eliminate all of the alum.

Results: Nuclei, motor organelles, and other cytoplasmic structures—blue black or black; general cytoplasm—transparent.

NOTE: In protozoans counterstains after iron hematoxylin rarely stain structures not already visible and frequency obscure some structures. Used along iron hematoxylin is the protozoological cytological stain of greatest precision. With protozoans in tissues where principal concern is to show effects of the protozoans on the tissues, use one of the staining combinations below rather than an iron hematoxylin.

Alcoholic Iron Hematoxylin

From Kofoid and Swezy (1915) and laboratory instructions of Kofoid

Designed to show: Nuclear detail, locomoter organelles, chromotoidal bodies, and plastids mainly in smear preparations.

Fixation recommended: Schaudinn, hot or at room temperature.

NOTE: Kofoid and Swezy warmed (to about 50°C) mordant and stain for a rapid, but rather precise cytological stain on trichomonads and intestinal amebas; for such pruposes use the shorter times. Alcoholic iron hematoxylin, providing a precise and reliable stain without the dissolving and dislodging effects of several aqueous changes, is the stain of choice for vulnerable albumen-affixed ("albumen smears" p. 284) materials. Longer times yield more precise cytological results.

Preparation of solutions:

A. Mordant solution: As Solution A in Heidenhain's method, but with 0.5–1% ferric alum dissolved in 50% alcohol; observe caveats as above.

B. Dilute 1 ml of ripened 10% hematoxylin in 95% alcohol with 9 ml 50% alcohol (Columbia dish portion) shortly before staining.

Staining schedule:

1. Transfer iodized material to 50% alcohol.
2. Mordant in Solution A 10–60 min.
3. Rinse in 3 changes of 50% alcohol.
4. Stain in Solution B 10–60 min.

NOTE: If much material is being stained, use new staining solution after each 3 or 4 batches of slides.

5. Rinse in 2 changes of 50% alcohol.
6. Differentiate in fresh Solution A under microscope.
7. Wash in at least 6 10–min changes of equal parts of tap water and 95% alcohol.
8. Dehydrate.
9. Clear and mount in resinous medium.

Results: Nuclei—black, cytoplasmic organelles—gray to black.

NOTE: Results sometimes a little less crisp than with Heidenhain's iron hematoxylin, but sharper than with alum hematoxylins. Precision is generally preportionate to care in carrying out staining, differentiation and washing.

Dilute Alum Hematoxylin

From laboratory instructions of E. G. Conklin

Designed to show: Nucleus and general structure progressively.

NOTE: This method is particularly suited to materials manipulated by the settling method because alum hematoxylins are complete stains, mordant and wash stages are avoided, the weak solutions do not overstain, and the difficult elimination of differentiating fluid is also avoided.

Method of fixation recommended: Fluid containing a mordant metal or picric acid; Schaudinn, Hollande, DaFano, etc.

Preparation of staining solution: Dilute ripened Delafield's hematoxylin stock solution (see p. 105) about 2 ml to 95 ml of distilled water, or Mayer's hemalum (see p. 106) about 10 ml to 95 ml of distilled water just before using.

Staining schedule:
1. Replace water on settled or centrifuged protozoans by pouring on at least 10 times their volume of diluted alum solution, suspending them by gentle shaking when they settle; stain 12 hr or until they have a grayish blue (with Delafield's or grayish violet with hemalum) tinge.
2. Decant or draw off fluid.
3. Wash in 2 changes of tap water, removing fluid after settling.
4a. Add glycerin nearly equal to volume of protozoans; let stand 1 hr.

NOTE: Material may remain in glycerin indefinitely.

5a. Mount in melted glychrogel and cover with No. 1 or 0 coverslip.

or

3b. Dehydrate.
4b. Clear and mount in resinous medium.

NOTE: See section on handling protozoans by settling (p. 283).

Results: Nuclei and cytoplasmic structures and fibrils—blue, general cytoplasm—very pale transparent blue.

SECTION F: MISCELLANEOUS METHODS

Basic Fuchsin-Picro-Indigo Carmine

Essentially from Langeron (1942), cited as
Trichrome of Ramon y Cajal as used by P. Masson
Similar to Borrel (1901) Stain in Kirby (1947)

Designed to show: Nuclei and basic form in whole mounts; protozoans and normal and altered tissues in sections of infected hosts.

Method of fixation recommended: Schaudinn for smears; fluids without chrome-osmium.

Embedded in: Paraffin.

Preparation of solutions:
A. Nuclear stain: dissolve about 0.3 gm of basic fuchsin in 100 ml of distilled water.
B. Counterstain: dissolve about 1.7 gm indigo carmine in 100 ml of distilled water. Mix 30 ml of indigo carmine solution with 60 ml of saturated aqueous picric acid. Filter this staining solution just before using.

Staining schedule:
1. Remove paraffin from sections in usual manner.
2. Stain in Solution A 10 min.
3. Wash in 3 changes of tap water.

4. Differentiate in 1% acetic acid until details in nuclei begin to be discernible.
5. Stain in Solution B 10 min.
6. Rinse in distilled water.
7. Wash in 0.5% acetic acid 2 min or until connective tissue is blue (*i.e.*, picric acid is removed).
8. Dehydrate and differentiate in several changes of absolute alcohol until connective tissue fibers ae sharply defined.
9. Stop the differentiation by transferring to toluene.
10. Clear in toluene and mount in resinous medium.

Results: Nuclei—red, cytoplasm—yellow or pink, collagen—blue, mucus—orange.

Mallory's (1900) Triple Stain

Modified for sections of cattle ciliates by Sharp (1914, p. 55)

Designed to show: Nuclei and cytoplasmic organelles of ciliates.

Method of fixation required: Zenker.

Embedded in: Paraffin.

Preparation of counterstaining solution: Dissolve 2 gm oxalic acid in 100 ml of distilled water, then add 0.5 gm anilin blue, w.s. and 2 gm orange G.

Staining schedule:
1. Treat with iodine as usual after Zenker fluid.
2. Stain 5–6 μm sections in 0.5% aqueous basic fuchsin 45 min.
3. Rinse in distilled water.
4. Immerse in 1% phosphomolybdic acid 60 min.
5. Rinse in distilled water; counterstain 60 min.
6. Wash in 3 changes of distilled water over 10 min.
7. Dehydrate in 95% alcohol 1 min and absolute alcohol 1 min.
8. Clear in carbol-xylene, xylene and mount in balsam.

Results: Macronucleus—orange brown, micronucleus and motor fibrils—bright red, ectoplasm (including skeletal plates)—blue-red, entoplasm-—pink, cilia—transparent.

NOTE: This modification is for sections of individual protozoans. Use regular Mallory's (anilin blue collagen stain, p. 117) stain, basic fuchsin-picro-indigo carmine, Masson trichrome or hematoxylin-zaure II-eosin for protozoan parasites in tissues.

Hematoxylin-Biebrich Scarlet-Fast Green

Modified from many sources
Based on Masson (1911)

Specially recommended for: Effects of parasitic protozoans on tissues of reptiles, lower vertebrates and invertebrates.

Designed to show: Mantle tissues, glandular, muscle, connective and nervous tissue.

Method of fixation recommended: Heidenhain's SUSA, Hollande or Bouin.

Embedded in: Paraffin.

> **NOTE:** Celloidin is required for some delicate invertebrates and for soft, delicate tissue next to tough structures (as a soft mass of fish sporozoans next to fibrous connective tissue).

Preparation of staining solution: Regaud's hematoxylin—Dissolve 1 gm hematoxylin in 10 ml 95% alcohol, 10 ml glycerin and 80 ml of distilled water.

Staining schedule:
1. Remove paraffin in the usual manner; if mercury-containing fixative is used, iodize.
2. Mordant in 5% ferric alum 30 min to 12 hr.
3. Wash in tap water; rinse in distilled water.
4. Stain in Regaud's hematoxylin at room temperature 12–24 hr or at 50°C for 30 min.

> **NOTE:** Heidenhain's hematoxylin used in place of Regaud's may give somewhat sharper stain of protozoan and other nuclei.

5. Differentiate in fresh ferric alum until nuclei are distinct, but still solidly black.

> **NOTE:** If differentiation proceeds too rapidly for good control dilute ferric alum with distilled water to about 2%.

6. Rinse in several changes of distilled water.
7. Stain in 1% biebrich scarlet in 1% acetic acid 5 min.
8. Rinse in several changes of distilled water.
9. Immerse in 1% phosphomolybdic acid 1 min.
10. Dehydrate to 95% alcohol.
11. Stain in 0.2% fast green FCF in 95% alcohol 30 sec.

> **NOTE:** Check connective tissue areas; if green is too pale, return to stain; if it is excessive (not common with invertebrate tissues), dilute fast green with 95% alcohol for further sections. If the complete connective tissue area does not stain or if a more transparent stain is desired, use 2% Luxol fast yellow TN in 95% alcohol in place of fast green FCF.

12. Dehydrate, clear and mount in resinous medium.

Results: Nuclei—blue black, cytoplasm—reddish, muscle and glands—different shades of red, connective tissue—green.

NOTE: This and similar combinations of hematoxylin, a red or red dyes, phosphomolybdic or phosphotungstic acid, and fast green or anilin blue have proved consistently more controllable than Mallory or Heidenhain AZAN in student laboratories and investigations with protozoan-infected reptile, amphibian and invertebrate tissues. The luxol fast yellow TN modification has been advantageous in sections of small crustaceans.

SECTION G: METHODS FOR STAINING PROTOZOANS IN BLOOD, MACROPHAGES OR BODY FLUIDS

Malaria parasites, Toxoplasma, piroplasms, trypanosomes, Leishmania and similar forms found in mammals and other vertebrates are commonly studied in dried smears by various Romanowksy methods, whether the protozoans are taken from the vertebrate or from an invertebrate alternate host. The custom has merit, even though all the cells are shrunken, because of the great accumulated knowledge of normal and pathological cells pre- pared this way. Most published Romanowsky methods will probably give somewhat useful results if "the initial quality of the stain is good," the methyl alcohol used as fixative is pure, and the technic is carried out carefully, especially with regard to pH control. Storage of stains, especially in tropical or particularly dirty environments, presents some difficulties. The following methods or modifications of them meet most of the needs of protozoologists. Certification of stains of the Romanowsky group is *not* made for the spectrum of parasites and certified batches, while staining blood cells characteristically, may not work well with some parasites.

Stains for Dried Thin Films

Giemsa Stain for Thin Films

From Conn (1940), p. 276; Lillie (1954, p. 387)
Original by Giemsa (1904)

Designed to show: Blood cells (including nucleated red cells), platelets and protozoan parasites.

Method of fixation: Methanol. This films prepared by spreading a drop of blood (known to be or suspected of being infected) over a scrupulously clean slide by means of a second slide or a coverslip; fix immediately in absolute methanol (hematological grade) 3–5 min.

Preparation of solutions:

A. Dissolve 0.75–0.8 gm Giemsa stain in 50 ml glycerin (reagent grade) and 50 ml absolute methanol in a mechanical shaker over 1–2 days (or shaking by hand occasionally during 3–4 days) with glass beads. Just before using, dilute 4 ml of Giemsa stock solution with 96 ml phosphate buffer.

B. Phosphate buffer, pH 6.8: Mix 50 ml $M/15$ Na_2HPO_4 (9.47 gm dried salt in 1000 ml of distilled water) and 50 ml $M/15$ KH_2PO_4 (9.08 gm dried salt in 1000 ml of distilled water).

Staining procedure:

1. Stain methanol fixed film in buffer-diluted stain 40–120 min.
2. Rinse quickly in distilled water.
3. Air dry. Examine under oil immersion objective.

NOTE: Rapidity of action varies; in general, parasites require longer stain than blood elements alone. Begin with 40 min stain, increasing staining time if colors are too pale.

Results: Nucleus or part of nucleus of protozoans—red or carmine, cytoplasm of protozoans—blue, undulating membrane—purple, chromatin of leukocytes—purple, basophil cytoplasm of agranulocytes—blue, eosinophil granule—pink to red, neutrophil granules—purple, red corpuscles—pink or bluish.

May-Grünwald: Giemsa Stain

From Langeron (1942, pp. 573–577)
Original by Pappenheim (1908)

Designed to show: Hematozoans for diagnosis.

Method of fixation: Methanol in undiluted May-Grünwald staining solution.

Preparation of solutions:

A. May-Grünwald stain: Mix 0.5 gm eosin Y and 0.5 gm methylene blue in 100 ml distilled water. Filter. Dry filtrate. Wash residue and dry. Dissolve in 50 ml absolute methanol; this is the stock solution.

NOTE: Jenner's stain, which has more eosin Y than methylene blue, but is otherwise similar, may be substituted for May-Grünwald stain.

B. Giemsa stain—see Giemsa stain for thin films.
C. Phosphate buffer—see under Giemsa stain for thin films.

Staining schedule:

1. Onto each fresh, dried film, pipette just enough May-Grünwald Solution A to cover blood (usually 10 drops or a little more) and let react 3 min.
2. Add an equal amount of distilled water, tilt to mix and stain 1 min.
3. Pour off fluid and, without washing, add about 10 drops of freshly buffer-diluted Giemsa fluid; stain 5–60 min.

NOTE: Time required varies with the batch of stain, age of stock solution, age of film (older ones staining more slowly and less precisely), and the kind of parasite. Try 10–15 min with first films.

4. Rinse with a quick, ample jet of distilled water.

NOTE: Some of stain remains as an unsightly deposit if one simply pours off the stain; however, do not prolong the rinse for the stain will be diminished.

5. Dry and examine under oil.

Results: Similar to preceding, but more intense, particularly with hematozoans.

NOTE: If coloration is too blue or intense, wash film with xylene to remove immersion oil, dry, and differentiate cautiously in 1% boric acid or in Wolbach's colophonium-acetone (15 gm rosin in 100 ml acetone).

Water-Soluble Blood Stain ("JSB Stain")

From Manwall (1961, pp. 587–588)
Original by Singh and Bhattacharji (1944)

Designed to show: Plasmodium in blood films, rapidly and reliably.

NOTE: The devisers sought an economical, reliable malaria-screening stain with better keeping quality in the tropics than Giemsa and other Romanowsky stains.

Method of fixation: Absolute methanol.

Preparation of staining solutions:

A. Dissolve 0.5 gm methylene blue in 500 ml of distilled water, then add 3.0 ml H_2SO_4. Mix thoroughly. Add 0.5 gm $K_2Cr_2O_7$. Heat product (with heavy purple precipitate) over a boiling water bath 3 hr, or until mixture turns from greenish to blue. Cool to room temperature. Filter out resultant precipitate and dissolve precipitate in 500 ml of $M/20$ Na_2HPO_4. Ripen 48 hr before using. (To "stock" JSB stain, dry ppt in a vacuum desiccator at room temperature; store dried powder in tightly stoppered bottle not much larger than required to contain the stain.)

B. Dissolve 1 gm eosin Y in 500 ml of distilled water.

Staining schedule:

1. Fix fresh film by dipping in absolute methanol.
2. Immerse in methylene blue (Solution A) 30 sec.
3. Wash in two jars of distilled water, buffered to pH 6.2–6.6, 5 sec each.
4. Dip in eosin Solution B.
5. Wash in a third jar of buffered (pH 6.2–6.6) water about 10 sec.
6. Immerse in Solution A again 30 sec.
7. Wash in 2 jars of buffered (pH 6.2–6.6) water 5 sec each.
8. Stand slides upright to dry.

Results: As with Giemsa for thin films.

Accelerated Giemsa Stain for Thick Films

From Lillie (1954, p. 388)

Essentially the same as Barber and Komp (1929)

Designed to show: Presence of Plasmodium in light infections.

No fixation.

Preparation of solutions:
  A. Giemsa solution—Stock as for Giemsa for thin films: just before using, mix 4 ml Giemsa stock solution, 3 ml acetone, 2 ml of phosphate buffer (pH 6.8) in 31 ml of distilled water.
  B. Phosphate buffer, pH 6.8—as in Giemsa for thin films.

Staining procedure:
  1. Spread 3–5 drops of mammalian blood suspected of having hematozoans on a clean slide in a disc about 15 mm in diameter; newsprint should be just visible through the drop.
  2. Dry film 1 hr.
  3. Stain in buffer-diluted Giemsa (Solution A) 5–10 min.
  4. Rinse in distilled water.
  5. Dry.

Results: Plasmodium chromatin—clear red, cytoplasm—clear blue, red corpuscles—laked, transparent.

  NOTE: Very brief exposure to water avoids detachment and loss of films which is otherwise not uncommon in thick film procedures.

Giemsa Stain on Saponin-lysed Thick Films

From Umlas and Fallon (1971)

Specially recommended for: Malaria screening for laboratories not handling many suspected cases.

Designed to: Lyse red corpuscle background especially cleanly.

Preparation of staining solution: Dissolve 0.3 gm Wright's stain or, preferred, 0.15 gm Wright's stain plus 1.5 gm Giemsa stain in 1000 ml absolute methanol. Allow to stand a day or more before using.

Staining schedule:
1. Prepare think film as for accelerated Giemsa stain.
2. Dry.
3. Lyse film with about 3 drops of 0.5% saponin solution about 5 sec.

  NOTE: Saponin solution for preparation of blood for white blood cell counts in electronic cell counters is suitable.

4. Drain off saponin solution and dry about 10 min.
5. Cover film with undiluted stain 2 min.
6. Dilute with twice as much distilled water; allow to react 3 min.
7. Rinse in distilled water and dry.

Results: Similar to those with Accelerated Giemsa, but with paler background.

Heidenhain's Hematoxylin for Wet Films

After Conn (1940; derived from Mayer (1899)
Orignal by Heidenhain (1892)

Specially recommended for: Hematozoans in non-mammalian blood; blood flagellates.

Designed to show: Cytological detail in undistored cells.

Method of fixation recommended: Schaudinn, Zenker or Hollande.

NOTE: Make films in a humid place, transferring to fixative *instantly. Films must not be permitted to dry at any stage. If,* as is likely with blood flagellates, some films are to be stained with Activated Protein Silver, fix in Hollande.

Staining schedule: As for sections (see p. 297).

Results: Nuclei and cytoplasmic organelles of protozoans and blood cells— blue-black or black, undistorted by major shrinkage.

Giemsa Stain for Wet Films or Sections

I. After Wolback (1911)
II. After Wenyon (1926)

Specially recommended for: Undistorted hematozoans for cytological studies complementing those done on dry films and on wet films with Activated Protein Silver.

Designed to show: Undistorted hematozoans and readily distinguished types of leucocytes in the same preparation.

Method of fixation: Schaudinn, Zenker or Hollande on clean coverslip or slide.

NOTE: *Films must not be permitted to dry at any time.*

Preparation of solutions:
A. Giemsa stain: As Giemsa stain for thin films.
B. Colophonium alcohol—Dissolve 10 gm of natural rosin in 100 ml of 95% alcohol. Dilute 2 drops to 10 ml 95% alcohol for differentiating solution.

Staining schedule: (proportions as carried out with coverslips in Columbia jars):
1. If a mercurial fixative has been used, iodize; rinse in 70% alcohol.
2. Wash in distilled water.
3. Stain in buffer-diluted Giemsa solution (p. 303) 12 hr.

4. Replace solution and stain another 12 hr.
5. Wash in distilled water.

I. Dehydration-differentiation by colophonium-alcohol

6. Put coverslip film downward on a drop of differentiating Solution B on microscope; control process under microscope.
7. When differentiation is nearly complete (only seconds may be required), rinse quickly in 2 dishes of absolute alcohol.
8. Clear in xylene and mount in cedar oil.

or

II. Dehydration in acetone-xylene

6. Transfer to 9.5 ml acetone: 0.5 ml xylene a few seconds.
7. Pass through 7 ml acetone: 3 ml xylene.
8. Pass through 3 ml acetone: 7 ml xylene.
9. Clear in xylene and mount in cedar oil.

Results: With chrome fixation, chrome fixation, chromatin—blue, cytoplasm—pink.

Alum Hematoxylin-Azure II-Eosin

Essentially from Lillie (1954)
Original by Maximow (1909)

Specially recommended for: Tissue stages of hematozoas and their effects on host tissues.

Designed to show: Tissue mast cells, early necrobiotic changes in cells, and protozoans, rickettsias and bacteria in tissues.

Method of fixation recommended: Mercuric-dichromate-formaldehyde (Helly, Bensley).

Embedded in: Paraffin or celloidin.
   NOTE: Maximow removed mercury precipitates from tissue with iodine alcohol before embedding, embedded in celloidin, and attached sections to slides with ether vapor before staining. Lillie embeds in paraffin and treats *sections* with iodine and thiosulfate.

Preparation of staining solutions:
   A. Alum hematoxylin: Prepare Delafield's hematoxylin (see p. 105).
   B. Azure II-eosin: Dilute 5 ml 0.1%xeosin Y with 40 ml of distilled water and add 5 ml 0.1% azure II.

Staining schedule:
1. Stain sections with alum hematoxylin (Solution A) 5 min.
2. Wash in water.
3. Stain in azure II-eosin (Solution B) 18–24 hr.

4. Differentiate sections individually in 95% alcohol until gross blue clouds stop coming into alcohol, and red cells and collagen are pink.
5. Dehydrate in 2 changes of absolute alcohol.
6. Clear in absolute alcohol/xylene, and 2 changes of xylene.
7. Mount in resin dissolved in xylene.

Results: Nuclei—blue, granules of basophils and mast cells—purple to violet, cytoplasm—blue to pink, secretion granules and eosinophil granules—pink.

Fluorescence Methods

A number of methods of examining protozoans with UV light sources have been employed, particularly with fluorescence of inclusions (as chlorophyll pigments in phytomastigophorans), introduction of a fluorochrome (as demonstration of trypanosomes vitally with 1:15,000 coriphosphine as they appear bright among nonstaining blood cells: Radna, 1938), and, most of all, with the rapidly expanding field of fluorescent antibodies. These are beyond the scope of the present effort. They are treated in Goldman (1968).

CHAPTER 11

IMMUNOENZYME METHODS FOR LOCALIZING SPECIFIC ANTIGENS IN TISSUES

S. S. Spicer

In the original immunoenzyme method, antibodies labeled with the enzyme, horseradish peroxidase (Nakane and Pierce, 1967), were substituted for the fluorescein-labeled antibodies employed in the prior immunofluorescent procedure (Coons, 1958) for localizing antigen. The immunoperoxidase method offers the advantage of viewing sections with visible light rather than for ultraviolet fluorescence through catalytic buildup of reaction product by enzyme activity. The initial immunoenzyme technic utilized horseradish peroxidase as the enzymatic label and employed antibody conjugated to this enzyme in either a direct or indirect immunostaining procedure. The direct method visualized antigen with labeled antibody to the antigen and the indirect localized antigen with a sequence of a primary antibody to the antigen followed by labeled antibody to the primary immunoglobulin. The horseradish peroxidase thus bound to the tissue antigen is visualized by incubation with the substrate 3, 3′-diaminobenzidine, which yields a brown to black reaction product (Graham and Karnovsky, 1966). Peroxidase can most conveniently be conjugated to the labeled antibody with glutaraldehyde (Avrameas, 1970).

The conjugation of peroxidase to antibody and its potential effect on the affinity of the antibody for the antigen can be avoided by an immunostaining procedure which links the labeling enzyme to the tissue antigen through a series of antigen-antibody reactions. This procedure, referred to as the immunoperoxidase bridge method (Mason et al., 1969), entails reacting the antigen first with a specific antibody generated against it in rabbits. Subsequent exposure of the section to a goat antiserum to rabbit globulin, then to antibody to horseradish peroxidase generated in rabbits and finally to the peroxidase, ties the labeling enzyme to the tissue antigen through a goat anti-rabbit globulin bridge. Horseradish peroxidase bound at the site of the tissue antigen can then be visualized by incubation in the diaminobenzidine substrate medium.

The application of rabbit anti-peroxidase and peroxidase in the immunoenzyme bridge procedure can be condensed into a single step by applying a preformed complex of peroxidase with antibody to peroxidase (Sternberger et al., 1970). This modification of the bridge requires prior preparation of the peroxidase-antiperoxidase (PAP) complex but offers the advantage of shortening the overall staining procedure by combining two steps into one. The PAP complex has provided an advantage for ultrastructural immunostaining in a postembedment technic by virtue of yielding recognizable pentameres with distinctive structure at the antigen site, but the method does not improve on results obtained at the light microscopic level.

The enzyme alkaline phosphatase has also been employed as an immunocytochemical label and, employed in sequence with a peroxidase immunostain, provides a means for localizing two antigens in different colors in the same tissue section. The latter objective can also be achieved by immunostaining a section sequentially for two antigens with the immunoperoxidase method and using different peroxidase substrates that yield different colors for visualizing the sites of binding of primary antibodies to the separate antigens.

As for the fluorescent immunostaining method, the validity of the immunoenzyme procedures depends primarily on the purity of the antigen against which the antibody was prepared and, hence, the immunospecificity of the resulting antiserum. Preferably antibodies are purified by absorption on an affinity column loaded with antigen of high purity and by subsequent elution of the antibody from the column.

Appropriate controls are required to differentiate staining of the specific antigen from that of sites that bind antibodies selectively but on a nonimmunospecific basis (Spicer et al., 1977). The latter sites are recognized by their reactivity in control sections stained by an immunoenzyme bridge in which a nonimmune normal serum—preferably from the rabbit subsequently used to produce the specific antiserum—replaces the specific antiserum as the primary step in the bridge. Nonimmunospecific staining occurs in different sites in tissues fixed by different procedures. Nonimmunospecific background is said to be minimized by employing a globulin not bound by goat anti-rabbit globulin (e.g. sheep or goat immunoglobulin) in a wash prior to the bridge and as an additive to each antiserum in the bridge sequence. Interfering background is also minimized in sections stained with high titer antisera which can be diluted 100-fold, or preferably more, for the staining procedure.

The following procedure is recommended for light microscopic immunocytochemistry:

   A. Tissue processing:

      (1) Fix specimens by an optimal procedure which should be deter-

mined for each antigen. This entails testing various fixative solutions for varying time intervals, ranging as a rule from ½ to 6 hr at 4°C. Solutions include: Carnoy's fluid, Bouin's fluid, 4% paraformaldehyde buffered with 2% calcium acetate; 0.5 to 2% glutaraldehyde, and mixtures of formaldehyde and glutaraldehyde.

For demonstrating antigens labile to the tested fixatives, prepare cryostat sections of frozen or sections of frozen dried specimens.

(2) Rinse fixed tissues in chilled 70% ethanol. Specimens can be kept at this stage for intervals up to a few days, allowing time for shipment between laboratories.

(3) *Dehydrate* fixed specimens rapidly through graded alcohols and *embed*, preferably in paraffin having a low melting temperature, and with a vacuum oven to shorten infiltration time.

B. Immunostaining:

(1) Hydrate paraffin sections taking to distilled water through graded alcohols. Rinse cryostat sections briefly in phosphate-buffered saline (PBS)* and blot.

(2) Treat 3–5 min in 3% hydrogen peroxide to inhibit erythrocyte pseudoperoxidase.

(3) Rinse in several changes of PBS.*

(4) Wipe excess PBS from slide and react with a 1:20 to 1:1000 dilution of the primary specific antiserum in PBS for 10 min.†

(5) Rinse 3 times with PBS.

(6) Wipe excess PBS from slide and react 10 min with a 1:20 dilution of goat anti-rabbit serum in PBS.

(7) Rinse 3 times with PBS.

(8) Wipe excess PBS from slide and react 10 min with a 1:20 dilution of rabbit antiserum to horseradish peroxidase.‡

(9) Wipe excess PBS from slide and react with 0.5 mg/100 ml horseradish peroxidase in PBS.

(10) Rinse 3 times for a total of at least 10 min.

(11) Wipe excess PBS from slide and react 10 min with 3,3'-diaminobenzidine substrate medium prepared fresh as follows:

---

* *Phosphate-buffered saline (5× concentrated stock): (1) 72.0 gm sodium chloride, (2) 14.8 gm sodium phosphate—dibasic (Na_2HPO_4), (3) 4.3 gm potassium phosphate—monobasic (KH_2PO_4), and (4) 2000.0 ml distilled water. Use this buffer diluted 1 part with 4 parts distilled water.*
† *Antibodies purified by affinity column chromatography offer advantages in increased sensitivity and lessened background nonimmunospecific staining.*
‡ *The antiperoxidase and peroxidase steps can be combined into one employing the peroxidase-antiperoxidase (PAP) complex (Sternberger). A recent variant employs Staphylococcus aureus "protein A" as a second step reagent for immunocytochemical localization of cell antigens on the basis of its selective affinity for the F_c component of immunoglobulin.*

(a) 10 ml of 0.05 M Tris buffer—pH 7.4 to 7.6.§

(b) 2–3 mg 3,3′-diaminobenzidine tetrahydrochloride (Sigma Chemical Co.).§

(c) Just prior to use add 0.02 ml 3% hydrogen peroxide.

(12) Rinse several times with running water, then distilled water.

(13) Add drops of 1% osmium tetroxide briefly over the section.

(14) Rinse several times with distilled water.

(15) Counterstain nuclei with methylene blue or a light hematoxylin as desired.

(16) Dehydrate slides through graded alcohols and mount in permount.

---

§ The Tris buffer is prepared by adding 0.606 gm Tris [tris-(hydroxymethyl)aminomethane] to 100.0 ml PBS pH 7.2–7.6. The diaminobenzidine is considered potentially hazardous for possible carcinogenic properties and must be handled with precautions to avoid contact with skin, ingestion or inhalation.

BOTANICAL SCIENCES

CHAPTER 12

PLANT ANATOMY AND GENERAL BOTANY

Henry Schneider

INTRODUCTION TO BOTANICAL SCIENCES

Staining procedures employed in plant microtechnic are presented in five chapters entitled: Plant Anatomy and General Botany; Plant Cytology; Plant Cytogenetics; Pollen and Pollen Tubes; and Plant Pathological Anatomy and Mycology. A chapter on stains for algae is not included but some procedures are applicable. Although staining procedures are grouped according to scientific disciplines, it is recognized that clear-cut distinctions cannot always be made. For instance, some of the staining procedures described later in this chapter are also valuable for cytogenetical studies and vice versa. Multipurpose staining procedures need not be repeated in every chapter unless some unique steps are involved.

Fixatives for electron microscopy such as glutaraldehyde followed by osmic acid give excellent results with some stains. These fixatives do not penetrate the large pieces of tissue sometimes used for light microscopy, but we recommend their trial where appropriate.

An effort is made to make the titles of each procedure descriptive and short. The stains and some other ingredients, such as mordants used to treat the sections, are included in the titles; and they may be separated by arrows which signify "followed by."

The term freehand sections is used for sections that are not embedded and glued to a slide with an adhesive. These include sections cut from tissues that are: (1) held with the fingers or between pieces of elder pith and cut with a razor; (2) clamped in a microtome and sectioned by a sliding knife with no embedding; (3) frozen on a microtome freezing stage; etc.

By tap water we mean water with a pH above 7; however, in some localities tap water is below pH 7 and a trace of $NaHCO_3$ should be added. Distilled water has a pH of 6 or below owing to CO_2 dissolved in it.

It is assumed that reasonable agitation will be used while staining, washing, and mordanting. With freehand sections, it is imperative that the sections be kept moving and floating and not allowed to lodge together and thus prevent their exposure to the stain or other reagents (see Schneider, 1960).

When a particular plant or plant part is mentioned, it is intended to convey that it is known that the procedure works on that material. If one encounters difficulty with another material, he should try the material suggested to test his technique.

Each investigator should combine primary and counterstains that give the best results for his material and for the staining he wishes to achieve. Therefore, in some schedules, several cross references are given for primary or counter stains.

In this chapter on Plant Anatomy and General Botany an attempt is made to group the staining procedures according to the mode of action of the primary dye or the principal dye in the procedure. We mention first the mordanted stains, hematoxylin and tannic acid; then the cationic or basic dyes. These are followed by the anionic or acidic dyes, the neutral dyes, the fat-soluble dyes, and histochemical tests.

SECTION A: MORDANTED DYES

Hematoxylin and tannic acid are widely used with ferric cations. They stain various plant structures and are among the most useful of the biological stains in the plant sciences. Once bound to a structure, they are not removed by solvents and acids employed for destaining counterstains, and by solvents employed in dehydration and clearing. Hence one has a free reign in the application of either cationic, anionic, or other counterstains. Ferric cations will often be referred to as iron.

Hematoxylin and its metal mordants may be premixed, and the resulting stain used as a single component dye. Examples are Weigert's premixed iron hematoxylin and Delafield's and Harris' premixed aluminum hematoxylin.

In other procedures with hematoxylin, the iron is applied first as ferric cations in a rather prolonged treatment, and the hematoxylin is applied later after removing the unbound iron by washing. The application of hematoxylin may be carried out in two different ways:

    I. *Progressively* by applying it dilute and with careful timing and microscopic observations so as to bring out the desired staining;

or

    II. *Regressively* by overstaining and then destaining with ferric cations or picric acid.

Progressive staining produces a general staining of walls, cytoplasm, and nuclei. Regressive staining is more specific because the dye is removed from all structures except those to which it is tightly bound, as for instance chromosomes.

When the use of hematoxylin for staining was first being developed, elaborate ripening procedures were devised. In the latest methods for testing stains submitted to the Biological Stain Commission for certification (Lillie (1977) pp. 609–610), use of hematoxylin without ripening is specified for Heidenhain's and Weigert's methods; and for Mayer's and Ehrlich's it is done quickly with the oxidizing agent $NaIO_3$. Short procedures will supplement long, older ones that are still preferred by some.

The application of tannic acid and then iron chloride as used by Foster (1934) is done in reverse order from the application of Fe^{3+} and then hematoxylin as in Heidenhain's, i.e., the tannic acid is applied first; then the unbound tannic acid is washed away and the ferric cations are applied to form colored compounds of Fe^{3+} and tannic acid. Apparently the tannic acid selects certain structures—especially meristematic cell walls—to which it binds, and then the iron reacts with the tannic acid to form colored compounds and no mordanting is involved. However, the procedure calls for repeating the staining if necessary; so, ferric ions apparently attach to other structures such as mature cell walls and then the tannic acid combines with the ferric ions to produce color. Thus for some structures mordanting apparently occurs. The $ZnCl_2$ used in Sharmen's safranin-orange G-tannic acid apparently acts as a mordant for tannic acid.

Premixed Iron Hematoxylin
Original by Weigert (1904)
Johansen (1940); Feder and O'Brien (1968)

Application: General plant tissue.

Designed to show: Cellulose walls, nuclei, cytoplasm.

Fixative: Any botanical fixative.

Embedded in: Paraffin or freehand sections.

Preparation of staining solutions:
  A. Hematoxylin: To 90 ml of 95% alcohol add 10 ml of a stock 10% alcoholic solution of hematoxylin.
  B. $FeCl_3$: To 90 ml of H_2O add 4 ml of 29% aqueous $FeCl_3$ anhydrous and 5 ml of 1:5 HCl.

Staining schedule:
  1. Remove paraffin and bring into water.
  2. Mix equal parts of A and B above and stain from 10 min to 1 hr until dark enough, checking with a microscope.
  3. Wash in several changes of tap water.
  4. Counterstains:
     a. Lacmoid (see p. 330).
     b. Safranin (see p. 324).

Results: Cellulose cell walls and nuclei are sharply stained by this easily used procedure.

Iron→Hematoxylin (Progressive Staining)

Rewritten by Henry Schneider
An old method given in many textbooks; Jeffrey (1917, pp. 457–458)

Application: Many plant tissues.

Designed to show: Cellulose walls, nuclei, chloroplasts, cytoplasm. A good stain for photomicrography.

Fixation: Any botanical fixative.

Embedded in: Paraffin, freehand, frozen, etc.

Preparation of solutions:
1. Stock solution: 10% hematoxylin in 95% alcohol.
2. Staining solution: ¼ ml of stock solution in 500 ml of water with a pH of 7.0 or above. (Add a trace of 0.5% $NaHCO_3$ if necessary.)

Staining schedule:
1. Remove paraffin and bring into water.
2. Mordant 30 min to overnight in 3% aqueous $NH_4Fe(SO_4)_2 \cdot 12H_2O$.
3. Wash in tap water, several changes, 10 min.
4. Stain in above solution. Examine a section every few minutes to see if staining has reached an optimum.
5. Wash in water, several changes, 5 min.
6. Counterstains if desired:
   a. Safranin (see p. 324).
   b. Lacmoid (see p. 330).
7. Dehydrate in alcohol series.
8. Clear in xylene or clove oil.
9. Mount in synthetic resin or balsam.

Results from hematoxylin: Middle lamellae, cellulose walls, nuclei, cytoplasm, chloroplasts, and p-protein—black. Lignified walls, callose, starch, cuticle—not stained.

NOTE: Use agitation for at least the first 5 min of all treatments.

Premixed Aluminum Hematoxylin

Delafield's procedure as communicated and refined by Prudden (1885)
Modified from Chamberlain (1932), Boke (1939) and Johansen (1940)

Application: General anatomical stain including meristematic tissue.

Designed to show: Cellulose cell walls, nuclei.

Fixation: Any botanical fluid.

Embedded in: Paraffin or freehand.

Preparation of staining solutions:

A. Short method: To 100 ml of a stock 5% aqueous solution of $NH_4Al(SO_4)_2 \cdot 12H_2O$, add 1.0 ml of a stock 10% alcoholic solution of hematoxylin. As an oxidation accelerator add 1 ml of 1% $NaIO_3$. (This is a further modification of a note in the second edition.)

B. Classical method: To 400 ml of aqueous 1% $NH_3Al(SO_4)_2 \cdot 12H_2O$ add 40 ml of 10% hematoxylin in absolute ethanol, and let stand exposed to air and light 1 wk. Filter, and then add 100 ml of glycerin and 100 ml of methyl alcohol. Let stand until solution becomes dark (6–8 weeks). This stock solution may be kept for a considerable period in a tightly stoppered bottle. Just before using, dilute with an equal volume of distilled water. (From Prudden, 1885.)

NOTE: Filter before using.

Staining schedule:

1. Remove paraffin and bring into water. Wash thoroughly.
2. Stain 2–10 min or longer in your choice of the above hematoxylin preparations. The exact time should be determined with the aid of a microscope.
3. Rinse in several changes of tap water (10–30 min).
4. Counterstain:
   a. Safranin (see p. 324).
   b. Lacmoid (see p. 330).
5. If no counter stain is used, treat for 5 min in 1% $NaHCO_3$.
6. Dehydrate through 30%, 50%, 70%, 95%, and 2 changes of absolute ethanol.

   NOTE: If precipitates are present, or sections are overstained, dip the slides twice into alcohol acidulated with 2 drops 1:5 HCl to 100 ml of 70% alcohol. Then place the slides into 70% alcohol to which one drop of ammonia has been added until a rich purple color appears.

7. Clear in 2 changes of xylene.
8. Mount in neutral balsam or a resin of your choice.

Results: Nuclear envelope; nucleolus, and chromatin—deep purple; cytoplasm and cellulose wall—lighter purple; sieve-tube slime—purple.

Tannic Acid→FeCl₃→Safranin→Fast Green

From Foster (1934) with modifications by Gifford (1966)
See review by Koch (1896); Zimmerman (1928)

Application: Meristematic tissues.

Designed to show: Cell walls; spindle fibers, chromosomes, and nucleoli may also be well stained.

Fixation: Formalin-acetic-alcohol, or one of the chromic acid fixatives (see p. 16, 17).

Embedded in: Paraffin.

Staining schedule:
1. Remove paraffin and bring into water. Wash thoroughly.
2. Place for 30 sec to 1 min in 1% aqueous tannic acid (reagent grade). The addition of 1% sodium salicylate prevents growth of molds and has no effect on the staining.
3. Wash thoroughly in distilled water (3 changes).
4. Place for 1–2 min in 3% aqueous $FeCl_3$.
5. Examine under microscope. If cell walls of meristematic tissue appear black or dark blue, proceed with the technic. If too weak, wash thoroughly in tap water and place again in the tannic acid solution. Alternate between Steps 2 and 4 until satisfactory differentiation is obtained, being sure to wash thoroughly in tap water each time the transfer is made.
6. Wash in distilled water.
7. Cationic counterstains:
   a. Safranin (go to Steps 8–19).
   b. Lacmoid (for phloem, see (p. 330).
8. Transfer from distilled water to 50% alcohol.
9. Place for 1–48 hr in 1% safranin in 50% alcohol. (Time may be cut down even to 20 min.)
10. Rinse in distilled water.
11. Wash in 50% alcohol and destain carefully in acidulated 70% alcohol. Destain until nucleoli can be seen. Observe with microscope.
12. Stop destaining in another jar of 70% alcohol.
13. 95% alcohol for 2 min.
14. 100% alcohol for 2 min.
15. Anionic counter stain: Fast green. If desired, otherwise go to Step 16. Dip slides in jar of absolute alcohol-xylene (1:1) to which is added only enough fast green from a stock bottle to form a slight green color. If walls were lightly stained with tannic acid-$FeCl_3$, continue dipping until walls are well differentiated (observe with microscope).
16. 1:1 absolute alcohol-xylene.
17. Xylene I.
18. Xylene II.
19. Mount (Harleco synthetic resin (HSR)).

Results: Cell walls—black, gray, or greenish; spindle fibers—blue; chromosomes and nucleoli—red; cytoplasm—pink or greenish.

ZnCl$_2$→Safranin→Orange G—Tannic Acid→Tannic Acid→NH$_4$Fe(SO$_4$)$_2$

From Sharman (1943)

Application: Shoot apex, other meristems, mature tissue.

Designed to show: Cell walls and protoplasts.

Fixations: Any of the usual botanical fixatives; but gives best results using formalin-acetic acid or formalin-aceto-alcohol. In place of acetic acid other fatty acids may be substituted (Ball, 1941).

Embedded in: Paraffin.

Preparation of solutions:
A. ZnCl$_2$: 2% aqueous. Filter off white precipitate.
B. Safranin: 1:25,000 aqueous (2 ml of a stock 2% aqueous solution to a liter of water).
C. Orange G—Tannic acid: Orange G, 2 gm; Tannic acid, 5 gm; 1:5 HCl, 0.3 ml; Thymol, few crystals; water to make 100 ml.
D. Tannic acid: tannic acid, 5 gm; thymol, few crystals; water to make 100 ml.
E. Clove oil-xylene: clove oil, 1 vol; xylene, 3 vol.

Staining schedule:
1. Take the slides down to tap water.
2. Treat 1 min in ZnCl$_2$.
3. Rinse about 5 sec in tap water.
4. Stain 5 min in safranin solution.
5. Rinse about 5 sec in tap water.
6. Stain 1 min in orange G-tannic acid.
7. Rinse about 5 sec in tap water.
8. Treat 5 min in tannic acid.
9. Rinse quickly (1–3 sec) in tap water.
10. Treat 2 min in 1% aqueous NH$_4$Fe(SO$_4$)$_2$·12H$_2$O.
11. Rinse 15 sec in tap water.
12. Place for 5 sec in 45% alcohol and for 5–10 sec in 90% alcohol. Wipe excess off slide.
13. Treat 10 sec in absolute alcohol. Wipe excess off slide.
14. Immerse for 20–30 sec in clove oil-xylene. Wipe slide. If cloudiness occurs, add a few drops of absolute alcohol.
15. Place for 20–30 sec in xylene (1 or 2 changes).
16. Mount in balsam.

NOTES: When placing slides in any solution, raise and lower them a dozen times.
The schedule may be simplified by substituting the following for Steps 6–8:

6A. Stain 5 min in the following: orange G, 0.5 gm; tannic acid, 10 gm; 1:5 HCl, 0.25 ml; thymol, a few crystals; water to make 100 ml. This is better for mature tissue than apical meristem.

Results: Cell walls—blue-black; nuclei—yellow to orange; cytoplasm—pale yellow, darker yellow to orange, or gray, depending on tissue; procambial cells—brilliant orange; protophloem sieve elements—dark blue or black; starch grains—black; phloem parenchyma—black walls with orange contents; young unlignified walls—blue-gray; partially lignified walls—violet, mauve or pink; lignified walls—cherry-red; mature sclerenchyma elements—red.

SECTION B: CATIONIC STAINS

The cationic or basic dyes are usually used in combination with anionic dyes or dyes used with a mordant. Among the most useful ones are safranin, crystal violet, acetocarmine, toluidine blue O, and thionin.

Bismarck Brown Y

An old stain for cellulose walls
Esau (1948), Jensen (1962)

Application: General plant stain.

Designed to show: Cellulose walls and plant mucins.

Fixation: Any botanical fixative.

Staining schedule:
1. Prestains: Hematoxylin applied regressively may be used (see p. 316).
2. Stain with a 1% aqueous Bismarck brown (5 min).
3. Wash with water, several changes.
4. Dehydrate in an alcohol series.
5. Clear in xylene.
6. Mount in balsam.

Results: Cell walls and tannins—orange brown; nuclei and cytoplasm— lightly stained.

> **NOTE:** Bismarck brown is only occasionally a choice for staining. It tends to stain everything uniformly orange brown with no contrast between structures. It is a cationic dye that tends to stain the same structures that are stained by anionic dyes or by hematoxylin used progressively. On occasions it has merit as indicated above for contrast with hematoxylin used regressively. Esau combined Bismarck brown with iodine green and resorcin blue for staining phloem.

Crystal Violet and Erythrosin B

From Conn (1953, p. 323)
Essentially the same as Lee (1937, p. 654)
Original by Jackson (1926)

Application: Woody tissue of vascular plants.

Designed to show: Vascular bundles; differentiation of lignified and non-lignified cell walls.

Fixation: Any botanical fixing fluid.

Embedded in: Paraffin, celloidin, or freehand.

Staining schedule:
1. Remove paraffin and bring into water.
2. Stain 15 min in 1% aqueous crystal violet.
3. Rinse quickly in distilled water.
4. Dehydrate quickly but thoroughly in 95% and absolute alcohol.
5. Stain 1–10 min in saturated solution of erythrosin B in clove oil.
6. Treat 1–2 min in equal volumes of absolute alcohol in xylene.
7. Clear in xylene and mount in balsam.

Results: Nonlignified walls—red; lignified walls, nuclei, cytoplasm—violet.

> NOTE: The principal variations of this method are in the lengths of time for which the stains are allowed to act. The xylem elements in some materials stain satisfactorily in a very few minutes; with others, ½ hr or more may be necessary. The action of the erythrosin must be carefully controlled, because it has a tendency to replace the violet in the lignified walls if allowed to act too long.

Periodic Acid-Leucobasic Fuchsin (Schiff's)→Amido Black 10B

Communications from W. A. Jensen and Donald B. Fisher
Jensen (1962); Fisher (1968)

Application: Any plant material.

Designed to show: Specific for total carbohydrates and proteins.

Fixation: Any botanical fixative, freeze substituted, or freeze dried.

Embedded in: Plastic, paraffin, or freehand.

Preparation of solutions: Leucobasic fuchsin (Schiff's reagent (see p. 74 or 350).

Staining procedure:
1. Bring sections into water.

2. Place slides in 0.5% aqueous periodic acid for 20 min.
3. Wash in running water for 10 min, and finally glass distilled water.
4. Stain in leucobasic fuchsin for 20 min.
5. Rinse the sections and place them in 2% sodium bisulfite for 1–2 min.
6. Wash in running tap water 5–10 min.
7. Stain with 1% Amido Black 10B (purchased from Matheson, Coleman and Bell as Aniline Blue Black C.I. 20470) in 7% acetic acid. (For epon sections warm to 50°C for about 10 min.)
8. Dip briefly in 7% acetic acid.
9. Mount in glycerol containing 5% acetic acid.
10. Ring with vas-par (see p. 367).

Results: Polysaccharides—red; mitochondria, plastids, nuclei and other protein-containing bodies—black.

Safranin with Various Counterstains

Henry Schneider

Application: Any plant material.

Designed to show: Lignified walls, cuticle, wound gum, cork cells, chromosomes, nucleoli.

Embedded in: Paraffin, carbowax, freehand sections.

1. Sections may be prestained with one of the following: Premixed iron hematoxylin, Weigert's (see p. 317); ferric cations followed progressively by hematoxylin, Heidenhain's (see p. 318); premixed aluminum hematoxylin, Delafield's (see p. 318) or Harris' (see p. 106).
2. Bring sections that are prestained or unstained into 30% alcohol.
3. Stain 30 min to overnight in 1% Safranin in 30% alcohol.
4. Wash in several changes of water.
5. Destain as needed (about 2 min) in 30% acid alcohol (1 ml of 1:5 HCl in 500 cc of 30% alcohol).
6. Rinse in 30% alcohol.
7. Rinse in 1% NaHCO$_3$ in 30% alcohol for 5 min.
8. Rinse in 50% and 70% alcohol.
9. Counter stains:
   a. Aniline blue (see p. 325).
   b. Fast green (see p. 329).
   c. Orange G (see p. 330).
10. If no counter stain is applied, dehydrate with 95% and absolute alcohol, clear in xylene and mount in neutral Canada balsam.

   NOTE: If freehand sections are too brittle and curled to mount, bring them through 2 changes of 95% alcohol, 2 changes of clove oil, mount on a slide and blot dry with bibulous blotting paper, and mount in balsam.

Safranin and Aniline Blue

From Darrow (1944)

Application: Meristematic tissue.

Designed to show: Chromosomes; cellulose walls.

Fixation: Randolph's modification of Navashin's (CRAF).

Embedded in: Paraffin.

Staining schedule:
1. Remove paraffin from sections in the usual manner.
2. Stain 15 min in 1% aqueous safranin O.
3. Rinse in distilled water.
4. Stain 2 min in 1% aniline blue w.s., in alcohol.
5. Dehydrate in absolute alcohol.
6. Clear and mount in balsam.

Results: Chromosomes—red; nucleoli—red; cellulose walls—blue; cytoplasm—nearly colorless.

Safranin and Fast Green

From Conn (1953, p. 313)
Similar to Haynes (1928)

Application: General plant tissue; especially meristematic tissue.

Designed to show: Chromatin; spindles; cellulose walls; lignified walls.

Fixation: Any CrO_3 fixative.

Embedded in: Paraffin.

Staining schedule:
1. Remove paraffin from sections as usual.
2. Stain 30–50 min in 1% aqueous safranin O.
3. Rinse in distilled water.
4. Differentiate with 0.2% fast green FCF in alcohol until the chromatin and nucleoli remain red.
5. Rinse in absolute alcohol.
6. Clear and mount in balsam.

Results: Chromosomes—bright red; nucleoli—bright red, lignified walls—bright red; spindles—green; cellulose walls—green; cytoplasm—green.

Safranin and Fast Green

From Johansen (1940)

Application: General plant tissue.

Designed to show: Chromosomes, spindle fibers, cuticle, lignified walls, cytoplasm, cellulose walls, etc. A general stain.

Fixation: Any botanical fixative.

Embedded in: Paraffin, celloidin, frozen or freehand sections.

Preparation of staining solution:
  A. Johansen's Safranin: After completely dissolving 4 gm of safranin in 200 ml of Methyl Cellosolve, add 200 ml of 50% ethanol. Then add 4 gm of sodium acetate and 8 ml of formalin.
  B. Fast green stock solution: Saturate a mixture of equal parts of methyl cellosolve and absolute ethanol with fast green FGF.
  C. Fast green staining solution: To 1 part of absolute ethanol and 3 parts of clove oil add enough of Solution B to give the desired intensity.

Staining schedule:
  1. Bring sections to 50% or 70% alcohol.
  2. Stain in safranin, Solution A, for 2 to 24 hr.
  3. Thoroughly wash out excess stain in H_2O for 5 min or longer.
  4. To differentiate and dehydrate, dip for 10 sec in an 0.5% picric acid solution in 95% ethanol. (Prepare in advance as picric acid dissolves slowly.) This solution may be reused.
  5. Rapidly transfer sections to 95% ethanol containing 4 or 5 drops of ammonia per 100 ml of ethanol for no more than 2 min.
  6. Transfer to 95% ethanol.
  7. Counterstain with the fast green staining Solution C for 15 sec or less.
  8. Rinse with clove oil.
  9. Clear in 2 parts of clove oil, 1 part absolute alcohol, 1 part xylol.
  10. Clear in 2 changes of xylol.
  11. Mount in balsam.

Results: Lignified and cutinized walls, chromosomes—brilliant red; cytoplasm and cellulose walls—brilliant green.

Safranin and Picro Aniline Blue

From Johansen (1940, pp. 62, 82)

Application: Woody tissue.

Designed to show: Differentiation of lignified and unlignified cellulose walls.

No fixation: Freehand sections.

Preparation of staining solutions:
  A. Dissolve 4 gm safranin in 200 ml Methyl Cellosolve; then add 100 ml

each of 95% alcohol and distilled water, followed by 4 gm sodium acetate and 8 ml formalin. (The purpose of the acetate is to intensify the stain.)

B. Prepare saturated solutions of picric acid and aniline blue, w.s., in alcohol; when ready to use, mix 78 vol picric acid with 22 vol aniline blue.

Staining schedule:

1. Stain sections at least 2 hr in Solution A.
2. Wash out excess stain with tap water.
3. Destain either in alcohol saturated with picric acid, or in 50% alcohol with a trace of HCl.
4. Stain 2 hr in Solution B.
5. Wash 10 sec in absolute alcohol.
6. Clear in clove oil.
7. Xylene.
8. Mount in balsam.

Results: Lignified or cutinized walls—bright red; cellulose walls—bright blue.

Thionin for Freehand Sections

An old stain (see p. 211).

Application: Temporary mounts of freehand sections.

Designed to show: General polychromatic staining.

No fixation: Frozen or freehand sections.

Staining schedule:

1. Stain sections in 0.25% aqueous thionin.
2. Mount in slightly alkaline tap water or 1% aqueous $NaHCO_3$.
3. Ring with vas-par.

Results: Cellulose walls and nucleoli—red; nuclei—purple; lignin—deep purple.

Methylene Blue-Azure A→Basic Fuchsin for Tissues in Resin

From Humphrey and Pittman (1974)
See also Hoefert (1968)

Application: Epoxy resin embedded tissues. Survey and monitor sections.

Designed to show: General polychromatic staining of tissue elements, and photomicrography.

Fixation: Glutaraldehyde and osmic acid.

Embedded in: Spurr's low viscosity epon resin or epoxy resin.

Preparation of staining solutions:

A. Methylene blue-azure A:

Methylene blue	0.130 gm
Azure A	0.020 gm
Glycerol C.P.	10 ml
Methanol C.P.	10 ml
Phosphate buffer, pH 6.9	30 ml
Distilled water	50 ml

Phosphate buffer (pH 6.9):

Anhydrous KH_2PO_4	9.078 gm
Anhydrous $NaHPO_4$	11.876 gm
Distilled water	1.0 l

B. Basic fuchsin. Stock solution:

Basic fuchsin	0.1 gm
50% ethanol	10 ml

Dilute 3 ml of stock solution to 60 ml with distilled water. The diluted solution was kept in a screw cap Coplin jar for up to 4 months with no deterioration.

Staining schedule:

1. Affix sections to slide (without removing resin) by floating them on a drop of 10% acetone in distilled water and heating on a hot plate at 122°C until solution has evaporated. Remove slide from hot plate immediately after becoming dry.
2. Immerse slides into preheated Solution A in a screw-capped Coplin jar at 65°C for 20 min. Times vary according to embedding media.
3. Rinse in 2 changes of distilled water.
4. Stain in Solution B at room temperature for 30 sec to 5 min.
5. Rinse in 3 changes of distilled water.
6. Air-dry the slide and apply 1 drop of xylene and 1 drop of mounting media and apply coverslip.
7. Dry slides at room temperature. Heat will cause the stains to fade.

Results: Cellulose walls, cuticle—red; nucleoli—dark blue-green; cytoplasm, chloroplasts—bluish gray; lignified walls—green.

SECTION C: ANIONIC DYES

The anionic dyes are often used as counterstains with cationic dyes and with metal mordanted dyes. Cationic and anionic dyes supplement each other very well. Mention should be made of fast green, orange G, aniline blue, chlorazol black E, and congo red, which are listed here or in other chapters where they are used in combination with other stains.

Chlorazol Black E

From Darrow (1940)
Original by Cannon (1937)

Application: General plant tissue, including that of cryptogams, for pathological material; also used in cytology.

Designed to show: Differentiation of various tissue elements.

Fixation: Bouin or Flemming.

Embedded in: Paraffin.

Staining schedule:
1. Remove paraffin from sections and bring down to 70% alcohol.
2. Stain 5–10 min in fresh, unfiltered, 1% chlorazol black E in 70% alcohol.
3. Drain off excess dye.
4. Dehydrate.
5. Clear and mount in balsam.

Results:
A. Vascular plants: Cell walls—jet black; cytoplasm—grayish green; nuclei—yellowish green; nucleoli—deep amber to dark green.
B. Fern leaf: Cell walls—definitely black; epidermis walls—heavy black; cytoplasm—light amber; nuclei—green; nucleoli—dark green; plastids—gray; suberized walls of midrib—dark amber; veins—dark amber.

Fast Green

Haynes (1928)

Application: Many plant tissues.

Designed to show: Cellulose walls, cytoplasm.

Fixation: Any botanical fixative.

Embedded in: Paraffin or freehand sections.

Staining schedule:
1. Prestain in safranin (p. 324); iron and hematoxylin used regressively (p. 353); or a stain of your choice.
2. Place in 0.2% alcoholic fast green FCF and control the degree of staining with the microscope starting with 15 sec.
3. Dehydrate in 2 changes of absolute alcohol.
4. Clear in xylene.
5. Mount in balsam.

Results: Cellulose walls and proteins—green.

Light Green

See Chamberlain (1932, p. 68)

Light green may be used similarly to fast green being made up as a 1% solution in alcohol or in clove oil. The stain fades and it has therefore been replaced by fast green for plant anatomy.

Orange G

Stoughton (1930), Johansen (1940), Jensen (1962)

Application: General plant tissue. Usually used as a counterstain.

Designed to show: A general stain for cellulose walls, starch granules, callose on sieve plates, cytoplasm, and nuclei.

Fixative: Any plant fixative.

Embedded in: Paraffin, Carbowax, or freehand.

Staining schedule:
1. Prestains:
    a. Iron and hematoxylin applied regressively (see p. 353).
    b. Thionin (as in Stoughton's, see p. 373).
    c. Used in Sharman's with tannic acid and safranin.
2. Stain in one of the following:
    a. If sections are in water, one may use 1% aqueous orange G for 30 sec or as required.
    b. After dehydration one may use a saturated solution of orange G in absolute alcohol, *or* 0.25% orange G in clove oil until dark enough.
3. For (2a) dehydrate in an alcohol series; then for both (2a) and (2b) rinse in 2 changes of xylene and mount in balsam.

Results: Intensity of staining of the above structures varies with maturity of tissues and with other stains used with the orange G.

SECTION D: NEUTRAL DYES

It is surmized that lacmoid is neutral in reaction.

Lacmoid Counterstain for Permanent Mounts

From Schneider (1952); Cheadle, Gifford, and Esau (1953); Schneider (1960)

Application: Phloem.

Designed to show: Callose on sieve areas of lateral walls and sieve plates of sieve tubes.

Fixation: Any botanical fixative.

Embedded in: Paraffin, Carbowax, or freehand.

Staining schedules:

I. For nonmeristematic tissue or meristematic tissue. (Schneider 1952, 1960, and personal communications.)

   1. Prestain with one of the following: premixed iron hematoxylin (Weigert's see p. 317). Ferric cations followed progressively by hematoxylin (Heidenhain's, see p. 318). Premixed aluminum hematoxylin (Dellafield's, see p. 318).
   2. Treat sections with frequent agitation for 30 min in a 1% aqueous solution of $NaHCO_3$.
   3. Stain for 1 hr to overnight in 0.25% lacmoid in 30% alcohol containing 0.5% $NaHCO_3$.
   4. Wash in several changes of tap water (pH over 7).
   5. Treat with 1% aqueous $NaHCO_3$ for 10 min.
   6. Dehydrate in 30% tertiary butyl alcohol (TBA) for 10 sec; 70% TBA 20 sec; 3 changes of 100% TBA for 20 sec, 1 min, and 1 min, respectively.

NOTE: Use 95% TBA if freehand section becomes too brittle to mount when 100% TBA is used.

   7. Rinse in clove oil from a bottle labeled 1; then place in a 2nd change of clove oil from a bottle labeled 2. The clove oil may be reused many times. Eventually discard the clove oil in bottle 1 and replace it with that in bottle 2 which in turn is replaced with new clove oil. Use a good grade of oil from flower buds that is straw colored.
   8. Mount freehand sections on slides. Remove excess clove oil from freehand or paraffin sections by placing a piece of lint free bibulous blotting paper on the sections and pressing gently with the pad of a finger or thumb.
   9. Mount in neutral Canada balsam and apply coverslip.

NOTE: None of the above dyes will be destained by the residue of clove oil. Both clove oil and balsam tolerate a little water; hence, the walls do not shrink as they do with complete dehydration, and the sections do not become so brittle.

Results: Lacmoid: callose and lignified walls—blue; wound gum—blue green; hematoxylin: See the schedule used.

II. For color photography (after (Schneider (1960)):

   1. Bring sections into water.
   2. Prepare a stock solution consisting of: lacmoid 0.1 gm; and congo red 0.1 gm in 100 ml of 30% alcohol. Stain overnight in a mixture of 10 ml of stock solution to 30 ml of 1% $NaHCO_3$ in water.

3. Go back to I, Steps 4–9 for details of washing, dehydration, and mounting.

Results: Cellulose walls—red; callose—blue.

III. Especially for primary phloem developing from procambial strands. After Cheadle, Gifford, and Esau (1953).
    1. Prestain with tannic acid and $FeCl_3$ according to Steps 1–5 in the procedure of Foster (see pp. 319–320).

NOTE: Mordanting with $ZnCl_2$ prior to treatment with tannic acid as in Sharman's procedure (p. 321) will increase the staining of cell walls.

    2. Place in a solution of 1% $NaHCO_3$ in 25% alcohol.
    3. Stain 12–18 hr in a 0.25% solution of lacmoid in 30% alcohol to which a few ml of 1% $NaHCO_3$ has been added.
    4. Place a few sec or min in 1% $NaHCO_3$ in 50% alcohol.

NOTE: Destaining occurs rapidly in ethanol, and a tertiary butyl alcohol series is suggested as a substitute.

    5. Wash 2–3 min in 80% and 95% alcohol.
    6. Wash 2–3 min in 2 changes of absolute alcohol.
    7. Wash 2–3 min in equal parts of absolute alcohol and xylene.
    8. Wash 2–3 min in 2 changes of xylene.
    9. Mount in synthetic neutral resin.

NOTE: If mucilage is present, the dyes and washes should be in 50% alcohol.

Results: Callose and lignified walls—greenish blue; cellulose walls, nuclei, slime and cytoplasm—light brown to grayish brown.

SECTION E: FAT-SOLUBLE DYES

Sudan IV

From Rawlins (1933, p. 47)

Application: Any plant material.

Designed to show: Suberized walls, cuticle, fat or oil globules.

Fixation: Either no fixation or any botanical fixative.

Embedded in: Paraffin or freehand sections.

Preparation of the staining and mounting solution: Make a saturated solution of Sudan IV in 95% alcohol. Add an equal volume of glycerin and filter; if the stain precipitates out on the sections, dilute further as necessary with a mixture of equal parts of glycerine and alcohol. Keep in a dropper bottle.

Staining schedule:
1. Bring sections into 30% alcohol.
2. Mount sections on a slide in a few drops of the staining and mounting solution.
3. Seal the edges of the coverslip with vas-par (see p. 367) for method of preparation).

Results: Cuticle, cutinized and suberized walls, and fat globules—orange.

CHAPTER 13

PLANT CYTOLOGY

Henry Schneider

In this chapter consideration is given to subcellular structures and to organelles other than chromosome morphology which is presented in the next chapter on cytogenetics.

Schedules that are presented in other chapters and which stain specific cell components are: periodic acid-leucobasic fuchsin (Schiff's) with Amido Black 10B (see p. 323); Azure B (pp. 370–371); and acridine orange with fluorescence microscopy (p. 369). Some of the procedures given for animal histology: "Miscellaneous Methods" are also applicable to plant cells. A schedule that is used for differentiation of DNA and RNA is methyl green with pyronin on page 199.

Aniline Blue WS and Fluorescence Microscopy

See also Chapter 3

From Currier and Strugger (1956)
Currier and Shih (1968)

Application: Bulb scales of *Allium cepa* L. or leaves of *Elodea canadensis* or *E. densa.*

Designed to show: Minute amount of pit callose.

Fixation: Formalin-acetic-alcohol or unfixed.

Embedded in: Cross or paradermal sections cut freehand or embedded in paraffin. Whole mounts of Elodea may be used.

Preparation of solutions: Aniline blue 0.005% or 0.01% in M/15 potassium phosphate buffer. (pH 10 for killed material; pH 8–9 for living tissue.) The dye becomes nearly colorless in this pH range.

Staining schedule:
1. Mount sections in the above staining solution. Sucrose (0.2 M) may be added. Stain for 10 min to 24 hr.
2. Examine with blue light of 390–430 mμ wave length.

Results: Pit callose fluoresces yellow.

Azure B for Differentiation of RNA and DNA

From Swift (1955)
Jensen (1962); original by Flax and Himes (1952)

Application: Plant and animal tissues, onion root tips.

Designed to show: DNA in chromatin and RNA in cytoplasm and nucleolus.

Fixation: Alcohol-acetic acid, 3:1 (12–24 hr).

Embedded in: Paraffin.

Preparation of staining solution: Dissolve 25 mg per 100 ml of methylene azure (azure B type) in McIlvaine's buffer at pH 4.0 (24.6 ml 0.1 M citric acid and 15.4 ml 0.2 M K_2HPO_4).

NOTE: Azure B is satisfactory.

Staining schedule:
1. Treat sections 2 hr at 37°C in the above staining solution.
2. Rinse slides in 3 changes of water and pass directly to tertiary butyl alcohol. Leave 12 hr.
3. Clear in xylene.
4. Mount in balsam or a synthetic medium.

Results: DNA of chromatin—blue-green; RNA in nucleolus and cytoplasm—purple; other basophilic substances (mucin, chondroitin)—reddish.

Fast Green for Nuclear Histones

From Alfert and Geschwind (1953) and communication with Ernest M. Gifford.
(See also Bloch and Godman (1955); Gifford and Dengler (1966))

Designed to show: Basic protein of chromatin.

Fixation: 10% neutral formalin 3–6 hr. Wash overnight in running water.

Embedded in: Paraffin.

Preparation of staining solution: A 0.1% solution of fast green FCF adjusted to pH 8.0–8.1 with a minimum of NaOH.

Staining schedule:
1. Remove paraffin and bring sections to water.
2. 5% solution of trichloroacetic acid at 90°C (15 min).
3. 70% alcohol wash 3 changes (10 min each).
4. Distilled water (pH 8.0. Adjust pH with 0.1 M NaOH); rinse for a few seconds.
5. Place in 0.1% solution of fast green (1–3 hr).
6. Wash in distilled water (pH 8.0) for 5 min, pour out and add new water each min of the 5.
7. Put directly into 95% alcohol (2 min).
8. Absolute alcohol (2 min).
9. 1:1 Absolute alcohol:xylene (2 min).
10. Xylene I (5 min).
11. Xylene II (5 min).
12. Mount in Harleco synthetic resin.

Control: At step 2, use a water control. Water at 90°C for 15 min.

Results: Chromosomes—green.

4′-6-Diamidino-2-phenylindole (DAPI), a Flurorescent Stain for Mycoplasma and DNA
Seemüller (1976)

Application: Mycoplasma-diseased phloem tissue. Apple with proliferation. Pear with pear decline.

Designed to show: Mycoplasma in sieve tubes; DNA.

Fixation: 5% glutaraldehyde in 0.1 M phosphate buffer pH 7 at 4°C.

Embedded in: Frozen sections.

Staining solution: 1 mg of DAPI in 1 liter of pH 7 phosphate buffer.

Staining schedule:
1. Stain for 20–30 min.

2. Wash and mount in pH 7 phosphate buffer.
3. Ring coverslip with nail polish.
4. Examine with 365 nm peak fluorescent or epi-fluorescent light.

Results: Fluorescent particles are irregularly distributed in the sieve tubes.

Phloroglucinol in HCl for Wound Gum

Schneider (1979)

Application: Diseased or injured plants. Psorosis diseased citrus trunks that are scaling. Pruning wounds on stone fruit trees.

Designed to show: Deposits of wound gum in vessels and throughout tissues.

Fixation: None or any botanical fixative.

Embedded in: Parafin, frozen, or freehand.

Staining schedule: Treat and then mount sections in a saturated solution of phloroglucinol in 18% HCl. Ring coverslip with vas-par to prevent HCl fumes from corroding microscope (see p. 367).

Results: Lignin—pink; wound gum—red. The color in wound gum may be more persistant than in lignin.

   NOTE: Lignin tests positive with the procedure of Mäule, but wound gum does not.

FeSO₄ for Tannins

From Schneider (1977)

Application: Diseased or healthy tissue.

Designed to show: Vacuolar and coacervated tannins.

Fixation: Living material.

Embedded in: Freehand sections.

Preparation of staining and fixing solution:

H_2O	89.0 ml
Glacial acetic acid	0.25 ml
37% formaldehyde	10.0 ml
$FeSO_4$	2.0 gm

Staining schedule:

1. Cut sections with a razor blade in water.
2. Place in the staining and fixing solution for 2–4 hr.
3. Gradually bring into glycerine.
4. Mount in glycerine.

Results: Tannins become orange to black.

NOTE: Formaldehyde precipitates tannin in their natural locations. Acetic acid slows precipitation of the $FeSO_4$ from solution but does not dissolve the tannins as inorganic acids do.

CHAPTER 14

PLANT CYTOGENETICS*

Ronald L. Phillips

The methods given in this chapter are commonly employed for chromo-some studies. As pointed out in the introduction to the section on Botanical Sciences (Chapter 12), some of the procedures given below may be used as general stains for plant tissue even though they are designed for chromo-somes.

Cytogeneticists commonly pretreat plant tissues prior to fixation for the purposes of arresting the processes of cell division in mitosis, contracting and condensing chromosomes for improving visibility, improving stainabil-ity, and improving separation and spreading in squash preparations. Of many pretreatment chemicals that have been introduced (Sharma and Sharma, 1965), colchicine (O'Mara, 1939), 8-hydroxyquinoline (Tjio and Levan, 1950), isopropyl-*N*-(3-chlorophenyl) carbamate (CIPC) (Storey and Mann, 1967), monobromonapthalene (Morrison, 1953), and *para*-dichloro-benzene (Meyer, 1945) are now used widely. Cytogeneticists working with cereal crops and many other species have been using a pretreatment in ice water or running cold tap water. The cold treatment has been used to induce banding patterns in somatic chromosomes (Darlington and LaCour, 1940).

Freshly prepared Farmer's acetic-alcohol solution (1 vol glacial acetic acid: 3 vol 95 or 100% alcohol) is the killing and fixing fluid most widely used for root tips, young leaves, apical meristems, and microsporocytes to be squashed for studying chromosomes in mitosis or meiosis. It is eminently satisfactory for most species of plants. However, Carnoy's fluid (see p. 12) is

* Helpful suggestions for the previous edition by W. B. Storey and Asim Esen, Department of Plant Sciences, University of California, Riverside, still are gratefully acknowledged. In addition, the following individuals made significant contributions to this edition: M. C. Albertsen, C. R. Burnham, R. Stavig, and J. Ruegemer, University of Minnesota; E. T. Bingham, University of Wisconsin; D. R. Dewey, Utah State University; J. E. Endrizzi, University of Arizona; B. S. Gill, University; J. N.Rutger and R. Snow, University of California, Davis; T. Tsuchiya, Colorado State University; C. H. Uhl, Cornell University; and M. H. Yu, U.S. Agricultural Research Station, Salinas, California.

preferable for some species. Penetration and fixation are rapid. Furthermore, the chloroform dissolves and removes oils, resins, tannins, gums, and lipid substances which tend to interfere with visibility of the chromosomes.

Many species of plants have cells containing raphides, druses, ceptoliths, and other kinds of crystals, as well as starch grains, glycosides, pigments, and various precipitated substances which interfere with chromosome visibility. Many of these can be removed from plant material being prepared for chromosome studies following squash procedures by washing thoroughly in water at 60–80°C following maceration in HCl. Maceration is conveniently done in a mixture of 1 part concentrated HCl to 1 part 95% ethanol. Hot water often softens the cells, also, enhancing separating and spreading of the chromosomes. Some good reference books on microtechnic include Berlyn and Miksche (1976), Darlington and LaCour (1975), and Jensen (1962).

Iron Acetocarmine

Original by Belling (1926)

Application: Microsporocytes in smears of anthers.

Designed to show: Chromosomes

Fixation: According to McClintock (1929), in acetic alcohol (1 vol glacial acetic acid to 3 vol absolute or 95% alcohol), or in Carnoy's fluid (see Chapter 1). If not to be used immediately, such material may be stored in 70% alcohol at 0°C after 1–3 days in the fixative at room temperature. This step is sometimes omitted, as the staining solution below is also a fixing fluid, but the results are often not as good as with prior fixation.

No embedding.

Preparation of staining solution: Add 0.5 gm carmine to 100 ml boiling 45% acetic acid and boil for 1–2 min or until there is a sudden change to a darker color. Cool and filter. For more consistent staining results, simmer the acetic acid and carmine in a flask with a reflux condenser. Refluxing can be done up to 6–8 hr. Cool and filter. Store in a brown bottle. For certain species such as maize, the solution is used at full strength. For barley and wheat, the solution should be diluted with 45% acetic acid; propionic acid may be substituted for acetic acid. This stain as prepared requires the use of slightly rusty needles to provide iron for optimal staining with some material. To prepare acetocarmine with iron already added, the following is suggested: to half of the cooled and filtered acetocarmine prepared as above, add a few drops of a solution of ferric chloride or ferric hydroxide in 45% acetic acid until the stain turns bluish-red, but without visible precipitate. Then add the rest of the untreated acetocarmine. Stainless steel needles should be employed when using acetocarmine with added iron.

Staining schedule:

1. Place anthers in a drop of staining fluid on a *clean* slide. After cutting them transversely, gentle pressure repeatedly applied with a curved needle will force the contents out into the drop.
2. Stir the drop vigorously with a rusty needle to separate the pollen mother cells and also to add iron to the stain. Note that the stage of meiosis is most easily recognized in cells at or near the edge of the drop. If not at the desired stage, the drop may be wiped off.
3. Remove the anther pieces by picking them up between the points of the curved needles.
4. Add a *clean* coverslip. The drop of stain should be no larger than sufficient to barely fill to the edge. For maize sporocytes no pressure is needed. For certain species, e.g. tomato, considerable pressure is needed to flatten and spread the cells. For these cases, squash the slide between blotting paper.
5. Heat the slide by passing it back and forth over an alcohol flame. Heat enhances the contrast between chromosomes and protoplasm. For maize, results are best when heated almost to boiling. Repeat the heating until no further improvement in contrast is noted. In maize, this degree of heating for fresh material or in fixative only a short time tends to remove the protoplasm.
6. The slide may be examined immediately under a microscope. Use light passed through a glass ground on both sides plus a green filter.
7. Seal with a temporary sealing mixture or make permanent.

Results: Cytoplasm—uncolored; chromatin—dark translucent red (appearing black if green filter is used).

> **NOTE:** These results lack permanency. See page 356–358 for method of making permanent.

The following technic has been found to be excellent for microsporocytes in tomato (B. S. Gill, personal communication)

Application: Microsporocytes of tomato or other plants with small chromosomes.

Designed to show: Pachytene identification of chromosomes and other stages.

Fixation: Fix flower buds stripped of calyx in 1:3 propio-alcohol saturated with ferric chloride crystals for 12–24 hr, change to 70% alcohol, 3 changes and store in a refrigerator until ready to use.

Staining: Macerate one anther completely in a drop of propio-carmine, remove large debris. Apply the coverslip. Heat the slide over steam on a water bath. The water bath top consists of concentric rings. Arrange the rings so that opening is large enough to heat the slide area underlying the coverslip. Press the slide between the folds of a filter paper. If any air bubbles, apply some stain, press gently and seal with wax. Make observations on temporary preparations. Slides stay good for many days when kept in a vapor chamber in the refrigerator. After necessary observations, slides can be made permanent by any of the standard procedures.

Results: Chromosomes well spread, cytoplasm clear.

Alcoholic Hydrochloric Acid-Carmine

From Snow (1963)

Application: Squashes of meiotic or mitotic cells of plants (and animals).

Designed to show: Chromosomes at various stages of nuclear division.

Fixation: Acetic alcohol (1 part glacial acetic acid, 3 parts 100% ethyl alcohol); Carnoy's solution (6 parts 100% ethyl alcohol, 3 parts chloroform, 1 part glacial acetic acid); or a mixture of 6 parts 100% methyl alcohol, 3 parts chloroform, 2 parts propionic acid. After fixation for 24 hr the material can be transferred to 70% alcohol for storage in the refrigerator.

No embedding.

Preparation of staining solution: To 15 ml of distilled water in a small beaker add 4 gm of certified carmine and 1 ml of concentrated HCl. Mix well and boil gently for 5–10 min while stirring frequently. Cool, add 95 ml of 85% alcohol, and filter.

Staining schedule:
1. Remove material from fixative or 70% alcohol, drain on absorbent paper, and place in a liberal quantity of the stain. Leave until well penetrated, from 24 hr to several days. The time must be learned by experience. Material can remain in the stain indefinitely, as it is approximately 70% alcohol. Staining can be greatly accelerated by placing the tightly capped vial in an oven at 60°C for a few hours or overnight.
2. Pour off used stain (which can be saved and used again until staining becomes too weak), rinse material with water or 70% alcohol, and prepare squashes in the same way as for the ordinary aceto-carmine method, except that plain 45% acetic acid is used instead of aceto-carmine. Maceration of the tissue, if desirable, should be done in 45% acetic acid at 60°C.

Results: Chromosomes—dark red (appearing black if a green filter is used); cytoplasm—clear or faintly reddish.

M. H. Yu (personal communication) also found excellent results using this stain for sugarbeets. He recommends (1) omitting the last step in preparation of the stain (filtering), and (2) staining the sugarbeet flower buds at room temperature for 4–7 days. The stain can be reused and is good for about 2 months.

Acetocarmine

From Tsuchiya (1971)

Application: Root tip squashes and smears of microsporocytes.

Designed to show: Mitotic or meiotic chromosomes of a wide range of cereal, ornamental, vegetable, and tree species.

Fixation:

1. Pretreat with 0.002 M 8-oxyquinoline at 18°C for 4 hr (cf. Tjio and Levan, 1950) or ice water for 12–72 hr (cf. Tsunewaki and Jenkins, 1960). Only for root or shoot tips.
2. Fix the material with 1:3 acetic-alcohol for 5 minutes or more; may store several months to about 2 years in the refrigerator.
3. Transfer materials to 2% iron alum mordant, especially for the study of nucleoli in mitotic and meiotic cells.

Preparation of staining solution: Prepare 0.5–1.0% acetocarmine stain as described for Iron Acetocarmine (pp. 342–343).

Staining schedule:

1. Stain in 0.5–1% acetocarmine several days or more.
2. Squash stained materials by the usual method: Use a mixture of 45% acetic acid and glycerine (10:1) for mounting (see below, d).
   a. Cut the meristematic region of the stained root tips at 0.1 mm thick or squeeze out the meristematic tissues with the tip of a forceps after removing the root cap. Anthers of certain species are used without cutting but larger anthers should be cut.
   b. Put a piece of cut or squeezed meristematic tissues or anthers on a slide.
   c. Apply a drop of mixture of concentrated acetocarmine (1 or 2%) and 1 N HCl (1:1 ratio) for 1–3 sec (depending on the stain), and remove it soon with a piece of blotting paper.
   d. Apply a drop of a mixture of 10 parts 45% acetic acid and 1 part glycerine (Rattenbury's fluid).
   e. Put on a coverslip.
   f. Heat the slide gently on a small alcohol flame or gas burner at about 80°C. Should not be boiled.
   g. Tap the coverslip gently after fixing the corner of the coverslip with a finger to avoid cover glass slipping.

h. Heat again for a second. Should not be boiled.
i. Apply a blotting paper on the cover glass and press with the finger after fixing the corner of the cover glass with another finger. Thus the preparation is finished and the excess fluid is removed by the covered blotting paper.
j. Observe the preparation.
k. If cells are crowded and/or the cytoplasm is still overstained, apply a small drop of the Rattenbury's fluid and gently heat again and repeat the procedures from f to i.

Results: Chromosomes—red (black if green filter is used); cytoplasm—uncolored.

Chlorazol Black E with Acetocarmine

From Nebel (1940)

Application: Squashes of freehand sections of flower buds; microsporocytes.

Designed to show: Chromosomes.

Fixation: Acetic alcohol (1 vol glacial acetic acid to 3 vol alcohol), preferably 12–24 hr, although short periods (10–15 min) can be used when speed is essential.

No embedding: Free-hand sections or bits of meristematic tissue, dissected out of buds, are placed in the fixing fluid, and subsequently washed, stained, etc.

Preparation of staining solution: Aceto-carmine: boil 45% acetic acid saturated with carmine and filter.

Staining schedule:
1. Wash in 3 changes of 70% alcohol.
2. Stain 5–25 min in filtered 1% chlorazol black E in 70% alcohol.
3. Rinse in 3 changes of 70% alcohol.
4. Transfer materials to a slide.
5. Cover with a drop of acetocarmine.
6. If necessary, dissect further under a binocular microscope.
7. Cover with cover glass, heat, flatten and seal, or make permanent.

   NOTE: For smears of sporocytes, chlorazol black E may be used alone; but for somatic tissue it is strictly an auxiliary stain and should not be used by itself.

Results: When chlorazol black E is used alone for sporocytes: Chromosomes—black; background—clear. When used with carmine: Chromosomes—deep reddish black; cytoplasm—fairly clear.

Iron Propiocarmine-Hematoxylin

From Endrizzi (personal communication)

Application: Microsporocytes in smears of anthers.

Designed to show: Chromosomes of microsporocytes, especially useful for cotton.

Fixation: Acetic alcohol (3 vol glacial acetic to 7 vol of 95% alcohol). Material stored overnight in refrigerator and changed to fresh fixative the following day. Store in refrigerator.

Preparation of staining solution:
A. Iron Propiocarmine. Add 1.5 gm carmine to 100 ml boiling 45% propionic acid and boil about 2 min. Cool and filter. 3% $NH_4Fe(SO_4)_2 \cdot 12H_2O$ in 45% acetic acid. To individual dropping bottles of stain, add drops of iron solution until stain becomes burgundy in color, but without precipitate.
B. 1% aqueous hematoxylin.

Staining procedure:
1. Chop anthers in a drop of stain with spear-shape, iron needle.
2. Observe under low power for desired stage.
3. Remove debris, and quickly mix ¼ drop of hematoxylin and put on coverslip.
4. Heat gently, being careful not to boil.

Results: Cytoplasm—uncolored; chromatin—dark red, appearing black if green filter used.

Crystal Violet-Iodine

Original by Newton (1925)
Modified by L. F. Randolph and R. Snow (personal communications)

Application: Sections of root tips or anthers.
Designed to show: Chromosomes in mitotic or meiotic divisions.
Fixation: Navashin's, Flemming's or Randolph's chrom-acetic-formalin (CRAF) fixatives (see Chapter 1).

Embedded in: Paraffin

Staining schedule:
1. Remove paraffin from the sections in the usual manner, bring down to water.
2. Stain in 1% aqueous crystal violet for a few minutes to several hours.
3. Rinse in distilled water.
4. Place in iodine solution (1% iodine and 1% KI in 70% alcohol) for 1–3 min, until the color changes from violet to brown.

5. Rinse in absolute alcohol, 1 or 2 changes.
6. Differentiate in clove oil until reaching the desired contrast.
7. Pass through 3 changes of xylene, allowing 2–5 min in each change.
8. Mount in a synthetic resin.

NOTES: 1. If material has been fixed in an osmic fluid, it may need bleaching. If so, after bringing slides down to water bleach in 1–3% hydrogen peroxide or treat 2–3 min with 1% $KMnO_4$, rinse, and place in 1–5% oxalic acid until bleached. Rinse in water after bleaching, then proceed with staining.

2. The stain will come out of the material quite rapidly when the slides are placed in absolute alcohol. This step can be used to partially differentiate the material. Final differentiation is done in the clove oil, which removes the stain more rapidly from the cytoplasm than the chromosomes.

3. If a counterstain is desired, it can be dissolved in clove oil and dropped on the slide after differentiation is completed.

4. It is very important that all traces of clove oil be removed before mounting, otherwise rapid fading is likely.

Lacto-Propionic-Orcein

From Dyer (1963)

Application: Root tips and pollen mother cells of a wide range of plants.

Designed to show: Root tip mitotic chromosomes and pollen mother cell meiotic chromosomes.

Fixation: Modified Carnoy's fixative [alcohol, acetic acid, chloroform, formalin (10:2:2:1)] for mitotic chromosomes. No fixative for meiotic chromosomes; prepare from fresh pollen mother cells.

No embedding.

Preparation of the staining solution: Stock staining solution; dissolve 2 gm of natural orcein (note: synthetic orcein may be substituted) in 100 ml of a mixture of equal parts of lactic and propionic acids at room temperature. Filter. To use, dilute the stock solution to 45% with water.

Staining schedule: For root tip mitotic chromosomes, use Steps 1 to 5; for pollen mother cell meiotic chromosomes, Steps 4 and 5 only.
1. Immerse detached root tips for 2 hr in 0.2% colchicine.
2. Fix 5 min or longer if material is not readily penetrated.
3. Macerate in 1 N HCl at 60°C for 5 min.
4. Tap out in lacto-propionic orcein, 2–10 min.
5. Squash under a coverslip (after either warming or leaving to harden whichever is needed).

If the coverslip has been albumenized or subbed before placing it over the material, the preparation can be successfully made permanent. The coverslip, with attached cells, is floated off the slide in 45% acetic acid, and then taken rapidly through 60, 80, 95% and absolute alcohol to xylene before mounting in a synthetic resin.

Results: Chromosomes—dark red; cytoplasm—colorless or nearly so; nucleolus—lightly stained.

Propionic Acid Orcein

From correspondence with Frank H. Smith, Corvallis, Ore.

Application: Root tip smears.

Designed to show: Chromosomes.

Fixation: Farmer's solution—1 vol glacial acetic acid and 3 vol absolute alcohol. Carnoy's solution B—100% alcohol, 90 ml; glacial acetic acid, 15 ml; chloroform, 45 ml.

Preparation of staining solution: Dissolve 1 gm orcein in 100 ml of hot 45% propionic acid; cool and filter.

Staining schedule:
1. Kill and fix in the above Farmer's solution (see Iron aceto-carmine technic, pp. 342–343).
2. Place 5–10 min or longer in a mixture of equal parts of alcohol and conc. HCl at 60°C.
3. Transfer for 5–20 min or longer to the above Carnoy's solution.
4. Spread slide with a thin film of albumen.
5. Place a small drop of the above propionic-orcein on slide.
6. Cut thin (less than 0.5 mm) cross sections of root tips in the stain and crush with a flat scalpel. Add cover glass. Press or tap on cover glass until cells are spread. Do not use too much stain or cells will float out to the edge of cover glass.
7. Heat gently over flame. Do not boil.
8. If slide is satisfactory, temporarily seal or make permanent.

Results: Chromosomes—red; nucleolus—lightly stained.

Modified Carbol Fuchsin

From Kao (1975a, 1975b)

Application: Squashes of somatic (root tip, tissue culture or protoplast) cells or meiotic cells.

Designed to show: Chromosomes, interphase chromatin.

Fixation: Acetic alcohol (1:3 by volume) or Carnoy's B solution.

Preparation of staining solution:
  A. Prepare a stock solution of 3 gm basic fuchsin in 100 ml of 70%
     ethanol.
  B. Add 9 ml of the stock solution to 81 ml of 5% phenol in distilled water.
     Then add 12 ml of glacial acetic acid and 12 ml of 37% formaldehyde.
  C. Add from 2 to 10 ml of the solution in Step B to 90–98 ml of 45%
     acetic acid and 1.8 gm sorbitol. This solution seems to be stable
     indefinitely in stoppered bottles at room temperature. Aging the stain
     at least 2 weeks improves nuclear staining.

Staining schedule:
  1. For soybeans and alfalfa, root tips may require hydrolysis (20–90 sec)
     in 1:1 95% ethanol:concentrated hydrochloric acid at room temperature
     (overlong hydrolysis reduces staining intensity and can disintegrate the
     tissue). For other species and tissue culture cells, hydrolyze in 1 N HCl,
     60°C for 3–15 min. The hydrolyzed root tips are then rapidly blotted
     dry and placed in 70% ethanol for several minutes.
  2. Remove root tips, anthers, or tissue culture cells from 70% ethanol or
     fixative, blot off the excess fluid, place in a drop of the stain, and
     macerate or gently mash as appropriate.
  3. Put on a coverslip and heat gently (without boiling) before squashing.
     If the coverslip is then ringed with paraffin, the preparations can
     remain useable for weeks.

Results: Magenta to purple chromosomes or nuclei against a colorless
  background. The color is more purple than in Feulgen procedures. Nucleoli and starch do not stain, but some cytoplasms do (especially in
  embryo sacs).

HCl→Leucobasic Fuchsin→Light Green

(DeTomasi's Feulgen Stain)

From DeTomasi (1936)
Original by Feulgen and Rossenbeck (1924)

Application: Root tips, or other meristematic tissue.

Designed to show: Chromosomes.

Fixation: Flemming or CRAF. Success is also obtained following various
  other fixation methods.

  NOTE: Formalin produces intense staining if material is to be used for
  microspectrophotometry.

Embedded in: Paraffin

Preparation of solutions:

A. Dissolve 0.5 gm basic fuchsin by pouring over it 100 ml boiling distilled water. Shake thoroughly and cool to 50°C. Filter through 2 layers of No. 1 filter paper and add 15 ml N HCl to the filtrate. Add 1.5 gm $K_2S_2O_5$ (potassium metabisulfite) and allow solution to stand in a well-stoppered bottle in the dark overnight or until a light straw or faint pink color appears. If not completely decolorized, add 500 mg charcoal and shake for 2 min; filter. Repeat charcoal if necessary. Store in an amber bottle, or bottle wrapped in foil, in refrigerator (Schiff reagent).

B. Mix 5 ml of 10% aqueous $K_2S_2O_5$ with 5 ml N HCl and 100 ml distilled water. Prepare just before using.

Staining schedule:

1. Remove paraffin from the sections in the usual manner.
2. Rinse sections in cold N HCl.
3. Place in N HCl at 60°C for 4–5 min. Optimal times and concentrations differ for different materials.

   NOTE: Itikawa and Ogura (1954) used 5 N HCl at room temperature for 60–90 min and obtained better results.

4. Rinse in cold N HCl.
5. Rinse in distilled water.
6. Stain 3–5 hr (usually 5 hr) in Solution A.
7. Drain and pass quickly to the first of a series of three Coplin jars, each containing Solution B. Allow the slide to remain 10 min in each of these jars in succession, keeping the jars closed.
8. Wash in tap water for 10 min.
9. Counterstain 30–60 sec with 0.1% alcoholic light green SF yellowish. *Optional.* Do not do this if stained for microspectrophotometry.
10. Dehydrate.
11. Clear and mount as usual.

Results: Cell walls and cytoplasm—green; chromosomes—reddish violet.

Feulgen Squash Method

From Darlington and LaCour (1960)

Application: Root tips or anthers.

Fixation: Any of the alcoholic fixatives suitable for the aceto-carmine method. Material may be pretreated to condense chromosomes and arrest mitosis.

No embedding.

Preparation of staining solution: See HCl-Leucobasic Fuchsin-Light Green method.

Staining schedule:

1. Remove material from 70% alcohol or fixative solution, place in water until alcohol or fixative has been soaked out of tissue.
2. Hydrolyze and macerate the tissue by placing in 1 N HCl at 60°C for 4–8 min. The best time must be learned by trial, and might be longer or shorter than that given here.
3. Rinse out HCl with water, place material in a small quantity of the stain in a tightly stoppered vial. Staining may take from a few minutes to several hours.
4. Rinse in several changes of tap water, then squash material in 45% acetic acid as described for the iron aceto-carmine method.
5. Slides may be made permanent by the Dry Ice or vapor chamber method.

Results:

Chromosomes—bright magenta red (appearing blackish if a green filter is used); cytoplasm—colorless.

Also see Palmer and Heer (1973) for a root tip squash technic for soybean chromosomes.

C-Banding, Giemsa

From Gill and Kimber (1974a, 1974b)

Application: Somatic and meiotic metaphase chromosomes.

Designed to show: Constitutive heterochromatin and identify somatic and meiotic metaphase chromosomes.

Fixation: 1:3 acetic-alcohol or glacial acetic acid; time 12–24 hr. Prefix root tips in saturated monobromonaphthalene solution or in any other standard prefixative for 2–3 hr.

Preparation of solutions:

A. Enzyme solution: 500 mg pectinase + 500 mg cellulase + 10 ml distilled water + 6 drops of 1 N HCl. Store in refrigerator.
B. Barium hydroxide: 100 ml distilled water + about 5 gm barium hydroxide, shake vigorously and use immediately.
C. 2× SSC: 17.4 gm sodium chloride + 8.8 gm sodium citrate, make 1 liter with distilled water, adjust to pH 7.0 with 1 N HCl.
D. Giemsa stock solution: 1 gm Giemsa powder + 66 ml glycerine + 66 ml methanol. Dissolve Giemsa powder in glycerine at 60°C for 1 hr with constant stirring. Add methanol, continue stirring at 60°C for 24 hr. Store in refrigerator.
E. Citrate buffer: 2.1 gm citric acid + 100 ml distilled water = A. 14.2 gm sodium phosphate + 500 ml distilled water = B. Mix 4.55 ml of A + 15.45 ml of B for use as stock citrate buffer.

F. Giemsa staining solution: 5.0 ml filtered Giemsa stock solution + 1.5 ml methanol + 1.5 ml stock citrate buffer + 60–100 ml distilled water. Prepare fresh and skim the top before use. Source of chemicals: Cellulase (200 units/gm)—E. Merck, Catalog No. 2329. Pectinase (1.1 units/gm polygalacturonase)—Sigma Chemical Co., Catalog No. P-4625. Source of other chemicals is not critical.

Staining schedule:
1. Soften root tips in the enzyme solution until meristematic tips appear dark brown (0.5–2 hr), wash in water. The root tips may be hydrolyzed with 1 N HCl for 6–7 min at 60°C and squashed directly or can be followed by enzyme softening. For anthers, softening in 45% acetic acid for a few minutes should suffice.
2. Macerate root tips completely in 45% acetic acid on a slide, apply coverslip and press gently. Remove the coverslip on Dry Ice, soak slides in 95% alcohol (5 min–2 hr), air-dry.
3. Treat slides with saturated barium hydroxide solution for 5 min in a staining dish, wash in water for 10 min, 3 changes.
4. Place in hot 2× SSC buffer for 1 hr at 60°C. Wash in water, 3 changes.
5. Stain in Giemsa for appropriate time (30 sec to few minutes). Wash in water, air-dry. Overstained slides can be decolorized in 95% ethyl alcohol. Place in xylene overnight. Mount in a synthetic resin.

Results: Constitutive heterochromatin—dark bands; euchromatin—light staining.

Iron→Hematoxylin→Iron (Regressive Staining)

Long Schedule

Original by Heidenhain (1892)
Options for variations as proposed by Johansen (1940), Hutner (1934), and Lang (1936)

Application: Sections of root tips, stem tips, or anthers.

Designed to show: Chromosomes, centrosomes, and pyrenoids, etc.

Fixation: Chromic-osmic-acetic fluids (e.g. Flemming's)' or some combination of formalin with other substances, such as Lewitsky's (formalin, 1:1, and 5% aqueous CrO_3 in ratio of 4:1 or 1:1), or CRAF; see Chapter 1.

Embedded in: Paraffin.

Preparation of staining solution. One of the following is recommended:
A. 0.5% aqueous hematoxylin. Ripen for a few days or weeks. May be reused if refrigerated.

B. A 10% stock alcoholic solution of the same dye, diluted 20 times with distilled water before use.

C. Johansen's solution, which is stable: 5 ml 10% alcoholic hematoxylin, 100 ml Methyl Cellosolve, 50 ml distilled water, 50 ml tap water which contains calcium salts.

Staining schedule:

1. Remove paraffin from sections and bring down to water.

   NOTE: Omit Step 2 unless necessary. See comments under NOTE below.

2. Bleach in equal parts of 3% H_2O_2 and distilled water; or proceed as follows (bleaching is needed only if material has been fixed in a mixture containing osmic acid).
   a. Treat 2–3 min with 1% aqueous $KMnO_4$.
   b. Rinse under running water.
   c. Treat 1 min or less with 5% aqueous oxalic acid, until bleached. Wash in water.

3. Mordant 2 hr in fresh 4% aqueous $NH_4Fe(SO_4)_2 \cdot 12H_2O$. (If 1% glacial acetic acid and 0.1% conc. H_2SO_4 are added to this solution, it is more stable, and often gives greater staining contrast). A mordant is a substance that combines with a dye to form an insoluble compound.

4. Wash 5–10 min in running water.

5. Rinse in distilled water.

6. Stain 1–24 hr in one of the above hematoxylin solutions.

7. Rinse in tap water.

8. Destain in 2% aqueous $NH_4Fe(SO_4)_2 \cdot 12H_2O$ checking by examination under the microscope.

   NOTE: An acidic destainer (e.g. picric acid) bleaches the cytoplasm more rapidly than the nucleus and gives greater contrast between them. [An oxidizer, e.g. 30% hydrogen peroxide (Merck's Superoxol), 1 vol in 2 vol alcohol, freshly mixed is preferable when it is desired to retain some stain in cytoplasmic structures.] The action of $NH_4Fe(SO_4)_2$ is intermediate between these. Bleaching tends to loosen the sections from the slides. Sections may be retained by bringing the slides to absolute alcohol and dipping them into a solution of 0.5% celloidin in equal parts ethyl ether and absolute alcohol. After dipping, expose to air and allow the slides to become almost dry and then dip into alcohol and proceed with staining. After staining, dehydrate in an alcohol series and absolute alcohol, then remove the celloidin with a mixture of absolute alcohol and ether and clear in xylene and mount in a synthetic resin.
   Picric acid caution: The crystalline substance is highly explosive and should not be stored on the shelf for any length of time. The solution is relatively safe.

9. Wash 1 hr in running water (incomplete washing results in fading).

10. Counterstains:
    a. Safranin (see p. 324).
    b. Fast green (see p. 329).
    c. Orange G (see p. 330).
11. Dehydrate in several grades of alcohol (25%, 50%, 75%, 95%), with 2 or 3 changes of absolute, allowing several minutes in each grade. Addition of a drop of ammonia in one of the alcohol baths may improve the color.
12. Transfer to equal parts of absolute alcohol and xylene.
13. Place slide in xylene, 2 or 3 changes.
14. Mount in a synthetic resin.

Results: Chromosomes—black or very dark blue; centrioles—black or very dark blue; cytoplasm—gray or colorless.

In situ rRNA/DNA Hybridization and Autoradiography

From Phillips et al. (1979)

Procedure:

1. Fix freshly collected microsporocytes in 3 parts 95% ethanol and 1 part glacial acetic acid for 1–3 days at room temperature.
2. Use one of the anthers in a floret to identify appropriate meiotic stage and keep remaining anthers in a vial containing fixative.
3. Squash 3–5 anthers on an acid-cleaned slide in a drop of 45% acetic acid and add an acid-cleaned coverslip.
4. Keep the slides moist in a petri dish and store in a refrigerator overnight (16 hr).
5. Place slides on Dry Ice for 5–10 min. Use a razor blade to flip off the coverslip.
6. Keep slides in 2× SSC (1 × SSC: 0.15 M NaCl and 0.015 M Na citrate, pH 7.4).

Denaturation of DNA:

1. Dissolve RNase in 2× SSC (0.2 mg/ml) and heat at 80°C for 10 min.
2. Digest the meiotic cells with the heated RNase for 2 hr at room temperature (or 1 hr at 37°C) to remove endogenous RNA which may compete with hybridizing rRNA.
3. Remove the RNase by washing 3 times with 2× SSC.
4. Treat the slides with 0.2 N HCl for 20 min at room temperature.
5. Wash the denatured slides 3 times with 2× SSC.

rRNA-DNA Hybridization

1. Place 10 μl of ^{125}I-rRNA on each slide (conc.: 0.25 μg/ml with 10 μg *Escherichia coli* RNA as carrier.

2. Cover the slides with an acid-cleaned cover glass and place in a pan containing 2× SSC and support the slides with glass rods.
3. Seal the pan with aluminum foil and keep in 60°C oven overnight (16 hr).
4. Float the cover glasses off the slides with 2× SSC and wash 3 times with 2× SSC.
5. Digest with 80°C heated RNase (0.2 mg/ml) for 2 hr at room temperature or at 37°C for 1 hr.
6. Wash the slides 3 times with 2× SSC and keep in 2× SSC.

Autoradiography:
1. Melt Kodak NTB-2 emulsion (112 ml) at 45°C and dilute with 2 vol of distilled water in absolute darkness.
2. Pour 30 ml of the diluted emulsion in a small container and save the remainder for future use.
3. Surface-dry the slides (cells are still wet) and dip into the emulsion and withdraw slowly. Place slide vertically in a rack.
4. Place the emulsion-coated slides in a light-tight plastic slide box, seal the box with black electrical tape, wrap with aluminum foil, and store in a desiccator at 4°C to expose.
5. After proper exposure time (2–4 days), develop the slides in Kodak D-19 developer for approximately 1 min and rinse 3 times with distilled water.
6. Fix slides in Kodak fixer for 2 min and wash with distilled water for 30 min through several changes of water.
7. Stain the slides with 5% Giemsa stain (Harleco) before drying (dilute the Giemsa solution with 0.01 M phosphate buffer at pH 6.8).

Results: Dark blue chromosomes with silver grains over nucleolus organizer region.

Method for Making Aceto- or Propiono-Carmine Smears Permanent

Modified by C. R. Burnham (1967) from McClintock

This method works best after the slides are a few days old.

The steps are:
1. Remove the temporary seal with a razor blade.
2. Immerse, right side up, in 10% acetic acid in a petri dish. The solution should soak under the coverslip and loosen it. With this method, most of the pollen mother cells are on the coverslip.
3. Run coverslip and slide through the following solutions in Coplin jars, keeping them in the same position relative to each other:
   a. 10 ml 95% alcohol plus 30 ml 45% acetic acid.
   b. 20 ml 95% alcohol plus 20 ml 45% acetic acid.

c.*20 ml absolute alcohol plus 20 ml glacial acetic acid.
d. 30 ml absolute alcohol plus 10 ml glacial acetic acid.
e.*36 ml absolute alcohol plus 4 ml glacial acetic acid.
f.*40 ml absolute alcohol.
g.*40 ml absolute alcohol.
h.*40 ml xylene (2 changes).
* These steps alone have been satisfactory for corn and barley.
4. Add a synthetic resin of proper consistency to the slide and quickly add the coverslip in its original position, if possible, to retain the original vernier readings that may have been recorded for certain cells.

Dry Ice Method for Making Squash Preparations Permanent

Slightly modified from Conger and Fairchild (1953)
Also see Bowen (1956)

Application: Squashes or smears of anthers or root tips.

Treatment:

1. Lay slide flat on a block of Dry Ice, coverslip up, or use liquid CO_2 (Bowen, 1956) until thoroughly frozen.
2. Remove any sealing material from around the edge of the coverslip with a razor blade (this usually can be done before freezing), then slip the edge of the blade under one corner and pop the slip off.
3. Immediately immerse the coverslip and the slide in 95% ethanol, followed by absolute ethanol and 2 changes of xylene. Leave slide and coverslip in each solution for several minutes.
4. Drain slide and slip briefly, place a drop of a synthetic resin on the slide, and carefully replace the coverslip.
5. Dry on a warm plate.

   NOTES: 1. If the coverslip is first albumenized *very lightly*, squashed material will adhere to it instead of to the slide. In this case only the coverslip need be saved for mounting.
   2. Instead of freezing with Dry Ice, the entire slide can be plunged into liquid nitrogen until frozen. Subsequent treatment is the same.

Vapor-Chamber Method for Making Squash Preparations Permanent

From Brooks, Bradley, and Anderson (1959)

Application: Squashes or smears of root tips or anthers.

Treatment:
1. Place slide in a tightly closed jar with 95% or 100% alcohol, which is lined with absorbent paper. Sufficient excess alcohol should be present to form a shallow layer on the bottom of the jar, but not so much as to

reach the lower edge of the coverslip. Allow slide to remain in the jar for 4–25 hr.

2. Remove slide from jar and carefully run a few drops of 100% alcohol under the coverslip while holding the slide at a tilt. Absorb excess alcohol with blotting paper.

3. Apply 1–3 drops of Euparal to two opposite edges of the slide and place slide upright on glass rods in a second vapor chamber. This chamber is conveniently made from a plastic sandwich box with absorbent paper on the bottom which has been dampened with 100% alcohol. (The atmosphere should be saturated with the alcohol vapor, but there should be no free liquid.) Leave slide until the Euparal has thoroughly diffused into the material under the coverslip.

4. Remove slide to warm, dry place to harden the mountant. Excess mountant can be removed from the slide with a rag dampened with absolute alcohol.

NOTE: This method is exceptionally useful when removal of the coverslip might result in excessive loss of materials or distortion of cells. The vapor exchange method does not work well for corn.

Applying Adhesives to Slides

Subbed slides: Mix 5.1 gm gelatin/1000 ml boiling distilled water and 0.5 gm chromalum. Dip slides in cooled solution and let dry in a dust-free area.

Mayer's albumen slide fixative: Mix 25 ml albumen, 25 ml glycerine, and 0.5 gm sodium salicylate. Filter before use (see also p. 24).

Pretreatments

It may be desirable to precede fixation with a chromosome constrictor and mitotic arresting chemical treatment. Chemicals include saturated solution of monobromonaphthalene, 0.01–0.2 % colchicine, 0.001–0.004 M 8-hydroxyquinoline, saturated solution of paradichlorobenzene, 0.01–0.02% cycloheximide. Cold water treatments (4°C) can be used with some material.

The concentration of the chemical solution and treatment time and the temperature during pretreatment should be modified depending on the materials to obtain best results. Pretreatment by ice cold water has been and is used for the same purposes (Tsunewaki and Jenkins, 1960) as monobromonaphthalene and/or for inducing differential staining of the chromosomes (Darlington and LaCour, 1938, 1940).

The duration of the pretreatment in the ice cold water also should be adjusted depending on the materials. For example, wheat, rye and similar cereals need pretreatment for 24 hr or more, while 15–17 hr pretreatment gives the best results for barley and 10–12 hr is enough for sugarbeets.

Temporary Seal

From Burnham (1967)

Dahl's varnish-beeswax-paraffin:
½ part Turtox Ringing Varnish
1 part beeswax
2 parts paraffin (parowax)
Apply to the 4 edges of the coverslip, using an L-shaped wire that replaces a dissecting needle and can be heated.

Enzyme Treatment

Some type of enzymatic treatment may be useful. Pectinase is often used. **To prepare:** 100 mg Bacto-peptone, 500 mg pectinase, 10 ml H_2O, adjust pH to 6, incubate 30 min at 30°C. Filter and freeze in small aliquots.

CHAPTER 15

POLLEN AND POLLEN TUBES

Ronald L. Phillips

This chapter is concerned with preparing pollen tubes for staining as well as the actual staining.

SECTION A: POLLEN

Methyl Green in Glycerin-Jelly

From personal correspondence with Clyde Chandler, N.Y. Botanical Garden
Similar to Wodehouse (1935, p. 107)

Application: Pollen grains.

Designed to show: Exine of pollen-grain wall.

No fixation or embedding.

Preparation of staining solution: Add saturated (about 3%) solution of methyl green in 50% alcohol, drop by drop, to melted glycerin-jelly (a good cŏmmercial grade is satisfactory), until the jelly becomes as dark as green ink.

Staining schedule:
1. Place dry pollen on slide.
2. Add drop of alcohol to dissolve the oily material of the pollen.
3. Wipe off oil rim from slide.
4. Add drop of the above methyl green-glycerin-jelly, melted.
5. Stir in pollen with a needle.
6. Keep jelly melted.
7. Warm a cover glass and place over specimen.

Results: Exine of pollen grain—green.

Cotton Blue (Aniline Blue)—Lacto Phenol

From Uhl (personal communication)

Application: Pollen grains.

Preparation of solution:
  A. Prepare lacto-phenol by mixing:
      1 part phenol crystals
      1 part lactic acid
      1 part distilled water
      2 parts glycerol (glycerin).
  B. Prepare a saturated solution of the dye in 95% ethanol and add 1 part
      of this to 3 parts of lactophenol. (Synonym of the dye is complicated.
      Aniline blue is probably a mixture of several related substances. The
      more purified products are marketed as water blue or methyl blue.
      Probably either of these would serve equally well.)

Staining schedule: Collect some fresh, dry pollen and add it to a small
drop of the staining solution on the slide. The whole anther may be
macerated with a needle or forceps, and the large bits of anther wall
discarded before adding the cover glass. The edges of the preparation may
be sealed with a plastic sealant, such as "Zut," and apparently keep
indefinitely in this condition. Thin "Zut" with amyl acetate, if necessary.

Results: Pollen grains—blue.

I_2-KI

From Burnham (1967)

Application: Pollen grains (starch); especially useful for determining percentage of pollen abortion.

Preparation of solution: Mix 0.3 gm I_2, 1.0 gm KI, and 100 ml H_2O. Stain
may be diluted several fold for pollen sterility classification. Glycerine
added to the stain will serve to partially immobilize the pollen grains.

Staining schedule:
  1. For maize, a 12–15 cm tassel section that includes a portion that has
      not extruded anthers but also a portion with extruded anthers, is
      removed from the plant and tagged with the proper identification. Store
      in 70% alcohol.
  2. Pollen from an anther is teased out into a drop of I_2-KI solution. For
      maize, pollen from one-third of an anther is teased out into a very small
      drop of stain and a small coverslip added. This may be ringed with
      mineral oil to prevent drying. A circular disc of note card cut to fit on
      the shelf in the ocular and with a rectangular opening cut out gives a
      field with parallel sides.

Results: Pollen—black; aborted pollen—empty (no staining) and smaller, or partially filled with black staining starch.

Certain technics presented in previous chapters also may be used with pollen, especially the Aniline Blue WS and Fluorescence Microscopy technic (Chapter 12).

SECTION B: POLLEN TUBES IN STYLES

Acetocarmine and Basic Fuchsin

From Chandler (1931)

Application: Pollen tubes in pistils, with either solid or hollow styles.

Designed to show: Cytoplasm of pollen tubes.

Fixation: Formalin, 6–7 ml; 70% alcohol, 100 ml. Pistils in this fluid may be examined promptly, or stored for a considerable period.

Method of dissection:
1. Place entire style on glass slide; split the tube longitudinally, with two very fine needles, along one side, to the central canal, beginning at the stigmatic end and cutting down.
2. Spread cut surfaces apart until they lie flat on the slide.

   NOTE: If pollen tubes are found near the base of the style, the ovary may be sectioned free-hand, by using a razor and a piece of pith. Each section may be placed on slide and stained.

Staining schedule:
1. Place a drop of aceto-carmine (warm 95% acetic acid saturated with carmine, approximately 0.2%).
2. Add drop of 3% aqueous basic fuchsin.
3. Draw off excess stain with blotting paper.
4. Destain with absolute alcohol by absorbing it at the basal end with filter paper, old linen cloth, or blotting paper.
5. Add several drops of glycerin and mount.

Results: Cytoplasm of pollen tubes—dark red; surrounding tissue— lightly stained.

Lacmoid and Martius Yellow

From Nebel (1931)

Application: Pollen tubes in the style or on agar.

Designed to show: Callose in pollen tubes.

No fixation or embedding.

Preparation of staining solution: Dissolve 5 mg resorcin blue (often called lacmoid) and 5 mg martius yellow in 10–15 ml distilled water. Add a few drops of 1:100 NH₄OH to bring the reaction to about pH 8, as shown by the solution assuming an olive shade.

Staining schedule:
1. Crush slender styles or ovaries, while still moist, between two slides; with larger styles or ovaries, section longitudinally by hand, and then crush.
2. Stain 2–5 min in the above staining fluid, either on the slide, or by immersing small pieces in the fluid in a small dish.
3. Mount in the stain, or in water.
4. Examine with a powerful light.

Results: Pollen tubes—blue; background—light yellowish green.

SECTION C: POLLEN TUBES IN CULTURE

Crystal Violet→I₂KI→Orange II

From Newcomer (1938)

Application: Pollen tubes in culture.

Designed to show: Chromosomes.

Method of growing pollen tubes: Boil 0.5 gm agar and an optimum quantity of sugar (2 gm for *Amaryllis belladonna*, 0.5 gm for *Begonia* sp.) in 25 ml tap water or Hoagland's solution. Cool to about 35°C and add 0.5 gm powdered gelatin. Stir till gelatin is melted. (If not used immediately, keep the solution on a warming plate.) Smear this with the finger in a thin film on a warmed slide and dust on the pollen. If the pollen is sticky and falls in clumps on the slide, it can be spread evenly by smearing it with another drop of the mixture. Place slides in moist chamber for germination.

Fixation: Kill in any suitable fixative. Killing for 8 hr or overnight and washing for 2–4 hr in cold running water seems adequate.

Preparation of solution: Iodine-potassium iodide: iodine, 1 gm; potassium iodide, 1 gm; 80% alcohol, 100 ml.

Staining schedule:
1. Place slides 2–3 min in 1% KMnO₄.
2. Rinse in tap water.
3. Bleach 1–3 min in 5% oxalic acid.
4. Wash 15 min in tap water.
5. Mordant 20 min in 1% chromic acid.
6. Rinse in tap water.

7. Rinse in 2 or 3 changes of distilled water.
8. Stain 4 hr in 1% aqueous crystal violet.
9. Treat 1–2 min with the above I-KI solution, or until the color of the pollen tubes changes from blue to brown.
10. Rinse in alcohol.
11. Counterstain 2–4 min with 1% orange II in clove oil.
12. Rinse in 2 changes of absolute alcohol.
13. Clear in xylene.
14. Mount in balsam.

Results: Chromosomes and nuclei—blue.

NOTE: This technic has been successfully used for germinating fungi spores and for cytological observations on mycelium.

Acetocarmine

From Swanson (1940)

Application: Pollen tubes of *Tradescantia* L. in culture.

Designed to show: Chromosomes.

Method of growing pollen tubes: Essentially the same as Newcomer (1938). The medium consists of 2 gm cane sugar, 0.5 gm agar, and 0.5 gm gelatin in 25 ml of tap water. Bring sugar, agar and water to a boil; and as medium cools somewhat, add the powdered gelatin slowly, stirring to prevent clumping of the gelatin.

Preparation of staining solution: Belling's aceto-carmine. Add 1 gm carmine to 100 ml boiling 45% acetic acid and boil about 2 min. Cool and filter. To half of this add a few drops of a solution of ferric hydroxide in 45% acetic acid until the liquid becomes bluish red, but without visible precipitate; then add the rest of the untreated aceto-carmine.

NOTE: If the iron is omitted, and preparations are made with clean steel needles, little aeration, and not much heat, initial staining is light, but darkens for 2–3 days and remains in good condition several weeks when stored at 2–4°C.

Growing the tubes:
1. Smear the medium on the slides while still hot, preparing 2 slides at a time. This allows the first slide to cool sufficiently to permit sowing of the grains by the time the second is prepared.
2. Place the slides in a horizontal type staining dish which was previously prepared by placing moist filter paper on the underside of the cover and on the bottom of the dish. (The paper should be *moist* and not wet.) Scatter acenaphthene crystals (Eastman Kodak No. 597—m.p. 93–4°) on the bottom of the dish. Then place staining dish in a moist chamber at room temperature. Length of time for generative nucleus

to reach metaphase varies: *Bellevalia* Lapeyr, 3½ to 4 hr; *Tradescantia*, 17–22 hr; *Anemone*, 38 hr.

Fixation:

3. Fix in acetic-alcohol (70 ml of alcohol to 30 ml glacial acetic acid); or other fixative.

Staining:

4. Stain in Belling's aceto-carmine, or other stain (orange G not recommended).
5. Apply firm pressure on the coverslip.

Results: Number of figures can be obtained with all chromosomes on the same focal plane.

CHAPTER 16

PATHOLOGICAL ANATOMY AND MYCOLOGY

Henry Schneider

The methods given in this chapter are different from those in the preceding chapters because they are designed to stain fungi, bacteria, and viruses. However, some schedules in other chapters are also useful for these disease causing agents. Among these are Weigert's premixed iron hematoxylin or iron and hematoxylin used progressively with a suitable counterstain. For phloem diseases a counterstain of lacmoid is desirable to show wound callose, while safranin selectively stains wound cork and wound gum. Thionin (see p. 327) or Crystal Violet-Iodine (p. 347) may also be used.

Vas-par is used in several schedules for sealing the edges of cover glasses. It is prepared by mixing equal parts of Vaseline and melted paraffin in a beaker. It may be applied with a metal rod about ⅛-in diameter bent L-shaped so that there is a 6-in handle and a 1-in applicator. Warm the applicator over a burner, then melt and pick up a few drops of vas-par and apply to coverslip edges.

Acid Fuchsin-Lactophenol

From Riker and Riker (1936)
(Davis 1924)

Application: Fungal infected plant tissue and fungi from culture.

Designed to: Stain fungi.

Fixation: Fresh or dried materials, or any botanical fixative.

Embedded in: Paraffin, freehand sections, or whole mounts.

Preparation of solutions: Lactophenol (Davis, 1924)
Distilled water   20 ml

Carbolic acid crystals (warm until melted)	20 ml
Lactic acid	20 ml
Glycerin	40 ml

Acid fuchsin-lactophenol staining solution: lactophenol with 1% acid fuchsin added.

Staining schedule:
1. Place material on a slide and add several drops of acid fuchsin-lactophenol. Warm gently until fumes are given off.
2. Pour off stain and put on clear lactophenol to remove excess stain. Warm gently as before.
3. Mount in lactophenol and seal cover glass with finger nail polish, lacquer, or vas-par.

Results: Fungus spores and hyphae stain bright red. Host protoplasm stains less intensely.

Acid Fuchsin-Lactophenol→Fast Green

From Myers and Fry (1978); Johansen (1940)

Application: Counterstain for the above Acid Fuchsin-Lactophenol.

Designed to: Differentiate between fungal structures on leaf surfaces and those that are subcuticular or intracellular.

Fixation: Absolute Alcohol (2 parts)—Glacial acetic acid (1 part).

Embedded in: Whole mounts.

Preparation of solutions:
I. Solvent for Fast Green (Johansen, 1940):

Absolute alcohol	100 ml
Methyl cellosolve	100 ml
Clove oil	100 ml

II. Fast Green staining solution: Solvent with 0.5% Fast green added.

Staining schedule:
1. Fix leaves 24–48 hr and then clear 24–48 hr in Lactophenol (see previous schedule).
2. Immerse tissue in Acid Fuchsin-Lactophenol in a small petri dish or other container (see previous schedule).
3. Remove tissue from acid fuchsin when fungal hyphae are pink-red and before plant tissue becomes overstained. (One hour was sufficient for sorghum leaves infected by various leaf-spotting fungi.)
4. Place tissue on bibulous paper and *gently* blot to remove excess stain. Dip tissue in lactophenol briefly to remove pools of stain remaining on the tissue surface.

5. Place tissue in fast green solution for 1–5 min. Examine at 30-sec intervals. Surface fungal growth should be stained green; however, a minimum area of plant tissue or internal fungal tissue may be stained green also.

6. Remove immediately from fast green staining solution. Blot quickly and carefully as before and dip in lactophenol *very* briefly to remove pools of stain on the tissue surface.

7. Mount on slides under a coverslip in a solution of glycerol, distilled water, and Tween 20 (90:9:1, v/v). Seal coverslip if desired.

Results

1. Subcuticular or intracellular fungal hyphae pink—red.
2. Host protoplasm a less intense pink or unstained.
3. Surface fungal structures including spores, germ tubes, and various types of appressoria—green to blue-green.
4. Penetration sites can be observed easily at low magnification.

Acridine Orange and Fluorescence Microscopy

See also Chapter 3
From Hooker and Summanwar (1964)
And personal communication with W. J. Hooker

Application: Virus infected plants.

Designed to show: RNA in virus infected tissue. Potato virus X, tobacco etch virus, potato virus Y, tobacco necrosis virus.

Fixation: 50% ethyl alcohol.

No embedding: Fresh cross sections of leaves or paradermal sections of the leaf epidermis.

Staining solution: Acridine orange stock solution: 0.1% in distilled water; $M/15$ phosphate buffer: pH 6.3 (see buffer tables, pp. 2–8).

Staining schedule:

1. Place freehand sections of turgid fresh tissue from lower surface of virus infected leaves into 50% alcohol for 20 min.
2. Rinse in 2 changes of distilled water (10 min each).
3. Dilute the 1% aqueous acridine orange stock solution 1:10 with the pH 6.3 phosphate buffer. Stain 5 min.
4. Two distilled water rinses of 3–4 min each.
5. Place in 0.1 M $CaCl_2$ (5 min).
6. Two distilled water rinses of 2–3 min each.
7. One change of phosphate buffer. Mount in buffer.
8. Examine with ultraviolet light between 350 and 480 mμ with a peak between 390 and 420 mμ. Use a yellow barrier filter (Kodak Wratten

K2 No. 8) between objective and eye pieces. Observe at once as stain intensity fades rapidly.

Results: x bodies and striate bodies—red; nucleus—greenish white; muco proteins—red.

Aniline Blue W.S.. (Cotton Blue)

From Rawlins (1933)

Application: Fungus spores and mycelium, bacteria, and organisms in tissue.

Designed to show: Fungus mycelium and fruiting bodies in relation to plant cells.

Fixation: AFA (100 ml, 50% alcohol; 6.5 ml, formalin; 2.5 ml, acetic acid).

Embedded in: Paraffin or freehand.

Preparation of staining solution:

Phenol	10 gm
Glycerin	10 ml
Lactic acid	10 ml
Distilled water	10 ml

Aniline blue (Cotton blue): 0.02 to 0.05 gm
Place in a dropper bottle.

Preparation of mounting medium: As for the staining solution, but omit the aniline blue.

Staining schedule:
1. Fix material 24–48 hr before sectioning, or 20 min afterward.
2. Wash in several changes of water for 20 min.
3. Stain in the above solution 10–15 min.

   NOTE: Warming intensifies staining.

4. Mount in the same solution or in the mounting solution.
5. Seal coverslip with vas-par.

Results: Fungus mycelium—blue; protoplasts of cells—blue.

Azure B For X Bodies in Plant Virus Infections

From Bald (1964) and personal communications

Application: Tobacco mosaic virus (TMV) in leaf tissues of *Nicotiana tabacum* L. "Turkish," and other virus diseased plants.

Designed to show: Plant-virus ribonucleoproteins.

Fixative: Stock solution:

Potassium iodide	4.5 gm
Calcium iodide	3.2 gm
Iodine	4.5 gm
MgCO$_3$	(in excess of saturation)
Water	100 ml

Embedded in: Freehand paradermal sections or strips from the lower surface of TMV affected leaves.

Preparation of solutions:

1. Polyvinyl alcohol (PVA), stock solution:

PVA (Histochemical grade[1])	40.0 gm
H$_2$O	100.0 ml
Propylene glycol	0.5 ml
Molybdic acid	0.05 gm

First mix the PVA and water at 80°C with a mechanical stirrer. Then add the other ingredients.

2. Molybdic acid: 2.5% aqueous.

Fixing and staining schedule: See Steps I–VII in table below. Use continuous agitation for all steps.

Step	Mixture used (ml of each)				Procedure and Time
	PVA	H$_2$O	Alcohol	Other Ingredients	
I	6	—	10	4 ml fixative	Fix 30 min
II	6	7	7	—	Rinse 5 min
III	6	7	5	2 ml molybdic acid; 0.1 gm Na$_2$S$_2$O$_3$	Rinse 15 min to remove I$_2$
IV	6	11	3	—	Rinse 10 min
V	6	12	2	0.1 gm Azure B[2]	Stain 5 min
VI	6	13	1	—	Rinse 1 min
VII[3]	6	7	—	—	Mount

1. See Altman (1971) and Polysciences catalog.
2. Place 0.1 gm of dry azure B (C.I. 52010) in a mortar and gradually add the PVA-H$_2$O-alcohol mixture, triturating all the while with a pestle.
3. Ring slides with vas-par or lacquer and store in a cool humid place to prevent drying out.

Results: X bodies—greenish changing to blue; nuclei—violet to purple; nucleoli—greenish to blue; muco proteins in apical hair cells—green.

NOTE: Only cells that are not damaged in the stripping process should be observed. PVA seems to support and stabilize the cellular organelles and cytoplasm.

Premixed Martius Yellow, Malachite Green and Acid Fuchsin (Pianese IIIB stain)

Details from L. W. Sharp (personal communication), adapted from Vaughn (1914)

Original (applied to cancer tissue) by Pianese (1896)

This schedule was dropped because of its vagueness as to the pathogenic fungi used and the kinds of tissues involved. Also it is likely that the malachite green used by Vaughn was different from currently available malachite green. We tried C.I. 42000. Cert. No. LMg-14 and it stained lignified walls and chromatin but not cellulose walls. The stain used by L. W. Sharp was said to stain both kinds of walls. When used alone, Martius yellow did not stain any part of the host tissue but it stained spores of *Melampsora lini* in pustules on flax leaves. When a mixture of the three stains was used, the yellow in spores was masked by acid fuchsin or malachite green. Martius yellow could affect the staining by malachite green or acid fuchsin; but more likely, it serves no purpose. Indications are that Plant Pathologists do not use Pianese IIIB. The procedure may be found in the 2nd and 3rd editions of *Staining Procedures*.

Safranin→Picro=Aniline Blue

From Rawlins (1933)

Original by Cartwright (1929)

Application: Plants with verticillium wilt, fusarium wilt, or other diseases with fungi in the xylem.

Designed to show: Fungi and bacteria in lignified tissue.

Fixation: Formalin-aceto-alcohol.

Embedded in: Paraffin, celloidin, or freehand.

Preparation of staining solution: Picro-aniline blue: mix 25 ml saturated aqueous aniline blue with 100 ml saturated aqueous picric acid.

Staining schedule:
1. Bring sections down to water.
2. Stain for approximately 1 min in 1% aqueous safranin.
3. Wash out excess stain in water.

4. Flood section with picroaniline blue solution. Heat slide over a flame until on the point of simmering.
5. Wash with distilled water until blue color is no longer apparent in sections.
6. Dehydrate in 30, 50, 70, and 95% alcohol.
7. Clear in clove oil.
8. Mount in balsam.

Results: Fungus mycelium stains blue, lignified host cell walls stain red.

Thionin and Orange G

From Conn (1953, p. 333)
Similar to Stoughton (1930)

Application: Plant pathological material.

Designed to show: Fungi and bacteria in tissue.

Fixation: Any of the usual botanical fixatives.

Embedded in: Paraffin.

Staining schedule:
1. Remove paraffin from the sections in the usual manner.
2. Stain 1 hr in 0.1% thionin in 5% aqueous phenol.
3. Dehydrate in successively stronger alcohols.
4. Differentiate in a saturated solution of orange G in absolute alcohol.
5. Wash in absolute alcohol.
6. Clear and mount in balsam.

Results: Parasites—violet-purple; cellulose walls—yellow or green; xylem and chromosomes—blue; spindles—purple.
See also:

4'-6-Diamidino-2-phenylindole (DAPI), a Fluorescent Stain for Mycoplasma and DNA
(See p. 337)

Phloroglucinol in HCl for Wound Gum
(See p. 338)

FeSO₄ for Tannins
(See p. 338)

MICROBIOLOGY

CHAPTER 17

STAINS FOR MICROORGANISMS IN SMEARS

James W. Bartholomew

The vast majority of the staining procedures applied to microorganisms involve the use of smears on glass slides. Usually the smears are fixed simply by passing the slide, smear side up, five or six times through the flame of a Bunsen burner, and such heat fixation is acceptable for most staining procedures. Heat fixation, however, is not adequate when fine cytological detail is needed or delicate cells are to be stained, such as for the demonstration of the nuclear apparatus of bacterial cells, or for the demonstration of *Mycoplasma* cells. When acceptable, heat fixed smears can be stored for a reasonable length of time before staining, but they cannot be stored indefinitely as the cells slowly deteriorate with time.

SECTION A: THE DIFFERENTIATION OF MICROORGANISMS IN SMEARS BY THE GRAM STAIN PROCEDURE

The Gram stain procedure differentiates cells into two groups, those which retain the primary dye and are called Gram-positive, and those which lose the primary dye and take the color of the counterstain, which are called Gram-negative. Any microorganism which stains Gram-positively at any time should be placed in the group of Gram-positive organisms even though it may stain Gram-negatively at some early or late stage of its growth cycle. Most eukaryotic cells are Gram-negative, with the important exception of yeasts which are Gram-positive; because of this the Gram characteristic is not of great taxonomic significance for eukaryotic cells as a group. The Gram characteristic is of great taxonomic importance, however, for prokaryotic cells where it is often the first step used in the identification of an unknown organism. Among the bacteria, all of the cocci (with the exception of *Neisseria* and *Veillonella*), all of the spore-forming bacilli, and a small group of non-spore-forming bacilli, are Gram-positive; all of the spirilla, rickettsiae, and chlamydiae, as well as a very large group of non-spore-forming bacilli, are Gram-negative.

The Gram-positive characteristic results from the ability of a cell to resist the loss of the primary stain during the decolorization step, as compared to Gram-negative cells which do not possess this ability. Since the vast majority of cell types are Gram-negative, the unique characteristic to be explained is the dye retention ability of the Gram-positive cells. Most researchers who have studied the mechanism of Gram stain differentiation have concluded that the mechanism is based on the distinctive chemistry and physical structure of the cell walls of Gram-positive organisms. However, the exact mechanism of Gram stain differentiation is still not known, and various different concepts exist (for a discussion see Bartholomew, Cromwell, and Gan, 1965).

Gram differentiation is a quantitative as well as a qualitative procedure. There are many points at which the technic used influences the final result, such as the thickness of the smear, the nature of the washing procedures, the concentration of the crystal violet and the iodine used, whether a wet or a dry smear is subjected to decolorization, as well as the nature of the decolorizer and the manner in which it is used (Bartholomew, 1962). Despite all of these variables, the procedure is sufficiently rugged to allow a considerable leeway in technics used and it will still achieve valid differentiations. It should not be forgotten, however, that allowable limits of technic variation do exist. One simple control as to the validity of a varied technic would be to include organisms of known Gram characteristic on the slide along with the unknown organisms.

It is not acceptable to make extensive changes in the usual Gram stain procedure, and then to subsequently term Gram-positive any organism which retains the color of the primary stain. For example, it would not be valid to use heat to enable crystal violet to stain acid-fast organisms, then to complete the Gram procedure and to conclude that all *Mycobacterium* species are Gram-positive since they retain the color of the crystal violet (Kretschmer, 1934). In such a case a test for validity of the apparent Gram-positive state would be to repeat the staining procedure but to omit the iodine step. If the organism still retains the color of the primary stain, then such dye retention reflects a cellular or staining characteristic other than the true Gram positive characteristic (Bartholomew, Mittwer, and Finkelstein, 1959). This is a consequence of the fact that in all known cases of true Gram-positivity the omission of the iodine step will result in the failure of the cell to retain the primary dye.

All Gram staining procedures require four reagents, a primary stain, iodine, a decolorizer, and a counterstain. Hucker's crystal violet is generally preferred as the primary stain since it is very stable and its 2% crystal violet and 1% ammonium oxalate aid differentiation and increase the validity of the results obtained (Hucker and Conn, 1927; Bartholomew, 1962). Other basic dyes have been tried as the primary stain but none has been shown to be superior to, or even equal to, crystal violet. The Burke formula for the iodine solution should be preferred since it contains a high level of iodine (1% as compared to 0.3% for Gram's iodine) which increases the validity of the procedure (Bartholomew, 1962) and the solution is very stable. A wet slide procedure is generally preferred over a dry slide procedure since it effectively standardizes an important variable and it does not involve vague instructions such as "blot dry," "air-dry," or "blot dry but do not air-dry." A large number of different decolorizers are acceptable; however, only one should be used routinely since the accumulation of experience with one decolorizer will improve the validity of the results obtained. Once determined, the decolorizer and the method and time of its application should become standardized. Any basic dye contrasting in color to the primary stain can be used as the counterstain. Although normally none has any great advantage over the 0.25% safranin usually used, some prefer a basic fuchsin counterstain which is less easily washed out of the cell and which tends to result in a deeper red color for Gram-negative organisms. Bacterial cells take up basic dyes rapidly and therefore exact staining times are not essential. Usually a 1-min time is used for the crystal violet, iodine, and safranin, although successful differentiation can be accomplished using only a few seconds for each (Paine, 1963). If the bacterial smears are prepared from acidic material, such as sour milk or stomach fluid, the acidity should be countered by covering the smear with a 1% solution of $NaHCO_3$, washing, and then proceeding with the staining steps.

Wet and Dry Procedures for the Gram Staining of Microorganisms in Smears

Bartholomew (1962)

Preparation of solutions:

A. Hucker's crystal violet. A stock solution is prepared by dissolving 2 gm of crystal violet chloride with 20 ml of 95% ethanol. For use, 20 ml of the stock solution is mixed with 80 ml of 1% ammonium oxalate. Both solutions are stable and will keep for months.

B. Burke's iodine. This should be prepared by placing 2 gm of potassium iodide (KI) in a mortar and then adding 1 gm of iodine (I_2) and mixing by grinding with a pestle. Then add 1 ml of distilled water, grind, then add 5 ml of distilled water and grind until all of the KI and I_2 are in solution. Add 10 ml of distilled water, mix, and pour into a 100-ml reagent bottle, and rinse the mortar and pestle with sufficient distilled water to bring the total volume to 100 ml. This solution is stable and can be stored for months.

C. Safranin. A stock safranin should be prepared by dissolving 2.5 gm of safranin O chloride with 100 ml of 95% ethanol. For use, 10 ml of this solution should be diluted with 90 ml of distilled water.

Procedure:

1. Prepare the usual heat fixed smear. The thickness of the smear will greatly influence the acceptable decolorization times.
2. Flood the slide with Hucker's crystal violet for 1 min.
3. Wash just enough to remove the excess dye. Washing for about 5 sec in a beaker with tap water running into it is recommended. If the slide is placed directly in the stream of tap water, expose for only 1–2 sec. The degree of washing at this step will influence the acceptable decolorization time.
4. Flood with Burke's iodine, drain, and flood again with the iodine for 1 min.
5. Wash. Overwashing is difficult at this stage.
6. Choose either the wet or dry procedure below.

Wet procedure	*Dry procedure*
7. Do not blot or dry following the wash but take dripping wet to the decolorization step.	7. Blot dry, then air-dry. Lightly heat in a flame to assure dryness. Cool.
8. Decolorize with 95% ethanol. Use a dropper bottle and allow the ethanol to run over the surface for about 5–15 sec for thin smears, and 15–60 sec for thick smears.	8. Decolorize with 95% ethanol. Use a dropper bottle and allow the ethanol to run over the surface for about 15–45 sec for thin smears, and about 1–3 min for thick smears.

9. Wash to stop the action of the decolorizer.
10. Flood with the 0.25% safranin for 1 min.
11. Wash very lightly. Overwashing removes the safranin from Gram-negative microorganisms, causing them to appear small and lightly stained.
12. Blot dry, and examine.

Results: Gram-positive microorganisms should be stained a blue black; Gram-negative microorganisms should be stained red.

Variations and comments: The two most critical points in the Gram procedure are the wash step following the crystal violet, and the method used for decolorization. The severity of the washing procedure can be controlled by running tap water into a 250-ml beaker in a sink at a rate of about 30 ml per sec. A 5-sec wash in the beaker is usually sufficient for all needed wash steps, and the water flow is sufficiently rapid to flush out reagents from the water in the beaker in the period of time between steps. The nature of the decolorization step, and the decolorizer used, are probably the items most often varied. The 95% ethanol can be replaced with acetone, acetone-ether, or other alcohols such as methanol, propanol, butanol, or pentanol. Such substitutions are valid if the decolorization time is properly matched to the decolorizer used. Generally methanol is considered to be too rapid, and butanol or pentanol too slow, in their decolorization effect. The most popular decolorizers are 95% ethanol, acetone, and a mixture of acetone (3 parts) and ether (1 part). One decolorizer that shows much promise, but with which there is very little experience in its use, is 95% propanol, which has been reported as being more foolproof than other decolorizers (Bartholomew, 1962). For the wet procedure, acetone decolorization times will vary from 15 to 60 sec, and 95% propanol decolorization times will vary from 1 to 5 min. Several methods can be used to more carefully standardize the decolorization procedure, such as using a buret which delivers the decolorizer at a constant drip rate, or placing the decolorizer in a series of 3 vertical (75 × 25 mm) Coplin staining dishes into which the slide is dipped for standard periods of time. The 3 Coplin dishes should be used in series to counter the build up of water in the decolorizer since water dilution (up to 30% water) of ethanol, propanol, or acetone, increases the speed of decolorization.

SECTION B: ACID-FAST STAINING PROCEDURES

Three types of staining procedures for the differentiation of acid-fast organisms in smears are useful: 1) The Ziehl-Neelsen procedure; 2) methods using fluorescent stains; and 3) "cold" procedures which are formulated to avoid the heat step normally used to speed the staining of acid-fast organisms with the primary dye. Of these, the Ziehl-Neelsen procedure is generally most used since it is the best known, it has long been established as a diagnostic procedure, and the results obtained can be interpreted in terms of a vast amount of clinical experience. Additional advantages are that the procedure is uncomplicated as to the steps involved and as to the equipment required. The procedures using fluorescent dyes as primary stains are excellent if the proper equipment for fluorescent microscopy is available and its use is well understood. The "cold" procedures are of interest, but they have not yet attained general acceptance as routine diagnostic procedures since some question exists as to their reliability.

The mechanism of acid-fastness is not known. Although the high lipid content of mycobacteria has been suspected as being related to the mechanism, especially the mycolic acid component, no definitive evidence has ever been presented which clearly establishes such a relationship. The concept is difficult to reconcile with the fact that all components of ruptured acid-fast cells are non-acid-fast (Kanai, 1962), and it has long been known that rather light mechanical injury to acid-fast cells will cause them to lose this characteristic. Apparently the intact cell is required for acid-fastness, and permeation through cell membranes might be an important part of the mechanism involved in the acid-fast characteristic.

The Ziehl-Neelsen Procedure

Ziehl (1882); Neelsen (1883)

Preparation of solutions:

A. Ziehl's carbol fuchsin. Liquid phenol (90%) is convenient for preparing phenolated stain solutions. Melt phenol crystals at 45°C, then add sufficient warm distilled water to give a phenol concentration of 90%. On cooling, this will remain liquid and can be dispensed with a pipette. Dissolve 0.3 gm of basic fuchsin chloride with 10 ml of 95% ethanol. After the dye is dissolved, add 100 ml of 5% phenol. This solution must be aged from 1 to 2 weeks before use, and it will keep for about 2 months. It should be discarded at the first sign of deterioration.

B. Acid alcohol. Add 3 ml of concentrated HCl to 97 ml of 95% ethanol.

C. Loeffler's methylene blue. Dissolve 0.3 gm of methylene blue chloride with 30 ml of 95% ethanol and add this to 100 ml of 0.1% aqueous KOH.

Procedure:

1. Prepare a 3–4 sq cm even and thin smear, with as little mechanical manipulation as possible. Unnecessary movement of the loop may injure bacterial cells which may cause acid-fast organisms to lose their acid-fast characteristic.

2. Heat fix by heating gently in the flame; or by placing on a temperature controlled hot plate at 65°C for 2 hr. Overnight fixation at 65°C is not harmful, and will fix to the slide material which might otherwise be inadequately fixed.

3. Cover the smear with a piece of filter paper just smaller than the slide and flood with Ziehl's carbol fuchsin. Gently heat to steaming and keep hot for 5 min. Do not allow the stain to boil, and above all, do not allow the solution to evaporate to the point at which dye will crystallize out. Always keep the filter paper flooded with stain. If controlled temperatures are available, the staining time decreases with increased temperatures. That is, if 90°C is reached, only 5 min of staining time is necessary; at 70°C, 10 min; at 55°C, 30 min; and at room temperature, overnight staining should be used.

4. After cooling, rinse briefly (5–10 sec) in tap water.

5. Decolorize with 3% HCl in 95% ethanol by running the acid alcohol over the smear. The decolorization time will vary with the thickness of the smear. Normally, thin smears should be decolorized for from 5 to 30 sec; thick smears for up to 2 min.

6. Rinse briefly in tap water.

7. Flood the smear with Loeffler's alkaline blue for 1–2 min.

8. Rinse briefly in tap water, blot dry, and examine.

Results: The acid-fast organisms should be stained a bright red; other organisms and the background material should be stained blue.

Diagnostic procedures: Sputum, gastric contents, urine, feces, or other body materials should be subjected to concentration procedures as given in *Diagnostic Standards and Classification of Tuberculosis*, published by the National Tuberculosis Association, New York. Heat fixed smears should be prepared following concentration, and stained by the Ziehl-Neelsen procedure. If one or more acid-fast bacteria are seen per microscope field, the slide should be examined for only a minute or so and the results recorded as "numerous acid-fast bacteria seen," or as +++. A slide to be reported as negative should be examined for a period of about 20 min, using both vertical and horizontal sweeps across the smear. If only 1 or 2 acid-fast organisms are seen during the entire search, record as such. If 3–9 are seen during the search, record as "rare" or as +; if 10 or more are seen, record as "few" or as ++. Information obtained from stained slides should always be backed up by culturing procedures which may be positive when the results for acid-fast stained smears are negative. Positive smears from gastric contents are open to suspicion, and the presence of tuberculosis organisms must be supported by culture information.

Comments: Reasonable heat fixation times do not influence results. The use of a controlled temperature staining rack will help to produce consistent results provided the temperature-time relationship used has been tested for its validity. Although many substitutes for the acid alcohol decolorization step have been suggested, acid alcohol both speeds the decolorization of non acid-fast material and retards the loss of primary dye from acid-fast cells, thus enhancing the differentiation obtained (Dickinson, 1963). No other decolorization procedure is known to be better, and most are greatly inferior. It has been suggested that the decolorization step could be omitted and that the dye replacement power of the counterstain could be used to both decolorize and to counterstain non-acid-fast material. While this may be technically true, it has been shown that the use of acid alcohol does in fact emphasize the differentiation obtained, and its use will give superior results. Although non-alkaline solutions of methylene blue could be used as the counterstain, the acid nature of the decolorizer, and the effect of acidity on the uptake of the basic dye counterstain, would make the choice of an alkaline counterstain most logical. Excessive washing following the carbol fuchsin may cause a heightened decolorization effect; excessive washing after the counterstain lightens the blue color of the non-acid-fast material. For a discussion of the variables influencing the results obtained by the Ziehl-Neelsen acid-fast stain procedure, see Dickinson (1963). More recently, it has been demonstrated that an oxidation step prior to staining with carbol fuchsin may increase the apparent acid-fastness of the cells, and may also result in an increase in the number of cells staining as acid-fast. Harada (1973) recommended a 30-min prior exposure to 1% potassium permanganate, while Nyka (1967) recommended a 4–24-hr prior exposure to 10% periodic acid.

Fluorescent Microscopy Acid-Fast Procedure

Truant, Brett, and Thomas (1962)

Fluorescent microscopy for the detection of acid-fast organisms should not be confused with technics which are based on the specific reaction of fluorescent antibodies with their corresponding antigens. In the procedure below, a fluorescent dye binds in a nonspecific manner to substrate. The technic for the demonstration of acid-fastness through fluorescence is similar to the usual Ziehl-Neelsen procedure except that a phenolated fluorescent dye is used instead of the usual carbol fuchsin. Many fluorescence procedures have been published, and the one given below is based on that of Matthaei (1950), as modified by Truant et al. (1962). The dyes used should be certified by the Biological Stain Commission for use in fluorescent staining procedures.

The advantage of a fluorescent procedure is that the acid-fast organisms appear as a bright fluorescence against a dark background and are so easy to detect that low power (25× to 45×) objectives can be used with their much greater field of view. The increased field of view makes it possible to search a much greater proportion of the smear area in a given time, and because of this, much time can be saved in establishing the positive or negative nature of the smear being observed. When properly carried out and interpreted, the validity of the results obtained are at least equivalent to that of the usual Ziehl-Neelsen procedure (Truant et al., 1962). The principal disadvantage of fluorescent microscopy is that the microscope equipment required is more complicated than that used for ordinary microscopy and this may result in false negative reports when the operation of the equipment is not completely understood. Despite this, fluorescent microscopy is now recommended by the Communicable Disease Center of the U.S. Public Health Service (Kubica and Dye, 1967) although they state also that confirmation with the Ziehl-Neelsen procedure is recommended.

Preparation of solutions:

A. Auramine O, rhodamine B, stain. Dissolve 3 gm of auramine O and 1.5 gm of rhodamine B with 150 ml of glycerol at room temperature by stirring. Add 20 ml of melted phenol (45°C), and then 100 ml of distilled water. After mixing, filter through glass wool. This solution will keep for several months at refrigerator temperature.

B. Decolorizer. Add 0.5 ml of concentrated HCl to 99.5 ml of 70% ethanol.

C. Counterstain. Dissolve 0.5 gm of potassium permanganate in 100 ml of distilled water.

Procedure:

1. Prepare heat fixed smears. See instructions for the Ziehl-Neelsen procedure.
2. Flood the smear with the auramine O-rhodamine B stain and heat to 60°C for 10 min; or 37°C for 15 min; or 25°C for 20 min.
3. Rinse briefly in tap water.
4. Decolorize with 0.5% HCl in 70% ethanol for 2–3 min, and then rinse thoroughly.
5. Flood the smear with 0.5% potassium permanganate for at least 2 min, but not longer than 4 min.
6. Rinse, blot dry, and examine by fluorescent microscopy. Use a 25× to 45× objective and a 10× eyepiece. Confirm the cellular nature of fluorescent areas using the 100× oil immersion objective.

Results: Acid-fast organisms should show fluorescence; non-acid-fast organisms and background material should have little or no fluorescence.

Variations: If rhodamine B is not available, the stain can be formulated as given using only auramine O. The function of the potassium permanganate is not clear, but it apparently serves as a gentle decolorizer and results in less fluorescence in the background material. Mansfield (1970) recommended a 15 min fixation in 10% formalin following heat fixation of the smear.

Comments: A well designed microscope system for fluorescent microscopy is a necessity. A suitable intense light source of ultraviolet light is needed such as an Osram HBO, 200 watt, mercury vapor lamp. Critical illumination should be used, which means that the condenser of the light source and the condenser of the microscope should be adjusted so as to provide a focused image of the light source, as well as a focused image of the lamp diphram, at the level of the specimen. The lamp diphram (field diaphram) should be opened so as to just cover the field of view, but the microscope condenser diaphram (aperature diaphram) should be used wide open since the more intense the excitation light, the stronger the fluorescence. A BG-12 exciter filter between the light source and the specimen will provide light of the proper wavelength for excitation of fluorescence, and an OG-1 barrier filter in the microscope body tube or ocular is required to produce a dark background as well as serving as a barrier to ultraviolet light, which might be injurious to the eye. While the detection of fluorescing cells can be accomplished using low power (25× to 45×) objectives, once found they should be confirmed using a 100× oil immersion objective. The higher magnification is required to distinguish between noncellular flou-

rescing particulates and fluorescing bacterial cells which might appear
similar at the lower powers of magnification. The problem of differentiating fluorescing bacterial cells from fluorescing particulates constitutes the
principle reason for the failure of acceptance of fluorescent microscopy as
a standard diagnostic procedure in many laboratories.

Matthaei (1950) reported that ordinary microscope slides, lenses, and
light sources could be used following staining with auramine O-rhodamine
B stain. He used a high intensity condensed filament projection lamp (12
V, 100 watt) with a large aperature biconvex condenser (50 mm diameter,
f = 50 mm) in a lamp housing which enabled him to achieve critical
illumination. An aqueous copper sulfate filter was used to supply the blue
light (about 496 mμ) and a Wratten G-15 gelatin barrier filter was used in
the ocular to remove all light over 510 mμ, thus giving a dark background
to the fluorescing bacterial cells.

Acid-Fast Staining at Room Temperature

Procedure modified from Aubert (1950); Gross (1952); Desbordes,
Fournier, and Guyotjeannin (1952)

Much interest has been shown in "cold" acid-fast procedures which omit
heating during primary staining since such a simplification of technic would
have obvious advantages. In general, the omission of the heating step is
made possible by increasing the concentration of the basic fuchsin and
adding small amounts of a detergent such as Tween-80 which serves to speed
the staining process for acid-fast organisms. Although the cold procedures
would be convenient, they have not received much acceptance even though
they have been in existence for almost as long as the Ziehl-Neelsen procedure
itself.

Preparation of solutions:
  A. Carbol fuchsin with Tween-80. Dissolve 4.0 gm of basic fuchsin
     chloride with 12 ml of phenol at 80°C with stirring. Cool to about
     45°C and add 25 ml of 95% ethanol, continue stirring, and then bring
     the volume to 300 ml with distilled water. Let stand for 1 or 2 weeks
     and filter through glass wool. Before use, add 10 drops of Tween-80
     (or similar detergent) for each 100 ml of solution, and filter just before
     use. The stain with detergent must be filtered at the start of each day
     of use.
  B. Decolorizer. Add 5 ml of concentrated HNO_3 to 95 ml of 70% ethanol.
  C. Counterstain. Loeffler's methylene blue, as used in the Ziehl-Neelsen
     procedure, may be used.

Procedure:
  1. Prepare heat fixed smears.
  2. Flood the smear with the primary stain for 5–10 min at room temper-
     ature.

3. Rinse with tap water.
4. Decolorize by running the 5% nitric acid in 70% ethanol over the smear for about 30 sec for thin smears, or up to 2 min for thick smears.
5. Rinse in tap water and counter stain with Loeffler's methylene blue for 1–2 min.
6. Rinse briefly, blot dry, and examine.

Results: Acid-fast organisms should be stained red; non acid-fast organisms and the background material should be stained blue.

Comments: The older Kinyoun stain (1915) used a 4% basic fuchsin with 8% phenol for the primary stain, which probably would result in a more rapid staining of acid-fast organisms at room temperature. Detergents other than Tween-80 have been shown to be effective when added to the primary stain, such as Tween-20, Anatarox A-400, sodium tetradecylsulfate, and Tergitol No. 7 (Desbordes et al., 1952; Gross, 1952).

SECTION C: STAINS FOR BACTERIAL CAPSULES AND SLIME

Bacterial capsules and slime are found external to the cell wall and are composed of polysaccharides, more rarely of polypeptides, and sometimes of a mixture of polysaccharides and polypeptides. If this high polymer viscous material shows a distinct boundary between itself and the background material, it is called a capsule; if the boundary is diffuse or highly irregular, the material is termed slime. Capsules and slime are demonstrated using either a negative stain or a positive stain. Negative stains color the cell and background material, leaving the capsule or slime colorless. Positive stains differentially stain the cell, capsule or slime, and the background material. If an organism has well developed capsules (0.5 μm or larger), their demonstration by staining methods can be convincing. However, if the capsule is not well developed (0.2 μm or smaller), then the results obtained by any capsule stain procedure are difficult to interpret since both the staining methods and the optical equipment used may produce capsule like artifacts. For example, when dry India ink or nigrosin negative staining procedures are used, the drying of the preparation may produce a cellular shrinkage which causes a thin clear line to appear around the cell. On observation it is impossible to determine if this clear area represents a thin capsule, or a cell shrinkage artifact, although normally such a result would be considered as negative for capsules. If a wet negative stain is tried, the problem will not be solved since optical halos may then appear which are also indistinguisable from thin capsules. The best answer to this problem, when it appears, is to culture the organism in a manner to accentuate capsule formation, such as growth in a serum enriched medium.

Most who have worked on capsule stains report that the preparation of smears from cells suspended in water is futile (Hiss, 1905; Howie and Kirkpatrick, 1934; Butt, Bonynge, and Joyce, 1936). The role of suspending organisms in serum or glucose solutions for the demonstration of capsules is not known, but it is a positive factor and represents more than just supplying a background for negative stains. Therefore, it can be recommended that, for any capsule stain procedure, the organisms should be suspended in serum or glucose solutions, or in 1% glucose in 10% serum. The best natural suspension medium is the serum from an animal infected with the organism, if it is available. Often in negative stains, the cells will be observed to be located off center and to the side of the apparent capsule region. This has not been explained, but it does not reduce the validity of the demonstration of the capsules.

In a survey of a variety of capsule stains, Duguid (1951) found that a wet film India ink method was preferable to all others; and that of the dry film negative stains, the best was the method of Butt et al. (1936). The Howie and Kirkpatrick (1934) eosin Y, serum, negative stain was reported as being good except for very thin capsules, and it was the only method which demonstrated slime as well as capsules. Five methods for staining capsules are given below, and it can be recommended that, when in doubt, both a positive and a negative staining procedure be employed.

Wet, Negative Capsule Stain Using India Ink

Duguid (1951)

Preparation of solutions: A high quality India ink is needed in which the carbon particles present are small (0.02–0.1 μm) and are not aggregated into clumps. Higgins waterproof India ink is acceptable providing that coarse particles are first centrifuged out. Peliken ink No. 541, Gunther Wagner, N.Y., is good, and an especially fine preparation for negative staining called carbochrome ink is available from Michrome Laboratories, East Sheen, London, England (E. Gurr, Director).

Procedure:
1. Place a large loop of India ink on the slide.
2. Add cells from a colony or slant growth, or from broth. A large number of cells should be added with as little dilution of the ink as possible.
3. Cover with a microscope cover glass, No. 1 or 0 thickness. Overlay with a blotting paper and press down to produce a film about one cell thick. The blotting paper will take up the expressed material. Trial and error is the only way to achieve the correct film thickness. If the film is too thick, the cells will float about; if it is too thin, the capsules will be distorted.
4. Observe with the oil immersion objective.

Results: Capsules should be clear, and the background dark. The bacterial cell will be visible due to its refractility. If only a small clear area (0.2 μm or smaller) is seen around the cell, the results should be interpreted as doubtful or negative. This clear area may be a diffraction halo which can be seen around all non capsulated organisms.

Dry, Negative Capsule Stain Using India Ink

Butt, Bonynge, and Joyce (1936)

Preparation of solutions:

A. India ink as described for wet negative capsule stains using India ink.
B. All aqueous solution containing 6% glucose.
C. An aqueous solution of 1% methylene blue chloride.

Procedure:

1. A drop of the 6% glucose solution is placed on a slide, and sufficient organisms are added to produce a thin suspension.
2. A drop of India ink is added and evenly mixed with the suspension.
3. Spread the mixture in a thin film on the slide using the edge of a glass slide as for the preparation of thin blood films (see the staining of protozoa in smears).
4. Let air-dry, then fix with absolute methanol. Wash, do not blot.
5. Counterstain with a cationic dye such as an aqueous 1% methylene blue.
6. Wash, dry without blotting, and examine.

Results: The capsules should be colorless, the background dark, and the cells should be stained blue.

Variation: The India ink, glucose mixture can be replaced with a solution of 1% azo blue in 6% glucose.

Comments: The use of the 6% glucose results in a larger clear area around the capsulated cells than if saline or water are used.

Dry, Negative Capsule Stain Using Congo Red

P. B. White (1947)

Preparation of solutions:

A. Dissolve 5 gm of Congo red with 100 ml of distilled water, then add 11 ml of serum (rabbit, horse, or human) and mix. Decant the supernatant, and discard the residue.
B. Add 0.5 ml of concentrated HCl to 95.5 ml of distilled water.
C. Dissolve 1 gm of methylene blue chloride with 100 ml of distilled water, and then add 5 drops of glacial acetic acid.

Procedure:

1. Place 1 drop of the Congo red, serum, solution on a slide and mix the organisms into it.

2. Spread in a film of varied thicknesses from thick to thin, and dry with gentle heating. Fix the smear in the flame. Cool.
3. Flood with the 0.5% HCl, drain, blot gently, and drive off excess fluid with gentle warmth.
4. Stain 15–20 sec with the acidulated aqueous 1% methylene blue.
5. Drain, do not wash, blot firmly, air-dry, and examine.

Results: The cells should be stained blue, the background should be orange to gold, and the capsule material should be colorless or light violet.

The Anthony Modification of the Hiss Capsule Stain

Anthony (1931)

This is a simplification of the Hiss (1905) capsule stain, which it has supplanted. The stain is a positive stain when the background material is sparse and the copper sulfate used sparingly. The procedure will produce a negative stain when the background material is abundant, and when the copper sulfate wash is excessive.

Preparation of solutions:
  A. Dissolve 1 gm of crystal violet chloride with 100 ml of distilled water.
  B. Dissolve 20 gm of copper sulfate ($CuSO_4 \cdot 5\ H_2O$) with 100 ml of distilled water.

Procedure:
  1. Prepare a relatively thick smear of the organisms. Air-dry, do not heat fix.
  2. Stain for 2 min with the aqueous 1% crystal violet.
  3. Wash with the 20% copper sulfate. The amount of washing must be determined empirically.
  4. Blot dry and examine.

Results: The capsules should be stained a lighter blue than either the cell or the background material. The copper sulfate removes dye faster from the capsule material than it does from either the cell or background material.

Variations: Good demonstrations of capsules are also obtained if a drop of 10% serum or milk are placed on the slide, the organisms added, and a thin film made. In this case the background and cells are stained, and the capsules are usually colorless.

Williamson (1956) recommended suspending the organisms directly in a drop of India ink, spreading in a thin film, staining with 0.1% crystal violet in 0.25% acetic acid for 5 min, draining, washing with 20% copper sulfate, blotting dry, and examining. The background should be dark, the cells blue, and the capsules colorless.

Hiss (1905) found that 0.25% potassium carbonate was as effective as copper sulfate as a wash agent. After washing in potassium carbonate, a drop was left on the slide, a cover glass added, and the preparation examined. A ring of dye-carbonate precipitate often indicated the outer limits of the capsule, the cells were blue, and the capsules ranged from blue near the cell to colorless at the outer limits of the capsule.

Positive Capsule Stain

Möller (1951)

Preparation of solutions:

A. The suspension solution for organisms. Mix in a test tube 0.65 ml of saline (0.85% NaCl), 0.2 ml of normal serum, and 0.15 ml of 20% glucose.

B. Fixative. Dissolve 9 gm of lead acetate in 280 ml of distilled water, then add 20 ml of formalin (37–40% formaldehyde solution). A precipitate should form which can be ignored. Store in a tightly stoppered bottle.

C. Stain. Dissolve 0.5 gm of crystal violet chloride with 10 ml of 95% ethanol, then add 90 ml of distilled water.

D. Prepare a saturated solution of copper sulfate ($CuSO_4 \cdot 5 \ H_2O$) by adding it to distilled water at room temperature until no more dissolves. The reported solubility is 31.6 gm per 100 ml of water at 0°C, and 203.3 gm at 100°C.

Procedure:

1. Place a loop of the saline, serum, glucose, solution of a clear glass slide.
2. Add organisms and spread in a thin film. Air-dry, do not heat fix.
3. Hold the slide on a slant and run the lead acetate, formalin, solution over the smear. After this, let the fixative stand over the smear for 15 sec, then drain and blot dry. Do not wash.
4. Stain for 1–3 min with the 0.5% crystal violet, and then drain.
5. Rinse with the saturated copper sulfate for 5–10 sec, then drain and carefully blot dry. Do not wash.

Results: The bacterial cells should be stained a dark bluish violet, the capsules a light reddish violet.

Comments: The lead acetate, formalin, solution is a fixative which is known to be effective for mucous polysaccharides in tissue sections. Variations in the time of application of the copper sulfate will vary the degree to which the capsules retain the stain; over rinsing with the copper sulfate solution will cause the capsules to be colorless.

SECTION D: STAINS FOR BACTERIAL SPORES

Bacterial spores demonstrate a very low rate of permeation to ordinary dyes in aqueous solution at room temperature. Staining can be achieved, however, by the use of long periods of staining such as overnight, by increasing the staining temperature, by increasing the concentration of the dye, or by applying some reagent which will counter or destroy the resistance of the spore to dye permeation. In most spore staining procedures heat is used to reduce the staining time to a total of 3–5 min. The heating step has been objected to as being messy, and some suggestions have been made for avoiding it, but usually with some loss of spore stain dependability. The comparatively simple Schaeffer and Fulton (1933) modification of the older Wirtz spore stain is very dependable and is probably the differential spore stain of choice. When cultivating organisms the purpose of demonstrating the presence of spores, the organisms ordinarily should be grown on a solid culture medium, and should be from 18–24 hr old.

Schaeffer and Fulton Modification of the Wirtz Spore Stain

Schaeffer and Fulton (1933)

Preparation of solutions:
  A. Dissolve 5 gm of malachite green chloride with 100 ml of distilled water. This solution is very stable.
  B. Dissolve 0.5 gm of safranin O chloride with 100 ml of distilled water. This solution is very stable.

Procedure:
  1. Prepare a smear of the organisms. Air-dry, then heat fix by passing 3 times through the flame of a Bunsen burner.
  2. Flood the smear with the 5% aqueous malachite green, and heat to steaming 3–4 times within 30 sec, then let cool.
  3. Wash in tap water for 30 sec.
  4. Flood with 0.5% safranin for 30 sec.
  5. Wash very lightly, blot dry, and examine.

Results: The spores should be stained green; the vegetative cells red.

Comment: The safranin counterstain serves two purposes. It replaces the malachite green in the vegetative cells by the process of dye exchange, but it does not replace the dye in the spore because it does not penetrate into this structure. It also serves to give a contrasting color to the vegetative cells. Safranin is easily washed from vegetative cells; therefore washing should be minimal.

A Modification of the Schaeffer, Fulton Spore Stain Which Omits the Heat Step

Bartholomew and Mittwer (1950)

Preparation of solutions:
  A. A saturated solution of malachite green chloride is prepared by

dissolving 10 gm of malachite green with 100 ml of distilled water. Filter to remove any undissolved dye.

B. A 0.25% aqueous solution of safranin O chloride is prepared. This is the same safranin solution as used for the counterstain in the Gram stain.

Procedure:

1. Prepare a smear of the organism and air dry.
2. Heat fix by passing the slide through the flame of a Bunsen burner about 20 times. Each passage should be similar to that used for normal heat fixation. Let the slide cool before proceeding.
3. Flood the slide with the saturated malachite green and let stand for 10 min at room temperature.
4. Rinse with tap water.
5. Flood the slide with 0.25% safranin, and let stand for 15 sec.
6. Rinse very lightly in tap water, blot dry, and examine.

Results: The spores should be stained green; the vegetative cells red.

Comments: Saturated malachite green cannot be used for the Schaeffer and Fulton spore stain since the heating step causes a dye precipitate which will prevent the differentiation of vegetative cells and spores. The 20 passages through the flame for fixation should reduce the resistance of the spore to permeation by malachite green, but should not result in a complete loss of permeation resistance. If over heat fixed, the safranin will replace the malachite green in the spores as well as in the vegetative cells.

Fluorescent Spore Stain, Without Heat

Bartholomew, Lechtman, and Finkelstein (1965)

If a good quality fluorescent microscope system is available, this procedure is the easiest and simplest method for the demonstration of spores in bacterial cultures.

Preparation of solutions:

A. Dissolve 0.1 gm of auramine O chloride with 100 ml of distilled water. Fresh dye solutions are recommended, for stored solutions show a rapid quenching of fluorescence on exposure to ultraviolet light.

B. Dissolve 0.25 gm of safranin O chloride with 100 ml of distilled water. This is the same safranin solution used as the counterstain in the Gram stain.

Procedure:

1. Prepare the usual heat fixed smears.
2. Flood the slide with 0.1% auramine O and stain at room temperature for 2 min.

3. Wash very lightly with tap water.
4. Counterstain with the 0.25% safranin for 1 min.
5. Wash very lightly, blot dry, and examine using an oil immersion objective and a good quality fluorescent microscopy system.

Results: Spores should be brightly fluorescent; vegetative cells should show no fluorescence.

Comments: The safranin serves to remove the auramine O from the vegetative cells. If the cells are rich in sudanophilic granules, the auramine O has a small affinity for this material and the granules may be dimly fluorescent as compared to the bright fluorescence of spores. Only one brightly fluorescing spore should be seen per cell, while many of the dimly fluorescent sudanophilic bodies may be seen.

Variation: If sudanophilic bodies are present, these can be identified by staining with 0.1% auramine O in 3% phenol for 2 min. Counterstain with the 0.25% safranin for 2 min, and observe with the fluorescent microscope system. Many brightly fluorescing bodies will be seen per cell.

Differential Spore Stain Using Inorganic Acids and No Heat

Lechtman, Bartholomew, Phillips, and Russo (1965)

Preparation of solutions:
  A. Stain. Dissolve 2 gm of crystal violet chloride in 26 ml of absolute ethanol, then add 64 ml of distilled water with stirring. Slowly add 10 ml of 10% phenol with constant stirring. Let stand 1 or 2 days before use.
  B. Counterstain. Dissolve 0.25 gm of safranin O chloride in 100 ml of distilled water. This is the same safranin solution as used in the Gram stain.
  C. Dilute 30 ml of reagent grade HNO_3 (16 N or 70%) with 80 ml of distilled water to obtain a 6 N solution of HNO_3. Fill a 75 × 25 mm screw cap vertical Coplin staining dish with this solution and keep tightly closed.
  D. Prepare 1 N NaOH by dissolving 4.0 gm of NaOH in distilled water and bringing to a final volume of 100 ml. Fill a 75 × 25 mm screw cap vertical Coplin staining dish with this solution and keep tightly closed.

Procedure:
  1. Prepare the usual heat fixed smear.
  2. Immerse the slide in the 6 N HNO_3 in the Coplin dish at room temperature for 10 sec. Wash immediately in tap water. Be careful to keep the HNO_3 away from skin tissue.
  3. Immerse the slide in 1 N NaOH in a Coplin dish for 30 sec, and then wash in tap water.

4. Flood the slide with the solution of 2% crystal violet in 1% phenol and 26% ethanol for 2 min at room temperature. Wash in tap water.
5. Counterstain with the 0.25% safranin for 1 min.
6. Wash lightly in tap water, blot dry, and examine.

Results: The spores should be stained blue, the vegetative cells red.

Variation: The HNO_3 causes a yellow discoloration of skin tissue. This can be avoided by replacing the nitric acid with 44 N H_3PO_4 (conc., 85%), and exposing the slide to this acid for 2 min, followed by washing and neutralization in 1 N NaOH. The use of nitric acid, however, results in the best differential spore stains.

SECTION E: THE STAINING OF LIPID

The sudan series of dyes have long been used as a histochemical test for the presence of lipid material in cells. Under the conditions of staining, the dye is more soluble in lipid material than in the solvent of the stain solution and therefore the dye moves into the lipid material. The sudanophilic bodies in bacteria are known to be composed mostly of poly-beta-hydroxy-butyric acid. This can be considered to be a lipid-like material since it is composed of a polymer of a C_4 fatty acid, but no glycerol is present. Curiously, while the granules are sudanophilic *in situ*, removal from the intact cell causes them to lose their sudanophilic characteristic (Williamson and Wilkinson, 1958). Only two staining methods will be given since the procedures are very dependable and additional ones would merely be redundant. Some organisms such as *Bacillus cereus* will possess large sudanophilic granules even when grown on a simple medium such as nutrient agar, while other organisms may require growth on an enriched medium. Usually, mature cells are richer in sudanophilic granules than young cells. However, 3-hr-old cultures of *B. cereus* on an enriched agar medium can be seen to possess well developed sudanophilic granules.

Dry Smear Differential Stain for Sudanophilic Granules

Burdon (1946)

Preparation of solutions:

A. Dissolve 0.3 gm of Sudan black B with 100 ml of 70% ethanol. This solution is very stable; however, it is near or at saturation with Sudan B and Burdon recommends that it be shaken before each use.
B. Dissolve 0.5 gm of safranin O chloride with 100 ml of distilled water.
C. Xylene.

Procedure:

1. Prepare the usual air dried, heat fixed, smear.
2. Flood the entire slide with Sudan black B and stain for 10 min. Add

stain solution during this period of time to compensate for evaoporation.

3. Drain off the excess stain and then blot dry. Do not wash with water.
4. Clear, by placing a few drops of xylene on the slide, drain, blot dry.
5. Counter stain with 0.5% safranin O for 5 to 10 sec.
6. Wash *very* briefly, blot dry, and examine.

Results: The sudanophilic granules should be stained a deep blue black; the vegetative cells should be stained red.

Comment: If the Sudan B stain dries on the slide, or if a Sudan B precipitate appears, it can be removed by running a few drops of xylene over the smear. Sudan B is very soluble in xylene, however, and this may reduce the intensity of staining of the sudanophilic granules.

Wet Mount Demonstration of Sudanophilic Granules

Hartman (1940)

Preparation of solutions: Use the same 0.3% Sudan B in 70% ethanol as for the dry differential stain above.

Procedure:
1. Place several loops of the stain on a clean glass slide and merge.
2. Add cells to the stain with a loop.
3. Cover with a No. 1 or 0 cover glass. Add a loop of the stain at each of the 4 edges of the cover glass to prevent drying out.
4. Observe with the oil immersion objective. The sudanophilic granules will be stained after about 2–3 min.

Results: The sudanophilic granules should become stained a deep blue black; the vegetative cells should remain unstained.

Comment: The sudanophilic granules will appear to be larger in wet mounts than in dry smears. A precipitate of the dye may form as drying of the wet mount progresses. The precipitate granules are small and must be differentiated from small sudanophilic granules.

SECTION F: STAINS FOR BACTERIAL CELL WALLS

The so-called cell wall stains for bacteria are used mostly in the teaching laboratory along with stains for other cytological structures such as bacterial spores, nuclei, and flagella. It is probably correct to state that there is no staining procedure which specifically stains bacterial cell wall material and nothing else (Girbardt and Taubeneck, 1955; Finkelstein and Bartholomew, 1958). This conclusion makes sense since, if the average bacterial cell wall thickness is from 0.01 to 0.08 μm as determined by electron microscopy, then such a thin structure, even when fully stained and folded on itself as in heat fixed smears, would either not be visible at all or it would be just barely visible with the optical microscope. Optical microscopes are limited in their resolution to points 0.2 μm or further apart, and to visualization of particles larger than 0.05 μm.

Cell wall staining procedures can be grouped into two types. Those which stain the developing cross wall regions and newly completed transverse septa well, but which result in only a very thin visible line representing the cell wall, such as the staining procedures using tannic acid as a mordant; and those which form an obvious precipitate on the outer surface of the cell resulting in a thick appearing cell wall, but which do not stain the area of developing cross walls or newly formed transverse septa, such as the Dyar cell wall stain. One of each type of cell wall stain will be given below. Although neither of these stains can be said to be specific for cell wall material, they do define an outer region of the cell, they may demonstrate the areas of developing cross walls and completed transverse septa, and they are pretty slides to look at.

Dyar Cell Wall Stain

Dyar (1947)

Preparation of solutions:
  A. Dissolve 0.34 gm of cetyl pyridinium chloride with 100 ml of distilled water.
  B. Prepare a saturated solution of Congo red by dissolving 5 gm of the dye with 100 ml of distilled water.
  C. Dissolve 0.5 gm of methylene blue chloride with 100 ml of distilled water.

Procedure:
  1. Prepare the usual air dried and heat fixed smear.
  2. Cover with 3 drops of the 0.34% cetyl pyridinium chloride.
  3. Add 1 drop of the saturated aqueous Congo red and mix with the cetyl pyridinium chloride.
  4. Wash with tap water.
  5. Counterstain with the 0.5% methylene blue for a few seconds.
  6. Wash, blot dry, and examine.

Results: The cell wall should be stained red; the cytoplasm blue. The region of developing cross walls is either not stained at all, or is stained poorly.

Variation: The methylene blue counterstain can be omitted.

Comments: A precipitate should form as you mix the Congo red into the cetyl pyridinium chloride. The staining mechanism here is probably that the cationic cetyl pyridinium chloride binds onto surface cell material, and on the addition of the anionic Congo red the dye is precipitated at the point of contact with the outer surface of the cell. When cells are washed away after the application of the Congo red, a red ring of precipitate can be seen on the slide in the original position of the cell. Cells stained by the Dyar method appear wider than those demonstrated with nigrosin or India ink negative stains.

SECTION G: THE STAINING OF BACTERIAL NUCLEAR MATERIAL

The bacterial genetic material consists of a double stranded DNA molecule 600–1200 μm in length with its ends joined to form a closed ring molecule. Neither a nuclear membrane nor histone type proteins are associated with this nuclear material. In properly fixed, imbedded, and ultrathin sectioned preparations, the electron microscope reveals a fibrous nuclear structure with the DNA molecule arranged much like the strands of yarn in a skein of knitting wool. Demonstrations with the optical microscope require young (3 hr) cultures, which must be digested with 1 N HCl at 60°C, which unmasks the DNA region of the cell by removing RNA from the cell faster than DNA, and which probably also intensifies the staining reaction of the DNA rich areas remaining. After HCl digestion, staining can be accomplished with almost any dilute (0.1%) solution of basic dye, or by dilute (1:10 to 1: 60) Giemsa stain. The nuclear morphology revealed is consistent with that seen with electron microscopy if you allow for the fact that what appears as a single stained bacterial cell with the optical microscope is in reality a cell in which the nuclear material may be well advanced into its second, or even third, division. The optical microscope is not capable of fully resolving the details present, but it can reveal areas in which the DNA material is concentrated, and the appearance is what one would expect from rapidly dividing nuclear material. A large number of methods have been published for staining bacterial nuclear material. Two are given below of which one is a simplified method taken from Cassel and Hutchinson (1955) and the other is taken from Robinow (1942, 1944).

A Simplified Method for Staining Bacterial Nuclear Material

Cassel and Hutchinson (1955)

Preparation of solutions:
 A. The fixative. Dissolve 4 ml of acetaldehyde in 95 ml of disstilled water, and then add 1 ml of formic acid. This fixative should be kept in a tightly closed 75 × 25 mm screw cap, vertical, Coplin staining dish. Acetaldehyde is a liquid at ordinary temperatures but it has a low boiling point. It should be stored in the refrigerator, and should be below 20°C (68°F) when opened. It is best to add the acetaldehyde to the distilled water under a hood with the exhaust fan on. The final fixative is safe at room temperatures, but it should be kept in a tightly closed vessel.

B. Staining solution. Dissovle 0.1 gm of basic fuchsin chloride with 100 ml of distilled water.

C. Fill a 75 × 25 mm screw cap, vertical, Coplin staining dish with 1 N hydrochloric acid and place in a 60°C water bath.

Preparation of cultures: A petri dish with brain-heart infusion agar is prepared. The whole surface is inoculated with the organism (such as *Bacillus cereus* or *Escherichia coli*) using a sterile glass spreader, and the culture incubated at 37°C for a period of 3 hr. Young cultures are essential.

Procedure:

1. Cut 3 blocks, each 1 sq cm square, from the agar culture and press, culture side down, on the surface of 3 clear glass slides. Press firmly down, then remove the agar blocks and discard them. These impression smears should be air dried.

2. Immerse the 3 slides in the formic-acetaldehyde fixative for 30 min.

3. Wash the 3 slides in tap water, and while still wet immerse in the 1 N HCl at 60°C.

4. Remove one slide after 8 min, one after 10 min, and one after 12 min, of digestion. Wash in tap water.

5. Flood the slides with the 0.1% basic fuchsin stain for 15 sec.

6. Wash, blot dry, and examine.

Results: The DNA rich areas should be stained red; the cytoplasm should be colorless. Usually 1 of the 3 slides will be superior to the other 2.

Comments and variations: The 3 digestion times are required since it is impossible to predict with certainty the exact digestion time which will give the best demonstration of nuclear material. For examination the oil can be placed directly on the smear, or a drop of water can be placed on the slide and covered with a No. 0 cover glass before examination with the oil immersion objective. This staining procedure has proven to be particularly useful under the conditions found in a crowded student laboratory.

Personal experience with this staining procedure has shown that following air drying of the impression smear, the fixation step in the formic-acetaldehyde solution can be omitted if the slide is immediately placed into the 1 N HCl at 60°C.

The Robinow Stain for Bacterial Nuclear Structures

Robinow (1942, 1944)

Preparation of solutions:

A. Fixative. Obtain a small screw cap cover glass staining dish which will accept an 18 mm square cover glass when laying flat. Fill the dish about ⅔ full with 3.0–4.0 mm glass beads, and then add 2% osmium tetroxide to a level of about 1 mm below the top surface of the beads. Keep tightly capped.

B. Schaudinn's fixative. Add to 66 ml of saturated mercuric chloride 33 ml of absolute ethanol and 1 ml of glacial acetic acid. The solubility of $HgCl_2$ at 20°C is listed as 6.9 gm per 100 ml of water.

C. Fill a screw cap cover glass staining dish with 1 N HCl and keep at 60°C in a water bath.

D. Dilute 95% ethanol to 70%, with distilled water.

E. Giemsa stain. Either buy ready prepared or prepare as follows. Add 0.5 gm of Giemsa stain (powder form) to a warm mortar and slowly add, with trituration, 33 ml of warm (55°C) glycerol. Cool to room temperature, and then add 33 ml of reagent grade absolute methanol and mix. Place in a tightly stoppered bottle and store overnight during which time a small amount of sediment will appear. Decant into small (30 ml) bottles leaving the sediment behind and tightly stopper. Store at least 2 weeks and filter before use. For use, the stock solution is diluted 1:10 with M/200 Sörensen's phosphate buffer of pH 6.7–7.0.

The Sörensen's buffer is prepared as follows. Prepare solution (A) by dissolving 0.675 gm of monobasic sodium phosphate ($NaH_2PO_4 \cdot H_2O$) in distilled water and bring the final volume to 1 liter. Prepare solution (B) by dissolving 0.710 gm of anhydrous dibasic sodium phosphate (Na_2HPO_4) in distilled water and bring the final volume to 1 liter. A buffer at pH 7.0 is prepared by mixing 26 parts of (A) with 24 parts of (B). A buffer of pH 6.9 is prepared by mixing 28 parts of (A) with 22 parts of (B).

Preparation of cultures: Young (3 hr) cultures grown on a solid medium in a petri dish at optimum (or near optimum) temperature for growth are essential.

Procedure

1. Cut a square block of agar from the petri dish culture. The block should be smaller than the 18 mm square cover glass to be used.

2. Place the agar block, culture side up, on an 18 mm square cover glass.

3. Expose the preparation of osmium vapor for 2–3 min by placing the cover glass, with the agar block up, in the small staining dish containing the osmium tetroxide fixative.

4. Place the agar block, culture side down, on the surface of a fresh, No. 0, 18-mm square cover glass, and press down firmly. Remove the agar and discard. Air-dry the impression smear.

5. Expose for 1–2 min to Schaudinn's fluid at 45–50°C. Rinse.

6. Place in 70% ethanol. The smears can be stored for long periods of time in 70% ethanol.

7. Remove from the 70% ethanol, rinse, and place in 1 N HCl at 60°C for 7–10 min.

8. Rinse in tap water, and 2 changes of distilled water.
9. Float, impression smear down, on Giemsa stain diluted 1:10 with Sörensen's M/200 phosphate buffer, pH 6.9–7.0. The best staining time should be arrived at empirically but should be between 20–40 min.
10. Rinse, and mount the cover glass, impression smear down, in water on a clear glass microscope slide. Seal with paraffin to prevent evaporation.
11. Examine with the oil immersion objective, and with an optical system which will allow the full use of the numerical aperture of the objective.

Results: The DNA rich areas of the cell should be stained; the cytoplasm should be colorless.

Variations: Robinow (1944) described the following rapid method which he reported as resulting in passably well defined nuclear structures. Immerse the unfixed, air dried, impression smear in N/5 HCl at boiling temperature for 5 sec. Rinse in tap water, and mount in 0.1% crystal violet.

Other staining solutions can be used. Loeffler's methylene blue is acceptable as is also a 0.1% aqueous solution of basic fuchsin.

SECTION H: STAINS FOR BACTERIAL FLAGELLA

Stains for bacterial flagella are noted for the variability in the results obtained, and attempts to execute flagella stains in the crowded atmosphere of a class laboratory are often unsuccessful. When the techniques are given individual attention, however, successful and reproducibly good results are frequent. Bacterial flagella are too thin (0.01 μm) to be seen with the optical microscope unless something is done to increase their apparent width. Flagella stains accomplish this by forming a precipitate at the surface of the flagella which so increases their width as to make them easily visible with the optical microscope. Flagella stains can be used to establish the polar or lateral distribution of flagella on a bacterial cell. However, they should not be used to count the number of flagella on a cell since several flagella in a tuft may appear as a single flagellum as the result of being mutually buried in a common precipitate. If an electron microscope is available, it should be used in preference to staining for the determination of both the number and distribution of flagella on bacterial cells. The technics are simple and the individual flagella are sharply resolved. Three flagella stains will be given below. The Leifson (1960) stain, and its variation as given by Clark (1976) can be recommended. The Gray (1926) modification of the Muir flagella stain has been a favorite since it results in slides with very clean backgrounds.

Preparation of slides: Clean slides are essential. Slides should be cleaned in a sulfuric acid-potassium dichromate cleaning solution at 70–80°C for at least 12 hr. The slides should then be washed in distilled water to remove all traces of the cleaning solution. A final treatment of the dried slides, surfaces exposed, in the hot air oven at 200°C for several hours is recommended. Gray (1926) recommended the storage of cleaned slides in ammonia alcohol. Slides stored in slide boxes for long periods of time must be recleaned.

Preparation of cultures: Leifson (1960) recommended that the cultures should be 12–18 hr old when grown on either a solid or a liquid medium. If possible, incubation should be at room temperature. Before staining, all cultures should be checked in wet mount preparations to confirm their motility. If visibly motile cultures do not show flagella on staining, the procedure will be at fault. However, nonmotile cultures may be shown to possess flagella. Suspend cells from a solid medium in distilled water to a light turbidity. Avoid placing any agar into the suspension. Centrifuge down the cells and carefully resuspend to a similar volume in 10% formalin (1 part of 37–40% formaldehyde solution to 9 parts of distilled water). Cells in liquid media should first be killed by adding 1 part of formalin to 9 parts of the culture. The cells should then be centrifuged out, the supernatant decanted, removing the last drop on the lip of the tube with distilled water. Resuspend the cells in 2 ml of distilled water, then dilute to the original volume. Recentrifuge out the cells and decant the supernatant as before, then resuspend the cells in distilled water to a light turbidity.

A simpler method of preparing the cells was suggested by Gray (1926). Cells from a solid medium were suspended in sterile distilled water in a test tube to a light turbidity and then left at room temperature for 20–30 min. During this time the cells could be checked for motility, and the motile cells apparently wash themselves free of slime and medium material.

In general, flagella can be shaken off of bacterial cells by rough handling. The best culture preparation procedure would be one that is very gentle, and which would also free the cells from any material which would produce undesirable precipitation of the staining solution in the background areas of the slide.

Preparation of smears: Leifson (1960) recommended the preparation of smears as follows. Draw a line with a wax pencil transversely across the middle of a clean slide. Be sure the line reaches both edges of the slide. A separate smear can be made on each side of this line. A large full loop of cell suspension is then placed on the distal end of the slide and tilted so that the suspension runs along the slide to the center line. Air-dry, and do not heat fix.

Gray (1926) used the following method. A large loop of the suspension is placed near the end of the glass slide. A strip of lens paper is cut narrower than the slide. One end is placed over the loop of suspension on the slide and the paper drawn gently toward the operator. A very thin film should be produced which will dry quickly in the air, or which can be gently dried in an air oven.

The selection of basic fuchsin: Basic fuchsin is supplied both as a chloride and as an acetate. These two salts of rosaniline have considerably different solubilities in both water and alcohol and this fact would influence the results obtained by flagella staining procedures. Different solubilities would give different precipitate-forming characteristics to the mordants used, and this would have a rather important effect on the results obtained. Unfortunately, both the catalog listings and the labels on the bottles purchased often fail to state whether the acetate or chloride salt of basic fuchsin is being supplied. The only answer to this problem is to use only basic fuchsin which has been specifically certified for flagella staining by the Biological Stain Commission; the use of any other product would be taking an unnecessary risk as to failure of the procedure.

Leifson's Bacterial Flagella Stain

Leifson (1960)

Preparation of solutions:

A. Dissolve 1.2 gm of basic fuchsin (Commission Certified specifically for use in flagella staining) with 100 ml of 95% ethanol. Allow 24 hr for the complete solution of the dye.
B. Dissolve 3 gm of tannic acid with 100 ml of distilled water. The solution should have a light yellow color. Mold growth can be retarded by adding 0.1 ml of liquid phenol.
C. Dissolve 1.5 gm of sodium chloride with 100 ml of distilled water.
D. The working stain solution is prepared by mixing equal volumes of Solutions B and C to A in a glass vessel which can be tightly stoppered (a ratio of 1:1:1). The stain is ready for use immediately. On storage a precipitate will form which should be left on the bottom of the vessel and not disturbed when the stain is removed from the vessel. The stain will keep about 1 week at room temperature, and from 1 to 2 months at refrigerator temperatures. As the stain ages the staining time becomes too long (10–15 min), at which point the staining solution should be discarded and replaced with a fresh preparation.

Staining procedure:

1. Prepare a slide as directed above.
2. Place the slide on a rack or black surfaced board which will slightly tilt the slides so that one end will be higher than the other.
3. Remove 1 ml of stain from the top of the working stain solution and cover the smear with it. One end of the smear should be covered by

more of the dye solution than the other due to the tilt. The stain must stand over the smear and not be running from the slide at any point.

4. The staining time should be from 5–15 min, but it will vary with many factors such as the age of the stain, the temperature, amount of air currents, or the depth of the staining solution over the smear. The principle is that the alcohol will evaporate faster than the water; and when the alcohol is reduced to 20–25%, a collodial precipitate should form which will settle onto the flagella. Close observance is needed to detect the change from a clear staining solution to the turbid or rust colored solution which is an indication of successful staining. A strong beam of reflected light against a black background is helpful in determining when the precipitate has formed.

5. After the precipitate has been observed, wash the slide in tap water. Before it is placed under the tap, do not allow any of the stain to run off the slide.

6. Drain off the water, and carefully blot dry. The use of a counterstain is optional, but if used it should be applied before the blot dry step.

7. Counterstain (when used) with a 1:10 dilution of Loeffler's methylene blue with a staining time of 1 min.

Results: The flagella should be visible, and if a counterstain was used, the vegetative cells should be blue.

A Simplified Leifson Flagella Stain

Clark (1976)

In essence this revision changes the basic fuchsin, tannic acid, sodium chloride mix of the Leifson staining solution. In addition, it includes the freezing of the stain which permits a long and indefinite storage period so that the stain is conviently available when needed. Clean slides are, of course, essential. Cells from solid media should be suspended in 3 ml of distilled water to produce a faint opalescence. Avoid carrying any agar over into the suspension. Cells in liquid media should first be formalinized by adding 0.05 ml of 37–40% formaldehyde solution per 1 ml of culture, then centrifuged, washed with distilled water, recentrifuged, and then resuspended in distilled water to give a faint opalescence.

Preparation of solutions:

A. Dissolve 1.2 gm of basic fuchsin in 100 ml of 95% ethanol. The dye recommended was a basic fuchsin special for flagella staining and obtained from Matheson Coleman & Bell, or from Eastman Kodak. Let the solution stand overnight at room temperature.

B. Dissolve 0.75 gm sodium chloride in 100 ml of distilled water, and then add 1.5 gm tannic acid.

C. Mix Solutions A and B. The final pH should be about 5.0. If it is not, it should be adjusted with 1 N NaOH.
D. Aliquots of this stain solution can be stored frozen in screw cap test tubes for an indefinite period of time. A thawed solution should be kept at 4°C until used and it will remain usable for about a month.

Procedure:
1. Pass the slide through a blue flame until very hot. Cool on a paper towel flamed side up. Place a thick wax pencil mark transversely across the center of the slide. Place a loop of cell suspension in the center of the slide next to the wax mark, and allow to run down the slide by tilting. An uneven smear is discarded since this indicates a dirty slide. Another smear can be made on the other side of the wax mark.
2. Do not heat fix. Place the slide on a rack and cover the smear with 1 ml of the stain solution at room temperature.
3. The optimum staining time will vary from 5 to 15 min and this must be determined by trial and error. Once determined, it should remain constant for that sample of the stain solution.
4. Wash with water, air-dry, and examine.

Gray's Modification of the Muir Flagella Stain

Gray (1926)

Preparation of solutions:
A. Mordant. Prepare a saturated aqueous solution of potassium alum (aluminum potassium sulfate, $AlK(SO_4)_2 \cdot 12 H_2O$). The solubility is given as 11.4 gm per 100 ml of water at 20°C, and as being infinitely soluble in hot water. Prepare also a saturated solution of mercuric chloride; the solubility is given as 6.9 gm per 100 ml of water at 20°C, and as 61.3 gm at 100°C. In addition, prepare a 20% solution of tannic acid. The actual mordant consists of a mixture of 5 ml of the saturated potassium alum, plus 2 ml of 20% tannic acid, and 2 ml of the saturated aqueous mercuric chloride.
B. Prepare a saturated solution of basic fuchsin in ethanol. The basic fuchsin should be Commission Certified specifically for flagella stains. The solubility should be about 6 gm per 100 ml of ethanol.
C. Preparation of the staining solution. Place 9 ml of mordant as prepared in Solution A in a test tube and add 0.4 ml of the saturated basic fuchsin in ethanol (Solution B). This solution is not stable and should be discarded after each day of use.

Staining procedure:
1. Prepare a smear as described above.
2. Cover the smear with about 0.5 ml of the mixed stain and let stand for 10 min at room temperature.

3. Wash in a gentle stream of distilled water until no more precipitate can be washed from the slide. The use of a counterstain is optional.
4. When used, counterstain with the carbol fuchsin used in the Ziehl-Neelsen acid-fast stain for 5–10 min.
5. Wash in tap water, drain dry, and examine.

Results: The flagella should be visible, and the vegetative cells should be stained red if the counterstain is used.

Comment: This has been a favorite flagella stain since it produces clean backgrounds.

SECTION I: STAINS FOR DIPHTHERIA ORGANISMS, DIPHTHEROIDS, VOLUTIN, OR METACHROMATIC GRANULES

Stains for diphtheria organisms should not only differentially demonstrate volutin granules (metachromatic granules) in the cells, but they should also reveal the characteristic barred staining patterns of the cytoplasmic region, both of which help to differentiate diphtheria organism and diphtheroids from other bacilli. The volutin granules of the diphtheria organisms do not differ in any important respect from those of other microorganisms, and therefore any staining procedure which differentially demonstrates volutin granules in diphtheria organisms can also be used for their demonstration in other microorganisms.

When Loeffler's alkaline methylene blue is used alone to stain diphtheria organisms or diphtheroids, the volutin granules should appear red; and the cytoplasmic area of the cells usually will stain blue and show the typical cytoplasmic barring. The red color of the volutin granules is due to the metachromatic staining property of methylene blue, a property which it shares with certain other dyes such as toluidine blue O. The light absorbing shift, which is responsible for the red appearance of the normally blue dye, is due to the very high affinity of volutin for the cationic molecule of methylene blue. When the dye molecules are tightly packed together in the volutin, the orbital paths of the electrons of the dye chromophore groups are changed, which also changes their light absorbing property (Bergeron and Singer, 1958). An important disadvantage of using Loeffler's methylene blue alone is that it often results in failure to differentiate small volutin granules, and cytoplasmic barring may also be absent. Such a staining response will make it impossible to differentiate diphtheria organisms from other bacilli. For this reason staining with methylene blue alone has largely been supplanted by methods which more clearly differentiate either the volutin granules, or the barred regions of the cytoplasm, or both. For example, volutin is chemically a very acidic polyphosphate material. Such acidic granules would be best differentiated if the pH of the cationic dye staining solution were around 2.8 (Morton and Francisco, 1942), and such an acidic staining solution will deeply color volutin granules but leave the cytoplasmic regions relatively colorless. The differentiation obtained is due to the fact that when a solution of a cationic dye is this acidic, it will be on the alkaline side of the more acidic volutin material but on the acidic side of the more basic cytoplasmic components. A cationic dye has a good affinity for a substrate with a lower pH than the pH of the solution in which it is dissolved, since under these conditions enough of the necessary anionic receptor sites are ionized in the substrate to accept significant quantities of the cationic dye molecule. Staining procedures, such as the Albert stain, which simultaneously differentiate both volutin granules and the barred regions of the cytoplasm, are based on more involved mechanisms and will be explained below.

Albert's Stain for Diphtheria Organisms or Volutin

Albert (1921)

This is a good staining procedure for the demonstration of volutin granules, but it is only a fair stain for the demonstration of the cytoplasmic barring found in diphtheria organisms and in diphtheroids. In an acidified mixture of the solutions of toluidine blue O and methyl green, the toluidine blue O preferentially stains volutin granules and the methyl green preferentially stains the barred regions of the cytoplasm. The dye in these substrates is then fixed (precipitated) in place with iodine and the preparation observed. The metachromatic property of the toluidine blue in volutin is not expressed in the Albert stain because of the effect of iodine on the dye molecule.

Preparation of solutions:
  A. Staining solution. Dissolve 0.15 gm of toluidine blue O and 0.2 gm of methyl green with 2 ml of 95% ethanol in 100 ml of distilled water. Then add 1 ml of glacial acetic acid.
  B. Iodine solution. Any iodine solution used in the Gram stain procedure is acceptable.

Procedure:
  1. Prepare the usual air-dried and heat fixed smear.
  2. Stain with the toluidine blue O, methyl green, staining solution for 1 min.
  3. Wash in water, blot, and air dry.
  4. Cover the smear with the iodine solution for 1 min, wash, blot, air dry, and examine with the oil immersion objective.

Results: Volutin granules should be stained black, and the cytoplasm light green. Diphtheria organisms, and diphtheroids, may show the typical internal cytoplasmic barred pattern which will be stained with various darker shades of green.

Comments: The most usual objection to the Albert stain is that the contrasting color given by the methyl green to the cytoplasm may be so light as to make both the cytoplasm and the barred appearance visible only with difficulty. Since the volutin granules are heavily stained, the nonvisibility of the cytoplasm may give these organisms an appearance suggesting cocci rather than bacilli. Two variations have been suggested to correct this deficiency of the Albert stain, and they are given below.

Variations:

1. Laybourn (1924) replaced the methyl green in the above Albert stain with a like amount of malachite green. This modification has been reported as giving better staining of the cytoplasmic region of the cells with a better demonstration of cytoplasmic barring. Briefly the procedure is to stain with the toluidine, malachite green, stain for 3–5 min, wash, apply Gram's iodine for 1 min, wash, blot dry, and examine. The volutin granules should be stained black, and the cytoplasmic region green, with barring appearing as deeper shades of green.

2. Christensen (1949) omitted the methyl green from the Albert (1921) stain formula and at the same time increased the glacial acetic acid from 1 to 5 ml. The iodine and safranin solutions used in Gram staining were used to fix the toluidine blue in place and to serve as the counterstain. The procedure is as follows. Stain with the acidic toluidine blue O stain for 1 min, wash, apply Grams iodine solution for 1 min, wash, then counterstain with safranin for 1 min, wash, blot dry, and examine. Volutin granules should be stained a deep black and the cytoplasm red. Many prefer this procedure over all other methods for the staining of diphtheria organisms.

Gohar's Stain for Volutin Granules

Gohar (1944)

Although this procedure was originally presented as a stain for diphtheria organisms, it can be recommended for the demonstration of volutin granules in any microorganism. It is an excellent stain for volutin, but it is relatively poor for the demonstration of the cytoplasmic barring typical for diphtheria organisms. In this procedure Loeffler's alkaline methylene blue is used to stain both the volutin granules and the cytoplasm. Differentiation of the volutin granules occurs when an acid decolorizer is used to destain the cytoplasm. The methylene blue in the volutin granules is then fixed in place with iodine, and the colorless cytoplasm then differentially stained with eosin.

Preparation of solutions:

A. Loeffler's alkaline methylene blue. Dissolve 0.3 gm of methylene blue chloride with 30 ml of 95% ethanol and add to this 100 ml of 0.01% aqueous potassium hydroxide.

B. Acid decolorizer. Add 0.1 ml of concentrated sulfuric acid to 100 ml of distilled water.

C. Iodine solution. Use the same iodine solution as for the Gram stain procedure.

D. Counterstain. Dissolve 1 gm of eosin Y with 100 ml of distilled water.

Procedure:

1. Prepare the usual air-dried and heat fixed smear.
2. Stain with Loeffler's alkaline methylene blue for 5 min, then wash.
3. Decolorize for only a few seconds with the dilute sulfuric acid, then wash.
4. Expose for 1 min to Gram's iodine solution, then wash.
5. Counterstain with the aqueous 1% eosin Y for 1 min, wash, blot dry, and examine.

Results: Volutin granules should be stained black, the cytoplasmic region pink.

Variation: A deeper colored counterstain, such as safranin or basic fuchsin, can be used.

Comment: Over decolorization could be a danger, and would result in a poor demonstration of volutin granules.

SECTION J: STAINS FOR SPIROCHETES IN SMEARS

The visualization of various species of spirochetes in smears is made difficult by the fact that many of them have a width which is near, or less than, the resolving power of the optical microscope. *Treponema* species are 0.1–0.2 μm wide and 5–15 μm long; *Leptospira* species are about 0.1 μm wide and 6–20 μm long; while *Borrelia* species are 0.2–0.5 μm wide and 10–20 μm long. Of these only *Borellia* can be comfortably visualized with the optical microscope using ordinary staining methods, and some form of augmentation of the width or optical contrast of the organisms must be used if *Treponema* or *Leptospira* are to be seen. Since such augmentation also may have an advantage for *Borrelia*, the same methods of augmentation can be recommended for all spirochetes.

The dark field microscope is an excellent method for the demonstration of all spirochetes and its use can be recommended whenever possible. Negative stains, such as nigrosin, have been used and are ample for the demonstration of the wider spirochetes, but cannot be recommended for those which are less than 0.2 μm in width. Most staining methods are based on the deposition of a precipitate on the surface of the organism to increase its apparent width in much the same manner as the staining methods used for the demonstration of bacterial flagella. In fact the Leifson flagella stain can also be used as a stain for spirochetes. Impregnation of the organism with silver salts traditionally has been one of the most popular staining methods since it yields preparations in which the width of the organisms is adequately increased and the organisms are an opaque black against a lightly colored background. The greatest objections to the various silver staining methods is that the preparation of the reagents is complicated, and the results often vary greatly in quality from day to day. Other staining methods include the use of various mordants such as tannic acid, sodium bicarbonate, or potassium permanganate, usually followed by staining with a basic dye which forms a precipitate in the presence of the mordant. Examples of these procedures will be presented below.

Fontana-Tribondeau Silver Stain

Fontana (1926)

This is an ammonical silver nitrate stain using tannic acid as the so-called mordant. Gilbert and Bartels (1924) found that the Fontana stain, and Tribondeau's variation of the Fontana stain, were superior to many other silver staining methods with which they were compared. Fontana (1926) combined the Fontana and the Tribondeau stains into the procedure given below. This procedure is recommended and it results in a striking contrast of the organisms in relation to the background material. Its principle drawback is the rather indefinite instructions for the preparation of the ammonical silver nitrate reagent.

Preparation of solutions:
- A. Ruge's solution. Add 1 ml of glacial acetic acid, and 2 ml of formalin (37–40% formaldehyde) to 100 ml of distilled water.
- B. Mordant. Dissolve 5 gm of tannic acid in 100 ml of distilled water, then add 1 ml of phenol (melted crystals).
- C. Ammonical silver nitrate. Ammonia water (ammonium hydroxide, technical, 26° Baumé) is diluted 1 part to 19 parts of distilled water. Dissolve 1 gm of silver nitrate in 100 ml of distilled water, then add the diluted ammonia water drop by drop until a coffee colored turbidity appears. If the correct point is exceeded, the turbidity will disappear; however, it can be reformed by adding more 1% silver nitrate. This solution will keep well.

Procedure:
1. Prepare a smear which is not too dense with material and air-dry.
2. The smear can be heat fixed, or heat fixation can be omitted.
⟋ 3. Flood the smear with Ruge's solution for 1 min. During this period renew the solution several times.
4. Wash in running water.

5. Cover the smear with a few drops of the tannic acid, phenol, mordant for 20 sec. During this time gently heat the slide to steaming. Wash.
6. Cover the smear with the ammonical silver nitrate solution for 30 sec. During this time heat gently.
7. Wash, dry, and examine.

Results: The spirochetes and other microorganisms will be dark brown to an opaque black.

Variations: According to DeLamater, Haanes, and Wiggall (1950) the ammonical silver nitrate of the above procedure can be replaced with a carbol crystal violet stain. One part of a saturated alcoholic solution of crystal violet is mixed with 10 parts of an aqueous 5% solution of phenol. The solubility of crystal violet in absolute ethanol has been given by Gurr (1960) as 8.75 gm per 100 ml. Stain for 2 min, dip once in distilled water, drain on filter paper, dip once in acetone, dry between filter paper, then gently heat to complete the drying process. The spirochetes should be stained a deep purple violet, with a lightly stained background.

Warthin and Starry Silver-Agar Stain

Warthin and Starry (1922)

Preparation of solutions:

A. Dissolve 2 gm of silver nitrate in 100 ml of distilled water. The solution should not be over 7 days old. It should be stored in a dark, tightly stoppered, bottle at refrigerator temperature.
B. Add 1.5 gm of agar to a clean 150-ml flask. Add 30 ml of distilled water and allow to soak until the agar is saturated with water. Pour off the excess water and wash with several changes of distilled water. Then add 100 ml of distilled water and melt the agar by bringing the temperature up to boiling. After the agar has melted, pour the resulting 1.5% agar colloid sol into a clean bottle which can be tightly stoppered. As the agar cools to 45–42°C, it will begin to gel. The gel must be broken up by shaking violently. The final product is kept warm on the top of a paraffin oven (such ovens are usually set at about 55°C). The final agar preparation should be just fluid enough to barely run when the bottle is inverted.
C. Place in a 50-ml flask, with constant stirring, 3 ml of a 2% silver nitrate solution; add 5 ml of a warm (35–45°C) aqueous 10% gelatin; add 5 ml of warm (35–45°C) glycerol; and then add 5 ml of the warm agar prepared in (B). Just before use, add with stirring 2 ml of an aqueous 5% solution of hydroquinone. This mixture is then placed in a cover glass staining dish so that any added cover glasses will be completely immersed. This solution must be prepared fresh just before use.

D. The following reagents will also be needed: absolute ethanol, an aqueous 5% solution of sodium thiosulfate, and possibly a concentrated solution of hydrogen peroxide. A supply of clean No. 1 or 0 cover glasses is essential.

Procedure:
1. Prepare the smear on the surface of a clean cover glass. Thoroughly dry in air. Smears dried for 3 or 4 weeks are still usable. Do not heat fix, or use any method of fixation other than absolute ethanol.
2. Immerse in absolute ethanol for 3–5 min. Wash in distilled water.
3. In some cases (usually when the smear is rather thick) it will be desirable to clear the background of the smear by immersing in concentrated hydrogen peroxide for a period of 5–10 min, then wash through 2 or 3 changes of distilled water.
4. Rinse in 2% silver nitrate. Cover the smear side of the cover glass with another clean cover glass which has also been rinsed, and is wet with, 2% silver nitrate. The adhering cover glass without a smear is essential to prevent an overly stained background. Immerse the adhering cover glasses in 2% silver nitrate in a small, brown, tightly stoppered, wide-mouth bottle. Keep the adhering cover glasses standing on edge against the side of the bottle and at least half covered with the silver nitrate solution. Be sure of some method of identifying the cover glass with the smear. Place in a 37°C incubator for 1–2 hr. Then remove the cover glasses from the solution and separate them.
5. The cover glass with the smear is then placed in a cover glass staining jar with the silver nitrate, gelatin, glycerol, agar, hydroquinone, developer mixture for a period of 30 sec to 2 min, or until the reaction is complete. The smear can be seen to turn a light brown; however, when the smear is very thin, no color change may be seen. The speed of this reaction can be increased or slowed by raising or lowering the amount of hydroquinone added to the mixture. This developer solution must be prepared fresh just before use.
6. Rinse in 5% sodium thiosulfate for a few seconds. Then rinse in distilled water.
7. After exposure to absolute ethanol, the smears can be dried and examined; or run through xylene, and mounted in Permount, balsam, or other suitable mounting medium.

Results: The spirochetes will be black against a light background.

Comments: These preparations may fade with time. The smears can be made permanent by toning in a solution of 6.25 gm of ammonium thiocyanate (NH_4SCN), 0.5 gm of tartaric or citric acid, 1.25 gm of sodium chloride, and 6.25 ml of a 1:100 solution of gold chloride, all in 250 ml of distilled water.

Sodium Carbonate Basic Fuchsin Stain

Ryu (1963)

Preparation of solutions:
A. Dissolve 5 gm of sodium bicarbonate ($NaHCO_3$) in 100 ml of distilled water.
B. Dissolve 0.75 gm of basic fuchsin (chloride) with 25 ml of 95% ethanol. Before use, dilute 1 part of this solution with 9 parts of distilled water.
C. Add 1 ml of formalin (37–40% formaldehyde) to 99 ml of distilled water.

Procedure:
1. Prepare a thin smear on a clean glass slide and air-dry.
2. Fix in 1% formalin for 1 min, then drain dry.
3. Place 1 drop of the 5% $NaHCO_3$ on the smear, then add 10 drops of the basic fuchsin, mix well, and stain for 3–5 min at room temperature. A visible precipitate should form.
4. Wash, dry, and examine.

Results: The spirochetes should appear red.

Variations: A similar stain, using 2.5% sodium carbonate (Na_2CO_3) and 1% crystal violet has been described by Levine (1952). In this procedure the smear is immersed in a mixture of 89 ml of distilled water, 1 ml of the 2.5% Na_2CO_3, and 10 ml of the 1% crystal violet, mixed in the order given just before use in a Coplin staining dish. Stain for 2 min, then wash, dry, and examine.

Potassium Permanganate Crystal Violet Stain

Harris (1930)

Preparation of solutions:
A. Dissolve 1 gm of potassium permanganate ($KMnO_4$) in 100 ml of distilled water.
B. Prepare an aqueous 2% solution of crystal violet.

Procedure:
1. Prepare a thin smear, air dry, and heat fix.
2. Cover the smear with the 1% potassium permanganate solution, 8–10 min, then wash.
3. Stain 8–10 min with the 2% crystal violet, then wash.

Results: The spirochetes should be stained black and should stand out clearly from background material.

Variation: When staining *Treponema pallidum*, one is advised to gently heat the crystal violet stain.

Negative Stain for Spirochetes

Dienst and Sanderson (1936)

Preparation of solutions: Dissolve 5 gm of nigrosin (water soluble) in 100 ml of distilled water, and then add 1.0 ml of formalin as a preservative.

Procedure:

1. Mix a loop of exudate material directly into a loop of nigrosin solution on a clean glass slide.
2. Spread the mixture with the loop, starting in the center of the material and spreading through a series of gradually increasing circular movements. The object is to prepare a smear with various degrees of thickness. This allows the selection of a field which best demonstrates the spirochetes.
3. Rapidly dry by heating. The oil can be placed directly on the nigrosin for observation with the oil immersion objective.

Results: The spirochetes should be colorless against a dark background. It should be remembered that some spirochetes are very thin and they will be difficult to see with the optical microscope using a negative stain.

Variation: A drop of exudate material can be allowed to dry on a slide and then stored for several weeks. The dried smear can be prepared for nigrosin staining by adding a loop of water, letting stand for a few sec, then adding a loop of nigrosin and proceeding as above.

SECTION K: STAINS FOR RICKETTSIAE, COXIELLA, AND CHLAMYDIAE

Electron microscopy has clarified the taxonomic position of the rickettsiae (typhus and spotted fevers) and the chlamydiae (psittacosis-ornithosis, trachoma, lymphogranuloma venereum, groups of infective agents). They are now considered to be small bacteria since they are prokaryotic cells possessing both RNA and DNA and they reproduce by the process of binary fission. They differ from all other bacteria by being obligate intracellular parasites. The chlamydiae differ from the rickettsiae by being completely dependent on the host cell for the process of oxidative phosphorylation, by possessing a definite life cycle from the infective elementary body to the reproductive form called the initial body with subsequent reversion to the infective form, and by the ability of the elementary body to pass through bacteriological filters.

The rickettsiae and chlamydiae are difficult to demonstrate with the optical microscope for several reasons. They are very small, 0.2–0.3 μm wide to 0.2–1.0 μm long, which is just above the resolution limit of the optical microscope; they are Gram-negative and therefore the Gram stain does not differentiate them from the cellular material with which they are usually associated; and no really good staining procedure is available to differentiate them from background material. The best of the differential stains available are the Macchiavello stain (1937), and the more recent, and possibly better, procedure of Giménez (1964). Both are given below. Roger and Roger (1958) indicated that the staining procedures applicable to the rickettsiae could also be applied to the chlamydiae. This makes sense since the organisms are very similar.

Macchiavello Stain for Rickettsiae

Macchiavello (1937)

Preparation of solutions:
A. The primary stain. Prepare 100 ml of 0.1 M phosphate buffer, pH 7.3–7.4 by adding 80 ml of 0.1 M anhydrous dibasic sodium phosphate (Na_2HPO_4) to 20 ml of 0.1 M monobasic sodium phosphate ($NaH_2PO_4 \cdot H_2O$). Then add 0.25 gm of basic fuchsin chloride. The solution must be prepared fresh each day of use and a visible precipitate should form. The precipitate must be removed by filtration, and it is recommended that the stain be passed through a coarse filter paper when applied to the slide.
B. Prepare a 0.5% solution of citric acid.
C. The counterstain. Add 10 ml of 95% ethanol to 1 gm of methylene blue chloride and mix, then slowly add 100 ml of distilled water, and 5 ml of phenol (melted crystals). For use, dilute this stock solution 1:10 with distilled water.

Procedure:
1. Prepare a smear from tissue or yolk sac material on a clean slide. Air-dry, then gently heat fix.
2. Cover the smear with the buffered basic fuchsin. This solution should be filtered through a coarse filter paper when applied to the slide. Stain for from 3–5 min; 3 min is recommended. Drain off the excess dye.
3. Rinse 1–3 sec with 0.5% citric acid.
4. Rinse in tap water.
5. Counterstain with the 1:10 dilution of the stock methylene blue solution for a period of 1–2 sec.
6. Rinse with tap water, blot dry, and examine.

Results: Rickettsiae and chlamydiae should be stained a brilliant red; cellular material should be stained a pale to deep blue.

Variations: Macchiavello's instructions for this procedure included a great flexibility in the reagents used and in the procedure to follow. The use of a phenol methylene blue solution as the counter stain was based on its availability in Macchiavello's laboratory. Although he recommended the use of the phenol methylene blue, he also reported that non phenolated dye solutions were equally effective. The most common variation of the above procedure is to replace the phenol methylene blue with a simple 0.1% aqueous solution of methylene blue and to increase the time of counterstaining to 1–2 min. Also widely used is 1.0% aqueous methylene blue with the time of counterstaining reduced to 10 sec. Macchiavello reported that several different cationic (basic) dyes of contrasting color to the primary stain could be used as the counterstain if suitable changes in the time of counterstaining were made.

Macchiavello stated that the basic fuchsin solution used as the primary stain could vary from 0.25 to 1.0%; however, most subsequent versions of this staining procedure recommended 0.25%. He also found that the use of a buffered basic fuchsin was not essential. If the phosphate buffer is omitted, this avoids the formation of a precipitate and allows the use of the primary stain without the filtration step.

Roger and Roger (1958) found that the citric acid decolorization step was not necessary for some smear preparations. Differentiation could be accomplished entirely through the principle of dye replacement by first staining with 0.025% basic fuchsin for 5 min, followed by 0.1% methylene blue for 2 min.

Comments: The mechanism of this staining procedure is based on the greater affinity of cationic (basic) dyes for rickettsial and chlamydial organisms, as compared to tissue cells when stained at pH values near neutrality or above. The acid decolorization step emphasizes this difference by decolorizing tissue cells faster than the microorganisms, and the difference is again emphasized by differences in the speed of dye replacement on the application of the counterstain. This explains the variability of the results obtained when using this procedure since the rickettsiae and chlamydiae cells from different sources undoubtedly vary in their affinity for cationic dyes for a variety of reasons. If poor differential results are obtained, it is very probable that improvement could be obtained by varying the strength and the time of application of the counterstain.

Giménez Stain for Rickettsiae and Psittacosis Agents in Yolk-Sac Cultures

Giménez (1964)

Preparation of solutions:

A. The stock solution of the primary stain. Add 10 ml of 95% ethanol to 1 gm of basic fuchsin chloride and mix by stirring. Add 65 ml of distilled water, and then 25 ml of 4% phenol, and continue stirring until the basic fuchsin is dissolved. Store 48 hr at 37°C before use.

The stock solution is stable and will keep for months. For use, dilute with phosphate buffer as described below.

B. The working solution of the primary stain. Prepare a 0.1 M phosphate buffer, pH 7.45, by mixing 17.6 ml of 0.1 M monobasic sodium phosphate ($NaH_2PO_4 \cdot H_2O$) with 82.4 ml of 0.1 M anhydrous dibasic sodium phosphate (Na_2HPO_4). The working stain is prepared by adding 4 ml of the stock stain solution to 10 ml of the 0.1 M phosphate buffer, pH 7.45. A precipitate will form and this solution must be immediately filtered, and filtered again before each use. The working solution of the primary stain should be prepared fresh on the day of use.

C. Dissolve 0.8 gm of malachite green oxalate with 100 ml of distilled water.

D. Dissolve 4 gm of ferric nitrate ($Fe(NO_3)_3 \cdot 9H_2O$) with 100 ml of distilled water.

E. For *Rickettsia tsutsugamushi* only, dissolve 0.5 gm of fast green FCF with 100 ml distilled water.

Procedure (for *Rickettsia typhi, Rickettsia prowazeki, Rickettsia rickettsiae, Rickettsia akari, Coxiella burnetii*, and for psittacosis agents):

1. Prepare a very thin smear from yolk-sac tissue from which the yolk has been drained as much as possible. Air-dry. The smear can be gently heat fixed, or heat fixation can be omitted.
2. Cover with a freshly filtered working Solution B of carbol fuchsin, and stain for 1–2 min.
3. Wash thoroughly with tap water.
4. Cover with the malachite green oxalate solution for 6–9 sec. Wash with tap water. Cover again with the malachite green for 6–9 sec. Wash in tap water.
5. Blot dry, and examine.

Results: The rickettsiae and psittacosis organisms should be stained a strong red; cells should be stained a greenish blue, and the background material should be stained a light green.

Procedure (for *R. tsutsugamushi*):

1. The procedure is the same through Step 3 above.
2. Cover the smear with 4–6 drops of 4% ferric nitrate, and then wash immediately and thoroughly in tap water. The red color of the smear will almost disappear.
3. Cover with 0.5% fast green FCF for 13–30 sec, then wash in tap water.
4. Blot dry, and examine.

Results: The rickettsiae should be stained a reddish black; the background material should be stained green.

Comment: Giménez accepts the dye replacement principle as the mechanism of this differential stain. He reported that one batch of basic fuchsin (Matheson, Coleman & Bell, B300, Cert. No. 51, dye content 94%) was excellent while another product with a dye content of 88% was not satisfactory. Filtration of the buffered carbol fuchsin just before use is essential or particles of red precipitate can confuse the reading of results.

SECTION L: STAINS FOR MYCOPLASMA (PLEUROPNEUMONIA AND PPLO ORGANISMS) AND L-PHASE VARIANTS OF BACTERIA

Mycoplasma, small bacteria without cell walls, are capable of growing on artificial culture media, and are not known to have originated from, or to be able to revert to, forms with cell walls. They are the smallest cells known which are capable of a free living existence, and because of the absence of cell walls they are highly pleomorphic. In the smallest form (0.125–0.15 μm) they are capable of passing through bacteriological filters. The fragility of these cells, and their small size, makes them difficult to demonstrate with the optical microscope. On the surface of solid culture media they form small colonies (10–600 μm) which are round with a granular center (the so called "fried egg" colony). The small size of the colonies and their transparency make them difficult to see without the aid of magnification and staining. The most useful stains for *Mycoplasma* colonies not only make the colonies more visible but also differentiate *Mycoplasma* colonies from those of other bacteria. The study of the morphology of individual cells of *Mycoplasma* can best be accomplished through the methods of electron microscopy.

The L-phase growth variant of bacteria are variants without cell walls, which are known to be produced by a wide variety of bacterial species. Such variants differ from *Mycoplasma* cells in that they are generally larger, and they can either spontaneously revert to their normal form with cell walls, or it is definitely known that the variant originated from a form with cell walls. Like *Mycoplasma* cells, growth on a solid culture medium produces a typical "fried egg" colony which is usually much larger than the colonies of *Mycoplasma*.

The problem of staining *Mycoplasma* cells and L-phase variants for observation with the optical microscope is the same. Because of the absence of a cell wall both are very fragile and the cells would be destroyed by the usual methods of smear preparation and heat fixation; both do not stain well as individual cells, and both are best stained through direct staining of young colonies on a solid culture medium. The easiest of the colony stains is that of Dienes as modified by Madoff (1960); however, the preparations are not permanent. Permanent, and better preserved cells can be obtained by the use of the more complex agar-fixation method of Klieneberger-Nobel (1962).

Most staining methods for these organisms call for freshly grown cells on the surface of a solid culture medium. A typical medium would contain a beef heart infusion base to which would be added peptone to 1%, horse serum to 10%, or ascitic fluid to 20%, one part of 25% yeast extract to 10 parts of medium, and the pH would be adjusted to 7.6–8.0. This medium can be used in the liquid form, or it can be made into a solid culture medium by adding agar to a level of 1%. Antibacterial agents such as penicillin (300–3000 units per ml) can be added to control bacterial contaminants. Incubation can be carried out aerobically in sealed plates; or if an anaerobic environment is wanted, a mixture of 5% CO_2 and 95% N_2 is acceptable. The cells are so delicate that inoculation of solid media from liquid media should be accomplished by simply laying a loop of material on the surface of the medium without streaking. Agar to agar transfers should be accomplished by cutting a block from the parent culture, placing it culture side down on to the surface of the new medium, and drawing the block along the surface of the new medium.

Dienes Stain for Colonies of Mycoplasma and L-Phase Variants of Bacteria

Madoff (1960)

Preparation of solutions: The stain. Dissolve in 100 ml of distilled water, 2.5 gm of methylene blue chloride, 1.25 gm of azure II, 10 gm of maltose, 0.25 gm of anhydrous sodium carbonate (Na_2CO_3), and 0.2 gm of benzoic acid.

Procedure:
1. Obtain clean microscope cover glasses of No. 1 or 0 thickness, and about 12 mm square.
2. Use a sterile cotton applicator to apply a thin even film of the stain to one side of several cover glasses, and allow to dry. The cover glasses with the film of stain can be stored in a sterile petri dish until used.
3. Cut an agar block about 6–8 mm square and 2–3 mm thick from a fresh agar culture of the organism. Place the agar, culture side up, on the surface of a clean glass slide.
4. Cover the block with one of the cover glasses with the dry film of stain, stain side next to the cells, leaving a projecting rim of the cover glass on all 4 sides of the block.
5. Fill the space around the agar block with a melted mixture of paraffin and 10% Vaseline. This hardens sufficiently to give an adequate support to the cover glass for observation with the oil immersion objective, as well as slowing down the process of dehydration.
6. After a few minutes, the surface of the agar can be examined for stained colonies. Begin with low magnifications to find the colonies, then use higher magnifications for closer observation.

Results: Colonies of *Mycoplasma* and bacteria can be differentiated as follows. At low magnifications the colonies of *Mycoplasma* show the "fried egg" appearance with the center of the colony stained deep blue, while the edge of the colony is only light blue. Bacterial colonies stain evenly and will lose color in a short time. At high magnifications (oil immersion) the typically pleomorphic appearance of *Mycoplasma* cells can be seen ranging from tiny granules in the center of the colony to large bodies 5–20 μm in diameter at the edges of the colony. The cells in colonies of bacteria, other than *Mycoplasma*, take on the usual appearance of bacterial cells with a constant size and shape.

Comments: These slide preparations can be stored for several days, but they cannot be considered to be permanent. The procedure is simple, however, and it can be recommended. The differentiation of *Mycoplasma* and L-phase variants is more difficult. L-phase variants usually produce larger colonies, and have larger individual cells.

Klieneberger-Nobel Agar-Fixation Stain for Colonies of Mycoplasma and L-Phase Variants

Klieneberger-Nobel (1962)

Preparation of solutions:
  A. Bouin's fixative. Add 75 ml of saturated aqueous picric acid to 25 ml of formalin (37–40% formaldehyde) and 5 ml of glacial acetic acid. The solubility of picric acid has been given as 1.4 gm per 100 ml of water at 20°C.
  B. Giemsa stain, stock solution. This can be purchased ready to use, or it can be prepared from the powder form as follows. Place 33 ml of glycerol in a beaker and bring to 55°C in a water bath. Slowly add 0.5 gm of Giemsa stain (powder form) with stirring. Hold at this temperature for 1½–2 hr with occasional stirring. Add 33 ml of methanol with stirring. Cool, and store overnight in a tightly stoppered bottle. Decant into 2 smaller (30 ml) bottles leaving no air space, and store at least 2 weeks before use. Filter before use.
  C. Phosphate buffer, 0.1 M pH 6.8. Add to 49 ml of 0.1 M anhydrous dibasic sodium phosphate (Na_2HPO_4) 5 ml of 0.1 M monobasic sodium phosphate ($NaH_2PO_4 \cdot H_2O$).

Procedure:
  1. Cut an agar block with visible colonies from a fresh agar culture and place the block, culture side down, on to a clean and sterile No. 1 or 0 cover glass. Better results can be obtained if the organism is allowed to

grow after placing the block on the cover glass. This can be done by pressing a block of agar with visible colonies, culture side down, on to the surface of a fresh agar medium in a Petri dish. The area of the new medium inoculated by the colonies can then be cut out and placed culture side down on to the sterile cover glass and the entire preparation incubated in a moist chamber for 2–3 days. The agar blocks of such preparations should not be over 2 mm thick.

2. Place the agar block-cover glass preparation in a watch glass, agar block up, and cover with the Bouin's fixative. Fixation is by diffusion through the agar, and it requires from 8 hr to overnight.

3. Peel off the agar block and discard. The cover glass should then be washed in water until the yellow of the picric acid has disappeared.

4. The cover glass is then stained for 8–24 hr in a Coplin cover glass staining dish filled with the following stain. Mix 1 ml of the stock Giemsa stain with 47 ml of distilled water and 2 ml of 0.1 M phosphate buffer with a pH of about 6.8. The degree of staining can be followed by placing the cover glass on a wet slide, culture side up, and observing with the high dry objective. When stained sufficiently, proceed.

5. Drain away any excess dye by standing the cover glass on its edge on a filter paper.

6. Before the cover glass can dry, pass it through the following solutions. The transfer from one solution to the next must be quick, allowing only enough time for the loss of water.
   a. Acetone 19 ml and xylene 1 ml.
   b. Acetone 14 ml and xylene 6 ml.
   c. Acetone 6 ml and xylene 14 ml.
   d. 2 or 3 changes in xylene

7. Mount in balsam and observe.

Results: The individual cells should be well preserved, and the relationships of the cells to colony appearance can be seen. These mounts are permanent.

Variation: Organisms from liquid cultures can be observed by placing a loop of the culture on a sterile cover glass, then covering with a block of semisolid culture medium, followed by incubation in a moist chamber. After incubation, fix and stain the agar-cover glass preparation as above. If permanent mounts are not required, the cover glass can be mounted in water and observed after Step 5.

Comment: This method is more complicated than that of Dienes (Madoff, 1960), but it is reported to give better preservation of cells, and the mounts are permanent.

Dienes Agar-Fixation Stain for Colonies of Mycoplasma and L-Phase Variants

Dienes (1967)

Preparation of solutions:

A. Prepare a 2% aqueous solution of toluidine blue O, or azure II, or thionin. Toluidine blue is recommended.

B. Prepare a 20% solution of tartaric acid. This is almost a saturated solution since tartaric acid is reported as having a solubility of 20.6 gm per 100 ml of distilled water at 20°C.

Procedure:

1. Fresh young cultures on an agar medium are required. Inoculate the surface of a suitable agar culture medium in a petri dish with organisms and incubate 1 or 2 days until visible colonies are seen. The agar should be at least 3 mm deep.

2. After visible colonies appear, invert the petri dish and place in the lid a piece of filter paper on which is then placed 8 drops of formalin (37–40% formaldehyde). Either seal the petri dish, or place in a closed container. The formaldehyde vapors will have produced sufficient fixation after an exposure period of from overnight to a few days.

3. Place on the surface of a No. 1 or 0 cover glass (24 × 40 mm) a loop (3 mm diameter) of stain, and mix in a loop of 20% tartaric acid. Spread the mixture evenly over the surface of the cover glass and let air dry. After drying, rub lightly with a piece of filter paper to accelerate the crystallization of the tartaric acid.

4. Thin agar blocks should be cut from the fresh cultures as follows:

   a. Cut a block of agar from the formaldyhyde fixed plate. The surface of the block should have one or more visible colonies.

   b. Place the block on the surface of a clean glass slide, culture side up. Place along one side of the agar block sufficient layers of glass slides to leave about 1 mm of agar higher than the slides. Then cut a 1 mm or thinner layer from the top of the agar block using a sharp razor blade and using the glass surface beside the block as a support, and a cutting action similar to a microtome. The tops of several blocks can be placed in a row and cut at the same time.

5. Staining should be carried out at a temperature of from 50–60°C, and the blocks should not be allowed to dry out. Dienes recommends the use of a 100 watt bulb in a gooseneck lamp at a suitable distance from the work material to heat it to this temperature.

6. Place the newly cut, 1 mm or less thick, agar block on to a prewarmed cover glass with a film of stain, the culture side against the stain.

Several blocks can be placed on a single cover glass. Work sufficiently fast so that no visible dehydration of the agar block appears.

7. Place the cover glass, blocks up, on the surface of a glass slide. Cut 2 filter paper strips the size of the slide. Moisten one with water and then blot it between 2 layers of filter paper, and place it over the blocks. Cover the moist filter paper with a dry strip of paper. Keep in place by weights placed on both ends of the slide.

8. Place the preparation in the lid of an inverted petri dish containing culture medium and preheated to 50°C, and then place the inverted petri dish in an oven at 50°C for a staining period of 30–60 min. The medium in the petri dish serves to prevent drying of the blocks. Better staining can sometimes be obtained at a temperature of 70–80°C.

9. In the heat of the 100 watt lamp, remove the preparation from the petri dish and remove the filter papers and the slide. Allow the cover glass agar block preparation to dry. If the medium is over 1.5% agar, or if the slice is not sufficiently thin, the dried slides may curl up and separate from the cover glass. This can be prevented by touching the dry edges of the slices with balsam before the center of the block becomes dry.

10. Mount in balsam and examine.

Results: *Mycoplasma*, L-phase variants, normal bacteria, fungi, and tissue cells, all are stained; however, the intensity of staining varies. Observation with the oil immersion objective allows the differentiation of these organisms on a morphological basis.

Variation: Organisms grown and stained on cover glass preparations can be achieved as follows. Inoculate the surface of an agar medium, either by placing a loop of liquid culture on the surface, or by pressing an agar block with a visible colony on to the surface and moving the block gently back and forth with a loop, and then removing the agar block. A 1 mm thin agar block is then cut from the inoculated area and placed, culture side against the glass, on a sterile cover glass. After incubation for 1–2 days in a moist chamber, the preparation is then fixed with formaldehyde vapors for from overnight to several days under conditions which prevent dehydration of the agar block. After fixation, the agar block on the cover glass is covered with a 1:1 mixture of the stain and 20% tartaric acid, and stained for a period of 5–10 min in a moist chamber at 50°C. After staining, the preparation is then covered with a moist filter paper and handled like the usual preparations.

SECTION M: THE STAINING OF BACTERIA IN DAIRY PRODUCTS

The direct microscope counting of bacteria in smears has been one of the methods used to determine the numbers of bacteria in dairy products for many years (Prescott and Breed, 1910; Breed and Brew, 1918; *Standard Methods for the Examination of Dairy Products*, 1978). Acceptable procedures have several things in common. They all require scrupulously clean slides; otherwise parts or all of the smear may be lost during processing which would result in nonvalidity of the bacterial count obtained. Some method of fixing the smear to the slide is incorporated, as well as a defatting step, and the staining procedure used visibly differentiates bacterial cells from the background milk material. Studies have shown that the methylene blue stain, recommended in the original Breed procedures, is a relatively poor stain (Olson and Black, 1951; Mantel and Robertson, 1954). Better procedures include the aniline methylene blue stain (North, 1945), the acid and water-free stain (Levine and Black, 1948), the polychrome methylene blue stain (Anderson, Moehring, and Gunderson, 1948), and a modified Newman (No. 2) single step stain (Levowitz and Weber, 1956). The three most recommended procedures will be given below.

Levowitz and Weber Modification of the Newman (No. 2) Single Solution Stain

Levowitz and Weber (1956)

Preparation of solutions: Preparation of the stain and defatting solution. Place 0.6 gm of methylene blue chloride in a beaker, and with stirring add 52 ml of 95% ethanol. Stir until the dye is dissolved. Continue stirring and add 44 ml of tetrachlorethane (technical). Place in a closed 100 ml container and let stand for 12–24 hr in a refrigerator at about 5°C. Bring to room temperature and add 4 ml of glacial acetic acid, and then filter through a Whatman No. 2 filter paper or equivalent. Store in a tightly stoppered container in a relatively cool and dark area, but do not refrigerate.

Preparation of slides: Clean glass slides are required since the rather thick smears produced from dairy products may peel from the slide during the staining procedure. If possible, use new slides. If old slides must be used, first clean with Bon-Ami, rinse in distilled water, immerse in dichromate-sulfuric acid cleaning solution overnight, rinse in distilled water, heat in an oven at 200°C for several hr with the glass surfaces exposed to the hot air, cool, and then store in 95% ethanol until required. If a quantitative determination of the numbers of organisms is required, a measured quantity (0.01 ml) of the dairy product (or a known dilution of it) is evenly spread over a 1 sq cm area of the glass slide. A calibrated microscope is required and the counts should be made as described in the latest edition of the *Standard Methods for the Examination of Dairy Products* (1978). Some dairy products should be diluted in 1.25% sodium citrate to assure the solution of casein.

Procedure:

1. The smears should be rapidly dried in a small hot air oven or on a thermostatically controlled hot plate at 45–50°C; the temperature should not rise over 51°C. Cool to room temperature.
2. Stain and defat for 2 min with the methylene blue, ethanol, tetrachlorethane solution. Use a closed staining dish which is capable of preventing the evaporation of the solution. The slide should be completely covered by the solution and it should be discarded at the first sign of any formation of precipitate.
3. Remove the slide from the staining dish and drain off the excess stain by resting the long edge on the surface of absorbant paper. Dry thoroughly, using forced air from a blower or fan.
4. Rinse the dried, stained, slides by passing through 3 changes of tap water at 38–43°C.
5. Dry rapidly and thoroughly, using forced air from a blower or fan, and examine.

Results: Normal bacterial cells are stained heavily, the background material should be evenly and lightly stained. Plasmolyzed bacterial cells should stain with various degrees of intensity depending on the degree of plasmolysis. Leucocytes should be stained with the cytoplasm slightly darker than the background and the nuclear areas more deeply stained which allows their differentiation into types.

Comments: The staining solution serves to fix and defat the smear as well as to stain the organisms. Levowitz and Weber found that this stain was the equivalent to North's aniline methylene blue stain, and also to the Levine and Black acid and water free stain, for the enumeration of bacteria in smears of dairy products.

Acid and Water Free Stain for Dairy Products

Levine and Black (1948)

Preparation of solutions:

A. Preparation of the stain. Dissolve 0.6 gm of methylene blue chloride with 100 ml of 95% ethanol. Store in a tightly stoppered container, and age 2 days before use.
B. Needed also are xylene, and 95% ethanol.

Procedure:

1. Prepare a smear on a clean glass slide by the procedure usually used for dairy products (see the Levowitz and Weber procedure). Dry rapidly and thoroughly on a thermostatically controlled hot plate at 54°C.
2. Fix the dried smear by immersion in 95% ethanol for 1–2 min.
3. Defat the smear by immersion in clear xylene for 1–2 min. Drain and air-dry. Do not use heat or forced air.

4. Immerse in 95% ethanol for 1–2 min, and then immerse in the staining solution for 1–2 min. Overstaining should not be a problem. Remove the slide and drain off the excess stain; and while the slide is still wet, gently rinse in tap water in a beaker (not in flowing water) by gently raising and lowering the slide in the water. Avoid any vigorous movements of the slide in any direction.
5. Thoroughly air-dry, and examine.

Results: The organisms should be stained deeply; the background material should be lightly stained.

Comment: This stain was reported as being superior to the 0.3% methylene blue in 30% ethanol stain as used in the Breed procedures. The authors claim the superiority is due to the lower surface tension of the staining solution which is low in water content. Some difficulty may be experienced due to the loss of parts of the smear during processing.

Aniline Methylene Blue Stain for Dairy Products

North (1945)

Preparation of solutions:
A. Methylene blue. Prepare a saturated solution of methylene blue chloride in 95% ethanol. The reported solubility of methylene blue in 95% ethanol varies considerably, ranging from 1.5–2.0% or higher.
B. Aniline, acid, solution. Add 3 ml of aniline to 10 ml of 95% ethanol in a 100-ml flask. Slowly add, with stirring, 1.5 ml of concentrated HCl.
C. The staining solution. Add 30 ml of the saturated solution of methylene blue in 95% ethanol to the aniline, acid, mixture in the 100-ml flask, then bring the total volume to 100 ml with distilled water. Store in a tightly stoppered bottle, and filter when necessary.

Procedure:
1. Prepare a smear on a clean glass slide by the procedures usually used for dairy products (see Levowitz and Weber stain above). Dry rapidly and thoroughly on a thermostatically controlled hot plate at 45–50°C.
2. Defat by immersing in xylene for at least 1 min.
3. Immerse in 95% ethanol for 1 min, and drain. Drying is not necessary.
4. Stain for 1 min, remove from the stain and rinse off the excess stain by dipping in tap water in a container with slowly running water.
5. Thoroughly air dry, and examine.

Results: The bacteria should be deeply stained against a lightly stained background.

SECTION N: THE STAINING OF BACTERIA IN SOIL

The staining of bacteria in soil is difficult because the background material is rich in organic material which also has an affinity for stains. However, available differential stains are usually based on the greater affinity of bacterial cells for the stain, as compared to background material, when the staining procedure is carried out at an acid pH. The dyes usually used are rose bengal (sometimes referred to as rose bengal 2B) or erythrosin Y for ordinary light microscopy, and acridine orange for fluorescent microscopy. In a study of a variety of fluorescein dyes for the staining of soil organisms, Conn and Holmes (1926) reported rose bengal to be the best for ordinary microscopy. For fluorescent microscopy, acridine orange is best since it results in a differential color of fluorescence for bacterial cells (green) as compared to background material (orange to red). While most of the published staining procedures for soil organisms can be modified in a manner to give quantitative results, one of the most successful attempts to accomplish this can be found in the papers of Jones and Mollison (1948) and Skinner, Jones, and Mollison (1952). Since the accuracy of quantitative procedures for soil organisms is often questioned, no quantitative procedure will be given below.

Staining of Bacteria in Soil

Conn (1928, 1929)

Preparation of solutions:

A. Gelatin fixative. Dissolve 0.015 gm of gelatin in 100 ml of distilled water by heating to 45–50°C. Distribute in test tubes, 4.5 ml per tube, stopper with cotton, and sterilize in the autoclave.

B. Stain. Dissolve 1 gm of rose bengal with 100 ml of 5% phenol. Add 0.01 gm of calcium chloride ($CaCl_2 \cdot 2 H_2O$, or anhydrous). The function of the calcium chloride is to form a relatively insoluble calcium salt of rose bengal which will be on the verge of precipitating out of solution. On addition of the calcium chloride, a faint precipitate should form which should not be removed but which should be resuspended by shaking at the time of use of the stain. The more calcium chloride added the deeper will be the color of the stained substrates. If too much calcium chloride is added, then the differential staining of bacteria and background material will be lost. Conn (1928) stated that the best amount of calcium chloride to use must be determined empirically but usually it will be between 0.1–0.0001%. Rose bengal stains bacterial cells a purple color; if a red to orange color would be helpful this can be obtained by replacing the rose bengal with erythrosin Y.

Procedure:
1. Place 0.5 gm of soil in a tube with 4.5 ml of 0.015% gelatin and mix thoroughly.
2. Place a loop, or drop, of the suspension on a clean glass slide, and spread evenly as for a thin film.
3. Allow to dry while being held level over a boiling water bath.
4. While still over the water bath, cover with the stain and let stand for 1 min.
5. Wash rapidly but gently in tap water.
6. Air-dry and examine.

Results: When proper differentiation is obtained, the bacteria should be stained a deep purplish shade while the background material should be stained only lightly.

Comments: This procedure can be quantitated by covering a 1 sq cm area of the slide with 0.01 ml of the soil-gelatin mixture, and then treating the stained slide as done for direct counts of bacteria in milk. The counts obtained should be interpreted as "organisms seen" rather than as the actual number of bacteria per gm of soil. Only about 1 in 50 of the stained bacterial cells observed will produce colonies when cultured on solid culture media (Skinner, Jones, and Mollison, 1952).

Fluorescent Staining of Bacteria in Soil

Strugger (1948a)

This is a good method for differentiating bacteria from the background material in soil. It was originally presented as a stain which could differentiate living bacteria, which show a green fluorescence, from dead organisms, which fluoresce red as does the background material (Strugger, 1948a). Certainly some dead bacterial cells show a red fluorescence, but it cannot be proven that all of the green fluorescing cells were actually in a living state at the time of the preparation of the slide.

Preparation of solutions: Stain. Five concentrations of acridine orange should be prepared in tap water as follows: 1/1000; 1/2000; 1/3000; 1/4000; and 1/5000. Dissolve 1 gm of acridine orange with 100 ml of tap water and prepare the dilutions as shown below.

1/1000 = Add 10 ml of the 1:100 (1%) solution to 90 ml of tap water.
1/2000 = Add 5 ml of the 1/100 solution to 95 ml of tap water.
1/3000 = Add 3 ml of the 1/100 solution to 97 ml of tap water.
1/4000 = Add 10 ml of the 1/2000 solution to 10 ml of tap water.
1/5000 = Add 10 ml of the 1/1000 solution to 40 ml of tap water.

Procedure:
1. Place 1 gm of soil in each of 5 test tubes.

2. Add to each of these 5 tubes 10 ml of one of the above dilutions of acridine orange. That is, add 10 ml of the 1/1000 to one tube, 10 ml of the 1/2000 to the next tube, etc.
3. Shake each tube to mix the soil and the dye, and let stand 5–10 min.
4. Choose the soil suspension with the best appearing supernatant:
   a. If the acridine orange was too weak, the supernatant will be colorless.
   b. The best tube to use will show a slightly colored supernatant.
   c. Tubes with heavily colored supernatants are useless since excess dye will result in failure to differentiate bacteria from background material.
   The correct amount of dye will vary according to the organic content of the soil. Once the correct amount of dye is known for a soil, only a single tube need then be used for staining.
5. Shake the suspension and place a small drop on a clean glass slide and cover with a cover glass.
6. Examine with a good quality ultraviolet microscope system.

Results: Humus particles should have a red fluorescence; bacterial cells should have a green fluorescence. Bacteria which have been heat killed before staining should have the same red color of fluorescence as the background material.

Variation: Casida (1962) prepared the stain and its dilutions in a potassium phosphate buffer, pH 6.0. After exposing soil to the stain, he then inoculated the stained soil material onto a thin layer of solid culture medium on a slide, and incubated 4 hr before observing with a ultraviolet microscope system, using 120–240× magnifications to observe small colonies.

SECTION O: THE STAINING OF BACTERIA ON MEMBRANE FILTERS

Most of the membrane filters used in microbiology are composed of cellulose acetate or cellulose triacetate, which fortunately have little affinity for the basic (cationic) dyes usually used to stain bacterial cells. Almost any normal bacterial staining procedure can be used, either to stain individual cells, or to visualize small bacterial colonies following incubation on a culture medium. Quantification of the number of bacteria in a suspension can be achieved by passing a known volume of the suspension through a 0.2 μm pore (or smaller) filter followed by staining, and making a direct count of the cells on the membrane. Or place the membrane on a culture medium; and after a relatively short incubation period, the microcolonies can be stained and counted with a low power objective.

Bacterial cells deposited on the surface of a cellulose acetate membrane usually adhere to it due to both mechanical and Van der Waals forces; thus fixation may not be necessary before staining. Fixation is possible, however, either by using heat such as a hot air oven set at 105°C and exposing for a period of 10–15 min, or chemical fixatives may be used if they are compatible with the chemical nature of the membrane. A safe chemical fixative would be 95% ethanol, and absolute methanol could be used if the fixation period was short. Ether-ethanol, or acetone-formaldehyde fixatives should be avoided.

After staining, it is necessary to clear the membrane for microscopy by transmitted light. This is easily done by exposure to immersion oil of the same refractive index (1.515) as the glass used for oil immersion objectives. A dry membrane can be cleared by floating on immersion oil in a petri dish, organisms up. On removal, the membrane can be drawn over the rim of the dish to remove excess oil from the bottom of the membrane. The membrane can then be placed on a glass slide and observed with the oil immersion objective. Xylene can also be used to clear the membrane and is useful if the membrane is to be observed with low power objectives, or if it is to be mounted in balsam or other mounting medium.

Procedure for the staining of bacteria on membrane filters:

1. Filter the suspension through a 0.2 μm filter.
2. Remove the filter and place it, organisms side up, on the surface of a glass slide. The membrane can be dried, and fixed in an oven at 105°C for 10–15 min, or dried and exposed to chemical fixatives, or both drying and chemical fixation can be omitted.
3. Stain with methylene blue, crystal violet, or other suitable bacterial stain.
4. Wash twice with water, and dry in a hot air oven at 60°C.
5. Clear by floating on immersion oil in a petri dish.
6. Place on a glass slide and observe with the oil immersion objective.

Procedure for the staining of bacterial colonies on membrane filters:

1. Filter the suspension through a 0.2 μm filter.
2. Remove the filter and place it, organisms up, on the surface of a suitable solid culture medium.
3. After incubation for 3–5 hr, either stain directly, or fix in the hot air oven at 105°C for 15 min. After fixation, the filter should be dry.
4. Clip the membrane to a glass slide and stain by immersion in 0.5% aqueous methylene blue chloride solution for 5 min.
5. Wash twice in water, and dry in the hot air oven at 60°C.
6. Clear by floating the membrane, colony side up, on the surface of immersion oil in a petri dish.
7. Transfer the membrane to a glass slide and examine. Observation can be made with the 16-mm objective. If the 4-mm objective is used, oil

should be placed on the membrane and it should be covered with a No. 0 or 1 cover glass.

Comments: Almost any bacterial staining procedure, such as the Gram stain, or the acid-fast stain, can be applied to these membranes. The membrane must be dehydrated before it can be cleared with oil. The best procedure is to dry the membrane in an oven, but if the decolorization action of alcohol would not be a problem, the wet membrane can be dehydrated by passing through alcohol and then cleared in oil or in xylene.

Procedure for the fluorescent staining of bacteria on membrane filters: The fluorescent staining of bacteria on membrane filters has proven to be of particular value where total counts of bacterial cells in aquatic environments are needed. Epifluorescent (incident light) microscopy is used since it allows the visualization of bacteria adhering to solid particles, and solves the problem of ultraviolet light transmission through the membrane filter. Good descriptions of the instrumentation have been published (Francisco, Mah, and Rabin, 1973; Jones and Simon, 1975; Hobbie, Daley, and Jasper, 1977).

Black, 0.22 μm pore size, cellulose ester filters are usually used for epifluorescent microscopy since such filters have no innate self fluorescence, and they produce a black background. Polycarbonate (Nucleopore) filters have some innate self fluorescence which can be sometimes overlooked, or prevented using the method of Hobbie et al. (1977). Jones and Simon (1975) have published a critical discussion of the cell counts obtained by epifluorescent microscopy as compared to the counts obtained by using the Petroff-Hauser counting chamber, the Coulter counter, and viable cell counts as obtained by plating out procedures.

Procedure for Staining Membranes for the Counting of Cells by Epifluorescent Microscopy

Jones and Simon (1975)

Preparation of the stain solution: Dissolve 10 mg of acridine orange (AO or E2GNX) in 100 ml of distilled water. The stain solution should be prefiltered before use since it may itself support the growth of microorganisms.

Procedure:

1. Add 1 part of the acridine orange stain solution to 99 parts of the water sample. Let stand for 5 min.
2. Prewet a 0.22 μm pore size, black, cellulose ester filter with a prefiltered sample of the water.

3. Filter enough of a measured amount of the sample to yield 10–30 organisms per microscope field.
4. Rinse with a prefiltered sample of the water.
5. Immediately place the membrane over a drop of immersion oil on a glass slide, then add immersion oil to the top of the membrane. Examine using epifluorescent microscopy.

Comments: The procedure of Francisco et al. (1973) is about the same except for their use of a 0.1% solution of acridine orange. Jones and Simon (1975) reported superior results when the acridine orange was reduced to 0.01%. Acridine orange fluoresces green when it is bound as a monomer to DNA, and red when it is bound as a dimer to RNA (see Hobbie et al., 1977). Strugger (1948b) has tried to use this fact to differentiate living from dead organisms present in the sample before filtering and staining. However, most are skeptical about such living-dead interpretations. For total counts, both red and green fluorescent cells should be counted.

SECTION P: THE DEMONSTRATION OF NEGRI BODIES OF RABIES IN SMEARS

The quickest and simplest method of demonstrating the presence of rabies in suspected animals is by the use of smear preparations. Using any of the acceptable staining methods, a positive or negative report can be made in from 30 min to 1 hr. The highly specific, but more technical, fluorescent antibody tests should be employed on smears wherever the proper equipment, reliable reagents, and experienced personnel are available. Instructions for fluorescent antibody procedures are not given below, but can be found in the World Health Organization's *Laboratory Techniques in Rabies* (1966). Smear techniques in the diagnosis of rabies are rapidly replacing the use of the more traditional but slower paraffin imbedded sections or frozen sections. Sections, however, are still used when permanent slides are desired for teaching or research purposes. For the diagnosis of rabies, impression smears or spread smears (see Sellers' stain below) should be made from tissue obtained from the Ammon's horn, olfactory portion of the cortex, cerebellum, and the medulla oblongata regions of the brain, and from tissue from the submaxillary salivary glands. Detection of Negri bodies or viral antigen in salivary gland material is of more immediate concern when animal bites are involved than the demonstration of rabies in brain material. Detailed instructions for obtaining specimens from animals can be found in the World Health Organization's publication No. 23, mentioned above.

Sellers' Stain for Negri Bodies in Smears

Sellers (1927); World Health Organization: *Laboratory Techniques in Rabies* (1966)

This is a popular and rapid staining procedure for the demonstration of

Negri bodies for the diagnosis of rabies. It is a modification of the earlier Williams and Van Gieson stains, which also use basic fuchsin and methylene blue, and it is recommended only for smears. Small transverse sections (2–3 mm thick) of brain material are used to prepare both the impression smears and spread smears.

Preparation of smears:

1. Impression smears. Impression smears are recommended and are prepared by placing the cut section on a wooden tongue depressor with the cut surface facing upward. Touch a clean microscope slide to this surface with enough pressure to spread the tissue on the slide. Several smears can be made on one slide. The smears must be stained while still moist.

2. Spread smear. Place a small section of tissue on one end of a slide. Use another slide to crush the section and to draw the section across the first slide to obtain the smear. The result should be a homogenous spread of material which should not be too thick. This method results in a copious amount of material being available for observation. Stain while still moist.

Preparation of solutions:

A. Stock solution of basic fuchsin. Dissolve 1 gm of basic fuchsin chloride (92% or more dye content) with 100 ml of absolute methyl alcohol (A.C.S., C.P. grade). When stored in a tightly capped container, this solution is very stable and will keep for long periods of time.

B. Stock solution of methylene blue. Dissolve 1 gm of methylene blue chloride (not less than 85% dye content) with 100 ml of methyl alcohol (A.C.S., C.P. grade). When stored in a tightly capped container, this solution is very stable.

C. Staining solution. Mix 1 part of the basic fuchsin solution into 2 parts of the methylene blue solution. Mix thoroughly, but do not filter. Let stand for at least 24 hr before use. When stored in a tightly stoppered container, this solution will keep indefinitely. Evaporation of the methyl alcohol will cause deterioration of staining ability. Some adjustment of the above proportions may be necessary. If the overall staining effect is reddish, increase the proportion of methylene blue; if the Negri bodies are a muddy maroon and the nerve cells deeply blue, increase the amount of basic fuchsin. Once adjusted, the staining characteristic should remain constant over a long period of time.

Procedure:

1. Prepare impression or streak smears as directed above. Do not fix. Stain while still moist with tissue juice. If the smear has dried, the Negri bodies will not be differentially stained.

2. Immerse the smear in the stain solution for 1–5 sec depending on the

thickness of the smear. Prolonged staining (20–30 sec) will result in poor differentiation.

3. Rinse quickly with distilled water containing M/150 phosphate buffer pH 7.0; tap water may be acceptable. Air dry without blotting since blotting may remove material from the slide.

4. Examine first with a low power objective to find areas where the nerve cells are well spread out, then examine with the oil immersion objective.

Results: The Negri bodies should be a bright cherry red and should stand out in strong relief, and they should contain well differentiated dark blue to black basophilic inner bodies. All parts of the nerve cell should stain blue, the interstitial tissue a rose pink, bacteria blue, muscle cells red, and erythrocytes should be a copper color. In smear preparations some of the Negri bodies may appear extracellular, but usually they appear intracytoplasmically in the nerve cell. Nonspecific (nonrabies) red bodies may be found in material especially if it is not fresh, and when present they can be distinguished from true Negri bodies in that only Negri bodies contain the differentiated basophilic inner bodies which stain with methylene blue.

Bond's Modification of Mann's Eosin-Methyl Blue Stain for Negri Bodies in Smears

From Mallory (1961)

This is also a very popular and rapid stain for Negri bodies in smears. It differs from the Sellers' stain by using a mixture of 2 acid (anionic) dyes rather than 2 basic (cationic) dyes.

Preparation of solutions:
   A. Stock eosin Y. Dissolve 1 gm of eosin Y with 100 ml of distilled water. The solution is very stable and should be aged for several weeks.
   B. Stock methyl blue. Dissolve 1 gm of methyl blue with 100 ml of distilled water. The solution is very stable. Do not confuse methyl blue with methylene blue.
   C. Staining solution. Just before use mix 1 part of the eosin Y solution with 6 parts of distilled water, and then add 1 part of the methyl blue solution. This solution should be discarded after use.

Procedure:
   1. Prepare impression or spread smears as directed above for the Sellers' stain. Partly air-dry, then fix in absolute methyl alcohol for 5–6 min.
   2. Wash in running tap water for 30 sec.
   3. Stain for 4–5 min in the eosin Y-methyl blue stain.
   4. Wash in running tap water for 30 sec.
   5. Blot, and air-dry.
   6. The smears may be observed directly, or if they are to be preserved

they should be dehydrated in absolute alcohol, cleared in xylene, and mounted in balsam or other suitable mounting medium.

Results: The Negri bodies will be stained red, with typical blue internal granules.

Harris Stain for Negri Bodies in Smears

Harris (1908)

This is one of the older, but previously popular, staining procedures for Negri bodies in smears. It differs from Sellers' and Mann's procedures in that it uses an acid (anionic) dye eosin Y, followed by a basic (cationic) dye methylene blue. It should be useful also for the demonstration of other eosinophilic inclusion bodies such as found in smallpox, cowpox, chickenpox, and shingles.

Preparation of solutions:

A. Eosin Y solution. Dissolve 2 gm of eosin Y (sodium salt) with 100 ml of 95% ethanol. This solution should be allowed to age for at least 2 months.

B. Methylene blue solution. Dissolve 1 gm of methylene blue chloride with 100 ml of distilled water, then add 1 gm of potassium carbonate. Let age 1 or 2 weeks. Just before use, dilute 1:10.

Procedure:

1. Prepare smears as directed for the Sellers' stain and fix in absolute methyl alcohol for 1 min.
2. Wash in water to remove the alcohol.
3. Immerse in the eosin Y for 1–3 min.
4. Wash 2 or 3 sec to remove excess eosin Y.
5. Immerse in freshly dilute (1:10) alkaline methylene blue for 5–15 sec.
6. Wash briefly in water.
7. Decolorize in 95% ethanol until only the nerve cells are blue, and the erythrocytes are bright red. It may be necessary to follow decolorization with the microscope. Over decolorization is preferable to under decolorization.
8. Either wash, blot and air-dry, and examine; or dehydrate in absolute ethanol, clear in xylene, and mount in balsam or other suitable mounting medium.

Results: The Negri bodies should be a light red; with bluish-red inner granules.

SECTION Q: THE STAINING OF PROTOZOA IN SMEARS AND WET MOUNTS

The clinically important protozoa are sought for in smears or wet mounts of blood, fecal, vaginal, and oral material. If the organisms are in fecal material, a concentration procedure may be used to increase the possibility of the finding of any protozoa which may be present. Concentration technics are effective for the demonstration of cysts, but are unreliable for the concentration of trophozoites. Smears or wet mounts may be prepared and may be examined without any staining procedure, or a simple iodine stain may be used. Staining with iodine, however, is usually necessary for the identification of organisms since it brings out the necessary details of the nuclear structure. If permanent preparations are desired, or if greater cytological detail is needed, the smears must be stained using procedures such as the iron hematoxylin procedure, or the Giemsa stain.

Direct Observation of Protozoa in Wet Mounts

This procedure is simple and is often used for the routine examination of specimens. Two side by side wet mounts in saline are usually made on a slide. For comparative purposes, one is left unstained and the other is stained with iodine.

Preparation of solutions:
A. Saline. This is an 0.85% solution of sodium chloride in distilled water.
B. Iodine stain. Various iodine solutions have been recommended for staining, ranging from Lugol's iodine (5% iodine), which has the disadvantage of being relatively unstable, to the stable D'Antoni 1.5% solution of iodine, which is usually preferred. The Burke iodine solution used in Gram staining is acceptable and it is very stable. Place 2 gm of potassium iodide in a mortar, add 1 gm of iodine, and grind with a pestle to mix. Add 1 ml of distilled water, grind, then add 5 ml of water and grind, then add an additional 10 ml of water and mix. All of the KI and I_2 should now be in solution. Pour into a 100 ml reagent bottle, and rinse the mortar and pestle with the water necessary to bring the volume to 100 ml. The equivalent of D'Antoni's iodine can be prepared by using 3 gm of KI and 1.5 gm of iodine in the above procedure.

Procedure:
1. Place a drop of saline on one side of the slide, and a drop of the iodine solution on the other.
2. Suspend fecal, vaginal, or oral material in both the drop of saline and the drop of iodine solution.
3. Cover each with a No. 1 or 0 cover glass and scan with the 16 mm objective. When interesting material is found it may be observed with the 4 mm and oil immersion objectives.

Results: The organisms can be detected in the saline wet mount, but the iodine mount is usually essential for the identification of the organisms since it reveals glycogen granules and nuclear structure.

Variation: Smears, air-dried and fixed in Schaudinn's fixative can also be examined in this manner. When fecal specimens are examined, it may be desirable to use a concentration technic which helps in the detection of cysts, but which is not very helpful for the detection of trophozoites.

Zinc sulfate concentration procedure:
1. Suspend 1 part of fecal suspension in about 10 times its volume of saline.
2. Strain 10 ml of this suspension through one layer of wet cheese cloth and into a tappered centrifuge tube and centrifuge in an International Clinical Centrifuge at about 2500 rpm for 1 min. Decant the supernatant. Resuspend in 2 or 3 ml of saline, centrifuge, and repeat 2 or 3 times.
3. After the last supernatant has been discarded, add 3–4 ml of 33% zinc sulfate solution (specific gravity 1.180), break up the sediment, and then fill the tube to a half inch from the top with the 33% zinc sulfate solution.
4. Centrifuge at top speed for 1 min.
5. A film of material will be seen floating on top of the zinc sulfate. Transfer several loops of this to a drop of the iodine solution on a glass slide, cover with a cover glass, and examine.

The MIF and MIFC Wet Mount Technics for the Preservation, Staining, and Concentration of Protozoa in Fecal Material

Sapero and Lawless (1953);
Blagg, Schloegel, Mansour, and Khalaf (1955)

The MIF (merthiolate, iodine, formalin) staining and preservation technic of Sapero and Lawless (1953) can be used directly for wet mounts, or it can be modified to the MIFC (C for concentration) technic of Blagg *et al.* (1955), which adds a concentration step. The advantages are that specimens can be preserved in tubes for long periods of time before examination, and the concentration procedure described was reported to be effective for many trophozoites as well as for all cysts of protozoa.

Preparation of solutions:
A. Obtain tincture of merthiolate (E. Lilly & Co., No. 99, 1:1000).
B. Lugol's iodine. Place 10 gm of potassium iodide in a mortar, then add 5 gm of iodine and mix with a pestle. Grind while slowly adding 5 ml of water, then 25 ml of water, then 30 ml of water. Pour into a 100 ml reagent bottle and wash the mortar and pestle with the water required to bring the final volume to 100 ml. This solution is not stable and must be prepared fresh every 2 or 3 weeks.
C. Stock MF solution (merthiolate, formaldehyde). Place 50 ml of distilled water in a 100 ml brown reagent bottle. Add 40 ml of the

merthiolate, 5 ml of formalin (37–40% formaldehyde), and 1 ml of glycerin. This solution is stable.

Procedure for wet mounts:
1. Place 0.775 ml of the merthiolate in a standard Kahn tube, add 0.1 ml of Lugol's iodine (less than 1 week old), then add 0.125 ml of formalin, and mix. In another Kahn tube place a similar volume of distilled water. Place a medicine dropper in each tube.
2. Place a small drop of the distilled water on a slide, add an equal volume of the merthiolate, iodine, formalin mixture.
3. Add the fecal sample to the mixture. It is important not to place too much feces into the drop. Cover with a No. 1 or 0 cover glass. The completed preparation should be thin enough to allow the slide to be tipped on edge without the cover glass sliding. The preparation may be sealed with petrolatum or other sealing material. Examine first with the 16-mm objective, then the 4-mm, and finally with the oil immersion objectives. The use of a blue filter will increase the contrast between cytoplasm and nuclear material.

Results: Fresh preparations will show the cysts and trophozoites stained yellowish green to yellow brown. Nuclear elements should be well defined; glycogen granules should appear reddish to dark brown. As the preparation stands, an eosin color will replace the yellowish green to yellow brown color of the cysts and trophozoites.

Procedure for material to be preserved and stored:
1. Place 2.35 ml of the stock MF solution into a standard Kahn tube and stopper with a cork. Place 0.15 ml of Lugol's iodine into another tube and close with a rubber stopper. These 2 tubes constitute the "collection" unit.
2. Collect the fecal specimen. Pour the MF solution into the iodine solution and immediately add a volume of feces about twice the size of a pea (about 0.25 gm) with an applicator stick and mix. Do not overload with fecal material. The top of the fecal material should be about 1–1.5 cm from the bottom of the tube. Tightly stopper, label, and store until time for examination.
3. To examine, use a medicine dropper to draw off a drop of a mixture of the supernatant and feces from the top of the layer of sediment and place on a clear glass slide. Mix thoroughly with a toothpick and cover with a No. 1 or 0 cover glass. The preparation should be thin enough to allow the slide to be tipped on edge without the cover glass sliding.
4. Examine first with the 16-mm objective, then the 4-mm objective. The oil immersion objective can be used to observe the nuclei. A blue filter will increase the contrast between the cytoplasm and the nuclear material.

Results: The trophozoites and cysts will be stained an eosin color.

A Concentration Procedure for MIF Preserved Specimens

Blagg, Schloegel, Mansour, and Khalaf (1955)

1. Mix the M.I.F. preserved specimen (Step 2 above) by shaking vigorously for 5 sec.
2. Strain through 2 layers of wet surgical gauze into a 15 ml tappered centrifuge tube.
3. Add 4 ml of cold (refrigerated) ether, insert a rubber stopper and shake vigorously. If ether remains on top after shaking, add 1 ml of water and reshake.
4. Remove the stopper and let stand for 2 min.
5. Centrifuge at 1600 rpm. Four layers will appear in the tube; a layer of ether on top, a plug of fecal material, an MIF layer, and a layer of sediment on the bottom which contains the protozoa.
6. Loosen the fecal plug with an applicator stick.
7. Quickly and carefully pour off all but the bottom layer of sediment.
8. Mix the remaining sediment, put a drop on a slide, cover with a cover glass, and examine.

The Demonstration of Protozoa in Fixed and Stained Smears

Fixed and stained smears are rarely used in the diagnosis of protozoan infections. They do, however, result in the better demonstration of cytological details; and since they are permanent they are especially helpful for teaching laboratories. The usually recommended fixative is Schaudinn's, and the best staining procedures use hematoxylin or Giemsa's stain.

Iron-Hematoxylin Stain for Protozoa in Fixed Smears

Preparation of solutions:

A. Schaudinn's fixative. Prepare 100 ml of a saturated aqueous solution of mercuric chloride. The solubility of mercuric chloride in water is listed as 6.9 gm per 100 ml at 20°C. Add to this, 50 ml of 95% ethanol. Just before use, add 7.5 ml of glacial acetic acid.
B. Iodine in alcohol. Dissolve 0.2 gm of iodine in 100 ml of 70% ethanol.
C. Ferric ammonium sulfate. Dissolve 2 gm of ferric ammonium sulfate ($FeNH_4(SO_4)_2 \cdot 12 H_2O$) with 100 ml of distilled water. This should be prepared fresh just before use.
D. Hematoxylin stain. Dissolve 0.5 gm of hematoxylin in 10 ml of 95% ethanol, then bring the volume to 100 ml with distilled water. Allow the stain to ripen several weeks before use.

Procedure:

1. Use an applicator stick to prepare a thin film of fecal material on a clean glass slide. It may be necessary to dilute the fecal material with a bit of saline. Do not let the smear dry before fixation.
2. Fix in Schaudinn's fixative for 1 hr at room temperature. Prolonged fixation is not harmful.
3. Wash and expose to the 0.2% iodine in 70% ethanol for 5 min.
4. Expose to 50% ethanol for 2-3 min.
5. Wash in tap water for 3-5 min.
6. Expose to the 2% ferric ammonium sulfate for 2-4 hr at room temperature.
7. Wash through 2 changes of distilled water for a total of 3-6 min.
8. Stain with hematoxylin at room temperature for 12-24 hr, or at 50°C for 5-10 min. The slides should be kept vertical in Coplin staining dishes to prevent dye precipitate from falling on to the smear.
9. Wash in 2 changes of tap water for a total of 3-6 min.
10. Destain in the 2% ferric ammonium sulfate. This step is critical and should be followed with the microscope to determine proper destaining. During this examination, keep the slide wet. Washing will stop the decolorization action which can be restored by re-exposure to the 2% ferric ammonium sulfate. The time of destaining will vary from 2 to 30 min depending on the thickness of the film and other factors.
11. Wash in tap water for 10-30 min or longer.
12. Dehydrate through 70%, 95%, and 2 changes of absolute ethanol, 3 min each.
13. Clear in xylene 3 min, then mount in a suitable mounting medium such as balsam or Permount. Examine with the 4 mm and oil immersion objectives.

Results: Protozoa trophozoites and cysts will be stained a brownish black and should show good internal differentiation of cytological details.

Variation: The 2% ferric ammonium sulfate used for destaining can be replaced with a saturated aqueous solution of picric acid (about 1.4% at room temperature). Some recommend this over ferric ammonium sulfate since over destaining is more difficult and it results in a nice blue stain rather than a brownish black.

The Demonstration of Malaria Organisms, Leishmania, Trypanosoma, and Microfilariae, in Blood Smears

Two types of blood films are used. The thick film is most valuable in diagnostic work since it enables the observation of a larger quantity of blood per microscope field. The thin film is less useful in diagnosis, but is superior for the observation of morphological details. Both can be prepared on the same slide.

Preparation of a thick blood film: Clean the skin of a finger tip with cotton moistened with 70% ethanol, and then let dry. Puncture the finger tip with a sterilized instrument. Press out a large drop of blood and touch it with the surface of a clean glass slide. Without touching the skin, move the slide so that a smear about the size of a penny is obtained. The smear should not be opaque in that you should be able to read newsprint through the smear. Let dry in a closed petri dish for at least 24 hr.

Preparation of a thin film: A small drop of blood from a finger tip should be placed on one end of a clean glass slide. A spreader slide should be touched to the drop of blood, and while held at an angle of about 30° it should be pushed forward over the surface of the slide. The end of the smear should have an irregular edge where the smear is at its thinnest.

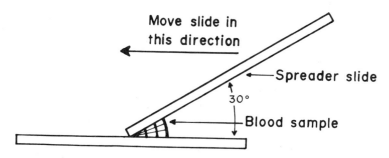

Move slide in this direction ←   ←—Spreader slide   30°   ←—Blood sample

This thinnest area is often the best area to search to find plasmodia. After making a good thin smear, it should be possible to make a thick smear on an unused area of the slide at one end. Let dry in a closed petri dish.

Preparation of solutions:

A. Giemsa stain. Either purchase ready prepared Giemsa stain, or prepare as follows: Heat 33 ml of glycerol to 55°C. Add 0.5 gm of Giemsa stain (powdered form) to a warm mortar and slowly add, with trituration, the 33 ml of warm glycerol. Cool to room temperature and add 33 ml of absolute methanol (reagent grade) and mix. Place in a tightly stoppered bottle and store overnight; a small amount of sediment will appear. Decant into small (30 ml) bottles, leaving the sediment behind, and tightly stopper. Store at least 2 weeks and filter before use.

B. Phosphate buffer, pH 7.0. Phosphate buffer, pH 7.0 can be prepared by dissolving 9.464 gm of anhydrous dibasic sodium phosphate (Na_2HPO_4) in distilled water and bring to a final volume of 1 liter; this is Solution A. Also dissolve 9.073 gm of monobasic potassium phosphate (KH_2PO_4) in distilled water and bring to a final volume of 1 liter; this is Solution B. For use, mix 61.1 ml of Solution A with 38.9 ml of Solution B and bring to a final volume of 1 liter.

Procedure:
1. Fix thin air-dried films in absolute methanol for 1–3 min. Thick films should not be fixed, but should be dried for at least 24 hr. If both a thick and thin films are on the slide, fix only the thin film in methanol, drain off the methanol, and let air-dry for 24 hr.
2. The staining solution should be in a Coplin staining dish. Fill with a solution of 2 parts of Giemsa stain in 98 parts of the diluted phosphate buffer of pH 7.0. Immerse the smear in the stain for 20–30 min for thin smears, and for 45 min for thick smears. If both types of smears are on the slide, stain for 45 min. The staining time may vary for different certification numbers on the dye label. The stain in this Coplin dish should be used only once.
3. Wash in the dilute phosphate buffer, pH 7.0, until properly differentiated. The thin part of the smear should be differentiated in about 1 or 2 dips in the buffer; the thick part of the smear may need decolorization for 3–5 min.
4. Drain dry, do not heat, do not blot. A warm lamp or small fan can be used to hasten drying.
5. The slide may be observed directly. If a slide is to be made permanent it can be cleared by xylene, and mounted in balsam or other suitable mounting medium.

Results: The cytoplasm of the protozoa should be stained blue; nuclear structures should be stained red; erythrocytes should be stained pink.

CHAPTER 18

STAINS FOR MICROORGANISMS IN SECTIONS

James W. Bartholomew

SECTION A: INTRODUCTION

Some of the staining procedures given in this chapter, such as the Gram and acid-fast stains, are based on the same differential principles as the procedures used for microorganisms in smears. The steps for staining microorgansism in tissue sections, however, are sufficiently different to warrant specific instructions in this chapter. Some general tissue stains, methyl green-pyronin B (p. 199), buffered azure eosinates (p. 209) and most Nissl stains (p. 206), all will also demonstrate microorganisms in sections if they are present. Since these procedures are presented elsewhere in this manual, they will not be repeated here.

Attachments of Sections to Slides, and the Removal of Paraffin

Mayer's albumin glycerol solution is often used to affix sections to glass slides. Albumin is obtained from fresh eggs, and 1 part of albumin is thoroughly mixed with 1 part of glycerol. The mixture is then filtered through cotton at 54°C; a preservative is usually added such as sodium salicylate to a concentration of 1%, or methiolate to a final concentration of 1:100,000 or 1:200,000.

The removal of paraffin from the sections before staining is essential. Warm the sections to just above the melting point of the paraffin used (usually 54–63°C) and then immerse in hot (70°C) xylene for 15 min, followed by a second immersion in fresh xylene. The sections should then be taken through 2 immersions of 1–2 min each in absolute, 95%, and 80% ethanol, and then taken to distilled water. The hot xylene may not be necessary if the time for each immersion is increased to 20–30 min. Nedzel (1951) reported that even after hot or cold xylene extraction, paraffin artifacts still could often be found which appeared as intranuclear birefringent crystals. Such artifacts, however, would not be important or apparent unless heat was used during the staining procedure for the sections. According to Nedzel the only way to completely remove such paraffin artifacts, if necessary, was through a xylene, alcohol, water, alcohol, xylene, alcohol, to water sequence.

If mercury or lead salts were a component of the fixation fluid used for the tissue (such as Zenker's, Schaudinn's, or Heidenhein's fixatives; or Lillie's (1965) lead nitrate formalin), then such salts must be removed from the section. This is accomplished following the 80% ethanol by immersing the sections in 70% ethanol containing 0.3–0.5% iodine for 5–10 min, rinsing in water, immersing in 5% sodium thiosulfate for 1–5 min to bleach out the iodine color, thoroughly washing, rinsing in distilled water, and then processing through the staining procedure.

SECTION B: GRAM STAIN DIFFERENTIATION OF MICROORGANISMS IN TISSUE SECTIONS

A Gram stain procedure for sections not only must differentiate the Gram-positive organisms from Gram-negative, but it must also differentially stain the Gram-negative microorganisms from the normally Gram-negative tissue cells and cellular components. Thus, the Gram staining procedure for sections must be more involved than that used for smears. A very good procedure for microorganisms in sections is that of Taylor (1966) which is in turn a modification of the older and widely used Brown and Brenn (1931) procedure.

Taylor's Modification of the Brown and Brenn Gram Procedure for Tissue Sections

Brown and Brenn (1931); Taylor (1966)

Preparation of solutions:
 A. Harris' alum-hematoxylin. Dissolve 60 gm of aluminum ammonium sulfate ($AlNH_4(SO_4)_2 \cdot 12H_2O$) or aluminum potassium sulfate ($AlK(SO_4)_2 \cdot 12H_2O$) in 800 ml of distilled water with the aid of heat. Dissolve 1 gm of hematoxylin in 40 ml of absolute ethanol with the aid of heat. Add the hematoxylin to the solution of aluminum potassium sulfate, and bring the mixture to a boil as rapidly as possible. Then slowly add 2 gm of red mercuric oxide (HgO). The solution will assume a dark purple color. As soon as this occurs, remove the vessel from the flame and cool by plunging into a basin of ice water. When cool, place in tightly stoppered bottles of about 100 ml each. The solution is now ready for use in staining and it will keep for years. However, it should be filtered just before use. If better nuclear staining of tissue cells is needed, add 32 ml of a 4% acetic acid solution to the 800 ml of distilled water used to prepare the stock solution.

B. Ammonium oxalate, crystal violet, stain. Dissolve 1 gm of crystal violet chloride with 10 ml of 95% ethanol. Add 2 ml of this to 18 ml of distilled water, and then add this to 80 ml of a 1% aqueous solution of ammonium oxalate. After standing for several days, this mixture should be filtered. The resulting stain is very stable and will keep for months. (*Note:* Do not substitute Hucker's crystal violet.)

C. Gram's iodine. Place 6 gm of potassium iodide and 3 gm of iodine in a mortar and mix by grinding with a pestle. Add 3 ml of distilled water, grind, then add 15 ml of distilled water, and grind until all of the KI-I_2 is in solution. Add 30 ml of distilled water, mix, and place in a 1-liter vessel with a mark at 900 ml. Rinse the mortar and pestle with distilled water and place the water in the liter vessel. Finally, add sufficient distilled water to bring the total volume to 900 ml. This solution is stable and will keep for months.

D. Basic fuchsin. The stock solution is prepared by dissolving 0.1 gm of basic fuchsin chloride with 100 ml of 95% methanol. The working solution is prepared by mixing 5 ml of the stock alcoholic basic fuchsin with 60 ml of distilled water.

E. Picric acid-acetone. Dissolve 0.1 gm of picric acid with 100 ml of acetone. If and when a greenish hue appears in this solution, it should be discarded.

F. Acetone-xylene I. Mix 1 part of acetone with 2 parts of xylene.

G. Acetone-xylene II. Mix 1 part of acetone with 3 parts of xylene.

H. Acid alcohol. Add 1 ml of hydrochloric acid (conc.) to 100 ml of 70% ethanol.

I. Saturated lithium carbonate. The solubility of lithium carbonate in water is listed as 1.33 gm per 100 ml at 20°C.

J. Ether-acetone decolorizer. Mix 1 part of ethyl ether with 3 parts of acetone.

Procedure:

1. Deparaffinize and hydrate to distilled water.
2. Stain in freshly filtered Harris' hematoxylin for 5 min. If greater nuclear detail is wanted, stain for 10 min.
3. Wash in running tap water to remove excess hematoxylin (about 1 min).
4. Differentiate by immersion in 1% HCl in 70% ethanol until the sections are brick red (about 1 dip).
5. Wash in running tap water for 3 min.
6. Wash in saturated lithium carbonate to intensify the blue color. From this point on, process only 1 slide at a time.
7. Wash in running tap water for 1 min.
8. Stain with the ammonium oxalate, crystal violet solution for 2 min.
9. Wash quickly in water.

10. Mordant for 1 min with Gram's iodine.
11. Wash with tap water. Blot, but do not allow to dry. In order to prevent drying, moisten a piece of filter paper with water and blot the slide.
12. Decolorize in ethyl ether-acetone until blue color no longer comes off the sections.
13. Blot, but do not allow to dry. Blot with ethyl ether-acetone moistened filter paper.
14. Stain in the working basic fuchsin solution for 3 min.
15. Wash in water. Blot with water moistened filter paper. Do not allow to dry.
16. Dip in acetone until the section begins to decolorize, about 10–15 sec.
17. Pass quickly to the picric acid-acetone solution and expose for 10–15 sec to decolorize and differentiate.
18. Pass through acetone-xylene I, then acetone-xylene II, then 2 changes of xylene.
19. Mount with balsam, Permount, or other suitable mounting medium.

Results: Gram-positive microorganisms should be blue to blue black, and Gram-negative microorganisms bright red. The cell nuclei should be stained brownish red and the cytoplasm brownish yellow. Leukocytes and connective tissue should be stained red. Necrotic tissue should be yellowish green, and red blood cells red to yellowish green.

Brown and Hopps Gram Stain for Sections

Brown and Hopps (1973)

This procedure is reliable for Gram differentiation of bacteria in 4–6-μm sections of tissues fixed in formal-Zenker's solution, Bouin's fluid, formalin, or glutaraldehyde. In this procedure the strength of the basic fuchsin counter stain is changed according to the fixative used.

Preparation of solutions:
A. Crystal violet. Dissolve 1 gm of crystal violet in 100 ml of distilled water.
B. Gram's iodine. Prepare as above directions for Taylor's modification of the Brown and Brenn Gram stain for sections.
C. Stock basic fuchsin. Dissolve 0.25 gm of basic fuchsin chloride with 100 ml of distilled water. Filter through No. 40 Whatman filter paper.
   a. For formalin, glutaraldehyde, and Bouin's fixed tissues, dilute 1 ml of stock basic fuchsin with 25 ml of distilled water.
   b. For formal-Zenker's fixed tissues, dilute 5 ml of the stock basic fuchsin with 25 ml of distilled water.
D. Gallego's differentiating solution. Add 1 ml of formalin (37–40% formaldehyde solution) to 50 ml of distilled water, then add 0.5 ml of glacial acetic acid.

E. Picric acid-acetone solution. Add 0.5 gm picric acid to 1 liter of acetone.
F. Acetone-xylene. Add 1 part of acetone to 1 part of xylene.

Procedure:

1. Deparaffinize and hydrate to distilled water.
2. Stain with 1% crystal violet for 2 min.
3. Wash in tap water to remove excess crystal violet.
4. Expose to Gram's iodine for 5 min.
5. Wash in tap water to remove excess iodine.
6. Blot off excess water, but do not dry.
7. Differentiate in acetone until the blue color ceases to run from the section. Vertically dip in an acetone filled Coplin jar at a rate of 2 dips per second. Total decolorization time should only be a few seconds.
8. Wash in tap water to remove acetone.
9. Stain with the proper basic fuchsin solution for 5 min.
10. Wash with tap water.
11. Expose to Gallego's solution for 5 min. Agitate the solution during this exposure. This can be done by blowing on the solution.
12. Wash in tap water, blot, but do not dry.
13. Give the slide 3 quick dips in acetone.
14. Then 3 quick dips in the picric acid-acetone solution.
15. Then 3 quick dips in acetone.
16. Then 5 quick dips in acetone-xylene.
17. Then 10 quick dips in xylene.
18. Then 2 dips in fresh zylene.
19. Mount in Permount.

Results: Gram-positive microorganisms should be blue, Gram-negative red, the background material should be yellow, nuclei and epithelium should be light red.

SECTION C: STAINS FOR ACID-FAST BACTERIA IN TISSUE SECTIONS

Blanco and Fite (1948) compared Zenker's, Bouin's, formalin (20 parts)-95% ethanol (80 parts), and formalin (20 parts)-water (80 parts), fixation fluids as to the results obtained when sections were subsequently stained by the Ziehl-Neelsen, or the Fite II, acid-fast procedures. Zenker's fixative was clearly superior, Bouin's fixative was poorest, and the formalin fixatives were in between. It can be recommended, therefore, that any tissue material to be used specifically for the demonstration of the presence of acid-fast organisms be fixed in Zenker's fixtive. If Zenker's fixative is used, following deparaffinization, mercury salts must be removed from the sections by extracting with alcoholic iodine, followed by thiosulfate. Zenker's fixative can be prepared by dissolving 25 gm of potassium dichromate in 950 ml of distilled water, then add 50 gm of mercuric chloride, and 50 ml of reagent grade glacial (99.5%) acetic acid. The fluid is stable over a period of months at room temperature. Procedures like the Ziehl-Neelsen are acceptable for the demonstration of tuberculosis organisms in tissue, but are not reliable for the demonstration of the presence of leprosy organisms. If leprosy organisms are to be demonstrated, use procedures involving oil-xylene deparaffinization such as the Fite II procedure (Fite, Cambre, and Turner, 1947), Wade's modification of the Fite I procedure (Wade, 1952, 1957), or Mansfield's (1970) modification of the Truant et al. (1962) fluorescence stain.

Modified Ziehl-Neelsen Procedure

Lillie (1944c, 1965)

This procedure is simple and is analogous to that used for smears, but it cannot be recommended for the demonstration of leprosy organisms in sections.

Preparation of solutions:
   A. Carbol fuchsin. Add 25 ml of melted phenol to 50 ml of 95% ethanol; use this to dissolve 5 gm of basic fuchsin chloride, then dilute to 500 ml with distilled water. This solution will form a dark red caked deposit which cannot be redissolved and which represents a weakening of the staining solution. Although the solution may be usable for years, it may become useless in a few months time, in which case it may take on a water-like appearance. Ziehl's carbol fuchsin can be used, as prepared for the acid-fast staining of smears.
   B. Acid alcohol. Add 2 ml of hydrochloric acid (conc.) to 90 ml of 95% ethanol, and bring to 100 ml volume with 95% ethanol.
   C. Counterstain. Add 20 ml of 95% ethanol to 79 ml of distilled water, and then add 1 ml of reagent grade glacial (99.5%) acetic acid. Use this to dissolve 1 gm of methylene blue chloride or janus green.

Procedure:
   1. Remove paraffin from the sections and take to water through the usual procedures. Treatment with alcoholic iodine and thiosulfate will be necessary if lead or mercuric salts were used in the fixative.
   2. Stain sections in the carbol fuchsin for 10 min at 70°C, 30 min at 55°C, 2 hr at 37°C, or 4–16 hr at room temperature.
   3. Rinse in water.
   4. Decolorize in acid alcohol for 20 sec or more.
   5. Rinse in water for 2–3 min.
   6. Counterstain with the acid methylene blue or acid janus green for 3 min.
   7. Rinse in water.
   8. Dehydrate by passing thorough 2 changes each of 95% acetone, 100% acetone, and then xylene.

9. Mount in balsam, Permount, Harleco synthetic resin (HSR), or other suitable mounting medium.

Results: The acid-fast bacilli will be red, non-acid-fast organisms and cell nuclei will take the color of the counterstain.

Comment: The acid counterstain is recommended by Lillie to help differentiate microorganisms and nuclei from cytoplasm and background material. Non-acid counterstains may be used.

Fite II Procedure for Acid-Fast Organisms

Fite, Cambre, and Turner (1947)

This procedure is very much like the Ziehl-Neelsen procedure except for the use of peanut-oil-xylene for the extraction of paraffin from sections rather than the usual xylene. The peanut-oil-xylene extraction is said to have a restorative effect of the acid-fast charateristic of older cells, and cells rendered weakly acid-fast by the tissue fixation, imbedding, and sectioning processes. Its use is especially important when the demonstration of leprosy organisms is involved; however, it can be recommended for the demonstration of any acid-fast organism in sections.

Preparation of solutions:
A. Deparaffinizing solution. Add 1 part of peanut oil to 2 parts of xylene. Although peanut oil is preferable, olive oil, cottonseed oil, or liquid petrolatum can be substituted.
B. Ziehl's carbol fuchsin. Prepare as given for the Ziehl-Neelsen procedure for the acid-fast staining of smears.
C. Decolorizer. Add 1 ml of hydrochloric acid (conc.) to 90 ml of 70% ethanol, and then bring to a volume of 100 ml with 70% ethanol.
D. Loeffler's methylene blue. Prepare as given for the Ziehl-Neelsen procedure for the acid-fast staining of smears.

Procedure:
1. Do not follow the usual method for the removal of paraffin from sections. Deparaffinize by immersing in 2 changes of the peanut oil-xylene mixture, allowing 3–6 min for each immersion.
2. Drain, wipe off excess oil, and blot to opacity. The residual oil in the section helps to prevent shrinkage and injury to the section.
3. If mercury or lead salts were used in the fixative, remove them by immersion in a strong iodine solution (5 gm of iodine and 10 gm of potassium iodine in 100 ml of distilled water) for 2 min, followed by immersion in 5% aqueous thiosulfate for 1–5 min. Rinse in tap water.
4. Stain at room temperature for 30 min in Ziehl's carbol fuchsin.
5. Decolorize with 1% HCl in 70% ethanol to a point where the sections are still faint pink. Usually about 1–2 min.
6. Rinse in tap water.

7. Counterstain with Loeffler's methylene blue for about 30 sec. Rinse in tap water.
8. Blot dry, then dry well in the air.
9. Dip in xylene before mounting.
10. Mount in Permount, HSR, or other suitable mounting medium.

Results: Acid-fast organisms will be stained red; other organisms and background material will be stained blue.

Comment: Nyka (1967) suggested a procedure which applied an oxidizing agent prior to the carbol fuchsin stain. This resulted in an increase in the intensity of the acid-fast stain, and in the number of bacterial cells appearing acid-fast. After the removal of the paraffin, rehydration, and washing, the sections were treated with 10% aqueous periodic acid for 4 hr or longer, and then subjected to an acid-fast stain procedure. If you want an acid-fast stain for tissue sections embedded in methacrylate, see the procedure of Pandolph and Poulter (1978).

Wade's Modification of the Fite I Acid-Fast Procedure

Wade (1957); Fite (1938, 1940)

This is a very good acid-fast stain for sections. Its disadvantages are that it is more complicated than the usual acid-fast procedure and that it results in a blue color for the acid-fast organisms rather than the traditional red color. The essential modification of the Fite I procedure is that it incorporates an oil-xylene deparaffinization step into the Fite I procedure which is to be recommended if leprosy organisms are to be demonstrated.

Preparation of solutions:
A. Fite's new fuchsin stain. Add 10 ml of absolute ethanol to 100 ml of distilled water and then add 5 ml of melted phenol. Use this solution to dissolve 0.5 gm of new fuchsin (magnetta III).
B. Modified van Gieson stain. Add 0.1 gm of picric acid to 100 ml of distilled water. Dissolve 0.01 gm of acid fuchsin with this solution.
C. Deparaffinizing solution. Add 1 part of heavy grade paraffin oil (liquid petrolatum) to 2 parts of rectified turpentine.
D. Reagent grade formaldehyde (37–40%). The use of a good quality reagent grade formaldehyde is essential.
E. Sulfuric acid solution. Add 5.0 ml of conc. sulfuric acid to 90 ml of distilled water, and then bring to a volume of 100 ml with distilled water.
F. Potassium permanganate solution. Add 1 gm of potassium permanganate to 90 ml of distilled water, and then bring to a volume of 100 ml with distilled water.
G. Oxalic acid bleach. Dissolve 2 gm of anhydrous reagent grade oxalic

acid in 90 ml of distilled water, and then bring to a volume of 100 ml with distilled water.

Procedure:

1. Do not follow the usual method for the removal of paraffin from sections. Deparaffinize by immersion in 2 changes of the turpentine-paraffin oil mixture for a total immersion time of 5 min.
2. Drain the slide, wipe off excess fluid on the back and edges of the slide, and blot with filter paper until the sections become opaque, and then place in water.
3. If mercury or lead salts were used in the fixative, remove them by extraction with an aqueous solution of 5% iodine in 10% potassium iodate for 2 min, followed by immersion in 5% aqueous thiosulfate for 1–5 min. Rinse in tap water.
4. Stain overnight (16–24 hr) in Fite's phenol-new fuchsin stain at room temperature. Rinse in tap water.
5. Immerse in reagent grade formaldehyde (37–40%) for 5 min. The bacilli become blue, but the sections may turn blue or remain a reddish color.
6. Wash in tap water for 2 min.
7. Decolorize 5 min in 5% sulfuric acid. Wash in tap water. The sulfuric acid has no visible effect but is important.
8. Expose to 1% potassium permanganate for 3 min to obtain a brown color. Wash.
9. Individually, bleach the secctions in 2% oxalic acid. Use agitation, and a period of time of from 30–60 sec. If a section does not bleach readily, use 5% oxalic acid.
10. Place in the modified van Gieson stain for 3 min. Do not wash.
11. Dehydrate rapidly in 95% ethanol, and then 100% ethanol, and clear in xylene.
12. Mount with Permount, HSR, or other suitable mounting medium.

Results: The acid-fast bacilli and free granules will stain a deep blue. Connective tissue elements, including reticulum, will stain red; other elements will stain yellowish.

Variations: Ziehl's carbol fuchsin can be used in place of the phenol-new fuchsin stain. Also, 1 part peanut oil, 2 parts xylene, can be substituted for the paraffin oil-turpentine mixture. Wade (1952) recommended the use of carbowax compounds, rather than paraffin, for sections to be examined for acid-fast bacilli, especially if leprosy organisms are to be demonstrated.

Comments: Wade (1952, 1957) showed that this procedure is superior to the usual Ziehl-Neelsen procedure when the staining of the leprosy bacillus is involved. It is believed that the paraffin oil-turpentine deparaffinizing treatment has a restorative effect on the acid-fast characteristic of cells weakly acid-fast, or which have lost some of their acid fast characteristic due to the procedures used in the fixation, imbedding and sectioning of tissues.

Fluorescence Microscopy for Acid-Fast Organisms in Tissue Sections

(A modification of the procedure of Truant, Brett, and Thomas (1962); and a procedure given in the Manual of Histologic Staining Methods, Armed Forces Institute of Pathology (1968); with a suggested modification of Mansfield (1970))

The use of fluorescence microscopy for the detection of acid-fast organisms in tissue sections has the same advantages and disadvantages as discussed for smear preparations. Its use, however, is becoming more and more popular.

Preparation of solutions:
 A. Auramine O-rhodamine B stain, the acid alcohol decolorizer, and the potassium permanganate counter stain are all prepared as given for the Truant, Brett, and Thomas (1962) procedure for fluorescence microscopy of acid-fast organisms in smears.
 B. Weigert's iron hematoxylin solution.
   Solution a. Dissolve 1.5 gm of anhydrous ferric chloride (or 2.5 gm of $FeCl_3 \cdot 6H_2O$) in 90 ml of distilled water, add 1 ml of hydrochloric acid (conc.), and then bring to a volume of 100 ml with distilled water.
   Solution b. Dissolve 10 gm of hematoxylin in 100 ml of absolute ethanol. Age at least 2 weeks before use. This stock solution will keep for months.
   Solution c. Just before use, add 10 ml of the 10% hematoxylin (b) to 90 ml of 95% ethanol.
   The working solution. Just before use add 1 part of the 1% hematoxylin (c) to 1 part of the acidic ferric chloride (a). The solution will blacken due to the formation of iron hematoxylin. The working solution is unstable and should be prepared fresh for each day of use.

Procedure:
 1. Remove paraffin from the sections and take to water through the usual procedure. Treatment with alcoholic iodine and thiosulfate may be necessary if lead or mercury salts were used in the fixative.
 2. Immerse in Weigert's iron hematoxylin for 10 min.
 3. Wash in running tap water for 10 min.
 4. Immerse in the auramine O-rhodamine B stain at 60°C for 10 min.
 5. Wash in tap water.
 6. Decolorize with the 1% HCl in 70% ethanol for about 2 min. Wash in tap water.

7. Immerse in 0.5% aqueous potassium permanganate. Wash in tap water, and then rinse in distilled water.
8. Dehydrate by passing through 2 changes each of 95% ethanol, 100% ethanol, and xylene. Some prefer to dehydrate by blot drying followed by thorough (overnight) drying in the air.
9. Mount in Permount, HSR, or other suitable mounting medium.
10. Observe with a suitable microscope system such as described for the fluorescence microscopy of acid-fast organisms in smears.

Results: Acid-fast organisms should show fluorescence, non-acid-fast organisms and background material should show little or no fluorescence.

Variation: Since deparaffinization of sections with peanut oil-xylene is known to greatly enhance the staining of leprosy organisms in sections stained by non-fluorescent stains, the question arises as to whether it would also benefit the results obtained using fluorescent stains. Mansfield (1970) compared non-fluorescent and fluorescent stains, both with and without the use of peanut oil-xylene as the deparaffinization agent. He found that deparaffinization with peanut oil-xylene was as beneficial for fluorescence procedures as it was for non-fluorescence procedures. Mansfield's recommended procedure was a combination of that of Hagemann (1938) and Truant, Brett, and Thomas (1962). Briefly, the recommended procedure was: Deparaffinize the sections in peanut oil (1 part)-xylene (2 parts), rinse in tap water, drain (do not blot), stain for 15 min at room temperature (0.3% auramine O, 4% phenol, 6% glycerol, in distilled water), rinse, decolorize in 0.5% HCl in 70% ethanol for 1 min, counter stain with 0.5% potassium permanganate for 3 min, rinse, drain, thoroughly air dry, and mount in Permount.

SECTION D: STAINS FOR SPIROCHETES AND LEGIONELLA ORGANISMS IN TISSUE BLOCKS AND SECTIONS

The silver staining of spirochetes in tissue blocks or sections presents many more technical problems than for staining spirochetes in smears. Good reviews and comparison of methods can be found in Campbell and Rosahn (1950) and in Bridges and Luna (1957). The procedures which follow have been known to produce disappointing results. It is recommended, therefore, that a known positive control slide be processed with the unknown material so that the effectiveness of the procedures can be evaluated.

Levaditi and Manouélian Silver Impregnation Stain for Spirochetes in Tissue Blocks

Levaditi and Manouélian (1906); as modified by Lillie (1965)

This procedure has had wide use for the demonstration of spirochetes in tissue blocks. Its chief disadvantage is that it is slow and time consuming, taking from 5–7 days. Generally, it has been replaced by more rapid procedures applied to tissue sections. The tissue blocks should be 1–2 mm thick and fixed in 10% formalin for 1–2 days.

Preparation of solutions:
  A. Silver nitrate. Dissolve 2 gm of silver nitrate with 100 ml of distilled water. The solution should be fresh and not over 1 week old.
  B. Developer solution. Dissolve 4 gm of pyrogallol (pyrogallic acid) with 100 ml of distilled water, then add 5 ml of formalin (37–40% formaldehyde solution).

Procedure:
  1. Rinse the formalin fixed block in distilled water and immerse in 95% ethanol for about 24 hr.
  2. Wash in distilled water and soak until the block sinks to the bottom of the container.
  3. Immerse in freshly prepared silver nitrate solution at 37°C for about 4 days; the solution should be changed each day.
  4. Wash in distilled water.
  5. Immerse in the solution of pyrogallol and formalin for 48 hr.
  6. Wash in several changes of distilled water.
  7. Dehydrate in 80%, 95%, and absolute ethanol, 2 changes each and for 15–30 min each change.
  8. Clear in oil of cedarwood for 2 changes, 1 hr each.
  9. Infiltrate with 2 changes of paraffin, 45 min each.
  10. Section at 5 µm, and mount sections on slides.
  11. Deparaffinize with xylene, 3 changes, and mount with Permount, balsam, or other suitable mounting medium.

Results: The spirochetes will be black, the tissue material will be various shades of yellow to brown.

Variation: Levaditi and Manouélian reported that the time of impregnation with silver nitrate could be reduced to 4–6 hr if a temperature of 50°C was used.

Warthin-Starry Procedure for the Silver Staining of Spirochetes in Sections

(Original by Warthin and Starry (1922); modifications from Kerr (1938); and Bridges and Luna (1957). This procedure also demonstrates Donovan bodies in sections)

This procedure has the advantage of being faster than the Levanditi-Manouélian procedure for tissue blocks. The tissue block should be fixed in 10% formalin, and there is no advantage in using buffered neutral formalin. While other tissue fixatives can be used, those with heavy metals must be avoided. The tissue block is processed through alcohols to xylene, and then imbedded in paraffin. Sections should be cut at 5–8 μm, and floated on warm, freshly distilled and boiled, water. Fresh distilled water is used to avoid microorganisms which might adhere to the sections. The sections are placed on acid-dichromate-cleaned cover glasses (No.1 or 0 thickness) or 1 × 3 inch microscope slids and dried in an oven up to 24 hr. The procedure given below is particularly recommended for *Leprospira*. Donovan bodies in sections are demonstrated when the acidulated water is used at a pH of 3.6–3.8.

Preparation of solutions:

A. Acidulated water. One liter of tripple distilled water (or other distilled water which will hold silver nitrate in solution for several hours) is acidified to a pH of 3.8–4.4 by adding a 1% solution of citric acid. The pH can be determined with a glass electrode or by using a colorimetric method.

B. Silver nitrate for impregnation. Dissolve 1 gm of silver nitrate (C.P., crystals) in 100 ml of acidulated water.

C. Silver nitrate for the developer. Dissolve 2 gm of silver nitrate (C.P., crystals) in 100 ml of acidulated water.

D. Hydroquinone solution. Dissolve 0.15 gm of hydroquinone (crystals, photographic quality) in 100 ml of acidulated water.

E. Gelatin solution. Dissolve 10 gm of a good grade of gelatin (sheet or granulated) in 200 ml of acidulated water. Gently heat to dissolve (30–40°C).

F. For use, Solutions C, D, and E should be held at 54–56°C in a water bath.

Procedure:

1. Paraffin sections are placed on acid-dichromate cleaned 1 × 3 inch glass slides, or 22-mm square cover glasses, and air-dried. Gently warm and then deparaffinize and take to water by placing in 2 changes of xylene, 2 changes of absolute ethanol, then 1 change each of 95%, 70% ethanol, and then to acidulated water.

2. Impregnate by immersion in the acidulated 1% silver nitrate solution heated in a water bath to 43°C, for 30 min. A vertical Coplin staining dish (75 × 25 mm) is recommended for glass slides.

3. Lay the slide (or cover glass) with the section across glass rods. The developer should be a temperature of between 54 and 56°C, and must be used as soon as it is mixed. Mix the following in a test tube in a water bath at 54–56°C; 1.5 ml of acidulated 2% silver nitrate, 3.75 ml of 5% gelatin, and 2.0 ml of the hydroquinone solution. Flood the slide with the hot developer and expose until the sections turn a light golden brown or yellow. The time will vary from 3 to 12 min, but will remain constant if the other conditions are not varied. On first runs it will be

desirable to use a series of developing times such as 3, 6, 9, and 12 min. A known positive section should be used as a control and should be checked in the microscope at this time; the spirochetes should be black and the background yellow. Overdevelopment results in a dark background and heavy appearing spirochetes with the spirals obscured. Under development results in a pale background and an attenuated blackness for the spirochetes.

4. Rinse quickly and thoroughly in hot tap water (56°C).
5. Rinse in distilled water at room temperature.
6. Dehydrate in 70%, 95%, and absolute ethanol.
7. Clear in 2 changes of xylene.
8. Mount in Permount, balsam, or other suitable mounting medium.

Results: Spirochetes should be black in properly developed sections, but may be yellow in under developed sections. The background will be pale yellow to light brown.

Variations: The allowable variations were studied by Bridges and Luna (1957). Faulkner and Lillie (1945) replaced the acidulated water with an acetone-acetic acid buffered water which they found to be more convenient since it did not require the use of a glass electrode or colorimetric methods for the determination of pH, and it avoided the problem of mold growth in the citric acid solution. To prepare the buffered acidulated water as recommended by them, dissolve 16.4 gm of sodium acetate in 1000 ml of distilled water. Dissolve 11.8 ml of glacial acetic acid in 1000 ml of distilled water. A buffer of pH 3.8 is prepared by mixing 17.6 parts of the acetic acid solution with 2.4 parts of the sodium acetate solution. The ratio for the pH of 3.6 is 18.5 to 1.5. For use, 20 ml of the buffer is added to 480 ml of distilled water. This buffered acidulated water replaces the citric acid acidulated water used in the above procedure.

Comments: This procedure usually requires some experience before good results can be expected. The following will help to avoid poor results: use only acid-dichromate-cleaned glassware, do not use old solutions, do not use the silver nitrate impregnation or developing solutions more than once, and do not shortcut the technical details of the procedure. Only a few sections should be processed at one time.

Krajian Silver Stain for Spirochetes in Sections

Krajian (1939); as Modified by Walter Smith, Isreal, and Gager (1969)

This is a more recent silver stain for spirochetes in sections and it is reported to be very dependable. It is fairly rapid and it is reported that 5 slides can be processed within an hour.

Preparation of solutions:

A. Fixative. Fix tissue blocks for 12–24 hr in 10% formalin. Then imbed in paraffin and cut sections at 5–8 μm.

B. Mordant solution. All chemicals should be C.P. or reagent grade. Place 30 ml of 95% ethanol in a clean 100-ml Erlenmeyer flask, and 30 ml of acetone, 15 ml of glycerine, and 9 ml of formic acid. Dissolve in this, 3 gm of uranium (uranyl) nitrate ($UO_2(NO_3)_2 \cdot 6H_2O$). This solution is stable and can be kept for months if stored in a brown glass bottle and kept in the dark.

C. Gum mastic (saturated alcoholic). Dissolve 25 gm of gum mastic with 35 ml of absolute ethanol. Shake several times a day for 3–5 days until the solution becomes clear. The solution is stable and can be stored for long periods of time. Use only the clear portion of the solution. It is not necessary to filter. However, a clear solution can be maintained by filtering through several layers of gauze.

D. Dilute gum mastic solution. Prepare fresh each day by adding 70 drops of the saturated alcoholic solution to 100 ml of 95% ethanol.

E. Silver nitrate stock solution. Dissolve 10 gm of silver nitrate with 100 ml of triple distilled water. Store in a brown glass bottle and in the dark in a refrigerator. The stock solution will keep for a week or so, but should not be kept for long periods of time. The 1% solution is prepared by mixing 1 part of the stock solution with 9 parts of triple distilled water. If only a few sections are to be processed, it would be best to prepare directly a 1% solution of silver nitrate on the day of use.

F. Developer. This should be mixed at the last possible moment. Just before starting the staining procedure, place 0.62 gm of hydroquinone in a 125-ml Erlenmeyer flask, and then add 0.2 gm of sodium sulfite, Complete the developer during the last 4 min of the time the sections spend in the 1% silver nitrate (see Steps 8, 9, and 10) by adding in the indicated order 5 ml of formaldehyde (37–40%), 5 ml of acetone, 5 ml of pyridine, 5 ml of the saturated alcoholic solution of gum mastic, and 30 ml of triple distilled water. Rotate the flask to mix and heat over an open flame to 60°C. Transfer to a vertical Coplin staining dish in a 60°C water bath. The developer solution can be used for only one set of slides.

G. Prepare a 5% solution of sodium thiosulfate in triple distilled water.

Procedure:

1. Place sections on 1 × 3 inch clean glass microscope slides. Prepare 5 slides for simultaneous processing. A known positive control should be used to demonstrate that the procedure has functioned as required.

2. Deparaffinize sections by passage through xylene, a series of absolute,

95% and 70% ethanol, and then to triple distilled water.

3. Place the prepared mordant Solution B (about 85 ml) in a vertical Coplin staining dish (75 × 25 mm), which will hold up to 10 slides. Be sure the fluid level will cover the sections. Bring the temperature to 60°C in a water bath. This solution can be used most of the day.

4. Immerse the sections in the mordant Solution B at 60°C for 10 min. The slides should be held in a Peel-A-Way plastic slide holder, 5 slides to the holder (Van Waters and Rogers, Cat. No. 69, No. 48440-002).

5. Rinse in distilled water.

6. Place one at a time in a vertical Coplin staining dish with dilute gum mastic for 5 sec.

7. Place the slides back in the Peel-A-Way holder, wash well in distilled water, and drain the slides on a paper towel.

8. Place sufficient 1% silver nitrate in a vertical Coplin staining dish to cover the sections and place in a water bath at 70–75°C. The temperature should never be below 70°C. Immerse the slides in the preheated silver nitrate and start timing. Keep the sections at this temperature for a total of 7 min.

9. When 4 min of the 7 min has elapsed, prepare the final mixing of the developer (Solution F, see instructions for its preparation) by pipetting the liquids into the flask with the hydroquinone and sodium sulfite. Bring the freshly mixed developer to 60°C and place in a vertical Coplin staining dish in a water bath at 60°C.

10. When the 7 min is up (Step 8), remove the slides from the 1% silver nitrate and dip the slides (still in the Peel-A-Way holder) into the hot developer and hold for 3–5 sec, withdraw and hold close (1–2 inches) to a 60 watt tungsten light source for 10–15 sec.

11. Repeat the dipping in the developer and the exposure to light for 5–7 times until the sections turn a medium brown or dark yellow color.

12. Wash the sections, still in the holder, by dipping 5 or 6 times in 95% ethanol to dissolve away the gum mastic.

13. Wash in distilled water.

14. Place the slides, still in the holder, for 5 min in aqueous 5% sodium thiosulfate in a vertical Coplin staining dish.

15. Rinse in distilled water for 1 min.

16. Dehydrate through 95% ethanol, and 2 changes of absolute ethanol, 30 sec each.

17. Clear in xylene for 15–30 sec, and then mount in permount, balsam, or other suitable mounting medium. Do not allow sections to stay in the xylene for any long period of time.

Results: The spirochetes should be stained black, the background material should be a brown to yellow color.

Dieterle Silver Impregnation Stain for *Legionella pneumophilia* in Sections (Legionnaire's Disease Organisms)

Dieterle (1927)

Chandler, Hicklin, and Blackmon (1977) compared the effectiveness of various stains for the organisms of Legionnaires' disease in sections. The Dieterle silver impregnation procedure was reported as being the method of choice. The attempted procedures and stains proving to be unsuitable were the Gomeri methenamine silver, acid-fast, Gridley fungus stain, periodic acid-Schiff, Giemsa, hematoxylin, and eosin. The Brown-Brenn or Brown-Hopps Gram stains gave highly variable results, but, when stained, the organisms were only faintly Gram-negative. The Giménez stain for rickettsiae and chlamydiae was effective for frozen sections of fresh or formalin-fixed human lung tissue, heat-fixed impression smears of human lung tissue, and yolk sac material. However, it was not suitable for sections which had been embedded in paraffin because the fuchsinophilic material in the organism was altered by the embedding process. Test of the Dieterle procedure were run on formalin-fixed and paraffin-embedded material, cut in sections 5 µm thick. The outstanding characteristic of the Dieterle procedure is that it uses uranium nitrate in an alcoholic solution to inhibit the uptake of silver by nerve fibers, collagen, and elastic tissue, and this results in a good selective stain for the *Legionella* organisms.

Preparation of solutions:
A. Ethanol, 95%.
B. Acetone.
C. A 1% solution of uranium nitrate in 70% ethanol.
D. A 10% gum mastic solution in absolute ethanol.
E. A 1% solution of silver nitrate in distilled water.
F. A developing solution:

Hydroquinone	1.5	gm
Sodium sulfite	0.25	gm
Neutral, 37–40% formaldehyde	10.0	ml
Acetone	10.0	ml
Pyridine	10.0	ml
Distilled water to make	90.0	ml

After dissolving the above, add about 10 ml of the 10% gum mastic in absolute ethanol, the solution should now appear "milky."

Procedure:
1. Deparaffinize with xylene, and bring the section to absolute ethanol by the usual method.

2. Immerse the section in 1% uranium nitrate in 70% ethanol, kept at 55°C, for 30 min.
3. Wash for a moment in distilled water.
4. Pass through 95% ethanol.
5. Immerse in 10% gum mastic in absolute ethanol for about 30 sec.
6. Immerse for an instant in 95% ethanol.
7. Wash in distilled water for a short period.
8. Immerse in the silver nitrate solution for 1 to 6 hr, at 55°C, and in the dark. The average suitable immersion time should be about 3 hr. The length of the staining period influences the intensity of staining.
9. Wash for a moment in distilled water.
10. Place in the developing solution for 5–15 min.
11. Wash for a moment in distilled water.
12. Immerse in 95% ethanol for a short time.
13. Immerse in acetone for a short time.
14. Clear in xylene, and mount in balsam or other suitable mounting medium.

Results: Tissue material should be golden yellow to golden brown. The *Legionella* organisms should be deep brown to black, and appear as small, blunt, bacilli, with occasional pleomorphic forms.

Comments: Stronger uranium nitrate solutions can be used (to 5%) to further reduce background staining. Smears of tissue material emulsified in one drop of physiological saline on a slide can be stained if first fixed for 1 hr in a solution of 10 ml formalin in 90 ml of 95% ethanol, rinsed in distilled water, and then taken to Step 2 above. Success was reported also for slide preparations of peritoneal exudate from inoculated guinea pigs.

References:
Chandler, F. W., Hicklin, M. D., and Blackmon, J. A. 1977. Demonstration of the agent of Legionnaires' disease in tissue. N. Engl. J. Med., **297**, 1218–20.
Dieterle, R. R. 1927. Method for demonstration of *Spirochaeta pallida* in single microscopic sections. Arch. Neurol. Psychiatry, **18**, 73–80.

SECTION E: STAINS FOR VIRAL INCLUSION BODIES IN SECTIONS

Staining procedures can be used for the demonstration in sections of the intracytoplasmic inclusion bodies of rabies (Negri bodies), smallpox and cowpox (Guarnieri bodies), and the intranuclear inclusion bodies of chickenpox and shingles (herpes zoster). Of these, the most important is the demonstration of Negri bodies, which is considered to be a specific diagnosis for the presence of a rabies infection. The traditional staining of sections for the diagnosis of rabies is rapidly being replaced by the faster smear procedures, especially when a fluorescent antibody procedure is used. Fluorescent antibody procedures are also replacing the traditional staining methods for sections, and are recommended wherever the proper equipment, reliable reagents, and experienced personnel are available (McQueen, Lewis, and Schneider, 1960; World Health Organization Publication No. 23, *Laboratory Techniques in Rabies*, 1973). Of the non-fluorescent staining procedures given below, the Schleifstein procedure is recommended for the demonstration of Negri bodies, and the Lendrum procedure is recommended for the inclusion bodies of smallpox, cowpox, chickenpox, and shingles.

Tissue for the diagnosis of rabies should be obtained from the Ammon's horn, olfactory portion of the cortex, cerebellum, and the medulla oblongata regions of the brain, and from tissue from the submaxillary salivary glands. Detection of Negri bodies or rabies virus antigen in salivary gland material is of more immediate concern when animal bites are involved than the demonstration of Negri bodies in brain material. Detailed instructions for obtaining animal specimens for the diagnosis of rabies and for fluorescent antibody techniques can be found in the World Health Organization's *Laboratory Techniques in Rabies* (1973).

Schleifstein Stain for Negri Bodies in Sections

Schleifstein (1937)

This is a rapid method which allows the fixation, imbedding, sectioning, and staining of sections to be completed in a period of about 8 hr. In principle it resembles the Sellers' stain for Negri bodies in smears. Its chief disadvantage is the toxicity of vapors of dioxan (diethylene oxide) used in the dehydration procedure and which must be handled with due caution and under a hood.

Blocks of tissue not more than 3 mm thick should be cut from several parts of the brain and fixed in Zenker's fixative for 4 hr at 37°C.

Preparation of solutions:
- A. Zenker's fixative. Add 25 gm of potassium dichromate to 950 ml of distilled water, then add 50 gm of mercuric chloride (HgCl₂) and 50 ml of glacial acetic acid. This solution is stable over long periods of time.
- B. Dioxan. This is diethylene oxide and is also called *p*-dioxane.
- C. Stock stain solution. Dissolve 1.8 gm of rosanilin chloride, and 1 gm of methylene blue chloride with 100 ml of absolute methanol, then add 100 ml of glycerol. This stock solution is stable and keeps indefinitely.
- D. Potassium hydroxide solution (1:40,000). Dissolve 0.025 gm of potassium hydroxide in 1 liter of distilled water.

Preparation of sections:
1. Fix in Zenker's fixative for 4 hr at 37°C.
2. Wash 30 min or longer in running tap water. The tissue can be left overnight in running tap water.

3. Place the following in a ground glass stoppered 100 ml bottle; anhy-drous calcium chloride to a depth of 1 cm, then add about 80 ml of dioxane. Dissolve in the dioxane a few flakes of iodine to assist in removing mercury salts from the tissue. Place a non-corrosive wire tripod with a fine mesh wire screen over the calcium chloride to prevent the tissue from having contact with the calcium chloride. The dioxane should rise above the screen so that when the tissue blocks are placed on the screen they are covered with dioxane. Keep the blocks on this screen, at 37°C for about 1 hr. Because the vapors are toxic, all vessels with dioxane should be kept under a hood.

4. Transfer to a mixture of equal parts of dioxane and paraffin, and hold at 56°C for 1 hr. Keep the blocks a few cm from the glass bottom of the vessel by use of a wire tripod screen.

5. Place the blocks in paraffin at 56°C for 1 hr. Then imbed in paraffin.

6. Section the blocks at 4 μm and attach to glass slides with Mayer's albumin fixative. Heat over a flame until the paraffin melts, then extract the paraffin with xylene and take to distilled water in the usual manner.

Staining procedure:

1. For use, dilute the stock stain solution by adding 1 drop to 2 ml of the 1:40,000 KOH solution in a test tube. This must be done just before each stain.

2. Remove the slides from the water and place on a hot plate (80–90°C) and flood the slide with the freshly diluted stain solution. Let steam for about 5 min, let cool, and wash in tap water.

3. Decolorize each section separately by swishing the slide in a jar with 90% ethanol until the section takes on a faint violet color. This step determines proper differentiation.

4. Rapidly pass through 95% ethanol, 100% ethanol, and clear in xylene.

5. Mount in balsam or other suitable mounting medium.

Results: Negri bodies should stain a deep magenta red and should contain the typical dark blue inner granules. Cytoplasm should stain a bluish violet, nucleoli should be bluish black, and the erythrocytes should be a copper color.

Variations: Lillie (1965) recommended that the toxicity of the dioxane vapors be avoided by substituting an acetone-benzene dehydration pro-cedure. For example, 4 passages through acetone for 20 min each, then acetone-benzene for 20 min, and 2 changes of benzene for 20 min each. Paraffin is then infiltrated at 55°C for 2 hr. Mercury salts should be removed from sections in the usual manner during the hydration procedure before staining.

The World Health Organization's *Laboratory Techniques in Rabies*

(1973) recommended the following as a rapid fixative; equal volumes of glacial acetic acid, acetone, and a saturated solution of mercuric chloride in absolute ethanol. Fix 1 mm thick tissue for 15–20 min, then transfer to absolute ethanol for 2 changes of 20–30 min each, then 2 changes of toluene 15 min each, followed by 2 impregnations with paraffin at 55°C for 15 min each. After sectioning, the mercury salts in the sections must be removed by the usual method before staining. That is, exposure during rehydration to iodine in 70% ethanol followed by 5% aqueous thiosulfate, and rinsing in distilled water.

Stoval and Black Stain for Negri Bodies in Acetone Fixed Tissue

Stoval and Black (1940)

This procedure uses ethyl eosin and methylene blue as stains. It employs an acidified ethyl eosin which greatly increases the affinity of ethyl eosin for substrate, and an acidified methylene blue as a counter stain which will not replace the ethyl eosin.

Preparation of solutions:
   A. Acidified ethyl eosin stain. Dissolve 1 gm of ethyl eosin with 100 ml of 95% ethanol. Adjust the pH to 3.0 with 0.1 normal HCl. The final pH must be near 3.0 and this should require about 5 ml of the HCl.
   B. Methylene blue stain. Dissolve 0.35 gm of methylene blue chloride with 100 ml of 22% ethanol. Adjust to a pH of 5.3 or below by adding acetate-acetic acid buffer. About 3–4 ml of the buffer should be required and it is made by dissolving 25 gm of sodium acetate and 5 ml of glacial acetic acid in 250 ml of distilled water. Adjustment of the pH to as low as 4.6 is acceptable.

Preparation of sections: Specimen blocks should be cut about 3–5 mm thick and placed between small squares of ordinary writing paper and placed, cut side up, in a Petri dish. The paper serves to prevent undue shrinking of the blocks which facilitates the demonstration of pyramidal cells which best show the Negri bodies. Cover the blocks with fresh acetone for 2½–6 hr depending on the size of the blocks. Overnight fixation in acetone is acceptable. Remove the tissue from the paper and place in paraffin at 60°C for 4 hr to overnight. The time can be reduced by going through a series of acetone-paraffin concentrations. Cut at 4.5–5 μm, float sections on slides, and warm gently to fix the sections on the slides and place in an oven at 60°C for 45 min. Take to water through xylene, absolute ethanol, 95% ethanol, 2 changes each. Then 70% ethanol and to distilled water. The sections can now be stained.

Procedure:
   1. Stain 2 min in the acidulated ethyl eosin.
   2. Rinse in distilled water.

3. Stain for 2 min in the acidulated methylene blue.
4. Rinse in distilled water.
5. rinse in distilled water adjusted to pH 3.0 by adding 13 drops of glacial acetic acid to 60 ml of water. The section should become a deep blue to brownish red color. This wash serves to remove excess methylene blue from the pyramidal cells and fixes the ethyl eosin in the Negri bodies.
6. Rinse in distilled water.
7. Rinse in 2 changes of 95% ethanol to differentiate and dehydrate.
8. Rinse in 2 changes of absolute ethanol to complete differentiation and dehydration. The degree of differentiation should be followed with the microscope.
9. Clear in 2 changes of xylene.
10. Mount in balsam or other suitable mounting medium. Examine with the oil immersion objective.

Results: The Negri bodies should be stained a reddish brown to bright red. Nucleoli and cytoplasm should be blue, and other material pink.

Mann's Method for Staining Negri Bodies in Sections

This is a classical method and has been popular for a very long time. Usually it takes some experience to produce good results. Sections should be cut from tissue fixed in Zenker's fixative, paraffin imbedded, sectioned, carried through absolute, 95%, and 80% ethanol, then 0.5% iodine in 70% ethanol, and to 5% sodium thiosulfate to remove mercury salts, rinsed in distilled water, and stained.

Preparation of solutions:
A. Stock eosin Y solution. Dissolve 1 gm of eosin Y with 100 ml of distilled water.
B. Stock methyl blue solution. Dissolve 1 gm of methyl blue (not methylene blue) with 100 ml of distilled water.
C. Alkaline ethanol. Add 0.4 ml of 0.1 normal sodium hydroxide to 100 ml of absolute ethanol.
D. Staining solution. At the time of use add to 49 ml of distilled water, 18 ml of the methyl blue solution, and 23 ml of the eosin Y solution.

Procedure:
1. Stain the section for 24 hr in the staining solution of methyl blue and eosin Y.
2. Wash rapidly with tap water, then with absolute ethanol.
3. Differentiate in the alkaline absolute ethanol until the section appears pink; this should take about 10 min.
4. Wash in water until the section is a light blue. If the blue color does not

appear, wash with 0.2% acetic acid for 1 min.
5. Dehydrate rapidly to absolute alcohol, clear with xylene, and mount in balsam or other suitable mounting medium.

Results: The Negri bodies should be stained a dark red and should show the typical internal blue granules. Erythrocytes should be pink, nucleoli violet, and cytoplasm blue.

Variation: The staining time for methyl blue, eosin Y stain can be reduced to 6–10 hr by heating to 38–40°C.

Lendrum Stain for Inclusion Bodies in Sections

Lendrum (1947)

This procedure is not usable for the demonstration of negri bodies, but is acceptable for the demonstration of a variety of other inclusion bodies such as the intracytoplasmic Guarnieri bodies of smallpox and cowpox, and the intranuclear inclusion bodies of chickenpox and shingles (herpes zoster); all are phloxinophilic when stained by this procedure (see also p. 208).

Preparation of solutions:
A. Mayer's hematoxylin. Dissolve 1 gm of hematoxylin with 100 ml of distilled water using gentle heating if necessary. Then add 0.2 gm of sodium iodate, and 50 gm of ammonium alum (aluminum ammonium sulfate, $AlNH_4(SO_4)_2 \cdot 12H_2O$) or potassium alum (aluminum potassium sulfate, $AlK(SO_4)_2 \cdot 12H_2O$) and shake until the alum is dissolved. Add 1 gm of citric acid (granular) and 50 gm of chloral hydrate ($Cl_3CCHO \cdot H_2O$). The color should be a reddish violet and this solution will keep for several months. Ripening is controlled by the concentration of sodium iodate and time. Over-ripening is indicated by prolonged staining times.
B. Phloxine solution. Dissolve 0.5 gm of phloxine B and 0.5 gm of calcium chloride with 100 ml of distilled water. This solution is very stable and can be kept for about a year. It can be reinvigorated by adding more calcium chloride.
C. Tartrazine solution. Dissolve 2.5 gm of tartrazine (sodium salt) with 100 ml of cellosolve (ethylene glycol monoethyl ether).

Preparation of sections: Tissue blocks should be fixed for 24 hr in a solution which consists of 9 parts of a saturated sodium of mercuric chloride and 1 part of formalin (37–40% formaldehyde solution). The solubility of mercuric chloride in water is listed as 6.9 gm per 100 ml of water at 20°C. The blocks should then be exposed to 70% ethanol with iodine to remove mercury salts, then dehydrated in 80%, 95%, and absolute ethanol, cleared in xylene, infiltrated and imbedded in paraffin as usual.

The 5–6 μm thick sections cut should be deparaffinized and brought to distilled water by the usual procedures, including iodine and thiosulfate treatment if the mercury salts have not already been removed.

Procedure:

1. Stain for 5–10 min in Mayer's hematoxylin.
2. Wash in running tap water for about 15 min. Wash until the section becomes blue in color.
3. Stain with the phloxine B, calcium chloride, solution for 30 min.
4. Rinse briefly in distilled water and drain.
5. Differentiated by flooding the slide with the saturated tartrazine in cellosolve, from a dropper bottle. The differentiation time will vary greatly with the nature of the material and should be as long as possible and still retain the red color of the inclusion bodies. The time can vary from a few minutes to several hours. The differentiation which occurs should be followed with the microscope. The yellow tartrazine replaces the red phloxine and the differentiation should be stopped when the inclusion bodies appear a bright red and before they begin to fade.
6. Wash briefly in water to remove enough tartrazine to produce a pleasing yellow background.
7. Dehydrate through 60%, 95%, and absolute ethanol, clear in 2 changes of xylene, and mount in balsam or other suitable mounting medium.

Results: The inclusion bodies should be stained a bright red; collagen and background material yellow. Nuclei should be stained with the hematoxylin.

SECTION F: THE STAINING OF RICKETTSIAE, CHLAMDIAE, AND MYCOPLASMA IN SECTIONS

The staining of rickettsiae, chlamydiae, and mycoplasma in sections is not often done in diagnostic procedures. Rickettsiae and chlamydiae organisms are usually demonstrated through smears of inoculated chicken embryo material, and mycoplasmas are usually cultured on an artificial culture medium, and then demonstrated by staining the entire colony, or by staining smears prepared direction from colonies.

Goodburn and Marmion Stain for Mycoplasma in Frozen Sections

Goodburn and Marmion (1962)

Preparation of sections: Inoculated chick embryos are incubated at 35°C for 6–7 days. The eggs are then opened, without chilling, the embryos removed, the lungs harvested and frozen in a bijou bottle at the temperature of solid carbon dioxide in ethanol, stored at −70°C, and sectioned at 5 μm. The sections are then placed on gelatin-coated glass slides, fixed in dehydrated acetone for 10 min, removed, and air-dried.

Preparation of solutions:

A. Giemsa stain. Either buy ready prepared, or prepare as follows. Heat 33 ml of glycerol to 55°C. Add 0.5 gm of Giemsa stain (powdered form) to a warm mortar and slowly add, with trituration, the 33 ml of warm glycerol. Cool to room temperature and add 33 ml of warm glycerol. Cool to room temperature and add 33 ml of absolute methanol (reagent grade) and mix. Place in a tightly stoppered bottle and store overnight; a small amount of sediment will appear. Decant into small (30-ml) bottles, leaving the sediment behind, and tightly stopper. Store at least 2 weeks and filter before use.

B. Staining solution. Dilute 1 part of the Giemsa stain with 25 parts of 0.1 M phosphate buffer pH 6.4. The buffer is prepared by dissolving 14.198 gm of anhydrous dibasic sodium phosphate (Na_2HPO_4) in distilled water and bringing the final volume to 1 liter; this is Solution A. Dissolve 13.801 gm of monobasic sodium phosphate ($N_2H_2PO_4 \cdot H_2O$) in distilled water and bring the final volume to 1 liter; this is Solution B. The pH 6.4 buffer is prepared by mixing 16 parts of Solution A (Na_2HPO_4) with 34 parts of Solution B ($NaH_2PO_4 \cdot H_2O$).

C. Potassium permanganate solution. Dissolve 1 gm of potassium permanganate with 100 ml of distilled water.

Procedure:

1. The acetone fixed sections of lung tissue on slides are covered with the 1% potassium permanganate for 2 min and then washed with distilled water.

2. Place in the buffered 1:25 dilution of the Giemsa stain for 18 hr. Wash in distilled water.

3. Differentiate with 0.5% glacial acetic until microscopical examination shows a clear differentiation has been obtained between the cytoplasm and nuclei of the epithelial cells in the mesobronchi.

4. Wash in distilled water.

5. The sections can either be mounted wet in a mixture of glycerol and phosphate buffer pH 6.4, and then covered with a cover glass and sealed with nail varnish; or more permanent preparations can be made by dehydrating the sections in acetone, clearing in xylene, and mounting in a suitable mounting medium. The wet mount is recommended.

Results: The mycoplasma cells are seen as minute pink and purple bodies, coccoid to coccobacillus in shape, and associated with the epithelial cells of the mesobronchi.

Wolback's Variation of the Giemsa Stain for the Demonstration of Rickettsiae and Chlamydiae in Sections

Wolbach, Todd, and Palfrey (1922)

Rickettsiae, the organisms of the trachoma, psittacosis-ornithosis, and lymphograuloma venereum groups are very small and difficult to demonstrate with the optical microscope. However, the following method can be recommended.

Preparation of solutions:

A. Giemsa stain solution. Prepare the stock Giemsa stains as described for the Goodburn and Marmion stain. For use, place 100 ml of distilled water in a 150-ml vessel and add 2.5 ml of the stock Giemsa stain, 2.5 ml of absolute methanol, and 0.25 ml of an aqueous 0.5% solution of sodium carbonate.

B. Colophonium alcohol. The stock is prepared by dissolving 10 gm of rosin (colophonium) in 100 ml of 95% ethanol. For use, add 0.5 ml of the 10% solution to 100 ml of 85% ethanol.

Preparation of sections: The tissue should be fixed in Schaudinn's fixative, or Zenker's fixative, cut, sections placed on glass slides, the paraffin removed with xylene, the mercury salt removed with an alcoholic iodine solution, taken to distilled water, exposed to sodium hyposulfate, and water washed as instructed elsewhere (see the introduction to this chapter).

Procedure:

1. Stain the section for 1 hr in the diluted Giemsa stain, methanol, and sodium carbonate, solution. After 1 hr replace the stain solution and repeat 1 hr later. Leave in the latter solution overnight.
2. Rinse in distilled water, then leave in distilled water for 10–15 min. The section should take on a pink color.
3. Differentiate in the 95% ethanol with 0.5 ml of rosin per 100 ml. Differentiation can be followed with the microscope using an oil immersion lens. After completion of the slide, some further differentiation can be obtained by prolonged exposure to direct sunlight. Mounting in cedar oil also results in further differentiation.
4. Dehydrate in absolute ethanol, clear in xylene, and mount in cedar oil. Do not mount in balsam.

Results: Rickettsiae and chlamydiae should stain an intense reddish purple unlike anything else stained in the section. Nuclei should stain blue or dark purple, cytoplasm should be a pale blue, collagen and muscle fibers pink, and erythrocytes an olive gray to pink.

SECTION G: THE DEMONSTRATION OF FUNGI IN TISSUE SECTIONS

Schneidau's Gram Stain for the Demonstration of Fungi in Sections

Schneidau (1963)

Theoretically, since all pathogenic fungi are Gram-positive and since all tissue cells are Gram-negative, and also since it would not be necessary to differentiate Gram-negative organisms from the Gram-negative tissue cells, any simple Gram stain procedure could be used for the differential demonstration of pathogenic fungi in tissue sections. However, the alcohol or acetone dehydration steps necessary before a section can be cleared in xylene and mounted in balsam, could well result in over decolorization of some fungi which are weakly Gram-positive, such as *Nocardia* and *Actinomyces*, and therefore failure to differentiate them from background material. Schneidau proposed the use of tertiary butanol, $(CH_3)_3COH$, for dehydration which only very slowly removes the dye-iodine complex from Gram-positive cells. The method is simple and can be recommended for some fungi, but not for all. Those reported to be well demonstrated are the eukaryotic fungi such as *Histoplasma capsulatum*, *Sporothrix schenckii*, and *Candida* species, as well as the prokaryotic *Nocardia asteroides*, *Nocardia brasiliensis*, and *Actinomyces bovis*. All of the staining solutions should be in vertical Coplin staining dishes since it was reported that flooding of the slide with the solutions was not satisfactory.

Preparation of solutions:

A. Hucker's crystal violet. Prepare as directed for the Gram stain procedure used for smears.

B. Gram's iodine. Prepare Burke's iodine as directed for the Gram stain procedure used for smears. Gram's iodine can be prepared from this by diluting 1 part of Burke's iodine with 2 parts of distilled water.

C. Saturated picric acid. The solubility of picric acid in water is given as 1.4 gm per 100 ml of distilled water at 20°C.

D. Tertiary butanol. The melting temperature of tertiary butanol is 25.5°C. It can be kept liquid for use at lower temperatures by adding *n*-butanol to a concentration of 2%.

Procedure:

1. Deparaffinize sections and bring to distilled water by the usual procedure.
2. Stain in Hucker's crystal violet for 2 min.
3. Rinse in distilled water to remove excess stain.
4. Place in Gram's iodine for 2 min.
5. Quickly rinse, and place in distilled water for 1 min.

6. Decolorize in absolute acetone in a vertical Coplin staining dish for about 5–10 sec. Actively move the slide up and down in the decolorizer. The section visually should appear to be unstained.
7. Quickly rinse, and place in distilled water for 1 min to stop the decolorization process.
8. Counter stain in the saturated aqueous picric acid solution for 2 min.
9. Rinse in distilled water to remove excess stain.
10. Place in a solution of 50% tertiary butanol and 50% distilled water for 2 min.
11. Place in 2 changes of tertiary butanol with 2% n-butanol, 3 min each change. Replace with fresh reagent at frequent intervals.
12. Clear in xylene for 2 min and mount in balsam, Permount, or other suitable mounting medium.

Results: *Candida, Histoplasma, Sporothrix, Nocardia,* and *Actinomyces* organisms should appear strongly Gram-positive against a background stained with the picric acid. Other fungi may not be well differentiated.

Gridley's Stain for Fungi in Sections

Gridley (1953)

Gridley observed that the Hotchkiss-McManus periodate, Schiff, staining procedure (Hotchkiss, 1948; McManus, 1948) stained deeply red the conidia or spore forms of fungi but that the hyphae were only lightly stained. On the other hand, Gomori's (1950) aldehyde-fuchsin stain for elastic tissue fibers in sections was observed to stain the hyphae of fungi deep blue with a good demonstration of septae and branching forms. It was found that these 2 procedures could be combined into a single procedure which demonstrated both conidia and hyphae at the same time. It was also found that the problem of positive tissue staining elements which confuse the results when periodic acid was used for hydrolysis followed by staining with the Schiff reagent could be avoided by hydrolysis with chromic acid, which was first used by Bauer (1933) for the histochemical demonstration of glycogen and polysaccharids in tissue sections. This procedure can be recommended for the demonstration of fungal forms in tissue sections, and for the screening of tissue for a suspected fungal infection. However, Luna (1964) reported that the Gridley procedure did not stain prokaryotic fungi in tissue such as *Nocardia* or *Actinomyces.*

Preparation of solutions:
A. Chromic acid solution. Dissolve 4 gm of chromic trioxide (CRO_3) in 100 ml of distilled water.
B. Coleman's Feulgen reagent. Dissolve 1 gm of basic fuchsin chloride in 200 ml of boiling distilled water, filter, cool, and add 2 gm of potassium metabisulfite ($K_2S_2O_5$), and 10 ml of 1 N HCl. Left bleach for 24 hr

and then add 0.5 gm of activated carbon (Norite) and shake for about 1 min, then filter through coarse filter paper. The filtrate should be colorless. This solution must be fresh for the best results.

C. Sodium metabisulfite, HCl, solution. Add to 100 ml of distilled water 6 ml of a 10% aqueous solution of sodium metabisulfite solution and 5 ml of 1 N HCl.

D. Aldehyde fuchsin, solution. Add to 200 ml of 70% ethanol, 1 gm of basic fuchsin chloride, 2 ml of paraldehyde, and 2 ml of conc. HCl. Let stand at room temperature for 3 days. The solution should turn a deep purple. Store in a refrigerator.

E. Metanil yellow solution. Dissolve 0.25 gm of metanil yellow with 100 ml of distilled water and then add 0.25 ml of glacial acetic acid.

Procedure:

1. Cut paraffin imbedded sections at 6 μm, affix to glass slides, deparaffinize, and bring to distilled water by the usual methods.
2. Place in the 4% chromic acid solution for 1 hr.
3. Wash in running water for 5 min.
4. Place in Coleman's Feulgen reagent for 15 min.
5. Rinse in 3 changes of the sodium metabisulfite, hydrochloric acid, solution using 3 min each change.
6. Wash for 15 min in running water.
7. Place in the aldehyde-fuchsin solution for 15–20 min.
8. Rinse off excess stain in 95% ethanol.
9. Wash well in water.
10. Counterstain in the 0.25% metanil yellow for 1–3 min.
11. Wash in water.
12. Dehydrate in 2 changes each of 95% and absolute ethanol. Clear in 2 changes of xylene, and mount in Permount or other suitable mounting medium.

Results: The hyphae and conidia of eukaryotic fungi are well stained. Hyphae should be stained a deep blue, conidia should be a rose to purple color. The capsules of yeasts should be a purple color. The background material should be yellow.

Comment: If the metanil yellow has stained the background too intently, some small fungal forms may be missed. If this happens, decrease the staining time and increase the time of the subsequent washing step.

Grocott's Modified Gomori Chromic Acid, Methenamine, Silver Nitrate, Stain for the Demonstration of Fungi in Tissue

Grocott (1955)

This procedure demonstrates prokaryotic fungi such as *Nocardia* and *Actinomyces* as well as all forms of eukaryotic fungi. It is simpler than the

Gridley method, and it results in contrasts which are much preferred if photomicroscopy is to be attempted. The procedure is based on Gomori's (1946) histochemical test for glycogen and mucin, and it is very popular. Since it results in black fungi against an almost colorless background, Luna (1964) reported that it is superior to the Gridley stain for the screening of tissue for a suspected fungal infection.

Preparation of solutions:

A. Chromic acid solution. Dissolve 5 gm of chromic trioxide (CrO_3) in 100 ml of distilled water.

B. Stock methenamine, silver nitrate, solution. Dissolve 3 gm of U.S.P. grade methenamine (hexamethylenetetramine, $C_6H_{12}N_4$) in 95 ml of distilled water and bring the total volume to 100 ml. Add 5 ml of a 5% aqueous solution of silver nitrate ($AgNO_3$). A white precipitate should . form which will disappear on shaking. This solution is stable and will keep for months in the refrigerator.

C. Sodium bisulfite solution.. Dissolve 1 gm of sodium bisulfite ($NaHSO_3$ in 95 ml of distilled water and bring to a final volume of 100 ml.

D. Borax solution. Dissolve 5 gm of U.S.P. grade borax (sodium borate, $Na_2B_4O_7 \cdot 10H_2O$) in 95 ml of distilled water and bring to a final volume of 100 ml.

E. Gold chloride solution. Dissolve 0.1 gm of gold chloride ($HAuCl_4 \cdot 3H_2O$) with 100 ml of distilled water. This solution may be used repeatedly.

F. Sodium thiosulfate solution. Dissolve 2 gm of sodium thiosulfate ($Na_2S_2O_3 \cdot 5H_2O$) in 95 ml of distilled water and bring the final volume to 100 ml.

G. Working solution of methenamine and silver nitrate. Just before use, mix 25 ml of the methanamine, silver nitrate, stock solution with 25 ml of distilled water and add 2 ml of the 5% solution of borax.

Procedure:

1. Deparaffinize sections and take to distilled water by the usual methods.

2. Place in the 5% chromic acid solution for 1 hr.

3. Wash in running tap water for 10 min.

4. Place in the sodium bisulfite solution for 1 min to replace any residual chromic acid.

5. Wash in tap water for 5 min, and then in 3 changes of distilled water.

6. Place in a freshly prepared working solution of the methanamine and silver nitrate solution (Solution G) in a water bath at 50°C for about 1 hr. The section should turn a yellowish brown color. Use paraffin coated forceps to add and to remove slides from this solution.

7. Rinse 2 or 3 times in distilled water.

8. Tone in the 0.1% gold chloride for 5 min, then rinse in distilled water.

9. Remove unreduced silver by placing in the 2% sodium thiosulfate for 2–3 min.
10. Wash thoroughly in tap water. A counterstain is not necessary.
11. If a counterstain is desired it should be applied at this time. Safranin can be used if a nuclear skin is wanted. Hematoxylin-eosin can be used if tissue detail is important. The Armed Forces Institute of Pathology recommends the use of 0.2% light green for 30–40 sec (Luma, 1964).
12. Mount in Permount or other suitable mounting medium.

Results: All fungal forms should be a deep black. The inner parts of hyphae may appear a rose color. Mucin should be a rose color. In the absence of a counter stain, the background material should be colorless.

SECTION H: ENDAMOEBA IN TISSUE SECTIONS

Gridley's Stain Endamoeba in Tissue Sections

Gridley (1954)

This procedure is relatively simple and it can be recommended.

Preparations of sections: Any good method of fixation can be used. Cut paraffin imbedded tissue at 6 μm and affix sections to glass slides. Deparaffinize and take to distilled water by the usual procedure.

Preparation of solutions:

A. Harris' alum-hematoxylin. Prepare as described for the Gram staining of tissue sections. Weigert's iron hematoxylin can be used if it is available.

B. Aniline, eosin Y, acetic acid, solution. Dissolve 1.5 gm of eosin Y with 100 ml of 80% ethanol. Add 3 ml of aniline (reagent grade) and 1 ml of glacial acetic acid.

C. Naphthol green B solution. Dissolve 1 gm of naphthol green B with 100 ml of distilled water and then add 1 ml of glacial acetic acid.

Procedure:

1. Stain for 5–10 min in Harris' hematoxylin, or for 2–3 min in Weigert's iron hematoxylin.
2. Rinse in water 3–5 min.
3. Destain with 1% HCl in 70% ethanol.
4. Rinse in water for 3–5 min.
5. Neutralize in dilute ammonia water (10 drops of ammonia in 100 ml of distilled water). The section should appear blue.
6. Wash well in running tap water.
7. Stain in the aniline, eosin Y, acetic acid, solution for 5 min.
8. Rinse well in water. The sections should be a deep rose color.

9. Stain in the naphthol green B solution for 5 min.
10. Differentiate in 2 changes in 95% ethanol until the erythrocytes are bright rose. Check with the microscope.
11. Dehydrate in absolute ethanol and clear in xylene, 2 changes each.
12. Mount in balsam, Permount, or other suitable mounting medium.

Results: The amoebae should be stained blue-green and the ingested erythrocytes a bright rose. The nuclei of the amoebae should be a darker blue-green, and connective tissue light green.

A Modified Methenamine, Silver Nitrate, Method for the Demonstration of Endamoeba in Sections

Luthringer and Glenner (1961)

Gomori's chromic acid, methenamine, silver nitrate, stain for glycogen and mucin (Gomori, 1946) has been modified by Grocott (1955) for the demonstration of fungi in sections and by Luthringer and Glenner (1961) for the demonstration of amoebae in sections.

Preparation of solutions: All solutions are prepared as given for the Grocott stain for fungi in sections (see p. 469) except the 5% chromic acid is replaced with a 0.5% solution of periodic acid (H_5IO_6). The counter stain stock solutions are 0.2% light green, or 0.2% eosin Y, to either of which is added 0.2% acetic acid. For use, the counter stain stock solution should be diluted by mixing 1 part of stock solution with 5 parts of distilled water.

Procedure:
1. Deparaffinize and take to distilled water by the usual procedure.
2. Place in 0.5% periodic acid for 10 min at room temperature (25°C).
3. Wash in running tap water for 5 min.
4. Place in 5% chromic acid for 1½ hr at room temperature.
5. Wash in running tap water for 10 min.
6. Place in 1% sodium bisulfite for 1 min to remove traces of chromic acid.
7. Wash in running tap water for 5 min, then rinse in distilled water.
8. Dilute 25 ml of the stock methenamine, silver nitrate, solution with 25 ml of distilled water and then add 2 ml of a 5% borax solution. Place the section in this diluted solution for 2½-3 hr.
9. Rinse in several changes of distilled water.
10. Tone in 0.1% gold chloride solution for 5 min. This solution can be used repeatedly.
11. Rinse thoroughly in 2% sodium thiosulfate (about 5 min) to remove unreduced silver.
12. Wash in running tap water for 10 min.

13. Counterstain with the dilute acidified light green or eosin Y for 30–45 sec.
14. Dehydrate in 95% ethanol, absolute ethanol, and clear in xylene, 2 changes each.
15. Mount in Permount or other suitable synthetic resin.

Results: Amoebae are found on the mucosal surface of colonic glands, in lymphatic channels, and in areas of mucosal inflammation. The trophozoite form of amoebae should contain fine dark black granules, or the whole endoplasm may be darkened. Cyst forms will be stained in the outer region only. The nuclei should not be stained. When eosin Y is used as a counter stain, the erythrocytes in *Endamoeba histolytica* are nicely demonstrated.

BIBLIOGRAPHY

Achúcarro, N. 1911. Neuvo método para del estudio de la neuroglia y el tejido connunctivo. Biol. Soc. Exp. Biol., **1**, 139–41.

Adams, C.W.M. 1956. A stricter interpretation of the ferric ferricyanide reaction with particular reference to the demonstration of protein bound sulph-hydryl and di-sulphide groups. J. Histochem. Cytochem., **4**, 23–35.

Adams, C.W.M. 1958. Histochemical mechanisms of the Marchi reaction for degenerating myelin. J. Neurochem., **2**, 178–86.

Adams, C.W.M. 1965. *Neurohistochemistry*, Elsevier Publishing, Amsterdam.

Adams, C.W.M., and Sloper, J.C. 1956. The hypothalamic elaboration of posterior pituitary principles in man, the rat and dog; histochemical evidence derived from a performic acid-alcian blue reaction for cystine. J. Endocrinol., **13**, 221–8.

Adams, L.R., and Kamentsky, L.A. 1971. Machine characterization of human leukocytes by acridine orange fluorescene. Acta Cytol., **15**, 289–91.

Adams, T.W., Thomas, R.W., and Davenport, H.A. 1948. Staining sections of peripheral nerves for axis cylinders and for myelin sheaths. Stain Technol., **23**, 191–6.

Aghajanian, G.K., Kuhar, M.J., and Roth, R.H. 1973. Serotonin-containing neuronal perikarya and terminals; differential effects of p-chlorophenylalaninc. Brain Res., **54**, 85–101.

Ajelis, V., Björklund, A., Falck, B., Lindvall, O., Lorén, I., and Walles, B. 1979. Application of the aluminum-formaldehyde (ALFA) histofluorescence method for demonstration of peripheral stores of catecholamines and indolamines in freeze-dried paraffin-embedded tissue, cryostat sections, and whole-mounts. Histochemistry, **65**, 1–15.

Albert, H. 1921. Modification of stain for diphtheria bacilli. J.A.M.A., **76**, 240.

Albrecht, M.H. 1954. Mounting frozen sections with gelatin. Stain Technol., **29**, 89–90.

Alfert, M., and Geschwind, I.I. 1953. A selective staining method for the basic proteins of cell nuclei. Proc. Nat. Acad. Sci. U.S.A., **39**, 991–9.

Allison, A.C., and Young, M.R. 1969. Vital staining and fluorescene microscopy of lysosomes. In: *Lysosomes in Biology and Pathology*, Vol. 2, J.T. Dingle and H.B. Fell (eds.), John Wiley & Sons, New York.

Altman, F.P. 1971. The use of a new grade of polyvinyl alcohol for stabilizing tissue sections during histochemical incubations. Histochemie, **28**, 236–41.

Amann, M.J. 1896. Conservierungsflüssigkeiten und Einschlussmedien für Moose, Chloro- und Cyanophyceen. Z. Wiss. Mikrosk., **13**, 18–21.

Anderson, C.W., Moehring, R., and Gunderson, N.O. 1948. A new method for fixing defatting and staining milk and cream films. J. Milk Food Technol., **11**, 352–64.

Anderson, D., and Greiff, D. 1964. Direct fluorochroming of rickettsiae. J. Histochem. Cytochem., **12**, 194–6.

Anderson, F.D. 1954. Alcoholic-thionin, counterstaining, following Golgi dichromate impregnation. Stain Technol., **29**, 174–84.

Anthony, E.E., Jr. 1931. A note on capsule staining. Science, **73**, 319.

Armstrong, J.A. 1956. Histochemical differentiation of nucleic acids by means of induced fluorescence. Exp. Cell Res., **11**, 640–3.

Armstrong, J.A., and Hopper, P.K. 1959. Fluorescence and phase contrast microscopy of human cell cultures infected with adenovirus. Exp. Cell Res., **16**, 584–94.

Armstrong, J.A., and Niven, J.S.F. 1957. Fluorescence microscopy in the study of nucleic acids. Nature, **180**, 1335–6.

Arndt-Jovin, D.J., and Jovin, T.M. 1978. Automated cell sorting with flow systems. Annu. Rev. Biophys. Bioeng., **7**, 527–58.

Ashley, C.A., and Feder, N. 1966. Glycol methacrylate in histopathology. Arch. Pathol., **81**, 381–97.

Atkins, K.N. 1920. A modification of the Gram stain. J. Bacteriol., **5**, 321–4.

Aubert, E. 1950. "Cold" stain for acid-fast bacteria. Can. J. Public Health, **41**, 31–2.

Austin, C.R., and Bishop, M.W.H. 1959. Differential fluorescence in living rat eggs treated with acridine orange. Exp. Cell Res., **17**, 35–43.

Avrameas, S. 1970. Immunoenzyme techniques: Enzymes as markers for localization of antigens and antibodies. Int. Rev. Cytol., **27**, 349.

Baird, T.T. 1936. A word of caution concerning dioxan. Stain Technol., **11**, 122–3.

Baker, J.R. 1946. The histochemical recognition of lipine. Q. J. Microsc. Sci., **87**, 441.

Baker, J.R. 1965. The fine structure produced in cells by fixatives. J. R. Microsc. Soc., **84**, 115–31.

Balamuth, W. 1941. Studies on the regeneration of ciliate protozoa; I. Microscopic anatomy of *Licnophora macfarlandi*. J. Morphol., **68**, 241–77.

Bald, J.G. 1964. Cytological evidence for the production of plant virus ribonucleic acid in the nucleus. Virology, **22**, 377–87.

Ball, E. 1941. Microtechnique for the root apex. Am. J. Bot., **28**, 233–43.

Bancroft, J.D., and Stevens, A. 1977. *Theory and Practice of Histological Techniques*, Churchill Livingstone, Edinburgh.

Bang, B.G., and Bang, F.B. 1972. Preparation of urn cells. Am. J. Pathol., **68**, 407–17.

Barber, M.A., and Komp, H.W. 1929. Method of preparing and examining thick films for the diagnosis of malaria. Public Health Rep., **44**, 2330–41.

Bargmann, W. 1950. Die elektive Darstellung einer marklosen diencephalen Bahn. Mikroskopie, **5**, 289–92.

Barka, T., and Anderson, P.J. 1963. *Histochemistry: Theory, Practice, and Bibliography*, Harper & Row, New York.

Barnard, J.E., and Welch, F.V. 1936. Fluorescence microscopy with high powers. J. R. Microsc. Soc., **56**, 361–4.

Bartholomew, J.W. 1962. Variables influencing results, and the precise definition of steps in Gram staining as a means of standardizing the results obtained. Stain Technol., **37**, 139–55.

Bartholomew, J.W., Cromwell, T., and Gan, R. 1965. Analysis of the mechanism of Gram differentiation by use of a filter-paper chromatographic technique. J. Bacteriol., **90**, 766–77.

Bartholomew, J.W., Lechtman, M.D., and Finkelstein, J. 1965. Differential spore and lipid staining at room temperature by use of fluorescent dye. J. Bacteriol., **90**, 1146–7.

Bartholomew, J.W., and Mittwer, T. 1950. A simplified bacterial spore stain. Stain Technol., **25**, 153–6.

Bartholomew, J.W., Mittwer, T., and Finkelstein, H. 1959. The phenomenon of gram-positivity; its definition and some negative evidence on the causative role of sulfhydryl groups. Stain Technol., **34**, 147–54.

Bauer, H. 1933. Mikroskopish-chemischer Nachweis von Glykogen und einigen anderen Polysacchariden. Z. Mikrosk. Anat. Forsch., **33**, 143–60.

Beck, R.C. 1938. *Laboratory Manual of Hematological Technique*, W.B. Saunders, Philadelphia.

Beech, R.H., and Davenport, H.A. 1933. The Bielschowsky staining technic. A study of the factors influencing its specificity for nerve fibers. Stain Technol., **8**, 11–30.

Belling, J. 1926. The iron-acetocarmin method of fixing and staining chromosomes. Biol. Bull., **50**, 160–2.

Benjaminson, M.A., and Katz, I.J. 1970. Properties of SITS (4-acetoamido-4-isothiocyanostilbene-2,2-disulfonic acid); fluorescence and biological staining. Stain Technol., **45**, 57–62.

Bennett, D., and Radimska, O. 1966. Flotation-fluid staining; toluidine blue applied to maraglas sections. Stain Technol., **41**, 349–50.

Bennett, H.S., Wyrick, A.D., Lee, S.W., and McNeil, J.H. 1976. Science and art in preparing tissues embedded in plastic for light microscopy, with special reference to glycol methacrylate, glass knives and simple stains. Stain Technol., **51**, 71–97.

Bennhold, H. 1922. Eine spezifische Amyloidfärbung mit Kongorot. Munch. Med. Wochenschr., **2**, 1537–8.

Bensley, C.M. 1930. Comparison of methods for demonstrating glycogen microscopically. Stain Technol., **14**, 47–52.

Berg, N.O. 1951. A histological study of masked lipids; stainability, distribution and functional variations. Acta Pathol. Microbiol. Scand. (Suppl.), **40**, 192.

Bergeron, J.A., and Singer, M. 1958. Metachromasy; an experimental and theoretical reevaluation. J. Biophys. Cytol., **4**, 433–57.

Berlyn, C.P., and Miksche, J.P. 1976. *Botanical Microtechnique and Cytochemistry*, Iowa State University Press, Ames.

Bertalanffy, F.D. 1960a. Fluorescence microscope methods for the detection of pulmonary malignancies. Can. Med. Assoc. J., **83**, 211–2.

Bertalanffy, F.D. 1960b. Cytological cancer diagnosis in gynecology by fluorescene microscopy. Mod. Med. Can., **15**, 55–65.

Bertram, E.G., and Ihrig, H.K. 1957. Improvement of the Golgi method by pH control. Stain Technol., **32**, 87–94.

Berube, G.R., Powers, M.M., and Clark, G. 1965. Iron hematoxylin chelates; I. The Weil staining bath. Stain Technol., **40**, 53–62.

Berube, G.R., Powers, M.M., Kerkay, J., and Clark, G. 1966. The gallocyanin-chrome alum stain; influence of methods of preparation on its activity and separation of active staining compound. Stain Technol., **41**, 73–81.

Best, F. 1906. Über Karminfärbung des Glykogens und der Kerne. Z. Wiss. Mikrosk., **23**, 319–22.

Betts, A. 1961. The substitution of acridine orange in the periodic acid-Schiff stain. Am. J. Clin. Pathol., **36**, 240–3.

Bielschowsky, M. 1904. Die Silberimprägnation der Neurofibrillen. J. Psychol. Neurol., **3**, 169–89.

Bielschowsky, M. 1909. Eine Modifikation meines Silberimpränations-verfahrens zur Darstellung der Neurofibrillen. J. Psychol. Neurol., **12**, 135–7.

Björklund, A., Lindvall, O., and Svensson, L.-A. 1972. Mechanisms of fluorophore formation in the histochemical glyoxylic acid method for monoamines. Histochemie, **32**, 113–31.

Blagg, W., Schloegel, E.L., Mansour, N.S., and Khalaf, G.I. 1955. A new concentration technique for the demonstration of protozoa and helminth egg in feces. Am. J. Trop. Med., **4**, 23–8.

Blanco, F.L., and Fite, G.L. 1948. The effect of fixatives on staining procedures for lepra bacilli in tissues. Int. J. Lepr. **16**, 367–8.

Blest, A.D. 1961. Some modifications of Holme's silver method for insect central nervous systems. Q. J. Microsc. Sci., **102**, 413–17.

Bloch, D., Fu, C.-T., and Chin, E. 1978. Flow-cytometric analysis of chromatin. Methods Cell Biol., **18**, 247–75.

Bloch, D.P., and Godman, G.C. 1955. A microphotometric study of the synthesis of desoxyribonucleic acid and nuclear histone. J. Biophys. Biochem. Cytol., **1**, 17–28.

Blum, F. 1893. Der Formaldehyd als Hartungsmittel. Z. Wiss. Mikrosk., **10**, 314–5.

Bodian, D. 1936. A new method for staining nerve fibers and nerve endings in mounted paraffin sections. Anat. Rec., **65**, 89–97.

Bodian, D. 1937. The staining of paraffin sections of nervous tissue with activated Protargol. The role of fixatives. Anat. Rec., **69**, 153–62.

Böhm, N., and Sprenger, E. 1968. A valuable method for the quantitative determination of nuclear Feulgen-DNA. Histochemie, **16**, 100–18.

Boke, N.H. 1939. Delafield's hematoxylin and safranin for staining meristemetic tissue. Stain Technol., **14**, 129–31.

Bordelon, M.R. 1977. Staining and photography for chromosome banding with the fluorescent dyes quinacrine mustard and Hoechst 33258. TCA Manual, **3**, 587–92.

Borrel, A. 1901. Les theories parasitaires du Cancer. Ann. Inst. Pasteur (Paris), **15**, 49–67.

Bosshard, U. 1964. Fluoreszenzmikroskopische Messungen des DNS Gehaltes von Zellkernen. Z. Wiss. Mikrosk., **65**, 391–408.

Botazzi, E.M., Schreiber, G., and Bowen, V.T. 1971. Acantharia in the Atlantic Ocean, their abundance and preservation. Limnol. Oceanogr., **16**, 677–84.

Bouin, P. 1897. Études sur l'evolution normale et l'involution du tube séminifère. Arch. Anat. Microsc., **1**, 225–63.

Bowen, C.C. 1956. Freezing by liquid carbon dioxide in making slides permanent. Stain Technol., **31**, 87–90.

Bradley, D.F., and Wolf, M.K. 1959. Aggregation of dyes bound to polyanions. Proc. Nat. Acad. Sci. U.S.A., **45**, 944–52.

Bradley, M.V. 1948. A method for making aceto-carmine squashes permanent without removal of the cover slip. Stain Technol., **23**, 41–4.

Bräutigam, F., and Grabner, A. 1949. *Beiträge zur Fluoreszenzmikroskopie,* Verlag Georg Fromme, Wien.

Brecher, G. 1949. New methylene blue as a reticulocyte stain. Am. J. Clin. Pathol., **19**, 895–6.

Breed, R.S., and Brew, J.D. 1916. Counting bacteria by means of microscope. N.Y. Agric. Exp. Stn. Tech. Bull., **49**.

Bresslau, E. 1921. Die Gelatinierbarkeit des Protoplasmas als Grundlage eines Verfahrens zur Schnellanfertigung gefärbter Dauer Präprate von Infusorien. Arch. Protistenkd., **43**, 467–80.

Brice, A.T. 1933. Three notes on biological stains; I. Combined peroxidase-Wright's stain on blood films. Am. J. Clin. Pathol., **3**, 381–3.

Bridges, C.B. 1937. The vapor method of changing reagents and of dehydration. Stain Technol., **12**, 51–2.

Bridges, C.H., and Luna, L. 1957. Kerr's improved Warthin-Starry technique. Study of the permissible variations. Lab. Invest., **6**, 357–67.

Brody, M.J., Hakanson, R., Owman, C., and Sundler, F. 1972. An improved method for the histochemical demonstration of histamine and other compounds producing fluorophors with o-phthaldehyde. J. Histochem. Cytochem., **11**, 945–8.

Brooks, R.M., Bradley, M.V., and Anderson, T.I. 1959. Plant micro-technique manual. Department of Pomology, University of California, Davis.

Brown, J.H., and Brenn, L. 1931. A method for the differential staining of gram-positive and gram-negative bacteria in tissue sections. Bull. John Hopkins Hosp., **48**, 69–73.

Brown, J.O., and Vogelaar, J.P.M. 1956. Amino-silver staining of nervous tissue. Stain Technol., **31**, 159–65.

Brown, R.C., and Hopps, H.C. 1973. Staining of bacteria in tissue sections; a reliable Gram stain method. Am. J. Clin. Pathol., **60**, 234–40.

Bruesch, S.R. 1942. Staining myelin sheaths of optic nerve fibers with osmium tetroxide vapor. Stain Technol., **17**, 149–52.

Burdon, K.L. 1946. Fatty material in bacteria and fungi revealed by staining dried, fixed slide preparations. J. Bacteriol., **52**, 665.

Burnham, C.R. 1967. Cytological smear techniques. Mimeograph. Department of Agronomy and Plant Genetics, University of Minnesota, St. Paul.

Burns, J., Pennock, C.A., and Stoward, P.J. 1967. The specificity of the staining of amyloid deposits with thioflavin T. J. Pathol. Bacteriol., **94**, 337–44.

Burtner, H.J., and Lillie, R.D. 1949. A five-hour variant of Gomori's methenamine silver method for argentaffin cells. Stain Technol., **24**, 225–7.

Butt, E.M., Bonynge, C.W., and Joyce, R.L. 1936. The demonstration of capsules about hemolytic streptococci with India ink or azo blue. J. Infect. Dis., **58**, 5–9.

Buttler, J.K. 1979. Methods for improved light microscope microtomy. Stain Technol., **54**, 53–69.

Cajal, S. Ramon y. 1903. Un Sencillo metodo de coloracion selectiva del reticulo protoplasmico. Trav. Lab. Rech. Biol. Univ. Madrid, **2**, 129–221.

Cajal, S. Ramon y. 1910. Las formulas del proceder del nitrato de plata reducio. Trav. Lab. Rech. Biol. Univ. Madrid, **8**, 1–26.

Cajal, S. Ramon y. 1913. Sobre un neuvo proceder de impregnación de la neuroglia y sus resultados en los centros nerviosos del hombre y animales. Trav. Lab. Inv. Biol. T. XI, Fasc. 3, Diciembre.

Cajal, S. Ramon y. 1916. El proceder del orosublimado para la coloracion de la neuroglia. Trav. Lab. Inv. Biol., T. XIV, Fasc. 3 y 4, Diciembre.

Cambrer, M.A., Wheeless, L.L., Jr., and Patten, S.F., Jr. 1977. A new post-staining fixation technique for acridine orange. Acta Cytol., **21**, 477–80.

Campbell, R.E., and Rosahn, P.D. 1950. The morphology and staining characteristics of the *Treponema pallidum*. Review of the literature and description of a new technique for staining the organisms in tissues. Yale J. Biol. Med., **22**, 527–43.

Cannon, H.G. 1937. A new biological stain for general purposes. Nature, **139**, 549.

Carnoy, J.B. 1886. La cytodiérèse de l'oeuf chez quelques nématodes. Cellue, **3**, 5–62.

Carnoy, J.B. 1887. Les globules polaries de l'Ascaris clavata. Cellule, **3**, 247–324.

Carnoy, J.B., and Lebrun, H. 1897. La fecondation chez l'Ascaris megalocephala. Cellule, **13**, 63–195.

Carrano, A.V., Gray, J.W., Langlois, R.G., Burkhart-Schultz, K.J., and Van Dilla, M.A. 1979. Measurement and purification of human chromosomes by flow cytometry and sorting. Proc. Nat. Acad. Sci. U.S.A., **76**, 1382–4.

Cartwright, K.St.G. 1929. A satisfactory method of staining fungal mycelium in wood sections. Ann. Bot., **43**, 412–3.

Casida, L.E. 1962. On the isolation and growth of individual microbial cells from soil. Can. J. Microbiol., **8**, 115–9.

Caspersson, T., Farber, S., Foley, G.E., Kudynowski, J., Modest, E.J., Simonsson, E., Wagh, V., and Zech, L. 1968. Chemical differentiation along metaphase chromosomes. Exp. Cell Res., **49**, 219–22.

Caspersson, T., Lomakka, G., and Rigler, R., Jr. 1966. Registrierender Fluoreszenzmikrospektrograph zur Bestimmung der Primär- und Sckundärfluoreszenz verschiedener Zellsubstanzer. Acta Histochem. Suppl., **6**, 123–6.

Caspersson, T., Zech, L., and Johannson, C. 1970. Differential binding of fluorochromes in human chromosomes. Exp. Cell Res., **60**, 315–9.

Caspersson, T., Zech, L., Modest, E.J., Foley

G.E., Wagh, U., and Simonsson, E. 1969. Chemical differentiation with fluorescent alkylating agents in *Vicia faba* metaphase chromosomes. Exp. Cell Res., **58**, 128–40.

Cassel, W.A., and Hutchinson, W.G. 1955. Fixation and staining of the bacterial nucleus. Stain Technol., **30**, 105–18.

Caventou, J.B. 1826. Ann. Chim., **31**, 358, cited in R.D. Lillie, 1954, p. 92.

Celarier, R.P. 1956. Tertiary butyl alcohol dehydration of chromosome smears. Stain Technol., **31**, 155–7.

Chamberlain, C.J. 1932. *Methods in Plant Histology*, Ed. 5, University of Chicago Press, Chicago.

Chandler, C. 1931. A method for staining pollen-tubes within the pistil. Stain Technol., **6**, 25–6.

Changaris, D.G., Combs, J., and Severs, W.B. 1977. A microfluorescent PAS method for the quantitative demonstration of cytoplasmic 1,2-glycols. Histochemistry, **52**, 1–15.

Chatton, E., and Lwoff, A. 1930. Imprégnation, par diffusion argentique, de l'infraciliature des Ciliés marins et d'eau douce, apres fixation cytologique et sans dessiccation. C.R. Soc. Biol. (Paris), **104**, 834–6.

Chatton, E., and Lwoff, A. 1936. Techniques pour l'etude des Protozoaires, spécialement de leurs structures superficielles (cinétome et argyrome). Bull. Soc. Franc. Microsc., **5**, 25–39.

Chayen, J., Bitensky, L., and Butcher, R.G. 1973. *Practical Histochemistry*, John Wiley & Sons, New York.

Cheadle, V.I., Gifford, E.M., Jr., and Esau, K. 1953. A staining combination for phloem and contiguous tissue. Stain Technol., **28**, 49–53.

Chen, T.R. 1977. In situ detection of mycoplasma contamination in cell cultures by fluorescent Hoechst 33258 stain. Exp. Cell Res., **104**, 255–62.

Chick, E.W. 1961. Acridine orange fluorescent stain for fungi. Arch. Dermatol. Syphilol., **83**, 305–9.

Chiffelle, T.L., and Putt, F.A. 1951. Propylene and ethylene glycol as solvents for Sudan IV and Sudan Black B. Stain Technol., **26**, 51–6.

Christensen, W.B. 1949. Observations on the staining of *Corynebacterium diphtheriae*. Stain Technol., **24**, 165–70.

Ciaccio, C. 1906. Sur une nouvelle espece cellulaire dan les glandes de Lieberkuhn. C.R. Soc. Biol. (1), **60**, 76–9.

Clara, M. 1935. Untersuchungen über die spezifische Färbung der Körnchen in den basal gekörnten Zellen des Darmepithels durch Beizenfarbstoffe. Z. Zellforsch., **22**, 318–52.

Clark, G. 1971. Unpublished material.

Clark, G. 1979a. Unpublished material.

Clark, G. 1979b. Displacement. Stain Technol., **54**, 111–9.

Clark, G. 1979c. Chromoxane cyanin R. Work in progress.

Clark, G., Berube, G.R., and Powers, M.M. 1971. Work in progress.

Clark, G., and Barnes, C.T. 1979. A simple time controlled connective tissue stain. J. Histotechnol., **2**, 19.

Clark, G., and Powers, J.M. 1976. A consistent phosphotungstic acid hematoxylin stain for glial fibers. Stain Technol., **51**, 227–9.

Clark, G., and Spicer, S.S. 1979. The assessing of acidophilia with Biebrich scarlet, ponceau de xylidine and wood stain scarlet. Stain Technol., **54**, 13–6.

Clark, S.L., and Ward, J.W. 1934. A variation of the Pal-Weigert method for staining myelin sheaths. Stain Technol., **9**, 53–5.

Clark, W.A. 1976. A simplified Leifson flagella stain. J. Clin. Microbiol., **3**, 622–34.

Cole, E.C. 1925. Anastomosing cells in the myenteric plexus of the frog. J. Comp. Neurol., **38**, 375–87.

Cole, E.C. 1936. A new methylene blue technic for permanent preparation. Stain Technol., **11**, 45–7.

Cole, M.B., and Sykes, S.M. 1974. Glycol methacrylate in light microscopy; a routine method for embedding and sectioning animal tissues. Stain Technol., **49**, 387–400.

Combs, J.D. 1936. Motor end-plates. Stain Technol., **11**, 147–8.

Comings, D.E. 1978. Methods and mechanisms of chromosome banding. Methods Cell Biol., **17**, 115–32.

Conger, A.D., and Fairchild, L.M. 1953. A quick freeze method for making smear slides permanent. Stain Technol., **28**, 281–3.

Conklin, M.E. 1934. Mercurochrome as a bacteriological stain. J. Bacteriol., **27**, 30–1.

Conn, H.J. 1928. On the microscopic method of studying bacteria in soil. Soil Sci., **26**, 257–9.

Conn, H.J. 1929. Use of the microscope in studying the activities of bacteria in soil. J Bacteriol., **17**, 399–405.

Conn, H.J. 1940. *Biological Stains*, Ed. 4, Biotech Publications, Geneva, New York.

Conn, H.J. 1953. *Biological Stains*, Ed. 6, Williams & Wilkins, Baltimore.

Conn, H.J., and Darrow, M.A. 1943. *Staining*

Procedures Used by the Biological Stain Commission, Biotech Publications, Geneva, New York.

Conn, H.J., Darrow, J.A., and Emmel, V.A. 1960. *Staining Procedures Used by the Biological Stain Commission*, Ed. 2, Williams & Wilkins, Baltimore.

Conn, H.J., and Holmes, W.C. 1926. Fluorescein dyes as bacterial stains. Stain Technol., **1**, 87–95.

Coons, A.H. 1958. *Fluorescent Antibody Methods*, Vol. 1, p. 399, J.F. Danielli (ed.), Academic Press, New York.

Corliss, J.O. 1953. Silver impregnation of ciliated protozoa by the Chatton-Lwoff technic. Stain Technol., **28**, 97–100.

Cornelisse, C.J., and Ploem, J.S. 1976. A new type of two-color fluorescence staining for cytology specimens. J. Histochem. Cytochem., **24**, 72–81.

Cottell, D.C., and Livingston, D.C. 1976. Fluorescent reagent for the periodic acid-Schiff and Feulgen reaction for cytochemical studies. J. Histochem. Cytochem., **24**, 956–8.

Cowden, R.R., and Curtis, S.K. 1974. Acridine orange as a supravital fluorochrome indicating varying degrees of chromatin condensation. Histochemistry, **40**, 305–10.

Cox, W.H. 1891. Imprägnation des centralen Nervensystems mit Quecksilbersalzen. Arch. Mikrosk. Anat., **37**, 16–21.

Crissman, H.A., Mullaney, P.F., and Steinkampf, J.A. 1975. Methods and application of flow systems for analysis and sorting of mammalian cells. Methods Cell Biol., **9**, 179–246.

Crissman, H.A., and Tobey, R.A. 1974. Cell-cycle analysis in 20 minutes. Science, **184**, 1297–8.

Crosier, R.H. 1968. An acetic acid dissociation, air-drying technique for insect chromosomes with aceto-lactic orcein staining. Stain Technol., **43**, 171–3.

Culling, C.F.A. 1963. *Handbook of Histopathological Techniques*. Butterworth, London.

Culling, C.F.A., and Vassar, P.S. 1961. Desoxyribose nucleic acid, a fluorescent histochemical technique. Arch. Pathol., **71**, 76–80.

Cumley, R.W., Crow, J.F., and Griffin, A.B. 1939. Clearing specimens for the demonstration of bone. Stain Technol., **14**, 7–11.

Cunningham, R.S. 1920. A method for permanent staining of reticulated red cells. Arch. Intern. Med., **26**, 405–9.

Currier, H.B., and Shih, C.Y. 1968. Sieve tubes and callose in elodea leaves. Am. J. Bot., **55**, 145–52.

Currier, H.B., and Strugger, S. 1956. Aniline Blue and fluorescence microscopy of callose in bulb scales of *Allium cepa* L. Protoplasma, **4**, 552–9.

Custer, R.P. 1933. Studies on the structure and function of bone marrow. Am. J. Med. Sci. (N.S.), **185**, 617–24.

Darlington, C.D., and LaCour, L.F. 1938. Differential reactivity of the chromosomes. Ann. Bot. (N.S.), **2**, 615–25.

Darlington, C.D., and LaCour, L.F. 1940. Nucleic acid starvation in chromosomes of *Trillium*. J. Genet., **40**, 185–213.

Darlington, C.D., and LaCour, L.F. 1960. *The Handling of Chromosomes*, Ed. 3, Macmillan, New York.

Darlington, C.D., and LaCour, L.F. 1975. *The Handling of Chromosomes*, Ed. 6, John Wiley & Sons, New York.

Darrow, M.A. 1940. A simple staining method for histology and cytology. Stain Technol., **15**, 67–8.

Darrow, M.A. 1944. Aniline blue as a counterstain in cytology. Stain Technol., **19**, 65–6.

Darrow, M.A. 1952. Synthetic orcein as an elastic tissue stain. Stain Technol., **27**, 329–32.

Dart, L.H., Jr., and Turner, T.R. 1959. Fluorescent microscopy in exfoliative cytology. Report of acridine orange examination of 5491 cases, with comparison by the Papanicolaou technic. Lab. Invest., **8**, 1513–22.

Davenport, H.A. 1929. Silver impregnation of nerve fibers in celloidin sections. Anat. Rec., **44**, 79–83.

Davenport, H.A. 1930. Staining nerve fibers in mounted sections with alcoholic silver nitrate. Arch. Neurol. Psychiatr., **24**, 690–5.

Davenport, H.A., McArthur, J., and Breusch, S.R. 1939. Staining paraffin sections with protargol. 3. The optimum pH for reduction. 4. A two-hour staining method. Stain Technol., **14**, 21–6.

Davenport, H.A., Windle, W.F., and Beech, R.H. 1934. Block staining of nervous tissue with silver; IV. Embryos. Stain Technol., **9**, 5–10.

Davis, W.H. 1924. Lactophenol. Bot. Gaz., **77**, 343–4.

Dawson, A.B. 1926. A note on the staining of skeleton of cleared specimens with alizarin red S. Stain Technol., **1**, 123–4.

DeBruyn, P.P.H., and Smith, N.H. 1959. Comparison between the in vivo and in vitro interaction of aminoacridines with nucleic acids and other compounds. Exp. Cell Res., **17**, 482–9.

DeGirolami, U., and Zvaigzne, O. 1973. Modification of the Achúcarro-Hortega stain for

BIBLIOGRAPHY

481

paraffin-embedded formalin-fixed tissue. Stain Technol., **48**, 48–50.

Deitrich, A. 1910. Zur Differenzial-diagnose der Fettsubstanzen. Verh. Dtsch. Pathol. Ges., **14**, 263–8.

DeLamater, E.D., Haanes, M., and Wiggall, R.H. 1950. Studies on the life cycle of spirochetes. II. The development of a new stain. Am. J. Syphilol., **34**, 515–8.

de Lerma, B. 1958. Die Anwendung von Fluoreszenzenlicht in der Histochemie. In: *Handbuch der Histochime*, Vol. 1, pp. 78–159, E. W. Graumann (ed.), Gustav Fisher Verlag, Stuttgart.

de Ment, J. 1950. Immersion liquids and mounting media for fluorescence microscopy. Trans. Am. Microsc. Soc., **69**, 357–8.

Demerec, M., and Kaufmann, B.P. 1957. *Drosophila Guide*, Ed. 6, Carnegie Institution of Washington, 1530 P St., Washington, D.C.

Dempsey, W.W., Bunting, H., Singer, M., and Wislocki, G.R. 1947. The dye-binding capacity and other chemo-histological properties of mammalian mucopolysaccharides. Anat. Rec., **98**, 417–29.

Desbordes, J., Fournier, E., and Guyotjeannin, C. 1952. Nouvelle application de l'usage des corps tensio-actifs en physiologie bactérienne. Coloration a froid des bacilles tuberculeux. Ann. Inst. Pasteur (Paris), **83**, 268–70.

DeTomasi, J.A. 1936. Improving the technic of the Feulgen stain. Stain Technol., **11**, 137–44.

Diagnostic Standards and Classification of Tuberculosis. 1961. National Tuberculosis Association, 1790 Broadway, New York, N.Y.

Dickinson, D.C. 1963. A study of the variables influencing differentiation during an acid-fast staining procedure. M.S. thesis, University of Southern California, Los Angeles, CA 90007.

Dienes, L. 1967. Permanent stained agar preparations of *Mycoplasma* and L forms of bacteria. J. Bacteriol., **93**, 689–92.

Dienst, R.B., and Sanderson, E.S. 1936. Use of nigrosine to demonstrate *Treponema pallidum* in syphilitic lesions. Am. J. Public Health, **26**, 910–2.

Doan, C.A., and Ralph, P. 1950. *Handbook of Microscopical Technique*, pp. 571–85, C.E. McClung (ed.), Paul B. Hoeber, New York.

Dohrmann, G.J., and Wick, K.M. 1971. Demonstration of the microvasculature of the spinal cord by intravenous injection of the fluorescent dye, thioflavin S. Stain Technol., **46**, 321–2.

Drawert, H. 1951. Beiträge zur Vitalfarbung

pflanzlicher Zellen. Protoplasma, **40**, 85–106.

Dryl, S. 1959. Antigenic transformation in paramecium aureba after homologous antiserum treatment during autogamy and conjugation. J. Protozool. Suppl., **6**, 25.

Duguid, J.P. 1951. The demonstration of bacterial capsules and slime. J. Pathol. Bacteriol., **63**, 673–85.

Dutrillaux, B. 1977. New chromosome techniques. In: *molecular Structure of Human Chromosomes*, pp. 233–65, J.J. Yunis (ed.), Academic Press, New York.

Dyar, M.T. 1947. A cell wall stain employing a cationic surface-active agent as a mordant. J. Bacteriol., **43**, 498.

Dyer, A.F. 1963. The use of lactopropionic orcein in rapid squash method for chromosome preparations. Stain Technol., **38**, 85–90.

Eggert, D.A. 1970. The use of morin for fluorescent localization of aluminum in plant tissues. Stain Technol., **45**, 301–3.

Ehinger, B., and Thunberg, R. 1967. Induction of fluorescence in histamine-containing cells. Exp. Cell Res., **47**, 116–22.

Ehrlich, P. 1885a. Das Sauerstoff-bedüriniss des Organismum. Eine farbenanalytische Studie. Dtsch. Med. Wochenschr., **11**, 295–7.

Ehrlich, P. 1885b. Zur biologischen Verwertung des Methylenblau. Centralbl. Med. Wiss., **23**, 113–7.

Ehrlich, P. 1886. Über die Methylenblaureaction der lebenden Nervensubstanz. Biol. Centralbl., **6**, 214–24.

Ehrlich, P. 1903. *Encyklopädie der Mikroskopischen Technik*. Urban u. Schwarzenberg, Berlin.

Einarson, L. 1932. A method for progressive selective staining of Nissl and nuclear substance in nerve cells. Am. J. Pathol., **8**, 295–307.

Ellinger, P. 1940. Fluorescence microscopy in biology. Biol. Rev., **15**, 323–50.

Ellinger, H.W., and Hirt, A. 1932. Eine Methode zur Beobachtung lebende Organe mit stärksten Vergrösserungen in Luminesgenzlicht (Intravitalmikroskopie). *Handb. Biol. Arbeitsmethoden* **5** (2/2), 1753–64.

Eng, L.K., and Cole, A.L.J. 1976. Tinopol AN in fluorescent microscopic detection of bacteria within plant tissues. Stain Technol., **51**, 277–8.

Eppinger, H. 1949. *Die Bedeutung der Fluoreszenzmikroskopie für die Pathologie*, Wien.

Esau, K. 1948. Phloem structure in the

grapevine, and its seasonal changes. Hilgardia, **18**, 217–96.

Eschrich, W., and Currier, H.B. 1964. Identification of callose by its diachrome and fluorochrome reactions. Stain Technol., **39**, 303–7.

Ewen, A.B. 1962. An improved aldehyde fuchsin staining technique for neurosecretory products in insects. Trans. Am. Microsc. Soc., **81**, 94–6.

Ezrin, C., and Murray, S. 1963. *Cytologie de L'adenohypophyse*, J. Benoit, and C. DaLage, (eds.), Editions Du.C.N.R.S., Paris.

Falck, B. 1962. Observations of the possibilities of the cellular localization of monoamines by a fluorescence method. Acta Physiol. Scand., **56**, Suppl. 197, 1–25.

Falck, B., Hillarp, N.A., Thieme, G., and Torp, A. 1962. Fluorescence of catecholamines and related compounds condensed with formaldehyde. J. Histochem. Cytochem., **10**, 348–54.

Falck, B., and Owman, Ch. 1965. A detailed methodologic description of the fluorescence method for the cellular demonstration of biogenic monoamines. Acta Univ. Lund, **II**, 5–17.

Faulkner, R.R., and Lillie, R.D. 1945. A buffer modification of the Warthin-Starry silver method for spirochaetes in single paraffin sections. Stain Technol., **20**, 81–2.

Feder, N., and O'Brien, T.P. 1968. Plant microtechnique; some principles and new methods. Am. J. Bot., **55**, 123–42.

Fenchel, T. 1965. On the ciliate fauna associated with the marine species of the amphipod genus *Gammarus* J.G. Fabricius. Ophelia, **2**, 281–303.

Feulgen, R., and Rossenbeck, H. 1924. Mikroskopische-chemischer Nachweis eine Nucleinsäure von Typus der Thymonucleinsäure und auf die darauf beruhende elektive Färbung von Zellkernen in mikroskopischen Präparaten. Z. Physiol. Chem., **135**, 203–48.

Fink, R.P., and Heimer, L. 1967. Two methods for selective silver impregnation of degenerating axons and their synaptic endings in the central nervous system. Brain Res., **4**, 369–74.

Finkelstein, H., and Bartholomew, J.W. 1958. An interpretation of several cell wall stains applied to heat-fixed *Bacillus subtilis* on glass slides. Stain Technol., **33**, 177–86.

Fischer, E. 1875. Über den Bau der Meissner'schen Tastkörperschen. Arch. Mikrosk. Anat., **12**, 364–90.

Fisher, D.B. 1968. Protein staining of ribboned epon sections for light microscopy. Histochemie, **16**, 92–6.

Fite, G.L. 1938. The staining of acid-fast bacilli in paraffin sections. Am. J. Pathol., **14**, 491–507.

Fite, G.L. 1940. The fuchsin-formaldehyde method of staining acid-fast bacilli in paraffin sections. J. Lab. Clin. Med., **25**, 743–4.

Fite, G.L., Cambre, P.J., and Turner, M.H. 1947. Procedure for demonstrating lepra bacilli in paraffin sections. Arch. Pathol., **43**, 624–5.

Flax, M.H., and Himes, M. 1952. Microspectrophotometric analysis of metachromatic staining of nucleic acids. Physiol. Zool., **25**, 297–311.

Flemming, W. 1884. Mittheilungen zur Färbetichnik. Z. Wiss. Mikrosk., **1**, 349–61.

Flemming, W. 1891. Über Theilung und Kernformen bei Leukocyten, und über denen Atractions-sphären. Arch. Mikrosk. Anat., **37**, 249–98.

Foley, J.O. 1936. Suggested counterstains for Davenport reduced silver preparations of peripheral nerves. Stain Technol., **11**, 3–8.

Foley, J.O. 1943. A protargol method for staining nerve fibers in frozen or celloidin sections. Stain Technol., **18**, 27–33.

Fontana, A. 1912. Verfahren zur intensiver und raschen Färbung des *Treponema pallidum* und anderer Spirochäten. Dermatol. Wochenschr., **55**, 1003–4.

Fontana, A. 1926. Uber die Silberdarstellung des *Treponema pallidum* und anderer Midroorganismen in Ausstrichen. Dermatol. Z., **46**, 291–3.

Foot, N.C. 1924. A technic for demonstrating reticulum fibers in Zenker-fixed paraffin sections. J. Lab. Clin. Med., **9**, 777–81.

Foot, N.C., and Menard, M.C. 1927. A rapid method for the silver impregnation of reticulum. Arch. Pathol. Lab. Med., **4**, 211–4.

Foster, A.S. 1934. The use of tannic acid and iron chloride for staining cell walls in meristematic tissue. Stain Technol., **9**, 91–2.

Fox, C.A., Ubeda-Purkiss, M., Ihrig, H.K., and Biagioli, D. 1951. Zinc chromate modification of the Golgi technic. Stain Technol., **26**, 109–14.

Francisco, D.E., Mah, R.A., and Rabin, A.C. 1973. Acridine orange-epifluorescence technique for counting bacteria in natural waters. Trans. Am. Microsc. Soc., **92**, 416–21.

Frey-Wyssling, A. 1947. Das Fluoreszenzmikroskop im Dienst der Pflanzenphysiologie. Vierteljahresschr. Naturforsch. Ges. Zür., **92**, 188–94.

Frost, H.M. 1963. Mean formation time of human osteons. Can. J. Biochem., **41**, 1307–10.

Frost, H.M. 1966. Bone dynamics in metabolic bone disease. J. Bone Joint Surg., **48**, 1192–1203.

Frost, H.M. 1968. Tetracycline bone labeling in anatomy. Am. J. Phys. Anthropol., **29**, 183–95.

Frost, H.M. 1969. Tetracycline-based histological analysis of bone remodeling. Calcif. Tissue Res., **3**, 211–37.

Frost, H.M., Villaneuva, A.R., and Roth, H. 1960. Tetracycline staining of newly forming bone and mineralizing cartilage in vivo. Stain Technol., **35**, 135–8.

Furness, J.B., and Costa, M. 1975. The use of glyoxylic acid for the fluorescence histochemical demonstration of peripheral stores of noradrenaline and 5-hydroxytryptamine in whole mounts. Histochemistry, **41**, 335–52.

Gage, S.H. 1890. Picric and chromic acid for the rapid preparation of tissues for classes in histology. Proc. Am. Soc. Microscopists, **12**, 120–7.

Galigher, A.E. 1934. *Essentials of Practical Microtechnique*, Published by the author, Berkeley, California.

Ganter, P., and Jollès, G. 1970. *Histochemie Normale et Pathologique*, Vol. 2, Gauthier-Villars, Paris.

Garven, H.S.D. 1925. The nerve endings in the panniculus carnosus of the hedgehog, with special reference to the sympathetic innervation of skeletal muscle. Brain, **48**, 380–441.

Garven, H.S.D., and Gairns, F.W. 1952. The silver diammine ion staining of peripheral nerve elements and the interpretation of the results. Q. J. Exp. Physiol., **37**, 131–42.

Gatenby, J.B., and Beams, H.W. 1950. *The Microtomist's Vade-Mecum (Bolles Lee)*, Ed. 11, pp. 533–4. Blakiston, Philadelphia.

Gatenby, J.B., and Cowdry, E.V. 1928. *Bolles Lees Micromists Vade-Mecum*. P. Blakiston, Philadelphia.

Gatenby, J.B., and Painter, T.S. (Eds.) 1937. *The Microtomist's Vade-Mecum (Bolles Lee)*, Ed. 10, J. & A. Churchill, London.

Gay, H., and Kaufmann, B.P. 1950. The corneal epithelium as a source of mammalian somatic mitoses. Stain Technol., **25**, 209–16.

Gelei, J. von. 1927. Eine neue Osmiumtoluidin methode für Protisten forschung. Mikrokosmos, **20**, 97–103.

Gendre, H. 1937. A propos des procédés de fixation et de detection histologique du glycogène. Bull. Hitol. Appl. Physiol. Pathol., **14**, 262–4.

German, W.M. 1939. Hortega's silver impregnation methods. Technic and applications. Am. J. Clin. Pathol. Suppl., **3**, 13–9.

Geyer, G., and Luppa, H. 1977. Eds. Fortschritte der Fluoreszenzund lipidhistochemie. Acta Histochem. Suppl. **14**.

Giemsa, G. 1904. Eine Vereinfachung und Vervolkommnung meiner Methylenazur-Methyl-enblau-Eosin-Färbemethode zur Erzielung der Romanowsky-Nochtschen Chromatinfärbung. Centralbl. Bakteriol., Abt. I, **37**, 308–11.

Gier, H.T. 1949. Differential stains for insect exoskeleton. J. Kans. Entomol. Soc., **22**, 79–80.

Gifford, E.M., Jr., and Dengler, P.E. 1966. Histones and alkaline fast green staining of onion roots. Am. J. Bot., **53**, 1125–32.

Gilbert, R., and Bartels, H.A. 1924. The staining of *Treponema pallidum* in dry smears. J. Lab. Clin. Med., **9**, 113–9.

Gill, B.S., and Kimber, G. 1974a. The Giemsa C-banded karyotype of rye. Proc. Nat. Acad. Sci. U.S.A., **71**, 1247–9.

Gill, B.S., and Kimber, G. 1974b. Giemsa C-banding and the evolution of wheat. Proc. Nat. Acad. Sci. U.S.A., **71**, 4086–90.

Gill, D. 1979. Inhibition of fading in fluorescence microscopy of fixed cells. Experientia, **35**, 400–1.

Gill, J.E., and Jotz, M.M. 1974. Deoxyribonucleic acid cytochemistry for automated cytology. J. Histochem. Cytochem., **22**, 470–7.

Gill, J.E., and Jotz, M.M. 1976. Further observations on the chemistry of pararosaniline-Feulgen staining. Histochemistry, **46**, 147–60.

Gill, J.E., Wheeless, L.L., Jr., Hanna-Madden, C., Marisa, R.J., and Horan, P.K. 1978. A comparison of acridine orange and Feulgen cytochemistry of human tumor cell nuclei. Cancer Res., **38**, 1893–8.

Giménez, D.F. 1964. Staining rickettsiae in yolk-sac cultures. Stain Technol., **39**, 135–40.

Giolli, R.A., and Karamanlidis, A.N. 1978. The study of degenerating nerve fibers using silver-impregnation methods. In: *Neuroanatomical Research Techniques*, R.T. Robertson (ed.), Academic Press, New York.

Girbardt, M., and Taubeneck, U. 1955. Zur Frage der Zellwandfärbung bei Bakterien. Zentralbl. Bakteriol., **162**, 310–3.

Glees, P. 1946. Terminal degeneration within the central nervous system as studied by a new silver method. J. Neuropath. Exp. Neurol., **5**, 54–9.

Glenner, G.G., and Lillie, R.D. 1957a. The histochemical demonstration of indole derivatives by the post coupled *p*-dimethylaminobenzylidene reaction. J. Histochem. Cytochem., **5**, 279–96.

Glenner, G.G., and Lillie, R.D. 1957b. A rhodocyan technic for staining the anterior pituitary. Stain Technol., **32**, 187–90.

Globus, J.H. 1927. The Cajal and Hortega glia staining methods. Arch. Neurol. Psychiatry, **18**, 263–71.

Globus, J.H. 1937. *Practical Neuroanatomy.* Williams & Wilkins, Baltimore.

Gohar, M.A. 1944. A staining method for *Corynebacterium diphtheriae.* J. Bacteriol., **47**, 575.

Golaboff, L.A., and Ezrin, C. 1969. Effect of pregnancy on the somatotroph and the prolactin cell of the human adeno-hypophysis. J. Clin. Endocrinol., **29**, 1533–8.

Goldetz, L., and Unna, P. 1909. Zur Chemie der Haut. Monatsschr. Prakt. Dermatol., **48**, 149–66.

Goldman, M. 1968. *Fluorescent Antibody Methods,* Academic Press, New York.

Golgi, C., 1873. Sulla sostanza grigia del cervello. Gazz. Med. Ital.; also in Opera Omnia (O. Hoepi. Milan), **1**, (1929), 91.

Golgi, C. 1875. Sulla fina struttura dei bulbi olfattorii. Riv. Sper. Freniatr., **1**, 405–25.

Golgi, C. 1878. Un nuovo processo di tecnica microscopia. Rendic. Ist. Lombardo Sc., 2d Ser., **12**, 5, ff.

Gomba, Sz., Szokoly, V., Soltész, M.B., and Endes, P. 1968. The effect of freezing and thawing on the granules of the juxtaglomerular cells. Acta Histochem., **30**, 346–48.

Gomori, G. 1939. Studies on the cells of the pancreatic islets. Anat. Rec. **74**, 439–60.

Gomori, G. 1941. Observations with differential stains on human islets of Langerhans. Am. J. Pathol., **17**, 395–406.

Gomori, G. 1946. A new histochemical test for glycogen and mucin. Am. J. Clin. Pathol. (Tech. Bull.), **16**, 177–9.

Gomori, G. 1948. Chemical characteristics of the enterochromaffin cells. Arch. Pathol., **45**, 48.

Gomori, G. 1950. Aldehyde-fuchsin; a new stain for elastic tissue. Am. J. Clin. Pathol., **20**, 665–6.

Gomori, G. 1952. *Microscopic Histochemistry,*

University of Chicago Press, Chicago.

Goodburn, G.M., and Marmion, B.P. 1962. A study of the properties of Eaton's primary atypical pneumonia organism. J. Gen. Microbiol., **29**, 271–90.

Goodrich, H.P. 1937. Protozoa. In *The Microtomist's Vade-Mecum (Bolles Lee),* Ed. 10, pp. 563–90, J.B. Gatenby and T.S. Painter (eds.), J. & A. Churchill, London.

Goodwin, R.H. 1953. Fluorescent substances in plants. Ann. Rev. Plant Physiol., **4**, 283–304.

Gottschewski, G.H.M. 1954. Die Methoden der Fluoreszenzund Ultraviolett-Mikroskopie und Spektroskopie in ihre Bedeutung für die Zellforschung. Mikroskopie, **9**, 147–67.

Gottschewski, G.H.M. 1958. Apparate und Einrichtungen für qualitative fluoreszenzmikroskopische Untersuchungen. In: *Handbuch der Histochime,* Vol. 1, pp. 160–70, W. Graumann (ed.), Gustave Fisher Verlag, Stuttgart.

Gower, W.C. 1939. A modified stain and procedure for trematodes. Stain Technol., **14**, 31–2.

Graham, R.C., Jr., and Karnovsky, M.J. 1966. The early stages of absorption of injected horseradish peroxidase in the proximal tubules of mouse kidney; ultrastructural cytochemistry by a new technique. J. Histochem. Cytochem., **14**, 291.

Gray, J.W., Langlois, R.G., Carrano, A.V., Burkhart-Schulte, K., and Van Dilla, M.A. 1979. High resolution chromosome analysis; one and two parameter flow cytometry. Chromosoma, **73**, 9–27.

Gray, P.H.H. 1926. A method of staining bacterial flagella. J. Bacteriol., **12**, 273–4.

Gregory, G.E. 1970. Silver staining of insect central nervous systems by the Bodian Protargol method. Acta Zool., **51**, 169–78.

Gridley, M.F. 1953. A stain for fungi in tissue sections. Am. J. Clin. Pathol., **23**, 303–7.

Gridley, M.F. 1954. A stain for *Endamoeba histolytica* in tissue sections. Am. J. Clin. Pathol., **24**, 243–4.

Grimelius, L. 1968. A silver nitrate stain for alpha-2 cells in human pancreatic islets. Acta Soc. Med. Ups., **73**, 243–70.

Grocott, R.G. 1955. A stain for fungi in tissue sections and smears. Am. J. Clin. Pathol., **25**, 975–9.

Gross, M. 1952. Rapid staining of acid-fast bacteria. Am. J. Clin. Pathol., **22**, 1034–5.

Gurr, E. 1951. Fluorescence microscopy. J. R. Nav. Med. Serv., **27**, 133–40.

Gurr, E. 1956. *A Practical Manual of Medica*

and Biological Staining Techniques, Interscience Publishers, New York.

Gurr, E. 1960. *Encyclopedia of Microscopic Stains*, Williams & Wilkins, Baltimore.

Hagemann, P.K. 1937. Fluoreszenzmikroskopische Untersuchungen über Virus and andere Mikroben. Zentralbl. Bakteriol. Parasitenkd., Abt. 1, **140**, 184–7.

Hagemann, P.K.H. 1938. Fluoreszenzfärbung von Tuberkelbakterien mit Auramin. Münch. Med. Wochenschr., **85**, 1066–7.

Haitinger, M. 1934. Methoden der Fluoreszenzmikroskipie. In: *Abderhaldens Handbuch der biologischen Arbeitsmethoden*, Abt. 2, Teil 3, pp. 3307–37, Urban & Schwarzenberg, Wien.

Haitinger, M. 1938a. *Fluoreszenzmikroskopie. Ihre Anwendung in der Histologie und Chemie*, Akad. Verlagsges, Leipzig.

Haitinger, M. 1938b. Die Fluoreszenzmikroskopie. In: *Handbuch der Virusforschung* Vol. 1, pp. 231–52, R. Doerr and C. Hallauer (eds.), Springer, Wien.

Haitinger, M. 1959. *Fluoreszenzmikroskopie*, Ed. 2, J. Eisenbrand and G. Werth (eds.), Akademie Verlag, Leipzig.

Hajduk, S.L. 1976. Demonstration of kinetoplast DNA in dyskinetoplastic strains of *Trypanosoma equiperdum*. Science, **191**, 858–9.

Hakanson, R., and Owman, C. 1966. Concomitant histochemical demonstration of histamine and catecholamines in enterochromaffin-like cells of gastric mucosa. Life Sci., **6**, 759–66.

Hamberger, A., and Hamberger. B. 1966. Uptake of catecholamines and penetrations of trypan blue after blood-brain barrier lesions. Z. Zellforsch., **70**, 386–92.

Hamm, J.J. 1966. A modified Azan staining technique for inclusion body viruses. J. Invertebr. Pathol., **8**, 125–6.

Hamperl, H. 1934. Die Fluoreszenzmikroskopie menschliche Gewebe. Virchows Arch., **292**, 1–51.

Hamperl, H. 1955. Fluoreszenzmikroskopie, Munch. Med. Wochenschr., **97**, 1121–25.

Harada, K. 1969. Staining juxtaglomerular granules with basic fluorescent stains. Stain Technol., **44**, 293–5.

Harada, K. 1973. Effect of prior oxidation on the acid-fastness of mycobacteria. Stain Technol., **48**, 269–73.

Harms, H. 1965. *Handbuch der Farbstoffe für die Mikroskopie*, Staufen Verlag, Kamp-Lintford.

Harrington, B.J., and Raper, K. 1968. Use of a fluorescent brightener to demonstrate cellulose in the cellular slime molds. Appl. Microbiol., **16**, 106–13.

Harris, D.L. 1908. A method for the staining of Negri bodies. J. Infect. Dis., **5**, 566–9.

Harris, H.F. 1900. On the rapid conversion of haematoxylin into haematein in staining reactions. J. Appl. Microsc. **111**, 777–81.

Harris, M.B.K. 1930. A simple method for staining spirochetes. Science, **72**, 275.

Hartman, T.L. 1940. The use of sudan black B as a bacterial fat stain. Stain Technol., **15**, 23–8.

Haynes, R. 1928. Fast green, a new substitute for light green SF yellowish. Stain Technol., **3**, 40.

Hawkes, S.P., and Bartholomew, J.C. 1977. Quantitative determination of transformed cells in a mixed population by simultaneous fluorescence analysis of cell surface and DNA in individual cells. Proc. Nat. Acad. Sci. U.S.A., **74**, 1626–30.

Hawkes, S.P., Meehan, T.D., and Bissell, M.J. 1976. The use of fluorescamine as a probe for labeling the outer surface of the plasma membrane. Biochem. Biophys. Res. Commun., **68**, 1226–33.

Heaton, L.H., and Pauley, G.B. 1969. Two modified stains for use in oyster pathology. J. Fish. Res. Board Can., **26**, 707–9.

Heidenhain, M. 1892. Ueber Kern and Protoplasma. *Festschr. f. A. v. Kolliker*, pp. 109–65, Wilhelm Englemann, Leipsig.

Heidenhain, M. 1915. Über die Mallorysche Bindegewebsfärbung mit Karmin und Azokarmin als Vorfarben. Z. Wiss. Mikrosk., **32**, 361–72.

Heidenhain, M. 1916. Über neuere Sublimatgemische. Z. Wiss. Mikrosk. Mikrosk. Technik., **33**, 232–4.

Heidenhain, R. 1870. Untersuchungen über den Bau der Labdrüsen. Arch. Mikrosk. Anat. **6**, 368.

Heimer, L. 1970. Selective silver-impregnation of degenerating axons. In *Contemporary Research Methods in Neuroanatomy*, W.J.H. Nauta, and S.O.E. Ebbesson (eds.), Springer-Verlag, New York.

Hellman, B., and Hellerström, C. 1960. The islets of Langerhans in duck and chicken with special reference to the argyrophil reaction. Z. Zellforsch., **52**, 278–90.

Helly, K. 1903. Eine Modification dev Zenkerschen Fixirungs-flussigkeit. Z. Wiss. Mikrosk., **20**, 413–5.

Henle, J. 1865. Über das Gewebe der Nebenniere und der Hypophyse. Z. Rat. Med., **24**, 143.

Herlant, M., and Pasteels, J.L. 1967. Histo-

486

physiology of human anterior pituitary. In: *Methods and Achievements in Experimental Pathology*, Vol. 3, pp. 250–305, E. Bajusz, and G.S. Jasmin (eds.), Karger, Basel.

Herxheimer, G. 1901. Uber Fettstoffe. Dtsch. Med. Wochenschr., 27, 607–9.

Herxheimer, G. 1903. Zur Fettfärbung. Centralbl. Allg. Pathol. Anat., 14, 841–2.

Heslop-Harrison, J., and Heslop-Harrison, Y. 1970. Evaluation of pollen viability by enzymatically induced fluorescence; intracellular hydrolysis of fluorescein diacetate. Stain Technol., 45, 115–20.

Hicks, J.D., and Matthaei, E. 1955. Fluorescence in histology. J. Pathol. Bacteriol., 70, 1–12.

Hicks, J.D., and Matthaei, E. 1958. A selective fluorescence stain for mucin. J. Pathol. Bacteriol., 75, 473–6.

Highman, B. 1942. A new modification of Perls's reaction for hemosiderin in tissues. Arch. Pathol., 33, 937–8.

Hilwig, I., and Gropp, A. 1972. Staining of constitutive heterochromatin in mammalian chromosomes with a new fluorochrome. Exp. Cell Res., 75, 122–6.

Hilwig, I., and Gropp, A. 1973. Decondensation of constitutive heterochromatin in L cell chromosomes by a benzimidazole compound ("33258 Hoechst"). Exp. Cell Res., 81, 474–7.

Hines, M. 1931. Studies on the innervation of skeletal muscle; III. Innervation of the extrinsic eye muscles of the rabbit. Am. J. Anat., 47, 1–53.

Hines, M., and Tower, S.S. 1928. Studies on the innervation of skeletal muscles; II. Of muscle spindles in certain muscles of the kitten. Bull. Johns Hopkins Hosp., 42, 264–307.

Hiss, P.H., Jr. 1905. A contribution to the physiological differentiation of pneumococcus and streptococcus, and to methods of staining capsules. J. Exp. Med., 6, 317–46.

Hobbie, J.E., Daley, R.J., and Jasper, S. 1977. Use of nucleopore filters for counting bacteria by fluorescence microscopy. Appl. Environ. Microbiol., 33, 1225–32.

Hoefert, L.L. 1968. Polychromatic stains for thin sections of Beta embedded in epoxy resin. Stain Technol., 43, 145–51.

Hökfelt, T.G.M., and Ljungdahl, A.S. 1972. Histochemical demonstration of neurotransmitter distribution. Res. Publ. Assoc. Res. Nerv. Ment. Dis. 50, 1–24.

Hollande, A.C. 1918. Enrichissement du liquide fixateur de Bouin en acide picrique, par addition d'acétate neutre de cuivre. C.R. Soc. Biol. (Paris), 81, 17–20.

Hollister, G. 1934. Clearing and dyeing fish for bone study. Zoologica, 12, 89–101.

Holmes, W. 1943. Silver staining of nerve axons in paraffin sections. Anat. Rec., 86, 157–87.

Hooker, W.J., and Summanwar, A.S. 1964. Intracellular acridine orange fluorescence in plant virus infections. Exp. Cell Res., 33, 609–12.

Hori, S.H., and Kitamura, T. 1972. The vitamin A content and retinol esterifying activity of a Kupfer cell fraction of rat liver. J. Histochem. Cytochem., 20, 811–6.

Hortega, P. del Rio. 1917. Notas técnios. Noticia de un neuvo y fácil método para la coloración de la neuroglia y del tejido conjunctivo. Trab. Lab. Invest. Biol., 15, 367–78.

Hortega, P. del Rio. 1928. Tercera aportación al concimiento morfologico e interpretación funcional de la oligodendroglia. Mem. Real Soc. Expan. His. Nat. 14, 5.

Horgeta, P. del Rio. 1932. Microglia. In: *Cytologia and Cellular Pathology of the Nervous System*, Vol. 2, edited by W. Penfield, P. B. Hoeber, New York.

Hortega, P. del Rio. 1962. *The Microscopic Anatomy of Tumors of the Central and Peripheral Nervous System*, Charles C Thomas, Springfield, Ill.

Hotchkiss, R.D. 1948. A microchemical reaction resulting in the staining of polysaccharide structures in fixed tissue preparations. Arch. Biochem., 16, 131–41.

Hoyt, R.F., Sorokin, S.P., and Bartlett, R.A. 1979. A simple fluorescence method for serotonin-containing endocrine cells in plastic-embedded lung, gut and thyroid gland. J. Histochem. Cytochem., 27, 721–7.

Howie, J.W., and Kirkpatrick, J. 1934. Observations on bacterial capsules as demonstrated by a simple method. J. Pathol. Bacteriol., 39, 165–9.

Hubschman, J.H. 1962. A simplified Azan process well suited for crustacean tissue. Stain Technol., 37, 379–80.

Hucker, G.J., and Conn, H.J. 1927. Further studies on the methods of Gram staining. N.Y. Agric. Exp. Stn. Tech. Bull. 128.

Hughes, J., and McCully, M.E. 1975. The use of an optical brightener in the study of plant structure. Stain Technol., 50, 319–29.

Humphrey, C.D., and Pittman, F.E. 1974. A simple methylene blue-azure II-basic fuchsin stain for epoxy-embedded tissue sections. Stain Technol., 49, 9–14.

Hutner, S.H. 1934. Destaining agents for iron alum hematoxylin. Stain Technol., 9, 57–9.

Hyde, B.B., and Gardella, C.A. 1953. A mordanting fixation for intense staining of small chromosomes. Stain Technol., **28**, 305–8.

Hyland, F. 1941. The preparation of stem sections of woody herbarium specimens. Stain Technol., **16**, 49–52.

Itikawa, O., and Ogura, Y. 1954. The Feulgen reaction after hydrolysis at room temperature. Stain Technol., **29**, 13–5.

Jackson, G. 1926. Crystal violet and erythrosin in plant anatomy. Stain Technol., **1**, 33.

Jalal, S.M., Clark, A.W., Hsu, T.C., and Pathak, S. 1974. Cytological differentiation of constitutive heterochromatin. Chromosoma, **48**, 391–403.

James, T.W., and Jope, C. 1978. Visualization by fluorescence of chloroplast DNA in higher plants by means of the DNA-specific probe 4'6-diamidino-2-phenylindole. J. Cell Biol., **79**, 623–30.

Janigan, D.T. 1965. Fluorochrome staining of juxtaglomerular cell granules. Arch. Pathol., **79**, 370–5.

Jarolim, K.L. 1975. A histochemical study of the developing enterochromaffin cell system of the mouse. Doctoral dissertation, University of Oklahoma, Norman, Oklahoma.

Jeffries, C.J., and Belcher, A.R. 1974. A fluorescent brightener used for pollen tube identification in vivo. Stain Technol., **49**, 199–202.

Jeffrey, E.C. 1917. *The Anatomy of Woody Plants*, University of Chicago Press, Chicago.

Jennings, J.B., and LeFlore, W.B. 1972. The histochemical demonstration of certain aspects of cercarial morphology. Trans. Am. Microsc. Soc., **91**, 56–62.

Jensen, R.H. 1977. Chromomycin A3 as a fluorescent probe for flow cytometry of human gynecologic samples. J. Histochem. Cytochem., **25**, 573–9.

Jensen, R.H., Langlois, R.G., and Mayall, B.H. 1977. Strategies for using a deoxyribonucleic acid stain for flow cytometry of metaphase chromosomes. J. Histochem. Cytochem., **25**, 954–64.

Jensen, W.A. 1962. *Botanical Histochemistry*, W.H. Freeman, San Francisco.

Johannisson, E., and Thorell, B. 1977. Mithramycin fluorescence for quantitative determination of deoxyribonucleic acid in single cells. J. Histochem. Cytochem., **25**, 122–8.

Johansen, D.A. 1940. *Plant Microtechnique*, Ed. 1, McGraw-Hill, New York.

Jones, J.G., and Simon, B.M. 1975. An investigation of errors in direct counts of aquatic bacteria by epifluorescence microscopy, with reference to a new method for dyeing membrane filters. J. Appl. Bacteriol., **39**, 317–29.

Jones, P.C.T., and Mollison, J.E. 1948. A technique for the quantitative estimation of soil micro-organisms. J. Gen. Microbiol., **2**, 54–69.

Jones, R.McC. (ed.) 1950. *McClung's Handbook of Microscopical Technique*, Ed. 3, Paul B. Hoeber, New York.

Jotz, M.M., Gill, J.E., and Davis, D.T. 1976. A new optical multichannel microspectrofluorometer. J. Histochem. Cytochem., **24**, 91–9.

Juhlin, L., and Shelley, W.B. 1966. Detection of histamine by a new fluorescent o-phthaldehyde stain. J. Histochem. Cytochem., **14**, 525–8.

Kanai, K. 1962. The staining properties of isolated mycobacterial cellular components as revealed by the Ziehl-Neelsen procedure. Am. Rev. Respir. Dis., **85**, 442–3.

Kao, K.N. 1975a. A nuclear staining method for plant protoplasts. In: *Plant Tissue Culture Methods*, Ch. 10, O.L. Gamborg and L.R. Wetter (eds.), National Research Council of Canada, Ottawa, Ontario.

Kao, K.N. 1975b. A chromosomal staining method for cultured cells. In: *Plant Tissue Culture Methods*, Ch. 11, O.L. Gamborg and L.R. Wetter (eds.), National Research Council of Canada, Ottawa, Ontario.

Kapil, R.N., and Tiwari, S.C. 1978. Plant embryological investigations and fluorescence microscopy; an assessment of integration. Int. Rev. Cytol., **53**, 291–331.

Karpechenko, G.D. 1924. Cytology of hybrids. J. Genet., **14**, 387–96.

Kashiwa, H.K. 1970. Calcium phosphate in osteogenic cells. A critique of the glyoxal bis(2-hydroxyanil) and the dilute silver acetate methods. Clin. Orthop., **70**, 200–11.

Kasten, F.H. 1958. Additional Schiff-type reagents for use in cytochemistry. Stain Technol., **33**, 39–45.

Kasten, F.H. 1959. Schiff-type reagents in cytochemistry; 1. Theoretical and practical considerations. Histochemie, **1**, 466–509.

Kasten, F.H. 1960a. The chemistry of Schiff's reagent. In Int. Rev. Cytol., **10**, 1–100.

Kasten, F.H. 1960b. Recent studies of the Feulgen reaction for deoxyribonucleic acid. Biochem. Pharmacol., **4**, 86–98.

Kasten, F.H. 1963. Schiff-type reagents in cytochemistry; 3. General applications. Acta Histochem. (Suppl.), **3**, 240–7.

Kasten, F.H. 1964. Schiff-type reagents. In: *Laboratory Techniques in Biology and Medicine*, Ed. 4, pp. 390–2, V.M. Emmel and

E.V. Cowdry (eds.), Williams & Wilkins, Baltimore.

Kasten, F.H. 1965. Loss of RNA and protein and changes in DNA during a 30-hour cold perchloric acid extraction of cultured cells. Stain Technol., **40**, 127–35.

Kasten, F.H. 1967a. Cytochemical studies with acridine orange and the influence of dye contaminants in the staining of nucleic acids. Int. Rev. Cytol., **21**, 141–202.

Kasten, F.H. 1967b. Histochemical methods in the study of nucleic acids. In: *Encyclopedia of Biochemistry*, pp. 408–10, R. Williams and E.M. Lansford, Jr. (eds.), Reinhold, New York.

Kasten, F.H. 1973. Acridine dyes. In: *Encyclopedia of Microscopy and Microtechnique*, pp. 4–7, P. Gray (ed.), Reinhold, New York.

Kasten, F.H., Burton, V., and Glover, P. 1959. Fluorescent Schiff-type reagents for cytochemical detection of polyaldehyde moieties in sections and smears. Nature, **184**, 1797–8.

Kasten, F.H., and Churchill, A.E. 1966. Cytochemistry of cytoplasmic and intranuclear inclusions induced by bovine parainfluenza 3 virus (SF-4) in human cell cultures. J. Histochem. Cytochem., **14**, 187–95.

Kasten, F.H., and Lola, R. 1975. The Feulgen reaction after glutaraldehyde fixation. Stain Technol., **50**, 197–201.

Kater, S.B., and Nicholson, C. (eds.) 1973. *Intracellular Staining in Neurobiology*, Springer Verlag, New York.

Katz, I. 1976. Vital and cytochemical staining of cell cultures with a fluorescent dye to determine viability. TCA Manual, **2**, 41–2.

Kerr, D.A. 1938. Improved Warthin-Starry method of staining spirochetes in tissue sections. Am. J. Clin. Pathol., Tech. Suppl., **8**, 63–7.

Kinyoun, J.J. 1915. A note on Uhlenhuths method for sputum examination for tubercle bacilli. Am. J. Public Health, **5**, 867–70.

Kirby, H. 1947. *Methods in the Study of Protozoa*, University of California Press, Berkeley.

Kirk, P.W., Jr. 1966. Morphogenesis and microscopic cytochemistry of marine pyrenomycete ascospores. Nova Hedwigia Suppl., **22**, 1–128. J. Cramer, Lehre.

Kirk, P.W., Jr. 1970. Neutral red as a lipid fluorochrome. Stain Technol., **45**, 1–4.

Klein, B.M. 1926. Ergebnisse mit einer Silbermethode bei Ciltaten. Arch. Protistenk., **56**, 243–79.

Klieneberger-Nobel, E. 1962. *Pleuropneumonia-like organisms (PPLO), Mycoplasmataceae*, Academic Press, New York.

Klüver, H., and Barrera, E. 1953. A method for the combined staining of cells and fibers in the nervous system. J. Neuropathol. Exp. Neurol., **12**, 400–3.

Klüver, H., and Barrerra, E. 1954. On the use of azaporphyrin derivatives (phthalocyanines) in staining nervous tissue. J. Psychol., **37**, 199–223.

Koch, L. 1896. Mikrotechnische Mittheilunger III. Jahrb. Wiss. Bot., **29**, 39–74.

Koenig. H., Groat, R.A., and Windle, W.F. 1945. A physiological approach to perfusion-fixation of tissues with formalin. Stain Technol., **20**, 13–22.

Kofoid, C.A., and Swezy, O. 1915. Mitosis and multiple fission in trichomonad flagellates. Proc. Am. Acad. Arts Sci., **5**, 289–378.

Kolmer, J.A., and Boerner, F. 1941. *Approved Laboratory Technic*, Ed. 3, Appleton-Century, New York.

Kornfield, H.J., and Werder, A.A. 1960. A differential nucleic acid fluorescent stain applied to cell culture systems. Cancer, **13**, 458–61.

Kornhauser, S.I. 1943. A quadruple tissue stain for strong color contrasts. Stain Technol., **18**, 95–7.

Kornhauser, S.I. 1945. A revised method for the "quad" stain. Stain Technol., **20**, 33–5.

Kosenow, W. 1952. Die Fluorochromierung mit Acridinorange, eine Methode zur Lebendbeobachtung gefärbter Blutzellen. Acta Haematol., **7**, 217–21.

Kossa, J.V. 1901. Über die in Organismus künstlich erzeugbaren Verkalkungen. Beitr. Pathol. Anat. Allg. Pathol., **29**, 163–202.

Kraemer, P.M., Deaven, L.L., Crissman, H.A., and Van Dilla, M.A. 1972. The paradox of DNA constancy in heteroploidy. In: *Advances in Cell and Molecular Biology*, pp. 47–108, E.S. DuPraw (ed.), Academic Press, New York.

Kraicer, J., Herlant, M., and Duclos, P. 1967. Changes in adenohypophyseal cytology and nucleic acid content in the rat 32 days after bilateral adrenalectomy and the chronic injection of cortisol. Can. J. Physiol. Pharmacol., **45**, 947–56.

Krajian, A.A. 1939. The clinical application of a twenty-minute staining method for *Spirochaeta pallida* in tissue sections. Am. J. Syphilol., **23**, 617–20.

Kramer, H., and Windrum, G.M. 1955. The metachromatic staining reaction. J. Histochem., **3**, 227–37.

Kramer, S. 1948. A staining procedure for the study of insect musculature. Science, **108**, 141–2.

Kretschmer, O.S. 1934. The Gram property of

the acid-fast form of the tubercle bacillus. J. Lab. Clin. Med., **19**, 350–8.

Krieg, A. 1953. Fluoreszenzanalyse und Fluorochromie in Biologie und Medizin. Klin. Wochenschr., **31**, 350–6.

Krug, H. 1979. (ed.). *Symposium on Quantitative Microscopy*, Leipzig.

Kubica, G.P., and Dye, W.E. 1967. *Laboratory Methods for Clinical and Public Health Mycobacteriology*. U.S.P.H. Serv. Publication No. 1547; Superintendent of Documents, U.S. Government Printing Office, Washington, D.C.

Kultschitzky, N. 1897. Sur Frage über den Bau des Darmkanals. Arch. Mikrosk. Anat. Forsch., **2**, 163.

Kuper, S.W.A., and May, R. 1960. Detection of acid fast organisms in tissue sections by fluorescence microscopy. J. Pathol. Bacteriol., **79**, 59–68.

Kuyper, Ch.M.A. 1957. Identification of mucopolysaccharides by means of fluorescent basic dyes. Exp. Cell Res., **13**, 198–200.

LaCour, L. 1935. Technic for studying chromosome structure. Stain Technol., **10**, 57–60.

LaCour, L. 1941. Acetic orcein: A new stain-fixative for chromosomes. Stain Technol., **16**, 169–74.

LaCour, L., and Farberge, A.C. 1943. The use of cellophane in pollen tube technic. Stain Technol., **18**, 196.

Lagunoff, D. 1972. The mechanism of histamine release from mast cells. Biochem. Pharmacol., **21**, 1889–96.

Lamm, M.E.E., Childers, L., and Wolk, M.K. 1965. Studies on nucleic acid metachromasy. 1. The effect of certain fixatives on the dye stacking properties of nucleic acids in solution. J. Cell Biol., **72**, 313–326.

Lang, A.G. 1936. A stable, high-contrast mordant for hematoxylin staining. Stain Technol., **11**, 149–51.

Langeron, M. 1934. *Précis de Microscopie*, Ed. 5; Ed. 6, 1942, Masson et Cie., Paris.

Langworthy, O.R. 1924. A study of the innervation of the tongue musculature with particular reference to the proprioceptive mechanism. J. Comp. Neurol., **36**, 273–97.

Larsell, O. 1921. Nerve terminations in the lung of the rabbit. J. Comp. Neurol., **33**, 105–31.

Lascano, E.F. 1958. A glycin HCl-formalin fixative for improved staining of neuroglia. Stain Technol., **33**, 9–14.

Laskey, A.M. 1950. A modification of Mayer's mucihematein technic. Stain Technol., **25**, 33–4.

Latt, S.A. 1976. Optical studies of metaphase chromosome organization. Annu. Rev. Biophys. Bioeng., **5**, 1–37.

Latt, S.A. 1977. Fluorescent probes of chromosome structure and replication. Can. J. Genet. Cytol., **19**, 603–23.

Laybourn, R.L. 1924. A modification of Albert's stain for the diphtheria bacilli. J.A.M.A., **83**, 121.

Lechtman, M.D., Bartholomew, J.W., Phillips, A., and Russo, M. 1965. Rapid methods of staining bacterial spores at room temperature. J. Bacteriol., **89**, 848–54.

Lee, B. 1890. *The Microtomist's Vade-Mecum*, Ed. 2, J. & A. Churchill, London; Ed. 10, 1937; Ed. 11, 1950, P. Blakiston's, Philadelphia.

Lehner, T. 1965. Juxtaglomerular apparatus staining with thioflavin T fluorochrome, and its confusion with amyloid. Nature, **206**, 738.

Leifson, E. 1951. Staining, shape, and arrangement of bacterial flagella. J. Bacteriol., **62**, 377–89.

Leifson, E. 1960. *Atlas of Bacterial Flagellation*, Academic Press, New York.

Lendrum, A.C. 1947. The phloxin-tartrazine method as a general histological stain for the demonstration of inclusion bodies. J. Pathol. Bacteriol., **59**, 399–404.

Levaditi, M.M., and Manouélian. 1906. Nouvelle méthode rapide pour la coloration des spirochétes sur coupes. C.R. Soc. Biol. (Paris), **60**, 134–6.

Levene, C., and Feng, P. 1964. Critical staining of pancreatic alpha granules with phosphotungstic acid hematoxylin. Stain Technol., **39**, 39–44.

Levine, B.S. 1952. Staining *Treponema pallidum* and other tryponemata. Public Health Rep., **67**, 253–7.

Levine, B.S., and Black, L.A. 1948. Newly proposed staining formulas for the direct microscopic examination of milk. Am. J. Public Health, **38**, 1210–8.

Levine, N.D. 1939. The dehydration of methylene blue stained material without loss of dye. Stain Technol., **14**, 29–30.

Levinson, J.W., Retzel, S., and McCormick, J.J. 1977. An improved acriflavine-Feulgen method. J. Histochem. Cytochem., **25**, 355–8.

Levowitz, K., and Weber, M. 1956. An effective "single solution" stain. J. Milk Food Technol., **19**, 121, 127–9.

Levy, H.A. 1943. Dioxane as an aid in staining insect cuticle. Stain Technol., **18**, 181–2.

Lillie, R.D. 1928. Eine Schell-methode zur Toluidinblau-Schleimfärbung. Z. Wiss. Mikrosk., **45**, 381.

Lillie, R.D. 1940a. Biebrich scarlet picro-anilin blue; a new differential connective tissue and muscle stain. Arch. Pathol., **29**, 705.

Lillie, R.D. 1940b. Further experiments with the Massen trichrome modification of Mallory's connective tissue stain. Stain Technol., **15**, 17–22.

Lillie, R.D. 1944a. Myelin staining by a fixed schedule for the occasional user. Arch. Pathol., **37**, 392–5.

Lillie, R.D. 1944b. Various oil soluble dyes as fat stains in the super-saturated isopropanol technic. Stain Technol., **19**, 55–8.

Lillie, R.D. 1944c. Acetic methylene blue counterstain in staining tissues for acid-fast bacilli. Stain Technol., **19**, 45.

Lillie, R.D. 1948. *Histopathologic Technic*, Blakiston, Philadelphia.

Lillie, R.D. 1951. Allochrome procedure; differential method segregating connective tissues collagen, reticulum and basement membranes into two groups. Am. J. Clin. Pathol., **21**, 484–8.

Lillie, R.D. 1954. *Histopathologic Technic and Practical Histochemistry*, Blakiston, New York.

Lillie, R.D. 1956a. A Nile blue staining technic for the differentiation of melanin and lipofuscins. Stain Technol., **31**, 151–3.

Lillie, R.D. 1956b. The mechanism of Nile blue staining of lipofuscine. J. Histochem., **4**, 377–81.

Lillie, R.D. 1965. *Histopathologic Technic and Practical Histochemistry*, Ed. 3, McGraw-Hill, New York.

Lillie, R.D. 1969. *Biological Stains*, Ed. 8, Williams & Wilkins, Baltimore.

Lillie, R.D. 1977. *Biological Stains*, Ed. 9, Williams & Wilkins, Baltimore.

Lillie, R.D., and Ashburn, L.L. 1943. Supersaturated solutions of fat stains in dilute isopropanol for demonstration of acute fatty degenerations not shown by the Herxheimer technic. Arch. Pathol., **36**, 432–5.

Lillie, R.D., and Fullmer, H.M. 1976. *Histopathologic Technic and Practical Histochemistry*, Ed. 4, McGraw-Hill, New York.

Lillie, R.D., Greco-Henson, J.P., and Cason, J.C. 1961. Azocoupling rate of enterochromaffin with various diazonium salts. J. Histochem. Cytochem., **9**, 11–21.

Lin, C.C., Biederman, B., and Jamro, H. 1968. Q-banding methods using quinacrine (QFQ) and Hoechst 33258 (QFH) for chromosome analysis of human lymphocyte cultures. TCA Manual, **4**, 937–40.

Lin, C.C., and van de Sande, J.H. 1975. Differential fluorescent staining of human chromosomes with daunomycin and adriamycin-the D bands. Science, **190**, 61–3.

Lindvall, O., and Björklund, A. 1974. The glyoxylic acid fluorescence histochemical method; a detailed account of the methodology for the visualization of central catecholamine neurons. Histochemistry, **39**, 97–127.

List, J.H. 1885. Zur Farbentechnik; I. Bismark-braun-Methylgrün. Z. Wiss. Mikrosk., **2**, 145.

Lockard, I., and Reers, B.L. 1962. Staining tissue of the central nervous system with Luxol fast blue and neutral red. Stain Technol., **37**, 13–6.

Love, L.D. 1979. Fluorescence microscopy of viable mast cells stained with different concentrations of acridine orange. Histochemistry, **62**, 221–5.

Love, R., Fernandes, M.V., and Koprowski, H. 1964. Cytochemistry of inclusion bodies in tissue culture cells infected with rabies virus. Proc. Soc. Exp. Biol. Med., **116**, 560–3.

Lorén, I., Bjorklund, A., Falck, B., and Lindvall, O. 1980. The aluminum formaldehyde (ALFA) histofluorescence method for improved visualization of catecholamines and indolamines. Application on the central nervous system. J. Neurosci. Meth. (in press).

Lubs, H.A., McKenzie, W.H., Patil, S.R., and Merrick, S. 1973. New staining methods for chromosomes. Methods Cell Biol., **6**, 345–80.

Luna, L.G. 1964. Evaluation of staining technique for pathogenic fungi. Am. J. Med. Technol., **30**, 139–46.

Lundvall, H. 1905. Weiteres über Demonstration embryonaler Skelette. Anat. Anz., **27**, 521.

Luthringer, D.G., and Glenner, G.G. 1961. The demonstration of endamoebae in tissue sections, by means of a modified methenamine-silver nitrate technique. Am. J. Clin. Pathol., **36**, 378–8.

Lynch, J.E. 1929. Eine neue Karminmethode für Total präparate. Z. Wiss. Mikrosk., **46**, 465–9.

Macchiavello, A. 1937. Estudios sobre Tifus Exantemático; III. Un neuvo método para tenier Rickettsia. Rev. Chil. Hig., **1**, 101–6.

Mackinnon, D.L., and Hawes, R.S.J. 1961. *An Introduction to the Study of Protozoa*, Oxford at the Clarendon Press.

MacNeal, W.J. 1922. Tetrachrome blood

stain; an economical and satisfactory imitation of Leischmann's stain. J.A.M.A., **78**, 1112.

Maddy, A.H. 1964. A fluorescent label for the outer components of the plasma membrane. Biochim. Biophys. Acta, **88**, 390–9.

Madoff, S. 1960. Isolation and identification of PPLO. Ann. N.Y. Acad. Sci., **79**, 383–92.

Magun, B.E., and Kelly, J.W. 1969. A new fluorescent method with pheneanthrenequinone for the histochemical demonstration of arginine residues in tissues. J. Histochem. Cytochem., **17**, 821–7.

Mallory, F.B. 1900. A contribution to staining methods; I. A differential stain for connective-tissue fibrillae and reticulum. J. Exp. Med., **5**, 15–20.

Mallory, F.B. 1938. *Pathological Technique*, W.B. Saunders Philadelphia. Reprinted 1961. Hafner, New York.

Manigault, P. 1955. Progrès récents de la microscopie par fluorescence. Bull. Micr. Appl. II Sér., **5**, 81–90.

Mann, G. 1894. Ueber die Behandlung der Nervenzellen für experimentell-histologische Untersuchunge. Z. Wiss. Mikrosk., **11**, 479–94.

Manocchio, I. 1964. Metachromasia e basophilia delle cellule insulari alfa nel pancreas di mammiferi depo metilazone e demetilazione. Arch. Vet. Ital., **15**, 3–7.

Mansfield, R.E. 1970. An improved method for the fluorochrome staining of mycobacteria in tissues and smears. Am. J. Clin. Pathol., **53**, 394–406.

Mantel, N., and Robertson, A.H. 1954. A comparison of six methods of preparing and using the methylene blue stain for bacterial counts by the direct microscopic method. J. Milk Food Technol., **17**, 179–84.

Manual of Histologic Staining Methods of the Armed Forces Institute of Pathology, Ed. 3, L.G. Luna (ed.), McGraw-Hill, New York.

Manwall, R.D. 1961. *Introduction to Protozoology*, St. Martin's Press, New York.

Mara, J.A., and Yoss, R.E. 1952. A rapid Marchi technic. Stain Technol., **27**, 325–8.

Marchi, V. 1886. Sulle degenerazioni consecutive allé estirpazione totale e parziale dell cervelletto. Riv. Sper. Freniatr., **12**, 50–6.

Marchi, V., and Algeri, G. 1885. Sulle degenerazioni discendenti consecutive a lesioni della corteccia cerebral. Riv. Sper. Freniatr., **11**, 492–4.

Margolena, L.A. 1935. Lugol's solution for the Flemming triple stain. Stain Technol., **10**, 35–6.

Martin, F.W. 1959. Staining and observing pollen tubes in the style by means of fluorescence. Stain Technol., **34**, 125–8.

Masin, F., and Masin, M. 1960. Difference in dye retention between proliferating atypical and malignant cells in smears from cervical epithelium. Cancer, **13**, 1221–9.

Mason, T.E., Phifer, R.F., Spicer, S.S., Swallow, R.A., and Dreskin, R.B. 1969. An immunoglobulin-enzyme bridge method for localizing tissue antigens. J. Cytol. Histochem., **17**, 563–9.

Masson, P. 1911. Le safran en technique histologique. C.R. Soc. Biol. (Paris), **70**, 573–4.

Masson, P. 1923. Diagnostics de Laboratoire. I. Tumeurs-Diagnostics histologiques. In: *Traité de Path. Medicale*. Maloine et Fils, Paris.

Masson, P. 1929. Some histological methods; trichrome stainings and their preliminary technique. J. Tech. Methods, **12**, 75–90.

Matta, J.F., and Lowe, R.E. 1969. A differential staining technique for a mosquito iridescent virus. J. Invertebr. Pathol., **13**, 457–8.

Matthaei, E. 1950. Simplified fluorescence microscopy of tubercle bacilli. J. Gen. Microbiol., **4**, 393–8.

Matzke, K.H., and Thiessen, G. 1976. The acridine dyes: Their purification, physiochemical, and cytochemical properties; 1. A purity test of some commercial acriflavine samples and the identification of their components. Histochemistry, **49**, 73–9.

Maximow, A. 1909. Über zweckmässige Methoden für cytologische und histogenetische Untersuchungen am Wirbeltier embryo, mit spezieller Berücksichtigung der celloidinschnittserien. Z. Wiss. Mikrosk., **26**, 177–90.

Mayall, B., and Gledhill, B.L. 1979. (eds.). Sixth Foundation Conference on Automated Cytology. J. Histochem. Cytochem., **27**, 1–641.

Mayer, P. 1891. Ueber das Färben mit Hämatoxylin. Mitt. Zool. Stat. Neapal, **10**, 170–86.

Mayer, P. 1896. Uber Schleimfärbung. Mitt. Zool. Stat. Neapel, **12**, 303–30.

Mayer, P. 1899. Ueber Hämatoxylin, Carmin, und verwandte Materien. Z. Wiss. Mikrosk., **16**, 196–220.

Mayor, H.D. 1961a. Cytochemical and fluorescent antibody studies on the growth of poliovirus in tissue culture. Texas Rep. Biol. Med., **19**, 106–22.

Mayor, H.D. 1961b. Acridine orange staining of purified polyoma virus. Proc. Soc. Exp. Biol. Med., **108**, 103–5.

Mayor, H.D., and Diwan, A.R. 1961. Studies on the acridine orange staining of two purified RNA viruses; poliovirus and tobacco mosaic virus. Virology, **14**, 74–82.

Mayor, H.D., and Hill, N.O. 1961. Acridine orange staining of a single-stranded DNA bacteriophage. Virology, **14**, 264–6.

McArdle, E.W. 1959. The preparation of ciliates for nuclear staining by embedding in albumen. J. Protozool., **6**(suppl.), 12–3.

McCarter, J.C. 1940. A silver carbonate method for oligodendrocytes and microglia for routine use. Am. J. Pathol., **16**, 233–5.

McClintock, B. 1929. A method for making aceto-carmine smears permanent. Stain Technol., **4**, 53–6.

McClung, C.E. 1937. *Microscopical Technique*, Ed. 2, Paul B. Hoeber, New York.

McClung, C.E. 1950. *Handbook of Microscopical Technique*, Ed. 3, Ruth McClung Jones (ed.), Paul B. Hoeber, New York.

McGee-Russell, S.M. 1958. Histochemical methods for calcium. J. Histochem., **6**, 22–42 (see p. 31).

McManus, J.F.A. 1948. Histological and histochemical uses of periodic acid. Stain Technol., **23**, 99–108.

McManus, J.F.A., and Mowry, R.W. 1960. *Staining Methods Histological and Histochemical.* Harper & Row, New York.

McNary, W.F., Jr. 1957. Dithizone staining of myeloid granules. Blood, **12**, 644–8.

McQueen, J.L., Lewis, A.L., and Schneider, N.J. 1960. Rabies diagnosis by fluorescent antibody; I. Its evaluation in a public health laboratory. Am. J. Public Health, **50**, 1743–52.

Meisel, M.N., Medvedeva, G.A., and Alekseeva, V.M. 1961. Discrimination of living, injured, and dead microorganisms. Mikrobiologiya, **30**, 699–704.

Meistrich, M.L., Göhde, W., White, R.A., and Schumann, J. 1978. Resolution of X and Y spermatids by pulse cytophotometry. Nature, **274**, 821–3.

Mellors, R.C., Keane, J.F., and Papanicolaou, G.N. 1952. Nucleic acid content of the squamous cancer cell. Science, **116**, 265–9.

Metcalf, F.L., and Patton, R.L. 1944. Fluorescence microscopy applied to entomology and allied fields. Stain Technol., **19**, 11–27.

Meyer, J.R. 1943. Colchicine-Feulgen leaf smears. Stain Technol., **18**, 53–6.

Meyer, J.R. 1945. Prefixing with paradichlorobenzene to facilitate chromosome study. Stain. Technol., **20**, 121–5.

Michaelis, L. 1900. Die vitale Färbung, eine Darstellungsmethode dez Zellgranula. Arch. Mikrosk. Anat., **55**, 558–75.

Milch, R.A., Rall, D.P., and Tobie, J.E. 1957. Bone localization of the tetracyclines. J. Nat. Cancer Inst., **19**, 87–93.

Milch, R.A., Rall, D.P., and Tobie, J.E. 1958. Fluorescence of tetracycline antibiotics in bone. J. Bone Joint Surg., **40**, 897–910.

Milch, R.A., Tobie, J.E., and Robinson, R.A. 1961. A microscopic study of tetracycline in skeletal neoplasms. J. Histochem. Cytochem., **9**, 261–70.

Moliner, E.R. 1957. A chlorate-formaldehyde modification of the Golgi method. Stain Technol., **32**, 105–16.

Moliner, E.R. 1958. A tungstate modification of the Golgi-Cox method. Stain Technol., **33**, 19–29.

Möller, O. 1951. A new method for staining bacterial capsules. Acta Pathol. Microbiol. Scand., **28**, 127–31.

Möller, W. 1899. Anatonische Beitrage zur Frage von des Sekretion und Resorption in der Dermschleimhaut. Z. Wiss. Zool., **66**, 69–135.

Mollier, G. 1938. Eine Vierfachfärbung zur Darstellung glatter und quergestreifter Muskulatur und ihre Beziehung zum Bindegewebe. Z. Wiss. Mikrosk., **55**, 472.

Moore, R.Y., and Loy, R. 1978. Fluorescence histochemistry. In: *Neuroanatomical Research Techniques*, Vol. 2, 115–39, R.T. Robertson (ed.), Academic Press, New York.

Morrison, J.W. 1953. Chromosome behaviour in wheat monosomines. Heredity, **7**, 203–17.

Morthland, F.W., DeBruyn, P.P.H., and Smith, N.H. 1954. Spectrophotometric studies on the interaction of nucleic acids with aminoacridines and other basic dyes. Exp. Cell Res., **7**, 201–14.

Morton, H.E., and Francisco, A. 1942. The staining of the metachromatic granules in *Corynebacterium diphtheriae*. Stain Technol., **17**, 27–9.

Mote, R.F., Muhm, R.L., and Gigstad, D.C. 1975. A staining method using acridine orange and auramine O for fungi and myobacteria in bovine tissue. Stain Technol., **50**, 5–9.

Mowry, R.W. 1956. Alcian blue techniques for the histochemical study of acidic carbohydrates. J. Histochem., **4**, 407.

Mowry, R.W. 1975. Special value of prestaining polyanions with basic dyes of contrasting color (Alcian blue) before staining with aldehyde fuchsin with particular reference to pancreatic islet B cells and the diagnosis of mesidoblastosis. J. Histochem. Cytochem., **23**, 322–3.

Müller, H. 1860. Anatomische Untersuchung eines Microphthalamus. Ver. Phys. Med. Ges. Wurzburg, **10**, 138–46.

Myers, D.F., and Fry, W.E. 1978. The development of gloeocercospora sorghi in sorghum. Phytopathology, **68**, 1147–55.

Nankane, P.K., and Pierce, G.B. 1967. Enzyme-labeled antibodies for the light and electron microscopic localization of tissue antigens. J. Cell Biol., **33**, 307–18.

Nassar, T.K., and Shanklin, W.M. 1951. Staining neuroglia with silver diaminohydroxide after sensitizing with sodium sulfite and embedding in paraffin. Stain Technol., **26**, 13–8.

Nauta, W.J.H., and Gygax, P.A. 1951. Silver impregnation of degenerating axon terminals in the central nervous system: (1) Technic (2) Chemical notes. Stain Technol., **26**, 5–11.

Nauta, W.J.H., and Gygax, P.A. 1954. Silver impregnation of degenerating axons in the central nervous system; a modified technic. Stain Technol., **29**, 91–3.

Nauta, W.J.H., and Ebbesson, S.W.E. 1970. *Contemporary Research Methods in Neuroanatomy,* Springer-Verlag, New York.

Nebel, B.R. 1931. Lacmoid-martius yellow for staining pollen-tubes in the style. Stain Technol., **6**, 27–9.

Nebel, B.R. 1940. Chlorazol black E as an aceto-carmine auxiliary stain. Stain Technol., **15**, 69–72.

Nedzel, G.A. 1951. Intranuclear birefringent inclusions, an artifact occurring in paraffin sections. Q. J. Microsc. Sci., **92**, 343–6.

Neelsen, F. 1883. Ein casuistischer Beitrag zur Lehre von der Tuberkulose. Centralbl. Med. Wiss., **21**, 497–501.

Newcomer, E.H. 1938. A procedure for growing, staining, and making permanent slides of pollen tubes. Stain Technol., **13**, 89–91.

Newton, W.C.F. 1925. Chromosome studies in *Tulipa* and some related genera. J. Linn. Soc. Bot., **47**, 339–54.

Nicolini, C., Belmont, A., Parodi, S., Lessin, S., and Abraham, S. 1979. Mass action and acridine orange staining; static and flow cytofluorometry. J. Histochem. Cytochem., **27**, 102–13.

Nocht. 1898. Zur Färbung der Malariaparasiten. Centralbl. Bakteriol., Abt. i, **24**, 839–43.

Nonidez, J.F. 1939. Studies on the innervation of the heart. Am. J. Anat., **65**, 361–413.

North, W.R. 1945. Anilin oil-methylene blue stain for the direct microscopic count of bacteria in dried milk and dried eggs. Assoc. Offic. Agr. Chem., **28**, 424–6.

Nyka, W. 1967. Method for staining both acid-fast and chromophobic tubercle bacilli with carbol fuchsin. J. Bacteriol., **93**, 1458–60.

Ogilvie, R.W., and Clark, G. 1972. In preparation.

Olson, J.C., and Black, L.A. 1951. A comparative study of stains proposed for the direct microscopic examination of milk. J. Milk Food Technol., **14**, 49–51; 64.

O'Mara, J.G. 1939. Observations on the immediate effects of colchicine. J. Hered., **30**, 35–7.

Ornstein, L., Mautner, W., Davis, B.J., and Tamura, R. 1957. New horizons in fluorescent microscopy. J. Mt. Sinai Hosp. N.Y., **24**, 1066–78, 1957.

Orth, J. 1896a. Ueber die Verwendung des Formaldehyd in pathologischen Institut in Göttingen. Berl. Klin. Wochenschr., **33**, 273–5.

Orth, J. 1896b. Ueber die Verwendung des Formaldehyd in pathologischen Institut in Göttingen. Z. Wiss. Mikrosk., **13**, 16–7.

Ostergren, G. 1948. Chromatin stains of Feulgen type involving other dyes than fuchsin. Hereditas, **34**, 510–1.

Pain, T.F. 1963. Gram staining without the clock. N. Engl. J. Med., **268**, 941.

Pal, J. 1886. Ein Beitrag zur Nervenfärbetechnik. Wien. Med. Jahrb., 1886, N.F. 1, 619–31. Abstracted in Z. Wiss. Mikrosk, **4**, (1887), 92–6.

Palmer, R.G., and Heer, H. 1973. A root tip squash technique for soybean chromosomes. Crop Sci., **13**, 389–91.

Pandolph, C., and Puolter, L.W. 1978. The staining of acid-fast bacilli in sections of glycerol methacrylate embedded tissues. Stain Technol., **53**, 173–6.

Papadimitrou, J.M., Van Duijn, P., Brederoo, P., and Streefkerk, J.G. 1978. A new method for the cytochemical demonstration of peroxidase for light, fluorescence and electron microscopy. J. Histochem. Cytochem., **24**, 82–90.

Papanicolaou, G.N. 1941. Some improved methods for staining vaginal smears. J. Lab. Clin. Med., **26**, 1200–5.

Pappenheim, A. 1899. Vergleichende Untersuchungen über die elementare Zuzammensetzung des rothen Knochenmarkes einige Säugethiere. Virchows Arch. Pathol. Anat., **157**, 19–76.

Pappenheim, A. 1908. Panoptische Universal-färbung für Blutpräparate. Med. Klin., **4**, 1244.

Pappenheim, A. 1912. Zur Blutzellfäfbung im klinischen Bluttrockempräparat und zur histologischen Schnittpräparatfärbung der hämatopoetischen Gewebe nach meinem Methoden. Folia Haematol., **13**, 338–44.

Parat, M. 1928. Contribution à l'étude morphologique et physiologique du cytoplasme, chondriome, vacuome (appareil de Golgi), enclaves, etc. pH. oxydase, peroxydase, rH de la cellule animale. Arch. Anat. Microsc., **24**, 73–357.

Paton, A.M., and Jones, S.M. 1973. The observation of microorganisms on surfaces by incident fluorescence microscopy. J. Appl. Bacteriol., **36**, 441–3.

Pauley, G.B. 1967. A modification of Mallory's aniline blue collagen stain for oyxter tissue. J. Invertebr. Pathol., **9**, 268–9.

Pearse, A.G.E. 1957. Solochrome dyes in histochemistry with particular reference to nuclear staining. Acta Histochem., **4**, 95–101.

Pearse, A.G.E. 1960. *Histochemistry, Theoretical and Applied,* Little, Brown, Boston.

Pearse, A.G.E. 1968. *Histochemistry, Theoretical and Applied,* Vol. 1, Ed. 3, Little, Brown, Boston.

Pearse, A.G.E. 1972. *Histochemistry, Theoretical and Applied,* Vol. 2, Ed. 3, Williams & Wilkins, Baltimore.

Peers, J.H. 1941. A modification of Mallory's phosphotungstic acid hematoxylin stain for formaldehyde-fixed tissue. Arch. Pathol., **32**, 446–9.

Penfield, W. 1928. A method of staining oligodendroglia and microglia (combined method). Am. J. Pathol., **4**, 153–7.

Perez, R.M. 1931. Contribution of l'étude des terminaisons nerveuses de la peau de la main. Trav. Lab. Recherches Biol. Univ. Madrid, **27**, 187–226.

Perls, M. 1867. Nachweis von eisenoxyd in Gewissen pigmenten. Virchows Arch. Pathol. Anat., **39**, 42–8.

Perner, E.S. 1957. Die Methoden der Fluoreszenzmikroskopie. In: *Handbuch der Mikrosk. i. d. Technik,* Vol. 1, H. Freund (ed.), Frankfurt a. M.

Peters, A. 1955a. Experiments on the mechanism of silver straining; I, II, III and IV. Q. J. Microsc. Sci., **92**: 84–115; 301–22.

Peters, A. 1955b. A general purpose method of silver staining. Q. J. Microsc. Sci., **96**, 323–8.

Peters, A. 1958. Staining of nervous tissue by protein-silver mixtures. Stain Technol., **33**, 47–53.

Petersen, H. 1924. Färben mit Säurealizarinblau. Z. Wiss. Mikrosc., **41**, 363–5.

Peterson, C.A., and Fletcher, R.A. 1973. Lactic acid clearing and fluorescent staining for demonstration of sieve tubes. Stain Technol., **48**, 23–7.

Phillips, R.L., Wang, A.S., Rubenstein, I., and Park, W.D. 1979. Hybridization of ribosomal RNA to maize chromosomes. Maydica, **24**, 7–21.

Phifer, R.F., III. 1971. Personal communication.

Pianese, G. 1896. Beitrag zur Histologie und Aetiologie des Carcinoms. Beitr. Pathol. Anat. Allg. Pathol., Suppl. **1**, 193.

Pickett, J.P., Bishop, C.M., Chick, E.W., and Baker, R.D. 1960. A simple fluorescent stain for fungi. Am. J. Clin. Pathol., **34**, 197–202.

Ploem, J.S. 1967a. Die Moglichkeit der auflichtfluoreszenzmethoden bei Untersuchungen von Zellen in Durchstromungskammern und Leightonrohren. Acta Histochem. Suppl., **7**, 339–43.

Ploem, J.S. 1967b. The use of a vertical illuminator with interchangeable dichroic mirrors for fluorescence microscopy with incident light. Z. Wiss. Mikrosk., **68**, 129–42.

Poirier, L.J., Ayotte, R.A., and Gauthier, C. 1954. Modification of the Marchi technic. Stain Technol., **29**, 71–5.

Polyak, S. 1942. *The Retina,* University of Chicago Press, Chicago.

Popper, H. 1941a. Histologic distribution of vitamin A in human organs under normal and under pathologic conditions. Arch. Pathol., **31**, 766–802.

Popper, H. 1941b. Visualization of vitamin A in rat organs by fluorescence microscopy. Arch. Pathol., **32**, 11–32.

Popper, H. 1944. Distribution of vitamin A in tissue as visualized by fluorescence microscopy. Physiol. Rev., **24**, 205–24.

Popper, H., and Szanto, P.B. 1950 *Fluorescence Microscopy,* Ed. 3, pp. 678–86, R.M. Jones (ed.), Hafner, New York.

Porter, R.W., and Davenport, H.A. 1949. Golgi's dichromate-silver method. 1. Effects of embedding. 2. Experiments with modifications. Stain Technol., **24**, 117–26.

Powers, M.M., and Clark, G. 1955. An evaluation of cresyl echt violet acetate as a Nissl stain. Stain Technol., **30**, 83–8.

Powers, M.M., and Clark, G. 1963. A note on Darrow red. Stain Technol., **38**, 289.

Powers, M.M., Clark, G., Darrow, M.A., and Emmel, V.M. 1960. Darrow red, a new basic dye. Stain Technol., **35**, 19–21.

Powers, M.M., Rasmussen, G.L., and Clark, G. 1951. The staining of hard tissues with silver. Anat. Rec., 111, 171–6.

Pratt, R., and Dufrenoy, J. 1948. Triphenyltetrazolium chloride, a valuable reagent in stain technology. Stain Technol., 23, 137–41.

Prenna, G. 1964. Caratteristiche del [2-(p-aminofenil)-6-methil-2,6-bibenzotia-SO₂], un nuovo reagente tipo Schiff zolo fluorescente a elevata sensibilita. Riv. Istochim. Norm. Pat., 10, 469–74.

Prenna, G. 1968. Qualitative and quantitative application of fluorescent Schiff-type reagents. Mikroskopie, 23, 150–4.

Prenna, G., and Zanotti, L. 1962. Reazioni di Feulgen fluorescenti e loro possibilita citofluorometriche quantitative; 1. Studio istochimico di alcuni reagenti tipo Schiff fluorescenti nella reazione di Feulgen. Riv. Istochim. Norm. Pat., 8, 427–46.

Prescott, S.C., and Breed, R.S. 1910. The determination of the number of body cells in milk by a direct method. J. Infect. Dis., 7, 632–40.

Price, G.P., and Schwartz, S. 1956. Fluorescence microscopy. In: Physical Techniques in Biological Research, Vol. 3, pp. 91–148, G. Oster and A.W. Pollister (eds.), Academic Press, New York.

Pringsheim, P. 1948. Fluorescence and Phosphorescence, Interscience, New York.

Proescher, F. 1927. Oil red O pyridin, a rapid fat stain. Stain Technol., 2, 60–1.

Proescher, F., and Arkush, A.S. 1928. Metallic lakes of the oxazines as nuclear stain substitutes for hematoxylin. Stain Technol., 3, 28–38.

Prudden, J.M. 1885. (A note without title.) Z. Wiss. Mikrosk., 2, 288.

Puchtler, H., and Sweat, F. 1960. Commercial resorcin-fuchsin as a stain for elastic fibers. Stain Technol., 35, 347–8.

Puchtler, H., and Sweat, F. 1962. On the binding of Congo red by amyloid. J. Histochem. Cytochem., 10, 355–64.

Radna, R. 1938. La microscopie de fluorescence comme mode de recherche du bacille de la lèpre et des trypanosomes. Ann. Soc. Belg. Med. Trop., 18, 623–8.

Rae, C.A. 1955. Masson's trichrome stain after Petrunkewitch or Susa fixation. Stain Technol., 30, 147.

Rahn, B.A., and Perren, S.M. 1970. Calcein blue as a fluorescent label in bone. Experientia, 26, 519–20.

Rahn, B.A., and Perren, S.M. 1971. Xylenol orange, a fluorochrome useful in polychrome sequential labeling of calcifying tissues. Stain Technol., 46, 125–9.

Rahn, B.A., Feisch, H., Moor, R., and Perren, S.M. 1970. The effect of fluorescent labels on bone growth and calcification in tissue culture. Eur. Surg. Res., 2, 137–8.

Ramanna, M.S. 1973. Euparal as a mounting medium for preserving fluorescence of aniline blue in plant material. Stain Technol., 48, 103–5.

Ramming, D.W., Hinrichs, H.A., and Richardson, P.E. 1973. Sequential staining of callose by aniline blue and lacmoid for fluorescence and regular microscopy on a durable preparation of the same specimen. Stain Technol., 48, 133–4.

Ramón-Moliner, E. 1957. A chlorate-formaldehyde modification of the Golgi method. Stain Technol., 32, 105–16.

Randolph, L.F. 1935. A new fixing fluid and a revised schedule for the paraffin method in plant cytology. Stain technol., 10, 95–6.

Ranson, S.W. 1911. Non-medullated nerve fibers in the spinal nerves. Am. J. Anat., 12, 67–87.

Ranvier, L. 1880. On the terminations of nerves in the epidermis. Q. J. Microsc. Sci., 20, 456–8.

Rattenbury, T.A. 1956. A rapid method for permanent aceto-carmine squash preparations. Nature (Lond.), 177, 1186.

Rawlins, T.E. 1933. Phytopathological and Botanical Research Methods, John Wiley & Sons, New York.

Regaud, C. 1910. Etudes dur la structure des tubes semiferes et sur la spermatogenese chez les Mifferes. Arch. Anat. Microsc., 11, 290–306.

Reichert, C. 1952. Fluorescence microscopy with fluorochromes. Phamphlet. Optische Werke C. Reichert, Wien.

Richards, O.W. 1941. The staining of acid-fast tubercle bacteria. Science, 93, 190.

Richards, O.W. 1955. Fluorescence microscopy. In: Analytical Cytology, pp. 5–37, R.C. Mellors (ed.), McGraw-Hill, New York.

Richards, O.W. 1960. Fluorescence microscopy. In: Medical Physics, Vol. 3, pp. 375–7, O. Glasser (ed.), Year Book Publ., Inc., Chicago.

Richmond, G.W., and Bennett, L. 1938. Clearing and staining of embryo for demonstrating ossification. Stain Technol., 13, 77–9.

Rigler, R. 1965. Mikroabsorptions- und Emissionsmessungen an Akridin orange-Nukleinsaure-Komplexen. Acta Histochem. Suppl., 6, 127–34.

Rigler, R., Jr. 1966. Microfluorometric characterization of intracellular nucleic acids and nucleoproteins by acridine orange. Acta Physiol. Scand., **67** (Suppl. 267), 122.

Rigler, R., Jr. 1969. Acridine orange in nucleic acid analysis. Ann. N.Y. Acad. Sci., **157**, 211–24.

Riker, A.J., and Riker, R.S. 1936. *Introduction to Research on Plant Diseases,* John S. Swift, St. Louis.

Ringertz, N. 1968. Cytochemical demonstration of basic proteins by dansyl staining. J. Histochem. Cytochem., **16**, 440–1.

Robbins, E., and Marcus, P.I. 1963. Dynamics of acridine orange-cell interaction; 1. Interrelationships of acridine orange particles and cytoplasmic reddening. J. Cell. Biol., **18**, 238–50.

Robbins, E., Marcus, P.I., and Ganatas, N.K. 1964. Dynamics of acridine orange-cell interaction; II. Dye-induced ultrastructural changes in multivesicular bodies (acridine orange particles). J. Cell Biol., **21**, 49–62.

Robertson, O.H. 1917. The effects of experimental plethora on blood production. J. Exp. Med., **26**, 221–37.

Robertson, R.T. 1978. *Neuroanatomical Research Techniques,* Academic Press, New York.

Robinow, C.F. 1942. A study of the nuclear apparatus of bacteria. Proc. R. Soc. Lond. (Biol.), **130**, 299–324.

Robinow, C.F. 1944. Cytological observations on *Bact. coli, Proteus vulgaris,* and various aerobic spore-forming bacteria with special reference to the nuclear structures. J. Hyg., **43**, 413–23.

Roger, F., and Roger, A. 1958. Les affinités tinctoriales des rickettsies et des gros virum visibles (groupe lymphogranulomatosepsitacose); I. Sur le méchanisme de la différenciation avec la coloration de Macciavello et sur quelques applications pratiques. Ann. Inst. Pasteur (Paris), **94**, 126–8.

Rogoff, W.M. 1946. The Bodian technic and the mosquito nervous system. Stain Technol., **21**, 59–61.

Romeis, B. 1968. *Mikroskopische Technik,* R. Oldenbourg Verlag, Munchen.

Rosselet, A., and Ruch, F. 1968. Cytofluorometric determination of lysine with dansylchloride. J. Histochem. Cytochem., **16**, 459–66.

Rotman, B., and Papermaster, B.W. 1966. Membrane properties of living mammalian cells as studied by enzymatic hydrolysis of fluorogenic esters. Proc. Nat. Acad. Sci. U.S.A., **55**, 134–41.

Ruch, F. 1964. Fluoreszenzphotometrie. Acta Histochem. (Suppl.), **6**, 117–21.

Ruch, F. 1966. Determination of DNA content by microfluorimetry. In: *Introduction to Quantitative Cytochemistry,* pp. 281–94, G. Wied (ed.), Academic Press, New York.

Ruch, F. 1970. Principles and some applications of cytofluorometry. In: *Introduction to Quantitative Cytochemistry,* Vol. 2, pp. 431–50, G.L. Wied and G.R. Bahr (ed.), Academic Press, New York.

Ruddell, C.L. 1971. Embedding media for 1-2 micron sectioning; 3. Hydroxyethyl methacrylate-benzoyl peroxide activated with pyridine. Stain Technol., **46**, 77–83.

Russell, W.C., Newman, C., and Williamson, D.H. 1975. A simple cytochemical technique for demonstration of DNA in cells infected with mycoplasmas and viruses. Nature, **253**, 461–2.

Ryu, E. 1963. A simple method for staining *Leptospira* and *Treponema.* Jpn. J. Microbiol., **7**, 81–5.

Sandritter, W., and Kasten, F.H., (eds.) 1964. *100 Years of Histochemistry in Germany,* F.K. Schattauer-Verlag, Stuttgart.

Sapero, J.J., and Lawless, D.K. 1953. The "MIF" stain-preservation technic for the identification of intestinal protozoa. Am. J. Trop. Med., **2**, 613–9.

Saunders, A.M. 1964. Histochemical identification of acid mucopolysaccharides with acridine orange. J. Histochem. Cytochem., **12**, 164–70.

Schabadasch, A. 1930. Untersuchungen zur Methodik der Methyleneblau-farbung des vegetativen Nervensystems. Z. Zellforsch. Mikrosk. Anat., **10**, 221–385 (4 papers).

Schabadasch, A. 1936. Histophysiologie des réactions réciproque entre le bleu de méthylène et le tissu nerveux. Bull. Histol. Appl. Physiol., **13**, 5–27, 72–89, 137–51.

Schaeffer, A.B., and Fulton, M. 1933. A simplified method of staining endospores. Science, **77**, 194.

Scharf, J.H. 1956a. Fluoreszenz und Fluoreszenzpolarisation den Nervenfaser nach Farbung mit Phenyloxyfluoronen. Versucheiner Interpretation. Teil. Mikrosk., **11**, 261–319.

Scharf, J.H. 1956b. Fluoreszenz und Fluoreszenzpolarisation den Nervenfaser nach Färbung mit Phenyloxyfluoronen. Versuch einer Interpretation. Teil. Mikrosk., **11**, 349–97.

Schaudinn, F. 1902. Studien uber kronkhertserregende Protozoen; II. *Plasmodium vivax* (Grossi and Felitti) der Erreger des tertian-

fibers heim Menschen. Arb. Kaiserl. Gesundh., **19**, 169–250.

Scheibel, M.E., and Scheibel, A.B. 1978. The methods of Golgi. In: *Neuroanatomical Research Techniques*, R.T. Robertson (ed.), Academic Press, New York.

Schlegel, J.U. 1949. Demonstration of blood vessels and lymphatics with a fluorescent dye in ultraviolet light. Anat. Rec., **105**, 433–43.

Schlegel, J.U., and Moses, J.B. 1950. A method for visualization of kidney blood vessels applied to studies of the crush syndrome. Proc. Soc. Exp. Biol. Med., **74**, 832–7.

Schleifstein, J. 1937. A rapid method for demonstrating Negri bodies in tissue sections. Am. J. Public Health, **27**, 1283–5.

Schmorl, G. 1899. Darstellung feinere Knochenstructuren. Centralbl. Allg. Pathol. Pathol. Anat., **10**, 745–9.

Schmorl, G. 1914. *Die pathologisch-histologischen Untersuchungsmethoden*, Ed. 7, F.C.W. Vogel, Leipzig.

Schmorl, G. 1928. *Die pathologisch-histologischen Untersuchungsmethoden*, Ed. 15, F.C.W. Vogel, Leipzig.

Schneidau, J.D. 1963. A simplified Gram stain for demonstrating fungi in tissues. Am. J. Clin. Pathol., **40**, 659–61.

Schneider, H. 1952. The phloem of the sweet orange tree trunk and the seasonal production of xylem and phloem. Hilgardia, **21**, 331–66.

Schneider, H. 1960. Sectioning and staining pathological phloem. Stain Technol., **35**, 123–7.

Schneider, H. 1977. Indicator hosts for pear decline: Symptomatology, histopathology, and distribution of mycoplasmalike organisms in leaf veins. Phytopathology, **67**, 592–601.

Schneider, H. 1980. Deposition of wound gum, callose, and suberin as response to diseases and wounding of citrus. Bull. Soc. bot. Fr., **127**, Actual. bot., (1) 193–98.

Schultze, M., and Rudneff, M. 1865. Weitere mittheilungen uber die Einwirkung der Überosmiumsäure auf thierische Gewebe. Arch. Mikrosk. Anat., **1**, 299–304.

Schümmelfeder, N. 1950a. Die Fluorochromierung des lebenden, überlebenden und toten Protoplasmas mit dem basischen Farbstoff Akridinorange und ihre Beziehung zur Stoffwechselaktivität der Zelle. Virchows Arch., **318**, 119–54.

Schümmelfeder, N. 1950b. Zur Morphologie und Histochemie nervöser Elemente; 1. Die

Fluorochromierung markhaltiger Nervenfasern mit Akridinorange. Virchows Arch., **319**, 294–320.

Schümmelfeder, N., Ebschner, K.-J., and Krogh, E. 1957. Die Grundlage der differenten Fluorochromierung von Ribo- und Desoxyribonucleinsäure mit Akridinorange. Naturwissenschaften, **44**, 467–8.

Schümmelfeder, N., and Stock, K.-F. 1956. Die Bestimmung des Umladebereiches (Isoelektrischer Punkt) von Gewebselementen mit dem Fluorochrom Akridinorange. Z. Zellforsch., **44**, 327–38.

Schwartz, D.S., Larsh, H.W., and Bartels, P.A. 1977. Enumerative fluorescent vital staining of live and dead pathogenic yeast cells. Stain Technol., **52**, 203–10.

Schweiger, D. 1976. Reverse fluorescent chromosome banding with chromomycin and DAPI. Chromosoma, **58**, 307–24.

Schwerdtfeger, W.K. 1978. The use of rhodamine B for the identification of injected horseradish peroxidase in the brain. Histochemistry, **58**, 237–9.

Seemüller, E. 1976. Fluoreszenzoptischer direktnachweis von mykoplasmaähnlichen organismen in phloem pear-decline-und triebsuchtkranker bäume? Phytopathol. Z., **85**, 368–72.

Sellers, T.F. 1927. A new method for staining Negri bodies of rabies. Am. J. Public Health, **17**, 1080–1.

Sengbusch, G.V., Couwenbergs, C. Kühner, J., and Müller, V. 1976. Fluorogenic substrate turnover in single living cells. Histochem. J., **8**, 341–50.

Setterfield, H.E., and Baird, T.T. 1936. Polarized light technic. A comparison with the Marchi method. Stain Technol., **11**, 41–4.

Seydel, H.G. 1966. The acridine orange and dye exclusion viability tests applied to irradiated Ehrlich tumour cells. Int. J. Radiat. Biol., **10**, 567–76.

Sharma, A.K., and Sharma, A. 1965. *Chromosome Techniques, Theory and Practice*, Butterworths, London.

Sharman, B.C. 1943. Tannic acid and iron alum with safranin and orange G in studies of the shoot apex. Stain Technol., **18**, 105–11.

Sharp, R.G. 1914. Diplodinium ecaudatum with an account of its neuromotor apparatus. University Calif. Pub. Zool., **13**, 43–122.

Shelley, W.B., Ohman, S., and Parnes, H.M. 1968. Mast cell stain for histamine in freeze-dried embedded tissue. J. Histochem. Cytochem., **16**, 433–9.

Sidman, R.L., Mottla, P.A., and Feder, N. 1961. Improved polyester wax embedding for histology. Stain Technol., **36**, 279–84.

Silver, M.L. 1942. Colloidal factors controlling silver staining. Anat. Rec., **82**, 507–29.

Simpson, C.F., Carlisle, J.W., and Mallard, L. 1970. Rhodanile blue; a rapid selective stain for Heinz bodies. Stain Technol., **45**, 221–4.

Sims, B. 1974. A simple method for preparing 1-2 μm sections of large blocks using glycol methacrylate. J. Microsc., **101**, 223–7.

Singh, J., and Bhattacharji, L.M. 1944. Rapid staining of malarial parasites by a water soluble stain. Indian Med. Gaz., **79**, 102–4.

Sjostrand, F. 1946. Cytological localization of riboflavin (vitamin B_2) and thiamine (vitamin B_1) by fluorescence microspectrography. Nature, **157**, 698.

Skinner, F.A., Jones, P.C.T., and Mollison, J.E. 1952. A comparison of a direct- and a plate-counting technique for the quantitative estimation of soil micro-organisms. J. Gen. Microbiol., **6**, 261–71.

Smith, F.H. 1934. The use of picric acid with the Gram stain in plant cytology. Stain Technol., **9**, 95–6.

Smith, G.L., Jenkins, R.A., and Gough, J.F. 1969. A fluorescent method for the detection and localization of zinc in human granulocytes. J. Histochem. Cytochem., **17**, 749–50.

Smith, J.L., and Mair, W. 1908. An investigation of the principles underlying Weigert's method of staining medulated nerve. With a note on the staining of fats by potassium dichromate and hematoxylin. J. Pathol. Bacteriol., **13**, 14–27.

Smith, M.C. 1956a. Observations on the extended use of the Marchi method. J. Neurol. Neurosurg. Psychiatry, **19**, 67–73.

Smith, M.C. 1956b. Recognition and prevention of artefacts of the Marchi method. J. Neurol. Neurosurg. Psychiatry, **19**, 74–83.

Smith, M.C., Strich, S.J., Sabina, J., and Sharp, P. 1956. The value of the Marchi method for staining tissue stored in formalin for prolonged periods. J. Neurol. Neurosurg. Psychiatry, **19**, 62–4.

Smith-Sonneborn, J. 1974. Acridine orange fluorescence; a temporary stain for paramecia. Stain Technol., **49**, 77–80.

Smithwick, R.W., and David, H.L. 1971. Acridine orange as a fluorescent counterstain with the auramine acid-fast stain. Tubercule, **52**, 226–31.

Smyth, J.D. 1951. Specific staining of eggshell material in Trematodes and Cestodes. Stain Technol., **26**, 255–6.

Snow, R. 1963. Alcoholic hydrochloric acid-carmine as a stain for chromosomes in squash preparations. Stain Technol., **38**, 9–13.

Solcia, E., Vassalo, G., and Capella, C. 1968. Selective staining of endocrine cells by basic dyes after acid hydrolysis. Stain Technol., **43**, 257–63.

Southgate, H.W. 1927. Note on preparing mucicarmine. J. Pathol. Bacteriol., **30**, 729.

Spicer, S.S., Frayser, R., Virella, G., and Hall, B.J. 1977. Immunocytochemical localization of lysozymes in respiratory and other tissues. Lab. Invest., **36**, 282.

Spicer, S.S., and Lillie, R.D. 1961. Histochemical identification of basic proteins with Biebrich scarlet at alkaline pH. Stain Technol., **36**, 365–70.

Spooner, G.H., Reed, C.S., and Clark, G. 1980. Work in progress.

Sprau, F. 1955. Pathologische Gewebeanderungen durch das Blattrollvirus bei der Kartoffel und ihr färbeachnischer Nachweis. Ber. Dtsch. Bot. Ges., **68**, 239–46.

Stadelmann, E.J., and Kinzel, H. 1972. Vital staining of plant cells. Methods Cell Physiol., **5**, 325–72.

Standard Methods for the Examination of Dairy Products, Ed. 14, W.G. Walker (ed.), 1978. American Public Health Association, New York.

Starr, T.J., Pollard, M., Tanami, Y., and Moore, R.W. 1960. Cytochemical studies with psittacosis virus by fluorescence microscopy. Texas Rep. Biol. Med., **18**, 501–14.

Stedman, H.F. 1960. *Section Cutting in Microscopy,* Charles C Thomas, Springfield, Ill.

Steedman, H.F. 1950. Alcian Blue 8GS: A new stain for mucin. Q. J. Microsc. Sci., **91**, 477–9.

Steinwall, O., and Klatzo, I. 1966. Selective vulnerability of the blood-brain barrier in chemically induced lesions. J. Neuropathol. Exp. Neurol., **25**, 542–59.

Stern, J.B. 1932. Neue Silberimpragnation versuche zur Darstellung der Mikro- und Oligodendroglia (An Celloidinserionschnitten anwendbare Methode). Z. Ges. Neurol. Psychiatr., **138**, 769–844.

Sternberger, L.A., Hardy, P.H., Cuculiss, J.J., and Meyer, H.G. 1970. The unlabeled antibody enzyme method of immunohistochemistry; preparation and properties of soluble antigen-antibody complex (horseradish peroxidase-antihorse-radish peroxidase) and its use in the identification of spirochtes. J. Histochem. Cytochem., **18**, 315–33.

Stevens, W., Lang, R.F., and Schneebeli, G. 1969. Acridine orange fluorescence for differentiating clean nuclei from nuclei with adherent cytoplasma in fractionated cells. Stain Technol., **44**, 211–2.

Stockinger, L. 1964. Vitalfärbung und Vitalfluorochromierun Tierischer Zellen. Protoplasmatologia, **2D1**, 1–96.

Storey, F.H., and Mann, J.D. 1967. Chromosome contraction by *o*-isopropyl-*N*-phenylcarbamate (IPC). Stain Technol., **42**, 15.

Stoughton, R.H. 1930. Thionin and orange G for the differential staining of bacteria and fungi in plant tissue. Ann. Appl. Biol., **17**, 163–6.

Stovall, W.D., and Black, C.E. 1940. The influence of pH on the eosin methylene blue method for demonstrating Negri bodies. Am. J. Clin. Pathol., **10**, 1–8.

Stoward, P.J. 1967a. Studies in fluorescence histochemistry; II. Mucosubstances with Pseudo-Schiff reagents. J. R. Microsc. Soc., **87**, 237–46.

Stoward, P.J. 1967b. Studies in fluorescence histochemistry; III. The demonstration with salicylhydrazide of the aldehydes present in periodate-oxidized mucosubstances. J. R. Microsc. Soc., **87**, 247–57.

Stretton, A.O.W., and Kravitz, E.A. 1968. Neuronal geometry; determination with a technique of intracellular dye injection. Science, **162**, 132–4.

Stretton, A.O.W., and Kravitz, E.A. 1973. Intracellular dye injection; the selection of procion yellow and its application in preliminary studies of neuronal geometry in the lobster nervous system. In: *Intracellular Staining in Neurobiology*, pp. 21–40, S.A. Kater and C. Nicholson (eds.), Springer-Verlag, New York.

Strugger, S. 1940. Fluoreszenzmikroskopische Untersuchunger über die Aufnahme und Speicherung des Akridinorang durch lebende und tote Pflanzenzellen. Jen. Z. Naturwiss., **73**, 97–134.

Strugger, S. 1948a. Fluorescence microscope examination of bacteria in soil. Can. J. Res. Sec. C., **26**, 188–93.

Strugger, S. 1948b. Fluorescence microscopy of trypanosomes in blood. Can. J. Res. Sec. E., **26**, 229–31.

Strugger, S. 1949. *Fluoreszenzmikroskopie und Mikrobiologie*, M. u. H. Schaper, Hannover.

Strugger, V.S. 1947. Die Vitalfluorochromierung des Protoplasmas. Naturwissenschaften, **34**, 267–73.

Sunberg, R.D., and Broman, H. 1955. The application of Prussian blue stain to previously stained films of blood and bone marrow. Blood, **10**, 160–6.

Sundler, F., and Hakanson, R. 1978. Fluorescence histochemistry of peptide hormone producing cells; observations on the pheneanthrenequinone method for the demonstration of arginine residues. Histochemistry, **56**, 221–7.

Swank, R.L., and Davenport, H.A. 1935. Chlorate-osmic-formalin method for staining degenerating myelin. Stain Technol., **10**, 87–90.

Swanson, C.P. 1940. The use of acenaphthene in pollen tube technic. Stain Technol., **15**, 49–52.

Sweat, F., Puchtler, H., and Rosenthal, S.I. 1964. Sirius red F3BA as a stain for connective tissue. Arch. Pathol., **78**, 69–72.

Sweat, F., Puchtler, H., and Woo, P. 1964. A light fast modification of Lillie's allochrome stain. Periodic acid-Schiff-picro-Sirius supra blue GI . Arch. Pathol., **78**, 73–5.

Swift, H. 1955. Cytochemical techniques for nucleic acids. In: *The Nucleic Acids*, E. Chargaff and J.N. Davidson (eds.), Academic Press, New York.

Szokol, M., and Gomba, Sz. 1971. Fluorescent staining of juxtaglomerular cells in frozen sections. Stain Technol., **46**, 102–3.

Taft, E.B. 1951. The problem of a standardized technic for the methyl green-pyronin stain. Stain Technol., **26**, 205–12.

Takaya, K. 1970. A new fluorescent stain with o-pthaldehyde for A-cells of the pancreatic islets. J. Histochem. Cytochem., **18**, 178–86.

Tanaka, C., and Cooper, J.R. 1968. The fluorescent microscopic localization of thiamine in nervous tissue. J. Histochem. Cytochem., **16**, 362–5.

Taylor, R.D. 1966. Modification of the Brown and Brenn Gram stain for the differential staining of gram-positive and gram-negative bacteria in tissue sections. Am. J. Clin. Pathol., **46**, 472–4.

Taylor, W.R. 1937. General botanical microtechnique. In: *Handbook of Microscopical Technique*, Ed. 2, C.E. McClung (ed.), Hoeber, New York.

Thaer, A.A., and Sernetz, M. (eds.) 1973. *Fluorescence Techniques in Cell Biology*, Springer-Verlag, New York.

Thiessen, G., and Thiessen, H. 1977. Microspectrophotometric cell analysis. Prog. Histochem. Cytochem., **9**, 37.

Tjio, J.H., and Levan, A. 1950. The use of oxyquinoline in chromosome analysis. An. Estac. Exp. Aula Dei, **2**, 21.

Toriumi, J., Inoue, T., and Ishida, M. 1959.

Histochemical studies on acid mucopolysaccharides. New fluorochromatic methods. Acta Pathol. Jpn., **9**, 343–50.

Torre, J.C., de la 1979. Standardization of the SPG histofluorescence method for monoamine transmitters. Society for Neuroscience, Bethesda, Md., Abstr. **5**: 333.

Torre, J.C., and Surgeon, J.W. 1976. A methodological approach to rapid and sensitive monoamine histofluorescence using a modified glyoxylic acid technique; the SPG method. Histochemistry, **49**, 81–93.

Traganos, F., Darzynkiewicz, Z., Sharpless, T., and Melamed, M.R. 1977. Simultaneous staining of ribonucleic and deoxyribonucleic acids in unfixed cells using acridine orange in a flow cytofluorometric system. J. Histochem. Cytochem., **25**, 46–56.

Truant, J.P., Brett, W.A., and Thomas, W. 1962. Fluorescence microscopy of tubercle bacilli stained with auramine and rhodamine. Henry Ford Hosp. Bull., **10**, 287–96.

Trujillo, T.T., and Van Dilla, M.A. 1972. Adaptation of the fluorescent Feulgen reaction to cells in suspension for flow microfluorimetry. Acta Cytol., **16**, 26–30.

Tsao, P.H. 1970. Applications of the vital fluorescent labeling techniques with brighteners to studies of saphrophytic behavior of Phytophthora in soil. Soil Biol. Biochem., **2**, 247–56.

Tsuchiya, T. 1971. An improved aceto-carmine squash method, with special reference to the modified Rattenbury's method of making a preparation permanent. Barley Genetics Newsletter, **1**, 71–2.

Tsunewaki, K., and Jenkins, B.C. 1960. Various methods of root tip preparation in screening wheat aneuploids. Cytologia, **25**, 373–8.

Tuan, H.C. 1930. Picric acid as a destaining agent for iron alum hematoxylin. Stain Technol., **5**, 135–8.

Tweedle, C.D. 1978. Single-cell staining techniques. In: *Neuroanatomical Research Techniques,* Vol. 2, pp. 141–74, R.T. Robertson (ed.), Academic Press, New York.

Uchida, I., and Lin, C. 1974. Quinacrine fluorescent patterns. In: *Human Chromosome Methodology,* Ed. 2, pp. 47–58, J.J. Yunis (ed.), Academic Press, New York.

Udenfriend, S. 1962. *Fluorescence Assay in Biology and Medicine,* Academic Press, New York.

Udenfriend, S., Stein, S., Bohlen, P., Dairman, W., Leimgruber, W., and Weigele, M. 1972. Fluorescamine; a reagent for assay of amino acids, peptides, proteins, and primary amines in the picamole range. Science, **178**, 871–2.

Ulrychova, M., Petru, E., and Pazourkova, Z. 1976. Permanent staining of callose in plant material by Ponceau S. Stain Technol., **51**, 272–5.

Umlas, J., and Fallon, J.N. 1971. New thick-film technique for malaria diagnosis. Am. J. Trop. Med. Hyg., **20**, 527–9.

Unna, P.G. 1890. Über die Tänzersche Färbung des elastischen Gewebe. Monatsschr. Prakt. Dermatol., **11**, 367.

Unna, P.G. 1891. Notiz, betreffend die Tänzersche Orceinfarbung des elastischen Gewebes. Monatsschr. Prakt. Dermatol., **12**, 394–6.

Van Gieson, I. 1889. Laboratory notes of technical methods for the nervous system. N.Y. Med. J., **50**, 57–60.

Van Ingen, E.M., Tanke, H.J., and Ploem, J.S. 1979. Model studies on the acriflavin Feulgen reaction. J. Histochem. Cytochem., **27**, 80–3.

Van Horne, R.L., and Zopf, L.C. 1951. Water-soluble embedding materials for botanical microtechnique. J. Am. Pharm. Assoc., (Sci. Ed.), **40**, 31–5.

Vassar, P.S., and Culling, C.F.A. 1959. Fluorescent stains with special reference to amyloid and connective tissue. Arch. Pathol., **68**, 487–98.

Vaughan, R.E. 1914. A method for the differential staining of fungus and host cells. Ann. Missouri Bot. Garden, **1**, 241–2.

Verhoeff, F.H. 1908. Some new staining methods of wide applicability; including a rapid differential stain for elastic tissue. J.A.M.A., **50**, 876–7.

Volk, B.W., and Popper, H. 1944. Microscopic demonstration of fat in urine and stool by means of fluorescence microscopy. Am. J. Clin. Pathol., **14**, 234–8.

von Bertalanffy, L., and Bickis, I. 1956. Identification of cytoplasmic basophilia (ribonucleic acid) by fluorescence microscopy. J. Histochem. Cytochem., **4**, 481–93.

von Muralt, A. 1943. Die sekundare thiochromfluorescenz des peripheren werven und ihre beziehung zu bethes polarisationsbild. Arch. Gesamte Physiol, **247**, 1.

von Querner, F.R. 1932. Die paraplasmatischen Einschlüsse der leberzellen im Fluoreszenzmikroskop und der Leuchtstoff X. Anz. Akad. Wiss. Wien, **69**, 172–5.

Wade, H.W. 1952. Demonstration of acid-fast bacilli in tissue sections. Am. J. Pathol., **28**, 157–70.

Wade, H.W. 1957. A modification of the Fite

formaldehyde (Fite I) method for staining acid-fast bacilli in paraffin sections. Stain Technol., **32**, 287–92.

Waggoner, A.S. 1979. Dye indicators of membrane potential. Annu. Rev. Biophys. Bioeng., **8**, 47–68.

Wagner, W.H., and Foerster, O. 1964. Die PAS-AO-Methode, eine spezial Färbung fur Coccidien in Gewebe. Z. Parasitenk., **25**, 28–48.

Waldrop, F.S.; Puchtler, H., and Valentine, L.S. 1972. Fluorescence microscopy of amyloid using mixed illumination. Arch. Pathol., **95**, 37–41.

Walter, E.K., Smith, J.L., Israel, G.W., and Gager, W.E. 1969. A new modification of the Krajian silver stain for *Treponema pallidum*. Br. J. Vener. Dis., **45**, 6–9.

Walton, W.R. 1952. On techniques for recognition of living foraminifera. Contrib. Cushman Found., **3**, 56–60.

Ward, D.C., Reich, E., and Goldberg, I.H. 1965. Base specificity in the interaction of polynucleotides with antibiotic drugs. Science, **149**, 1259–63.

Warthin, A.S., and Starry, A.C. 1920. A more rapid and improved method of demonstrating spirochetes in tissues. Am. J. Syphilol. Gonor. Vener. Dis., **4**, 97.

Warthin, A.S., and Starry, A.C. 1922. The staining of spirochetes in cover-glass smears by the silver-agar method. J. Infect. Dis., **30**, 592–600.

Weigert, K. 1884. Ausfuhrliche Beschreibung der in no. 2 dieser Zeitschrift erwahnten neuen Färbungsmethod fur das Centralnervensystem. Fortschr. Med., **2**, 190–1.

Weigert, K. 1885. Eine Vergesserung der Haematoxylin Blutlaugensalzmethode für das Centralnervensystem. Fortschr. Med., **3**, 236–9.

Weigert, K. 1891. Zur Markscheidenfärbung. Dtsch. Med. Wochenschr., **17**, 1184–6.

Weigert, K. 1898. Über eine Methode zur Färbung elastischer Fasern. Centralbl. Allg. Pathol. Pathol. Anat., **9**, 289–92.

Weigert, K. 1904. Eine kleine Verbesserung der Hamatoxylin-van-Gieson-Methode. Z. Wiss. Mikrosk., **21**, 1–5.

Weil, A. 1928. A rapid method for staining myelin sheaths. Arch. Neurol. Psychiatry, **20**, 392–3.

Weil, A., and Davenport, H.A. 1930. Eine Methode zur Silberimpragnierung von Gliomen. Z. Ges. Neurol. Psychiatr., **154**, 228–38.

Weil, A., and Davenport, H.A. 1933. Staining of oligodendroglia and microglia in cel-

loidin sections. Arch. Neurol. Psychiatry, **30**, 175–8.

Weinblatt, F.M., Shannon, W.A., Jr., and Seligman, A.M. 1975. A new fluorescent method for the demonstration of macromolecular aldehydes. Histochemistry, **41**, 353–9.

Weir, D.M. 1967. *Handbook of Experimental Immunology*, F.A. Davis, Philadelphia.

Weisblum, B., and de Haseth, P. 1972. Quinacrine—a chromosome stain specific for deoxyadenylate-deoxythymidylate-rich regions of DNA. Proc. Nat. Acad. Sci. U.S.A., **69**, 629–32.

Weissmann, C., and Gilgen, A. 1956. Die fluorochromierum lebender Ehrlich-Ascitescarcinomzellen mit Akridinorang und der Einfluss der Glycolyse auf das Verhalten der Zelle. Z. Zellforsch., **44**, 292–326.

Wenrich, D.H., and Diller, W.F. 1950. Methods of protozoology. In: *McClung's Handbook of Microscopical Technique*, pp. 432–74, R.McC. Jones (ed.), Paul B. Hoeber, New York.

Wenyon, C.M. 1926. *Protozoology, a Manual for Medical Men, Veterinarians and Zoologists*, Balliere, Tindall & Cox, London.

Wessenberg, H. 1972. Personal communication.

West, S.S. 1969. Fluorescence microspectrophotometry of supravitally stained cells. In: *Physical Techniques in Biological Research*, pp. 253–316, A.W. Pollister (ed.), Academic Press, New York.

Weste, S.M., and Penington, D.G. 1972. Fluorometric measurement of deoxyribonucleic acid in bone marrow cells. The measurement of megakaryocyte deoxyribonucleic acid. J. Histochem. Cytochem., **20**, 627–33.

White, P.B. 1947. A method for combined positive and negative staining of bacteria. J. Pathol. Bacteriol., **59**, 334–5.

Widholm, J.M. 1972. The use of fluorescein diacetate and phenosafranine for determining viability of cultured plant cells. Stain Technol., **47**, 189–94.

Wied, G.L., and Manglano, J.I. 1962. A comparative study of the Papanicolaou technic and the acridine-orange fluorescence method. Acta Cytol., **6**, 554–68.

Wilder, H.C. 1935. An improved technique for silver impregnation of reticulum fibers. Am. J. Pathol., **11**, 817–9.

Williams, T.W. 1941. Alizarin red S and toluidine blue for differentiating adult or embryonic bone and cartilage. Stain Technol., **16**, 23–5.

Williamson, C.K. 1956. Morphological and physiological considerations of colonial variants of *Pseudomonas aeruginosa*. J. Bacteriol., **71**, 617–22.

Williamson, D.H., and Wilkinson, J.F. 1958 The isolation and estimation of the poly-B-hydroxybutyrate inclusions of *Bacillus* species. J. Gen. Microbiol., **19**, 198–209.

Williamson, D.H., and Fennell, D.J. 1975. The use of fluorescent DNA-binding agent for detecting and separating yeast mitochondrial DNA. Methods Cell Biol., **12**, 335–51.

Windle, W.F. (ed.) 1957. *New Research Techniques of Neuroanatomy*, Charles C Thomas, Springfield, Ill.

Windle, W.F., Rhines, R., and Rankin, J. 1943. A Nissl method using buffered solutions of thionin. Stain Technol., **18**, 77–86.

Wismar, B.I. 1966. Quad-type stain for the simultaneous demonstration of intracellular and extracellular tissue components. Stain Technol., **41**, 309–13.

Wittekind, D. 1959. Über die derzeitige Beduntung der Vitalaörbung speziell der Vitafluorochromierung in der Cytologie und über Möglichkeiten ihrer weiteren Anwendung, insbesondere in Kombination mit anderen Methoden. Mikroskopie, **14**, 9–25.

Wodehouse, R.P. 1935. *Pollen Grains*, Ed. 1, McGraw-Hill, New York.

Wolbach, S.B. 1911. The use of colophonium in differentiating the eosin-methylene blue and other stains. J.A.M.A., **56**, 345–6.

Wolbach, S.B., Todd, J.L., and Palfrey, F.W., 1922. *The Etiology and Pathology of Typhus*, Harvard University Press, Cambridge, Mass.

Wolf, M.K., and Aronson, S.B. 1961. Growth, fluorescence and metachromasy of cells cultured in the presence of acridine orange. J. Histochem. Cytochem., **9**, 22–9.

World Health Organization: 1973. *Laboratory Techniques in Rabies*, Ed. 3., M.M. Kaplan and H. Koprowski (eds.).

Wright, J.H. 1902. A rapid method for the differential staining of blood films and malarial parasites. J. Med. Res., **7**, 138–44.

Yasaki, Y. 1959. A new histological method for the detection of Crytococcus neoformans and its histopathological study. Acta Pathol. Jpn., **9**, 351–60.

Yataghanas, X., Gahrton, G., and Thorell, B. 1969. Microspectrofluorometry of a periodic acid-Schiff reaction in blood cells. Exp. Cell Res., **56**, 59–68.

Yeager, J.F. 1938. A modified Wright's blood-staining procedure for smears of heat-fixed insect blood. Ann. Entomol. Soc. Am., **31**, 9–14.

Zanker, V. 1952. Über den Nachweis definierter reversibler Assoziate ("reversible Polymerisate") des akridinorange durch absorptions- und Fluoreszenzmessungen in wassriger Losung. Z. Physiol. Chem., **199**, 225–58.

Zeiger, K., and Harders, H. 1951. Über die vitale Fluorochromaörbung des Nervengewebes. Z. Zellforsch., **36**, 62–78.

Zelenin, A.V. 1967. *Fluorescence Cytochemistry of Nucleic Acids*. Nauka Publ., Moscow.

Zenker, K. 1894. Chromkali-Sublimat-Eiseeig als Fixirungsmittel. Munch. Med. Wochenschr., **42**, 532–4.

Ziehl, F. 1882. Zur Färbung des Tuberkelbacillus. Dtsch. Med. Wochenschr., **8**, 451.

Zimmermann, W.A. 1928. Histologische Studien am Vegetationspunkt von Hypericum uralum. Jhrb. Wiss. Bot., **68**, 296–7.

Zirkle, C. 1940. Combined fixing, staining, and mounting media. Stain Technol., **15**, 139–53.

Zon, L. 1936. The physical chemistry of silver staining. Stain Technol., **11**, 53–67.

INDEX